The Science of Vehicle Dynamics

Massimo Guiggiani

The Science of Vehicle Dynamics

Handling, Braking, and Ride of Road and Race Cars

Third Edition

 Springer

Massimo Guiggiani
Dipartimento di Ingegneria Civile e
Industriale
Universita di Pisa
Pisa, Italy

ISBN 978-3-031-06463-0   ISBN 978-3-031-06461-6   (eBook)
https://doi.org/10.1007/978-3-031-06461-6

1st edition: © Springer Science+Business Media Dordrecht 2014
2nd edition: © Springer International Publishing AG, part of Springer Nature 2018, corrected publication 2018
3rd edition: © The Editor(s) (if applicable) and The Author(s), under exclusive license to Springer Nature Switzerland AG 2023, corrected publication 2023

This Springer imprint is published by the registered company Springer Nature Switzerland AG
The registered company address is: Gewerbestrasse 11, 6330 Cham, Switzerland

Preface to the Third Edition

Close interactions with leading race engineers gave me motivation to further improve the book. Handling and braking of race cars have been completely reformulated, and expanded with some very interesting original results. Almost all other parts have undergone a deep revision, like the section on the axle characteristics. The analysis of limited slip differential mechanisms has been also improved.

The third edition is intended to serve as a textbook while in college, and afterward as a reference book for professionals, including race engineers. Indeed, any topic is addressed starting from very basic concepts, easy to understand. The analysis is then improved up to a very high level, always paying attention to develop sound mathematical models.

Acknowledgments I wish to thank Tito Amato, Lorenzo Bartali, Maurizio Bocchi, Daniele Calderini, Nazzareno Ciccorossi, Lucia Conconi, Leontina Di Cecco, Timothy Drotar, Andrea Ferrarelli, Ernesto Desiderio, Shaid Farzand, Daniele Giordano, Eugeniu Grabovic, Damian Harty, Nicolas Lalande, Andrea Landi, Basilio Lenzo, David Loppini, Gene Lukianov, Stylianos Markolefas, Alessandro Moroni, Federico Sánchez Motellón, William J Oberlies, Sandro Okutuga, Matteo Pergoli, Antonino Pizzuto, Andrea Quintarelli, Luigi Romano, Carlo Rottenbacher, Francesco Senni, Loic Serra, Matteo Togninalli, and Andrea Toso.

I am most grateful to Alessio Artoni and Marco Gabiccini for their valuable suggestions.

Pisa, Italy
March 2022

Massimo Guiggiani

Preface to the Second Edition

This second edition pursues, even more than the first edition, the goal of approaching Vehicle Dynamics as a scientific subject, with neat definitions, clearly stated assumptions, sound mathematics, critical analysis of classical concepts, step-by-step developments. This may sound theoretical, but it is actually very practical.

Indeed, some automotive companies have drastically changed their approach on some topics according to some (apparently) theoretical results presented in the first edition of this book.

These achievements, along with the willingness to better explain some issues, have been the motivations for writing a new edition.

All chapters have been thoroughly revised, with the inclusion of some new results. Several parts have been expanded, like the section on the differential mechanism. Moreover, worked-out exercises have been included to help clarify the matter, particularly for students.

In several parts the book departs from commonly accepted explanations. Somehow, the more you know (classical) vehicle dynamics, the more you will be surprised.

Pisa, Italy   Massimo Guiggiani
February 2018

Preface to the First Edition

Vehicle dynamics should be a branch of Dynamics, but, in my opinion, too often it does not look like that. Dynamics is based on terse concepts and rigorous reasoning, whereas the typical approach to vehicle dynamics is much more intuitive. Qualitative reasoning and intuition are certainly very valuable, but they should be supported and confirmed by scientific and quantitative results.

I understand that vehicle dynamics is, perhaps, the most popular branch of Dynamics. Almost everybody has been involved in discussions about some aspects of the dynamical behavior of a vehicle (how to brake, how to negotiate a bend at high speed, which tires give best performance, etc.). At this level, we cannot expect a deep knowledge of the dynamical behavior of a vehicle.

But there are people who could greatly benefit from mastering vehicle dynamics. From having clear concepts in mind. From having a deep understanding of the main phenomena. This book is intended for those people who want to build their knowledge on sound explanations, who believe equations are the best way to formulate and, hopefully, solve problems. Of course along with physical reasoning and intuition.

I have been constantly alert not to give anything for granted. This attitude has led to criticize some classical concepts, such as self-aligning torque, roll axis, understeer gradient, handling diagram. I hope that even very experienced people will find the book interesting. At the same time, less experienced readers should find the matter explained in a way easy to absorb, yet profound. Quickly, I wish, they will feel not so less experienced any more.

Pisa, Italy
October 2013

Massimo Guiggiani

Contents

About the Author

Massimo Guiggiani is professor of Applied Mechanics at the Università di Pisa, Italy, where he also teaches Vehicle Dynamics in the MSc degree program in Automotive Engineering.

He has achieved important results also in other fields, such as Guiggiani's algorithm for the evaluation of singular integrals in the Boundary Element Method (ASME Journal of Applied Mechanics, 1990 and 1992), and the invariant approach for gear generation (Mechanism and Machine Theory, 2005 and 2007).

Chapter 1
Introduction

Vehicle dynamics is a fascinating subject, but it can also be very frustrating without the tools to truly understand it. We can try to rely on experience, but an objective knowledge needs a scientific approach. Something grounded on significant mathematical models, that is models complex enough to catch the essence of the phenomena under investigation, yet simple enough to be understood by a (well trained) human being. This is the essence of science, and vehicle dynamics is no exception.

The really important point is the mental attitude we should have in approaching a problem. We must be skeptical. We must be critical. We must be creative. Even if something is commonly accepted as obviously true, or if it looks very reasonable, it may be wrong, either totally or partially wrong. There might be room for some sort of improvement, for a fresh point of view, for something valuable.

Vehicle dynamics can be set as a truly scientific subject, it actually needs to be set as such to achieve a deep comprehension of what is going on when, e.g., a race car negotiates a bend.

When approached with open mind, several classical concepts of vehicle dynamics, like, e.g., the roll axis, the understeer gradient, even the wheelbase, turn out to be very weak concepts indeed. Concepts often misunderstood, and hence misused. Concepts that need to be revisited and redefined, and reformulated to achieve an objective knowledge of vehicle dynamics. Therefore, even experienced readers will probably be surprised by how some topics are addressed and discussed here.

To formulate vehicle dynamics on sound concepts we must rely on clear definitions and model formulations, and then on a rigorous mathematical analysis. We must, indeed, "formulate" the problem at hand by means of mathematical formulas [5]. There is no way out. Nothing is more practical than a good theory. However, although we will not refrain from using formulas, at the same time we will keep the analysis as simple as possible, trying to explain what each formula tells us.

To help the reader, the Index of almost all mathematical symbols is provided at the end of this book. The Index shows in which context each symbol is introduced and defined.

© The Author(s), under exclusive license to Springer Nature Switzerland AG 2023 1
M. Guiggiani, *The Science of Vehicle Dynamics*,
https://doi.org/10.1007/978-3-031-06461-6_1

Fig. 1.1 Vehicle expected
behavior when negotiating a
curve

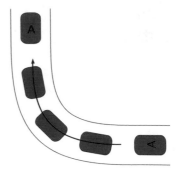

Fig. 1.2 Acceptable
behaviors for a road vehicle

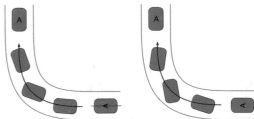

1.1   Vehicle Definition

Before embarking into the development of mathematical models, it is perhaps advisable to discuss a little what ultimately is (or should be) a *driveable road vehicle*.
Since a road is essentially a long, fairly narrow strip, a vehicle must be an object with
a clear *heading direction*.[1] For instance, a shopping cart is not a vehicle since it can
go in any direction. Another common feature of road vehicles is that the driver is
carried on board, thus undergoing the same dynamics (which, again, is not the case
of a shopping cart).

Moreover, roads have curves. Therefore, a vehicle must have the capability to be
driven in a fairly precise way. This basically amounts to controlling simultaneously
the *yaw rate* and the magnitude and direction of the *vehicle speed*. To fulfill this task
a car driver can act (at least) on the brake and accelerator pedals and on the steering
wheel. And here it is where vehicle dynamics comes into play, since the outcome of
the driver actions strongly depends on the vehicle dynamic features and state.

An example of proper turning of a road vehicle is something like in Fig. 1.1. Small
deviations from this target behavior, like those shown in Fig. 1.2, may be tolerated.
On the other hand, Fig. 1.3 shows two unacceptable ways to negotiate a bend.

All road vehicles have wheels, in almost all cases equipped with pneumatic tires.
Indeed, also wheels have a clear heading direction. This is why the main way to steer
a vehicle is by turning some (or all) of its wheels.[2]

---

[1] Usually, children show to have well understood this concept when they move by hand a small toy
car.

[2] Roughly speaking, wheels location does not matter to the driver. But it matters to engineers.

Fig. 1.3 Unacceptable
behaviors for a road vehicle

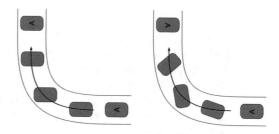

To have good directional capability, the wheels in a vehicle are arranged such that their heading directions almost "agree", that is they do not conflict too much with each other. However, tires do work pretty well under small slip angles and, as will be shown, some amount of "disagreement" is not only tolerated, but may even be beneficial.

Wheel hubs are connected to the chassis (vehicle body) by means of suspensions. The number of possible different suspensions is virtually endless. However, suspension systems can be broadly classified into two main subgroups: dependent and independent [7, 9]. In a dependent suspension the two wheels of the same axle are rigidly connected together. In an independent suspension they are not, and each wheel is connected to the chassis by a linkage with "mainly" one degree of freedom. Indeed, the linkage has some compliance which, if properly tuned, can enhance the vehicle behavior.

1.2 Vehicle Basic Scheme

A mathematical model of a vehicle [6] should be *simple*, yet *significant* [1, 2]. Of course, there is not a unique solution. Perhaps, the main point is to state clearly the assumptions behind each simplification, thus making clear under which conditions the model can reliably predict the behavior of a real vehicle.

There are assumptions concerning the *operating conditions* and assumptions regarding the *physical model* of the vehicle.

Concerning the *operating conditions*, several options can be envisaged:

Performance: the vehicle goes straight on a flat road, possibly braking or accelerating (nonconstant forward speed);

Handling: the vehicle makes turns on a flat road, usually with an almost constant forward speed;

Ride: the vehicle goes straight on a bumpy road, with constant forward speed.

Obviously, real conditions are a mixture of all of them.

A significant, yet simple, *physical model* of a car may have the following features:

1. the vehicle body is a single rigid body;
2. each wheel hub is connected to the vehicle body by a one-degree-of-freedom linkage (independent suspension);
3. the steering angle of each (front) wheel is mainly determined by the angular position δ_v of the steering wheel, as controlled by the driver;
4. the mass of the wheels (unsprung mass) is very small if compared to the mass of the vehicle body (sprung mass);
5. the wheels have pneumatic tires;
6. there are springs and dampers (and, maybe, inerters) between the vehicle body and the suspensions, and, likely, between the two suspensions of the same axle (anti-roll bar). Front to rear interconnected suspensions are possible, but very unusual;
7. there may be aerodynamic devices, like wings, that may significantly affect the downforce.

The first two assumptions ultimately disregard the elastic compliances of the chassis and of the suspension linkages, respectively, while the third assumption leaves room for vehicle models with compliant steering systems.

A vehicle basic scheme is shown in Fig. 1.4, which also serves the purpose of defining some fundamental geometrical parameters:

1. the vehicle longitudinal axis x, and hence the vehicle heading direction $\mathbf{i}$;
2. the height h from the road plane of the center of gravity G of the whole vehicle;
3. the longitudinal distances a_1 and a_2 of G from the front and rear axles, respectively;
4. the lateral position b of G from the longitudinal axis x;
5. the wheelbase $l = a_1 + a_2$;
6. the front and rear tracks t_1 and t_2;
7. the geometry of the linkages of the front and rear suspensions;
8. the position of the steering axis for each wheel.

All these distances are positive, except possibly b, which is usually very small and hence typically set equal to zero, like in Fig. 1.4.

It must be remarked that whenever, during the vehicle motion, there are suspension deflections, several of these geometrical parameters may undergo small changes. Therefore, it is common practice to take their reference value under the so called *static conditions*, which means with the vehicle moving straight on a flat road at constant speed, or, equivalently if there are no wings, when the vehicle is motionless on a horizontal plane.

Accordingly, the study of the performance and handling of vehicles is greatly simplified under the hypothesis of small suspension deflections, much like assuming very stiff springs (which is often the case for race cars).[3] Yet, suspensions cannot

---

[3] However, handling with roll will be covered in Chap. 9, although at the expense of quite a bit of additional work.

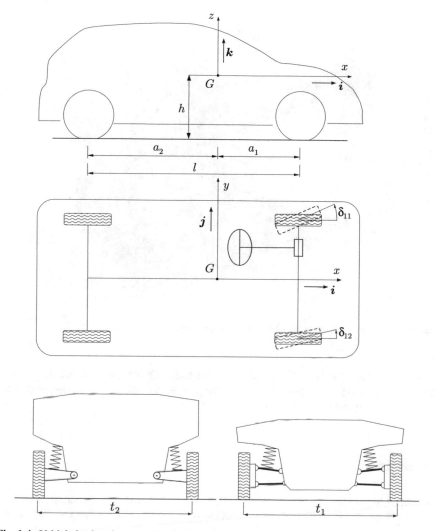

Fig. 1.4 Vehicle basic scheme and body-fixed reference system

be completely disregarded, at least not in vehicles with four or more wheels. This aspect will be thoroughly discussed.

The vehicle shown in Fig. 1.4 has a swing arm rear suspension and a double wishbone front suspension. Perhaps, about the worst and one of the best kind of independent suspensions [3, 4]. They were selected to help explaining some concepts, and should not be considered as an example of a good vehicle design. An example of a double wishbone front suspension is shown in Fig. 1.5.

As shown in Fig. 3.2, it is useful to define the *body-fixed reference system* $S =$ $(x, y, z; G)$, with unit vectors $(\mathbf{i}, \mathbf{j}, \mathbf{k})$. It has origin in the center of mass G and axes

Fig. 1.5 Example of a double wishbone front suspension [8]

fixed relative to the vehicle. The horizontal x-axis marks the forward direction, while the y-axis indicates the lateral direction. The z-axis is vertical, that is perpendicular to the road, with positive direction upward.

References

1. Arnold M, Burgermeister B, Fuehrer C, Hippmann G, Rill G (2011) Numerical methods in vehicle system dynamics: state of the art and current developments. Veh Syst Dyn 49(7):1159–1207
2. Cao D, Song X, Ahmadian M (2011) Editors' perspectives: road vehicle suspension design, dynamics, and control. Veh Syst Dyn 49(1–2):3–28
3. Genta G, Morello L (2009) The automotive chassis, vol 1. Springer, Berlin
4. Genta G, Morello L (2009) The automotive chassis, vol 2. Springer, Berlin
5. Guiggiani M, Mori LF (2008) Suggestions on how not to mishandle mathematical formulæ. TUGboat 29:255–263
6. Heißing B, Ersoy M (eds) (2011) Chassis handbook. Springer, Wiesbaden
7. Jazar RN (2014) Vehicle dynamics, 2nd edn. Springer, New York
8. Longhurst C (2013) https://www.carbibles.com/guide-to-car-suspension/
9. Schramm D, Hiller M, Bardini R (2014) Vehicle dynamics. Springer, Berlin

Chapter 2
Mechanics of the Wheel with Tire

All road vehicles have wheels and almost all of them have *wheels* with *pneumatic tires*. Wheels have been around for many centuries, but only with the invention, and enhancement, of the pneumatic tire it has been possible to conceive fast and comfortable road vehicles [5].

The main features of any tire are its *flexibility* and *low mass*, which allow for the contact with the road to be maintained even on uneven surfaces. Moreover, the rubber ensures *high grip*. These features arise from the highly composite structure of tires: a carcass of flexible, yet almost inextensible cords encased in a matrix of soft rubber, all inflated with air.[1] Provided the (flexible) tire is properly inflated, it can exchange along the bead relevant actions with the (rigid) rim. Traction, braking, steering and load support are the net result.

It should be appreciated that the effect of air pressure is to increase the structural stiffness of the tire, not to support directly the rim. How a tire carries a vertical load F_z if properly inflated is explained in Fig. 2.1.[2] In the lower part the radial cords encased in the sidewalls undergo a reduction of tension because they no longer have to balance the air pressure p_a acting on the contact patch [10, p. 279]. The net result is that the total upward pull of the cords on the bead exceeds that of the downward pull by an amount equal to the vertical load F_z [26, p. 161]. A very clear explanation can also be found in [31].

---

[1] Only in competitions it is worthwhile to employ special (and secret) gas mixtures instead of air. The use of nitrogen, as often recommended, is in fact almost equivalent to air [18], except for the cost.

[2] As pointed out by Jon W. Mooney in his review, in Noise Control Engineering Journal, Vol. 62, 2014, the explanation and the figure provided in the first edition of this book were *not correct*. A similar (incorrect) explanation has appeared in [9, Fig. 1.19], published in 2017.

© The Author(s), under exclusive license to Springer Nature Switzerland AG 2023
M. Guiggiani, *The Science of Vehicle Dynamics*,
https://doi.org/10.1007/978-3-031-06461-6_2

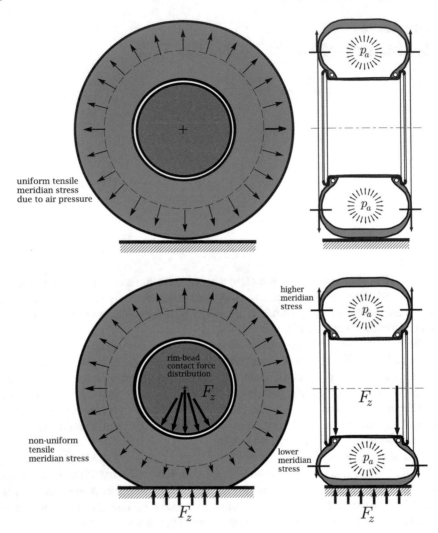

Fig. 2.1 How a tire carries a vertical load if properly inflated

The *contact patch*, or footprint, of the tire is the area of the tread in contact with the road. This is the area that transmits forces between the tire and the road via pressure and friction. To truly understand some of the peculiarities of tire mechanics it is necessary to get some insights on what happens in the contact patch.

Handling of road vehicles is strongly affected by the mechanical behavior of the wheels with tire, that is by the *relationship* between the *kinematics* of the rigid rim and the *force* exerted by the road. This chapter is indeed devoted to the analysis of experimental tests. The development of simple, yet significant, tire models is done in Chap. 11.

2.1 The Tire as a Vehicle Component

A wheel with tire is barely a wheel, in the sense that it behaves quite differently from a rigid wheel.[3] This is a key point to really understand the mechanics of wheels with tires. For instance, a rigid wheel touches the (flat) road at one point C, whereas a tire has a fairly large contact patch. Pure rolling of a rigid wheel is a clear kinematic concept [17], but, without further discussion, it is not obvious whether an analogous concept is even meaningful for a tire. Therefore, we have to be careful in stating as clearly as possible the concepts needed to study the mechanics of wheels with tire.

Moreover, the analysis of tire mechanics will be developed with no direct reference to the dynamics of the vehicle. This may sound a bit odd, but it is not. The goal here is to describe the *relationship* between the *motion* and *position* of the rim and the *force* exchanged with the road through the *contact patch*:

$$\text{rim kinematics} \quad \Longleftrightarrow \quad \text{force and moment}$$

Once this description has been obtained and understood, then it can be employed as one of the fundamental components in the development of suitable models for vehicle dynamics, but this is the subject of other chapters.

Three basic components play an active role in tire mechanics:

1. the *rim*, which is assumed to be a rigid body;
2. the flexible *carcass* of the inflated tire;
3. the *contact patch* between the tire and the road.

2.2 Carcass Features

The tire carcass C is a highly composite and complex structure. Here we look at the tire as a vehicle component [19] and therefore it suffices to say that the inflated carcass, with its flexible sidewalls, is moderately compliant in all directions (Figs. 2.1 and 2.2). The external belt is also flexible, but quite inextensible (Fig. 2.3). For instance, its circumferential length is not very much affected by the vertical load acting on the tire. The belt is covered with tread blocks whose elastic deformation and grip features highly affect the mechanical behavior of the wheel with tire [13–15].

Basically, the carcass can be seen as a nonlinear elastic structure with small hysteresis due to rate-dependent energy losses. It is assumed here that the carcass and the belt have negligible inertia, in the sense that the inertial effects are small in comparison with other causes of deformation. This is quite correct if the road is flat and the wheel motion is not "too fast".

---

[3] A rigid wheel is essentially an axisymmetric convex rigid surface. The typical rigid wheel is a toroid.

Fig. 2.2 Radial flexibility of
the tire carcass [13]

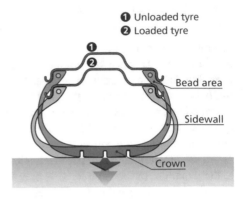

Fig. 2.3 Structure of a radial
tire [13]

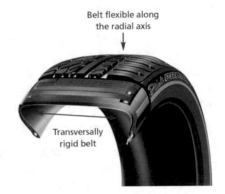

2.3 Contact Patch

Tires are made from rubber, that is elastomeric materials to which they owe a large
part of their grip capacity [25]. Grip implies contact between two surfaces: one is
the tire surface and the other is the road surface.

The contact patch (or footprint) $\mathcal{P}$ is the region where the tire is in contact with
the road surface. Most tires have a tread pattern, with lugs and voids, and hence the
contact patch is the union of many small regions (Fig. 2.4). It should be emphasized
that the shape and size of the contact patch, and also its position with respect to the
rim, depend on the tire operating conditions.

Grip depends, among other things, on the *type* of road surface, its *roughness* and
whether it is *wet or not*. More precisely, grip comes basically from road roughness
effects and molecular adhesion.

Road roughness effects, also known as indentation, require small bumps measuring
a few microns to a few millimeters (Fig. 2.5), which dig into the surface of the rubber.
On the other hand, *molecular adhesion* necessitates direct contact between the rubber
and the road surface, i.e. the road must be dry.

Percentage of contact with the road and pressure in the contact patch				
Inflation pressure	2 bars 150 cm²		8 bars 500 cm²	
Rubber/void percentage	30 %		30 %	
Mean pressure in the contact patch	3 bars		11 bars	
Percentage of rubber in contact with the road (load bearing surface)	on very rough surfaces 7%	on slightly rough surfaces 60%	on very rough surfaces 7%	on slightly rough surfaces 60%
Local pressure on rough spots (mean value)	43 bars	5 bars	157 bars	18 bars

Fig. 2.4 Typical contact patches with tread pattern (1 bar = 0.1 MPa = 14.5 psi) [13]

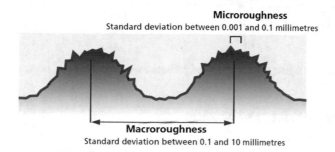

Microroughness
Standard deviation between 0.001 and 0.1 millimetres

Macroroughness
Standard deviation between 0.1 and 10 millimetres

Fig. 2.5 Road roughness description [13]

Two main features of road surface geometry must be examined and assessed when considering tire grip, as shown in Fig. 2.5:

Macroroughness: this is the name given to the road surface texture when the distance between two consecutive rough spots is between 100 microns and 10 millimeters. This degree of roughness contributes to indentation, and to the drainage and storage of water. The load-bearing surface, which depends on road macroroughness, must also be considered since it determines local pressures in the contact patch.

Microroughness: this is the name given to the road surface texture when the distance between two consecutive rough spots is between 1 and 100 microns. It is this degree of roughness that is mainly responsible for tire grip via the road roughness effects. Microroughness is related to the surface roughness of the aggregates and sands used in the composition of the road surface.

In Fig. 2.6 the contact patch is schematically shown as a single region.

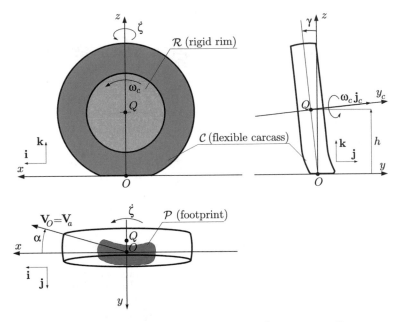

Fig. 2.6 Wheel with tire: nomenclature and reference system $S_w = (x, y, z; O)$

2.4 Rim Position and Motion

For simplicity, the *road* is assumed to have a hard and *flat* surface, like a *geometric plane*. This is a good model for any road with high quality asphalt paving, since the texture of the road surface is not relevant for the definition of the rim kinematics (while it highly affects grip [13]).

The *rim* $\mathcal{R}$ is assumed to be a *rigid body*, and hence, in principle, it has six degrees of freedom. However, only two degrees of freedom (instead of six) are really relevant for the rim *position* because the road is *flat* and the wheel rim is *axisymmetric*. Let Q be a point on the rim axis y_c (Fig. 2.6). Typically, although not strictly necessary, a sort of midpoint is taken. The *position* of the rim with respect to the flat road depends only on the *height h* of Q and on the *camber angle* γ (i.e., the inclination) of the rim axis y_c. More precisely, h is the distance of Q from the road plane and γ is the angle between the rim axis and the road plane.

In [32] and [22] the distance h is called loaded tire radius. In our opinion, the word "radius" may be misleading. There is no circle with radius h.

Now, we can address how to describe the rim velocity field.

The rim, being a rigid body, has a well defined angular velocity $\boldsymbol{\Omega}$. Therefore, the velocity of any point P of the (space moving with the) rim is given by the well known equation [12, p. 124]

$$\mathbf{V}_P = \mathbf{V}_Q + \boldsymbol{\Omega} \times QP \tag{2.1}$$

where $\mathbf{V}_Q$ is the velocity of Q and QP is the vector connecting Q to P. The three components of $\mathbf{V}_Q$ and the three components of $\boldsymbol{\Omega}$ are, e.g., the *six* parameters which completely determine the rim *velocity* field.

2.4.1 Reference System

A moving reference system $\mathbf{S}_w = (x, y, z; O)$ is depicted in Fig. 2.6. It is defined in a fairly intuitive way. The y-axis is the intersection between a vertical plane containing the rim axis y_c and the road plane. The x-axis is given by the intersection of the road plane with a plane containing Q and normal to y_c. The intersection between axes x and y defines the origin O as a point on the road. The z-axis is vertical, that is perpendicular to the road, with the positive direction upward.[4] The unit vectors marking the positive directions are ($\mathbf{i}$, $\mathbf{j}$, $\mathbf{k}$), as shown in Fig. 2.6.

An observation is in order here. The *directions* ($\mathbf{i}$, $\mathbf{j}$, $\mathbf{k}$) have a physical meaning, in the sense that they clearly mark some of the peculiar features of the rim with respect to the road. As a matter of fact, $\mathbf{k}$ is perpendicular to the road, $\mathbf{i}$ is perpendicular to both $\mathbf{k}$ and the rim axis $\mathbf{j}_c$, $\mathbf{j}$ follows accordingly. However, the *position* of the Cartesian axes (x, y, z) is arbitrary, since there is no physical reason to select a point as the origin O. This is an aspect whose implications are often underestimated.

The selected point O is often called *center of the footprint*, or center of the wheel.

2.4.2 Rim Kinematics

The moving reference system $\mathbf{S}_w = (x, y, z; O)$ allows a more precise description of the rim kinematics. On the other hand, a reference system $\mathbf{S}_f = (x_f, y_f, z_f; O_f)$ fixed to the road is not very useful in this context.

Let $\mathbf{j}_c$ be the direction of the rim spindle axis y_c

$$\mathbf{j}_c = \cos \gamma \, \mathbf{j} + \sin \gamma \, \mathbf{k} \tag{2.2}$$

where the *camber angle* γ of Fig. 2.6 is positive. The total *rim angular velocity* $\boldsymbol{\Omega}$ is

$$\begin{aligned}
\boldsymbol{\Omega} &= \dot{\gamma}\,\mathbf{i} + \dot{\theta}\,\mathbf{j}_c + \dot{\zeta}\,\mathbf{k} \\
&= \dot{\gamma}\,\mathbf{i} + \omega_c\,\mathbf{j}_c + \omega_z\,\mathbf{k} \\
&= \dot{\gamma}\,\mathbf{i} + \omega_c \cos \gamma \, \mathbf{j} + (\omega_c \sin \gamma + \omega_z)\,\mathbf{k} \\
&= \Omega_x\,\mathbf{i} + \Omega_y\,\mathbf{j} + \Omega_z\,\mathbf{k}
\end{aligned} \tag{2.3}$$

---

[4] $\mathbf{S}_w$ is the system recommended by ISO (see, e.g., [20, Appendix 1])

where $\dot{\gamma}$ is the time derivative of the camber angle, $\omega_c = \dot{\theta}$ is the angular velocity of the rim about its spindle axis $\mathbf{j}_c$, and $\omega_z = \dot{\zeta}$ is the yaw rate, that is the angular velocity of the reference system S_w about the vertical axis $\mathbf{k}$.

It is worth noting that there are *two* distinct contributions to the *spin velocity* $\Omega_z \mathbf{k}$ of the rim,[5] a *camber* contribution $\omega_c \sin \gamma$ and a *yaw rate* contribution ω_z

$$\Omega_z = \omega_c \sin \gamma + \omega_z \qquad (2.4)$$

Therefore, as will be shown in Fig. 2.19, the same value of Ω_z can be the result of different operating conditions for the tire, depending on the amount of the *camber angle* γ and of the *yaw rate* ω_z.

By definition, the position vector OQ is (Fig. 2.6)

$$OQ = h(-\tan \gamma \, \mathbf{j} + \mathbf{k}) \qquad (2.5)$$

This expression can be differentiated with respect to time to obtain

$$\begin{aligned}
\mathbf{V}_Q - \mathbf{V}_O &= \dot{h}(-\tan \gamma \, \mathbf{j} + \mathbf{k}) + h \left(\omega_z \tan \gamma \, \mathbf{i} - \frac{\dot{\gamma}}{\cos^2 \gamma} \mathbf{j} \right) \\
&= h \, \omega_z \tan \gamma \, \mathbf{i} - \left(\dot{h} \tan \gamma + h \frac{\dot{\gamma}}{\cos^2 \gamma} \right) \mathbf{j} + \dot{h} \, \mathbf{k}
\end{aligned} \qquad (2.6)$$

since $d\mathbf{j}/dt = -\omega_z \mathbf{i}$. Even in steady-state conditions, that is $\dot{h} = \dot{\gamma} = 0$, we have $\mathbf{V}_Q = \mathbf{V}_O + h \, \omega_z \tan \gamma \, \mathbf{i}$ and hence the velocities of points Q and O are not exactly the same, unless also $\gamma = 0$. The camber angle γ is usually very small in cars, but may be quite large in motorcycles (up to 60 deg).

The velocity $\mathbf{V}_o = \mathbf{V}_O$ of point O has, in general, longitudinal and lateral components (Fig. 2.6)[6]

$$\begin{aligned}
\mathbf{V}_o &= V_{o_x} \mathbf{i} + V_{o_y} \mathbf{j} \\
&= V_{o_x} (\mathbf{i} - \tan \alpha \, \mathbf{j})
\end{aligned} \qquad (2.7)$$

where α is the wheel *slip angle*.

As already stated, the selection of point O is *arbitrary*, although quite reasonable. Therefore, the velocities V_{o_x} and V_{o_y} do not have much of a physical meaning. A different choice for the point O would provide different values for the very same motion. However, a wheel with tire is expected to have longitudinal velocities much higher than lateral ones, that is $|\alpha| < 12°$, as will be discussed with reference to Fig. 11.33.

Summing up, the position of the rigid rim $\mathcal{R}$ with respect to the flat road is completely determined by the following six degrees of freedom:

$h(t)$ distance of point Q from the road (often, improperly, called loaded radius);

---

[5] In the SAE terminology, it is $\omega_c \, \mathbf{j}_c$ that is called spin velocity [6, 16].

[6] The two symbols $\mathbf{V}_o$ and $\mathbf{V}_O$ are equivalent. Using $\mathbf{V}_o$ is just a matter of taste.

$\gamma(t)$ camber angle;
$\theta(t)$ rotation of the rim about its axis y_c;
$x_f(t)$ first coordinate of point O w.r.t. S_f;
$y_f(t)$ second coordinate of point O w.r.t. S_f;
$\zeta(t)$ yaw angle of the rim.

However, owing to the *circular* shape of rim and the *flatness* of the road, the kinematics of the rigid rim $\mathcal{R}$ is also fully described by the following six functions of time:
$h(t)$ distance of point Q from the road;
$\gamma(t)$ camber angle;
$\omega_c(t)$ angular velocity of the rim about its axis y_c;
$V_{o_x}(t)$ longitudinal speed of O;
$V_{o_y}(t)$ lateral speed of O;
$\omega_z(t)$ yaw rate of the moving reference system S_w.

The rim is in steady-state conditions if all these six quantities are constant in time. However, this is not sufficient for the wheel with tire to be in a stationary state. The flexible carcass and tire treads could still be under transient conditions.

Now, there is an observation whose practical effects are very important. If we are interested only in the truly kinematic (geometric) features of the rim motion, we can drop the number of required functions from six to five:

$$ h, \quad \gamma, \quad \frac{V_{o_x}}{\omega_c}, \quad \frac{V_{o_y}}{\omega_c}, \quad \frac{\omega_z}{\omega_c} \tag{2.8} $$

Essentially, we are looking at the relative values of speeds, as if their magnitude were of no relevance at all. This is what is commonly done in vehicle dynamics, as we will see soon. Again, we emphasize that a vehicle engineer should be aware of what he/she is doing.

2.5 Footprint Force

As well known (see, e.g., [27]), any set of forces or distributed loads is statically equivalent to a force–couple system at a given (arbitrary) point O. Therefore, regardless of the degree of roughness of the road, the distributed normal and tangential loads in the footprint yield a resultant force $\mathbf{F}$ and a resultant couple vector $\mathbf{M}_O$

$$ \mathbf{F} = F_x \mathbf{i} + F_y \mathbf{j} + F_z \mathbf{k} $$
$$ \mathbf{M}_O = M_x \mathbf{i} + M_y \mathbf{j} + M_z \mathbf{k} \tag{2.9} $$

The resultant couple $\mathbf{M}_O$ is simply the moment about the point O, but any other point could be selected. Therefore it has no particular physical meaning. However,

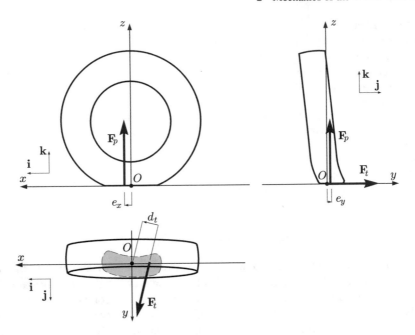

Fig. 2.7 Forces acting on the tire from the road

if O is somewhere within the footprint, the magnitude $|\mathbf{M}_O|$ is expected to be quite "small" for the wheel with tire to resemble a rigid wheel.

Traditionally, the components of $\mathbf{F}$ and $\mathbf{M}_O$ have the following names:

F_x longitudinal force ($F_x < 0$ when braking);
F_y lateral force;
F_z vertical load or normal force;

M_x overturning moment;
M_y rolling resistance moment;
M_z self-aligning torque, called vertical moment here.

The names of the force components simply reaffirm their directions with respect to the chosen reference system S_w, and hence with respect to the rim. On the other hand, the names of the moment components, which would suggest a physical interpretation, are all quite questionable. Their values depend on the arbitrarily selected point O, and hence are arbitrary by definition.

For instance, let us discuss the name "self-aligning torque" of M_z, with reference to Fig. 2.7 and Eq. (2.11). The typical explanation for the name is that "M_z produces a restoring moment on the tire to realign the direction of travel with the direction of heading", which, more precisely, means that M_z and the slip angle α are both clockwise or both counterclockwise. But the sign and magnitude of M_z depend on the position of O, which could be anywhere! The selected origin O has nothing special, nothing at all. Therefore, the very same physical phenomenon, like in Fig. 2.7, may be described with O anywhere and hence by any value of M_z. The inescapable conclusion is that the name "self-aligning torque" is totally meaningless and even misleading.[7] For these reasons, here we prefer to call M_z the *vertical moment*. Similar considerations apply to M_x and M_y.

It is a classical result that any set of forces and couples in space, like $(\mathbf{F}, \mathbf{M}_O)$, is statically equivalent to a unique wrench [27]. However, in tire mechanics it is more convenient, although not mandatory, to represent the force–couple system $(\mathbf{F}, \mathbf{M}_O)$ by *two* properly located *perpendicular forces* (Fig. 2.7): a vertical force $\mathbf{F}_p = F_z \mathbf{k}$ having the line of action passing through the point with coordinates $(e_x, e_y, 0)$ such that

$$M_x = F_z e_y \quad \text{and} \quad M_y = -F_z e_x \tag{2.10}$$

and a tangential force $\mathbf{F}_t = F_x \mathbf{i} + F_y \mathbf{j}$ lying in the xy-plane and having the line of action with distance $|d_t|$ from O, properly located according to the sign of d_t

$$M_z = \sqrt{F_x^2 + F_y^2}\, d_t = |\mathbf{F}_t|\, d_t \tag{2.11}$$

We remark that the two "displaced" forces $\mathbf{F}_p$ and $\mathbf{F}_t$ (Fig. 2.7) are completely equivalent to $\mathbf{F}$ and $\mathbf{M}_O$.

These forces are transferred to the rigid rim (apart for a small fraction due to the inertia and weight of the tire carcass and belt). Indeed, the equivalence of the distributed loads in the contact patch to concentrated forces and/or couples makes sense precisely because the rim is a rigid body.

2.5.1 Perfectly Flat Road Surface

To perform some further mathematical investigations, it is necessary to completely discard road roughness (Fig. 2.5) and to assume that the road surface in the contact patch is *perfectly flat*, exactly like a geometric plane (Figs. 2.6 and 2.7).[8] This is a fairly unrealistic assumption whose implications should not be underestimated.

Owing to the assumed flatness of the contact patch $\mathcal{P}$, we have that the *pressure* $p(x, y)\,\mathbf{k}$, by definition normal to the surface, is always vertical and hence forms a *parallel* distributed load. Moreover, the flatness of $\mathcal{P}$ implies that the *tangential stress* $\mathbf{t}(x, y) = t_x \mathbf{i} + t_y \mathbf{j}$ forms a *planar* distributed load. Parallel and planar distributed loads share the common feature that the resultant force and the resultant couple vector are perpendicular to each other, and therefore each force–couple system at O can be

---

[7] What is relevant in vehicle dynamics is the moment of $(\mathbf{F}, \mathbf{M}_O)$ with respect to the steering axis of the wheel. But this is another story (Fig. 3.1).

[8] More precisely, it is necessary to have a mathematical description of the shape of the road surface in the contact patch. The plane just happens to be the simplest.

further reduced to a *single* resultant force applied along the *line of action* (in general not passing through O). A few formulas should clarify the matter.

The resultant *vertical* force $\mathbf{F}_p$ and horizontal couple $\mathbf{M}_p^O$ of the distributed pressure $p(x, y)$ are given by

$$\mathbf{F}_p = F_z \mathbf{k} = \mathbf{k} \iint_{\mathcal{P}} p(x, y)dxdy$$

$$\mathbf{M}_p^O = M_x \mathbf{i} + M_y \mathbf{j} = \iint_{\mathcal{P}} (x\mathbf{i} + y\mathbf{j}) \times \mathbf{k}\, p(x, y)dxdy \tag{2.12}$$

where

$$M_x = \iint_{\mathcal{P}} y\, p(x, y)dxdy = F_z e_y, \qquad M_y = -\iint_{\mathcal{P}} x\, p(x, y)dxdy = -F_z e_x \tag{2.13}$$

As expected, $\mathbf{F}_p$ and $\mathbf{M}_p^O$ are perpendicular. As shown in (2.13), the force–couple resultant $(\mathbf{F}_p, \mathbf{M}_p^O)$ can be reduced to a single force $\mathbf{F}_p$ having a vertical line of action passing through the point with coordinates $(e_x, e_y, 0)$, as shown in Fig. 2.7.

The resultant *tangential* force $\mathbf{F}_t$ and vertical couple $\mathbf{M}_t^O$ of the distributed tangential (grip) stress $\mathbf{t}(x, y) = t_x \mathbf{i} + t_y \mathbf{j}$ are given by

$$\mathbf{F}_t = F_x \mathbf{i} + F_y \mathbf{j} = \iint_{\mathcal{P}} (t_x(x, y)\mathbf{i} + t_y(x, y)\mathbf{j})dxdy$$

$$\mathbf{M}_t^O = M_z \mathbf{k} = \iint_{\mathcal{P}} (x\mathbf{i} + y\mathbf{j}) \times \big(t_x(x, y)\mathbf{i} + t_y(x, y)\mathbf{j}\big)dxdy \tag{2.14}$$

$$= \mathbf{k} \iint_{\mathcal{P}} \big(x\, t_y(x, y) - y\, t_x(x, y)\big)dxdy = \mathbf{k}\, d_t \sqrt{F_x^2 + F_y^2}$$

where

$$F_x = \iint_{\mathcal{P}} t_x(x, y)dxdy, \qquad F_y = \iint_{\mathcal{P}} t_y(x, y)dxdy \tag{2.15}$$

$$d_t = \frac{M_z}{\sqrt{F_x^2 + F_y^2}} \tag{2.16}$$

Also in this case $\mathbf{F}_t$ and $\mathbf{M}_t^O$ are perpendicular. As shown in (2.14), the force–couple resultant $(\mathbf{F}_t, \mathbf{M}_t^O)$ can be reduced to a tangential force $\mathbf{F}_t$, lying in the xy-plane and having a line of action with distance $|d_t|$ from O (properly located according to the sign of d_t), as shown in Fig. 2.7.

Fig. 2.8 Example of
distributed tangential stress
in the contact patch, along
with the corresponding
resultant tangential force $\mathbf{F}_t$.
Reference system as in
Fig. 2.7 (bottom)

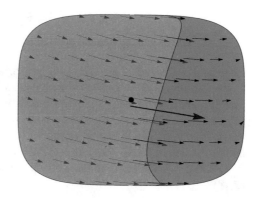

Obviously, the more general (2.9) still holds

$$\mathbf{F} = \mathbf{F}_p + \mathbf{F}_t$$
$$\mathbf{M}_O = \mathbf{M}_p^O + \mathbf{M}_t^O \tag{2.17}$$

An example of distributed tangential stress is shown in Fig. 2.8. It was obtained
by means of the tire brush model, a topic developed in Chap. 11.

2.6 Global Mechanical Behavior

The analysis developed so far provides the tools for quite a precise description of
the global mechanical behavior of a *real* wheel with tire interacting with a road.
More precisely, as already stated on Sect. 2.1, we are interested in the *relationship*
between the *motion* and *position* of the rim and the *force* exchanged with the road in
the *contact patch*.

We assume as given, and constant in time, both the wheel with tire (including its
inflation pressure and temperature field) and the road type (including its roughness).
Therefore we assume all grip features as given and constant in time.

2.6.1 Tire Transient Behavior

Knowing the mechanical behavior means knowing the relationships between the six
kinematical parameters $(h, \gamma, \omega_c, V_{o_x}, V_{o_y}, \omega_z)$ that fully characterize the position
and the motion of the rigid rim and the force–couple resultant $(\mathbf{F}, \mathbf{M}_O)$. We recall
that the inertial effects of the carcass are assumed to be negligible.

Owing mostly to the flexibility of the tire structure, these relationships are of
differential type, that is there exist *differential* equations

$$\mathbf{f}(\dot{\mathbf{F}}, \mathbf{F}, h, \gamma, \omega_c, V_{o_x}, V_{o_y}, \omega_z) = \mathbf{0}$$
$$\mathbf{g}(\dot{\mathbf{M}}_O, \mathbf{M}_O, h, \gamma, \omega_c, V_{o_x}, V_{o_y}, \omega_z) = \mathbf{0}$$

(2.18)

In general, differential equations of higher order may be needed.

The identification of these differential equations by means solely of experimental tests is a formidable task. The point here is not to find them, but to appreciate that the transient behavior of a wheel with tire does indeed obey differential equations, maybe like in (2.18). Which also implies that *initial conditions* have to be included and the values of $(\mathbf{F}, \mathbf{M}_O)$ at time t depend on time history.

In Chap. 11, suitable models will be developed that allow to partially identify (2.18).

2.6.2 Tire Steady-State Behavior

If all features are constant (or, at least, slowly varying) in time, the overall system is in steady-state conditions. Mathematically, it means that there exist, instead of (2.18), the following *algebraic* functions

$$\mathbf{F} = \bar{\mathbf{F}}(h, \gamma, \omega_c, V_{o_x}, V_{o_y}, \omega_z)$$
$$\mathbf{M}_O = \overline{\mathbf{M}}_O(h, \gamma, \omega_c, V_{o_x}, V_{o_y}, \omega_z)$$

(2.19)

which relate the rim position and steady-state motion to the force and moment acting on the tire from the footprint. In other words, given the steady-state kinematics of the rim, we know the (constant in time) forces and couples (but not viceversa).[9]

The algebraic functions in (2.19) are, by definition, the equilibrium states of the differential equations (2.18)

$$\mathbf{f}(\mathbf{0}, \bar{\mathbf{F}}, h, \gamma, \omega_c, V_{o_x}, V_{o_y}, \omega_z) = \mathbf{0}$$
$$\mathbf{g}(\mathbf{0}, \overline{\mathbf{M}}_O, h, \gamma, \omega_c, V_{o_x}, V_{o_y}, \omega_z) = \mathbf{0}$$

(2.20)

Equations (2.19) can be split according to (2.17)

$$\mathbf{F}_p = F_z \mathbf{k} = \bar{\mathbf{F}}_p(h, \gamma, \omega_c, V_{o_x}, V_{o_y}, \omega_z)$$
$$\mathbf{F}_t = F_x \mathbf{i} + F_y \mathbf{j} = \bar{\mathbf{F}}_t(h, \gamma, \omega_c, V_{o_x}, V_{o_y}, \omega_z)$$
$$\mathbf{M}_p^O = M_x \mathbf{i} + M_y \mathbf{j} = \overline{\mathbf{M}}_p^O(h, \gamma, \omega_c, V_{o_x}, V_{o_y}, \omega_z)$$
$$\mathbf{M}_t^O = M_z \mathbf{k} = \overline{\mathbf{M}}_t^O(h, \gamma, \omega_c, V_{o_x}, V_{o_y}, \omega_z)$$

(2.21)

---

[9] We remark that, as discussed in Chap. 11, steady-state kinematics of the rim does not necessarily implies steady-state behavior of the tire.

Fig. 2.9 Flat roadway testing machine (Calspan's Tire Research Facility)

2.6.3 Simplifications Based on Tire Tests

Typical *tire tests* (like those in Figs. 2.9 and 2.10) are aimed at investigating some aspects of these functions. It arises that the pressure-dependent forces and torques can be simplified drastically, since they are functions of h and γ only

$$\mathbf{F}_p = F_z(h, \gamma)\,\mathbf{k}$$
$$\mathbf{M}_p^O = M_x(h, \gamma)\,\mathbf{i} + M_y(h, \gamma)\,\mathbf{j} \tag{2.22}$$

Actually, quite often the *vertical load* F_z takes the place of h as an independent variable, as discussed in Sect. 2.9. This is common practice, although it appears to be rather questionable in a neat approach to the analysis of tire mechanics. As already stated, a clearer picture arises if we follow the approach "impose the whole kinematics of the rim, measure all the forces in the contact patch" [20, p. 62].

2.6.3.1 Speed Independence (Maybe)

Moreover, tire tests suggest that the *grip force* $\mathbf{F}_t = F_x\,\mathbf{i} + F_y\,\mathbf{j}$ and moment $\mathbf{M}_t^O = M_z\,\mathbf{k}$ are *almost speed-independent*, if ω_c is not too high. Essentially, it means that

$$F_x = \overline{F}_x(h, \gamma, \omega_c, V_{o_x}, V_{o_y}, \omega_z)$$
$$F_y = \overline{F}_y(h, \gamma, \omega_c, V_{o_x}, V_{o_y}, \omega_z) \tag{2.23}$$
$$M_z = \overline{M}_z(h, \gamma, \omega_c, V_{o_x}, V_{o_y}, \omega_z)$$

Fig. 2.10 Drum testing machine [13]

can be replaced by the following functions of only *five* variables, as anticipated in (2.8):

$$F_x = \widetilde{F}_x \left(h, \gamma, \frac{V_{o_x}}{\omega_c}, \frac{V_{o_y}}{\omega_c}, \frac{\omega_z}{\omega_c} \right)$$

$$F_y = \widetilde{F}_y \left(h, \gamma, \frac{V_{o_x}}{\omega_c}, \frac{V_{o_y}}{\omega_c}, \frac{\omega_z}{\omega_c} \right) \qquad (2.24)$$

$$M_z = \widetilde{M}_z \left(h, \gamma, \frac{V_{o_x}}{\omega_c}, \frac{V_{o_y}}{\omega_c}, \frac{\omega_z}{\omega_c} \right)$$

In other words, we assume that the grip-dependent forces and moments depend on the *geometrical* features of the rim motion (i.e., the trajectories), and not on how fast the motion develops in time. Therefore, we are discarding all inertial effects and any influence of speed on the phenomena related to grip. Of course, this may not be true at very high speeds, like in competitions.

Actually, as will be discussed in Sect. 2.9, it is convenient to employ the *tire slips* as independent kinematic variables. Therefore, (2.24) is usually replaced by (2.79). But to do that we need first to define the *pure rolling* condition for tires, as done in (2.25), (2.26), and (2.27).

2.7 Definition of Pure Rolling for Tires

Pure rolling, in case of *rigid* bodies in *point contact*, requires two kinematical conditions to be fulfilled: *no sliding* and *no mutual spin*. However, the term pure rolling is somehow ambiguous, since absence of sliding does not exclude the transmission of a tangential force, lower in magnitude than the limiting friction. In [10, p. 242], the terms *free rolling* and *tractive rolling* are used therefore to describe pure rolling of rigid bodies in which the tangential force is zero and non-zero, respectively.

These concepts and results are not, however, immediately applicable for the definition of pure rolling of a wheel with tire. As a matter of fact, there are no rigid surfaces in contact and the footprint is certainly not a point (Fig. 2.4). Therefore, even if it is customary to speak of pure rolling of a wheel with tire, it should be clear that it is a *different concept* than pure rolling between rigid bodies.

A reasonable definition of *pure rolling* for a wheel with tire, in steady-state conditions[10] and moving on a flat surface, is that the grip actions **t** have no *global* effect, that is

$$F_x = 0 \tag{2.25}$$
$$F_y = 0 \tag{2.26}$$
$$M_z = 0 \tag{2.27}$$

These equations do not imply that the local tangential stresses **t** in the contact patch are everywhere equal to zero, but only that their force–couple resultant is zero (cf. (2.14) and see Fig. 11.52). Therefore, the road applies to the wheel only a vertical force $\mathbf{F}_p = F_z\,\mathbf{k}$ and a horizontal moment $\mathbf{M}_p^O = M_x\,\mathbf{i} + M_y\,\mathbf{j} \simeq M_y\,\mathbf{j}$.

Therefore, in our analysis, pure rolling of a wheel with tire means *torque rolling*. However, owing to the small values of the rolling resistance coefficient f_r (defined in Sect. 2.13), there is not much quantitative difference between torque rolling and tractive rolling for a wheel with tire.

The goal now is to find the *kinematical conditions* to be *imposed* to the rim to fulfill Eqs. (2.25)–(2.27), that is to have the just defined pure rolling conditions. In general, the six parameters in Eqs. (2.21) should be considered. However, it is more common to assume that *five* parameters suffice, like in (2.24) (as already discussed, it is less general, but simpler, to assume that the speed is not relevant)

$$\widetilde{F}_x\left(h, \gamma, \frac{V_{ox}}{\omega_c}, \frac{V_{oy}}{\omega_c}, \frac{\omega_z}{\omega_c}\right) = 0 \tag{2.28}$$

$$\widetilde{F}_y\left(h, \gamma, \frac{V_{ox}}{\omega_c}, \frac{V_{oy}}{\omega_c}, \frac{\omega_z}{\omega_c}\right) = 0 \tag{2.29}$$

$$\widetilde{M}_z\left(h, \gamma, \frac{V_{ox}}{\omega_c}, \frac{V_{oy}}{\omega_c}, \frac{\omega_z}{\omega_c}\right) = 0 \tag{2.30}$$

2.7.1 Zero Longitudinal Force (Rolling Radius)

First, let us consider Eq. (2.28) alone

---

[10] We have basically a steady-state behavior even if the operating conditions do not change "too fast".

Fig. 2.11 Longitudinal pure
rolling of a cambered wheel.
Definition of $c_r < 0$ and of
point C

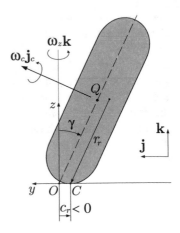

$$\widetilde{F}_x \left(h, \gamma, \frac{V_{o_x}}{\omega_c}, \frac{V_{o_y}}{\omega_c}, \frac{\omega_z}{\omega_c} \right) = 0 \qquad (2.31)$$

which means that $F_x = 0$ if[11]

$$\frac{V_{o_x}^r}{\omega_c} = f_x \left(h, \gamma, \frac{V_{o_y}}{\omega_c}, \frac{\omega_z}{\omega_c} \right) \qquad (2.32)$$

Under many circumstances, there is experimental evidence that the relation above
almost does not depend on V_{o_y}. Moreover, it can be recast in the following, more
explicit form

$$\frac{V_{o_x}^r}{\omega_c} = r_r(h, \gamma) + \frac{\omega_z}{\omega_c} c_r(h, \gamma) \qquad (2.33)$$

that is

$$V_{o_x}^r = \omega_c \, r_r(h, \gamma) + \omega_z \, c_r(h, \gamma) \qquad (2.34)$$

where c_r is a (short) signed length, as shown in Fig. 2.11. This equation means that
there exists a *special point* C of the y-axis such that

$$OC = c_r(h, \gamma) \, \mathbf{j} \qquad (2.35)$$

Like O, also point C belongs to the moving reference system $\mathbf{S}_w$.
    Therefore, (2.33) can be rearranged to get

$$V_{c_x}^r(\omega_c, h, \gamma) = \omega_c r_r(h, \gamma) = V_{o_x}^r - \omega_z c_r(h, \gamma) \qquad (2.36)$$

---

[11] As a general rule, a subscript or a superscript r means "pure rolling".

This is quite a remarkable result and clarifies the role of point C: the condition $F_x = 0$ requires point C to have a longitudinal velocity $V^r_{C_x} = \omega_c r_r(h, \gamma)$, regardless of the value of the yaw rate ω_z.

The function $r_r(h, \gamma)$ in (2.33) can be seen as a sort of longitudinal *rolling radius* [30, p. 18], although this name would be really meaningful only for a rigid wheel. In [22, p. 3] r_r is called *effective rolling radius* .

Point C would be the point of contact in case of a rigid wheel (Fig. 2.11). For a wheel with tire, we can call C the *point of virtual contact*.

If $\gamma = 0$, the origin O of the reference system S_w (Fig. 2.6) coincides with C. That is

$$c_r(h, 0) = 0 \tag{2.37}$$

and equations become much simpler.

The value of $r_r(h, \gamma)$ for given (h, γ) can be obtained by means of the usual indoor testing machines (Figs. 2.9 and 2.10) with $\omega_z = 0$. In most practical cases, particularly for radial tires, the rolling radius is quite insensitive to (reasonable) variations of h [22, p. 461]. Therefore

$$r_r(h, \gamma) \simeq r_r(\gamma) \tag{2.38}$$

Moreover, car tires operate at low values of γ and hence have almost constant r_r. This is not true for motorcycle tires.

An additional, more difficult, test with $\omega_z \neq 0$ is required to obtain also $c_r(h, \gamma)$ and hence the variable position of the point of virtual contact C with respect to O. Only for large values of the camber angle γ, that is for motorcycle tires, the distance $|c_r|$ can reach a few centimeters (Fig. 2.14).

A rough estimate shows that the ratio $|\omega_z/\omega_c|$ is typically very small, ranging from zero (straight running) up to about 0.01. It follows that usually $|(\omega_z/\omega_c)c_r|$ is negligible and points O and C have almost the same velocity.[12] However, particularly in competitions, it could be worthwhile to have a more detailed characterization of the behavior of the tire which takes into account even these minor aspects.

2.7.2 Zero Lateral Force

We can now discuss when the lateral force and the vertical moment are equal to zero. According to (2.29), we have that $F_y = 0$ if

$$\widetilde{F}_y \left(h, \gamma, \frac{V_{o_x}}{\omega_c}, \frac{V_{o_y}}{\omega_c}, \frac{\omega_z}{\omega_c} \right) = 0 \tag{2.39}$$

---

[12] However, in the *brush model*, and precisely on Sect. 11.1.5, the effect on C of the elastic compliance of the carcass is taken into account.

which means

$$\frac{V_{o_y}}{\omega_c} = f_y\left(h, \gamma, \frac{\omega_z}{\omega_c}\right) \tag{2.40}$$

where, as suggested by the experimental tests, there is no dependence on the value of V_{o_x}. Nevertheless, it seems that (2.40) does not have a simple structure like (2.33).

2.7.3 Zero Vertical Moment

Like in (2.30), the vertical moment with respect to O is zero, that is $M_z = 0$ if

$$\tilde{M}_z\left(h, \gamma, \frac{V_{o_x}}{\omega_c}, \frac{V_{o_y}}{\omega_c}, \frac{\omega_z}{\omega_c}\right) = 0 \tag{2.41}$$

which provides

$$\frac{V_{o_y}}{\omega_c} = f_z\left(h, \gamma, \frac{\omega_z}{\omega_c}\right) \tag{2.42}$$

where, like in (2.40), there is no dependence on the value of V_{o_x}. Also in this case, it is not possible to be more specific about the structure of this equation.

2.7.4 Zero Lateral Force and Zero Vertical Moment

However, the fulfilment of *both* conditions (2.40) and (2.42) together, that is $F_y = 0$ *and* $M_z = 0$, yields these results

$$V_{o_y}^r(h, \gamma) = V_{c_y}^r(h, \gamma) = 0 \tag{2.43}$$

$$\omega_z^r = -\omega_c \sin\gamma\left(1 - \varepsilon_r\right) \qquad\qquad \Omega_z^r = \omega_c \sin\gamma\,\varepsilon_r(h, \gamma) \tag{2.44}$$

which have a simple structure. Sometimes $\varepsilon_r(h, \gamma)$ is called the *camber reduction factor* [20, p. 119], [21]. Usually, ε_r is almost constant for a given tire. Therefore, it does not really depend on h and γ

$$\varepsilon_r(h, \gamma) \simeq \varepsilon_r \tag{2.45}$$

A car tire has $0.4 < \varepsilon_r < 0.6$, while a motorcycle tire has ε_r almost equal to 0.

   Equation (2.43) requires the lateral velocity of point O, and hence also of point C, to be equal to zero.

   Equation (2.44) is equivalent to

Fig. 2.12 Yaw rate ω_z to compensate the camber induced spin ($\gamma > 0$, $\omega_z < 0$)

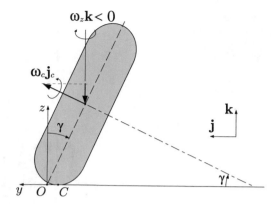

$$\frac{\omega_z^r}{\omega_c} = -\sin\gamma\left(1 - \varepsilon_r\right) \qquad (2.46)$$

that is, the camber effects have to be compensated by the proper amount of yaw rate (Fig. 2.12). As a special case, to have pure rolling, the yaw rate ω_z must be equal to zero only when $\gamma = 0$.

The physical interpretation of (2.46) is that, to have $F_y = 0$ *and* $M_z = 0$, a cambered wheel with tire must go round as shown in Fig. 2.12 and in Fig. 2.16, with no lateral velocity and with a precise combination of ω_c and ω_z. Since no condition is set by (2.46) on the speed V_{c_x}, the radius of the circular path traced on the road by point C does not matter, unless we also want $F_x = 0$.

It is worth remarking that (2.43) alone does not imply zero lateral force (Fig. 2.13). A cambered wheel can yield a lateral force even if it has no lateral velocity. Similarly, (2.44) alone does not imply zero vertical moment.

2.7.5 Pure Rolling Summary

Summing up, we have obtained the following kinematic conditions for a wheel with tire to be in what we have defined *pure rolling* in (2.25)–(2.27):

$$F_x = 0 \quad \Longleftrightarrow \quad V_{o_x}^r = \omega_c\, r_r(h, \gamma) + \omega_z c_r(h, \gamma)$$

$$\begin{cases} F_y = 0 \\ M_z = 0 \end{cases} \quad \Longleftrightarrow \quad \begin{cases} V_{o_y}^r = 0 \\ \omega_z^r = -\omega_c \sin\gamma\,(1 - \varepsilon_r) \end{cases} \qquad (2.47)$$

or, equivalently

Fig. 2.13 A cambered
wheel under two different
working conditions (see also
Fig. 2.19)

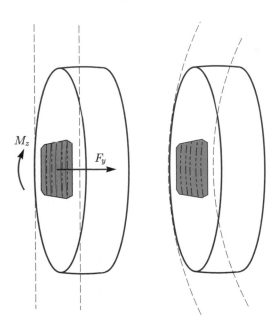

$$F_x = 0 \qquad \Longleftrightarrow \qquad \frac{V^r_{c_x}}{\omega_c} = r_r(h, \gamma)$$

$$\begin{cases} F_y = 0 \\ M_z = 0 \end{cases} \qquad \Longleftrightarrow \qquad \begin{cases} V^r_{c_y} = 0 \\ \dfrac{\omega^r_z}{\omega_c} = -\sin\gamma\,(1 - \varepsilon_r) \end{cases} \qquad (2.48)$$

These equations provide a sort of *reference condition* for the behavior of a wheel
with tire (Fig. 2.14). Moreover, they are of key relevance for the subsequent definition
of *tire slips*.

The complete characterization of pure rolling conditions essentially means obtaining the following functions (Fig. 2.14)

$$c_r(h, \gamma), \qquad r_r(h, \gamma) \simeq r_r(\gamma), \qquad \varepsilon_r(h, \gamma) \simeq \varepsilon_r \qquad (2.49)$$

Of them, the rolling radius r_r is the most important, followed by the camber reduction
factor ε_r. Of course, everything becomes much simpler if there is no camber: $c_r = 0$,
and ε_r becomes irrelevant.

The fulfillment of only the first condition in (2.47) or (2.48) corresponds to longitudinal pure rolling. The fulfillment of only the last two conditions in (2.47) or
(2.48) corresponds to lateral pure rolling.

It is worth recalling the main *assumptions* made (which are not always verified in
real life):

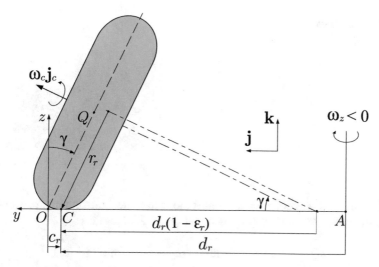

Fig. 2.14 Pure rolling of a cambered wheel with tire ($\gamma > 0, \omega_z < 0, c_r < 0, \varepsilon_r > 0$)

- negligible inertial effects (five instead of six parameters);
- grip features unaffected by speed;
- point O defined as in Fig. 2.6;
- point C not affected by ω_z;
- lateral velocity not affecting $F_x = 0$;
- longitudinal velocity not affecting $F_y = 0$ and $M_z = 0$.

2.7.6 Rolling Velocity and Rolling Yaw Rate

Point C and the first two equations in (2.48) provide the basis for the definition of the *rolling velocity* $\mathbf{V}_r$ (Fig. 2.14)

$$\mathbf{V}_r = \omega_c\, r_r\, \mathbf{i} = V_r\, \mathbf{i} = V_{c_x}^r\, \mathbf{i} \qquad (2.50)$$

Similarly, the third equation in (2.47) leads to the definition of the *rolling yaw rate* ω_r of the reference system S_w

$$\omega_r\, \mathbf{k} = -\omega_c \sin\gamma\,(1 - \varepsilon_r)\, \mathbf{k} = \omega_z^r\, \mathbf{k} \qquad (2.51)$$

Therefore, for a wheel with tire to be in total pure rolling it is necessary (according to (2.48)) that

$$\mathbf{V}_c = \mathbf{V}_r \qquad \text{and} \qquad \omega_z\, \mathbf{k} = \omega_r\, \mathbf{k} \qquad (2.52)$$

Fig. 2.15 Pure rolling of a
cambered rigid wheel
($\varepsilon_r = 0$)

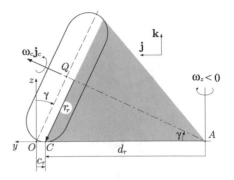

To fulfill *both* these conditions we must move the wheel on a circular path centered
at A, with radius $AC = d_r(h, \gamma)\,\mathbf{j}$ such that (Figs. 2.14 and 2.16)

$$\omega_z d_r = -\omega_c r_r \quad \text{with} \quad \omega_z = -\omega_c \sin \gamma \, (1 - \varepsilon_r) \tag{2.53}$$

which yields, for given γ, the radius d_r of the circular path for total pure rolling

$$d_r = \frac{r_r}{\sin \gamma (1 - \varepsilon_r)} \tag{2.54}$$

Typically the tire rolling radius r_r is slightly bigger than the distance of point C
from the rim axis (Fig. 2.14). We recall that a car tire has $0.4 < \varepsilon_r < 0.6$, while
a motorcycle tire has ε_r almost equal to 0. Therefore, if we take a car tire and a
motorcycle tire with the same rolling radius r_r and the same camber angle γ, to have
total pure rolling we must move the car tire on a circle about twice bigger than that
of the motorcycle tire.

To help, hopefully, better understand Fig. 2.14, we also provide in Fig. 2.15 its
counterpart in case of a rigid wheel. We can see that it behaves like a rolling rigid
cone.

It is often stated that a free-rolling tire with a camber angle would move on a
circular path [28, p. 163], [29, p. 128]. This statement is clearly incorrect. It should
be reformulated as "a tire with camber must be moved on a definite circular path
to have pure/free rolling" (Fig. 2.16). We are not doing dynamics here, but only
investigating the (almost) steady-state behavior of wheels with tire. Therefore, we
can say nothing about what a wheel would do by itself.

2.8 Definition of Tire Slips

Let us consider a wheel with tire under real operating conditions, that is *not* necessarily in pure rolling. The velocity of point C (defined in (2.35)) is called the *travel
velocity* $\mathbf{V}_c$ of the wheel (Fig. 2.14)

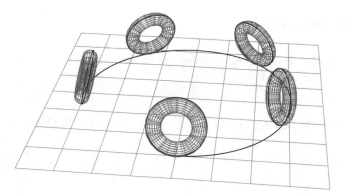

Fig. 2.16 Cambered wheel forced to move on a circular path (courtesy of M. Gabiccini)

$$\mathbf{V}_c = V_{c_x}\,\mathbf{i} + V_{c_y}\,\mathbf{j} = (V_{o_x} - \omega_z c_r)\,\mathbf{i} + V_{o_y}\,\mathbf{j} \tag{2.55}$$

The components of $\mathbf{V}_c$ also have specific names: V_{c_x} is the *forward velocity* and V_{c_y} is the *lateral velocity*. In almost all practical cases, $\omega_z \simeq 0$, and hence

$$\mathbf{V}_o \simeq \mathbf{V}_c \tag{2.56}$$

To describe any steady-state conditions of a wheel with tire we need at least two parameters plus three kinematical quantities, as in (2.24). However, it is more informative to say how "distant" these three quantities are from pure rolling. It is therefore convenient to define the *slip velocity* [24] $\mathbf{V}_s$

$$\mathbf{V}_s = \mathbf{V}_c - \mathbf{V}_r \tag{2.57}$$
$$= (V_{c_x}\,\mathbf{i} + V_{c_y}\,\mathbf{j}) - \omega_c\,r_r\,\mathbf{i} \tag{2.58}$$
$$= [(V_{o_x} - \omega_z c_r)\,\mathbf{i} + V_{o_y}\,\mathbf{j}] - \omega_c\,r_r\,\mathbf{i} \tag{2.59}$$
$$= V_{s_x}\,\mathbf{i} + V_{s_x}\,\mathbf{j} \tag{2.60}$$

as the difference between the actual travel velocity (2.55) and the rolling velocity V_r. Similarly, it is useful to define what can be called the *slip yaw rate* ω_{s_z}

$$\omega_{s_z} = \omega_z - (-\omega_c \sin\gamma\,(1 - \varepsilon_r))$$
$$= \omega_z - \omega_r \tag{2.61}$$

as the difference between the actual yaw rate ω_z of the reference system $\mathbf{S}_w$ and the rolling yaw rate ω_r.

2.8.1 Theoretical Slips

Consistently with the assumed speed independence as in (2.24), it is meaningful to divide (2.57) and (2.61) by

$$V_r = \omega_c r_r \qquad (2.62)$$

which leads to the definition of the well known (wheel with) tire *slips* σ_x, σ_y, and φ:

$$\sigma_x = \frac{V_{c_x} - \omega_c r_r}{\omega_c r_r} \qquad = \frac{V_{c_x} - V_r}{V_r} \qquad = \frac{V_{s_x}}{V_r} \qquad (2.63)$$

$$\sigma_y = \frac{V_{c_y}}{\omega_c r_r} \qquad = \frac{V_{c_y}}{V_r} \qquad = \frac{V_{s_y}}{V_r} \qquad (2.64)$$

$$\varphi = -\frac{\omega_z + \omega_c \sin \gamma \, (1 - \varepsilon_r)}{\omega_c r_r} \qquad = -\frac{\omega_z - \omega_r}{V_r} \qquad = -\frac{\omega_{s_z}}{V_r} \qquad (2.65)$$

where the slip angle α was introduced in Fig. 2.6 and in (2.7). Car tires operate with very small camber angles. Therefore, we have $c_r \simeq 0$, that is $V_{o_x} \simeq V_{c_x}$ and $V_{o_y} \simeq V_{c_y}$.

These quantities have the following names [20, 21]:

σ_x theoretical longitudinal slip ($\sigma_x > 0$ means braking);
σ_y theoretical lateral slip ($\sigma_y > 0$ means a right turn);
φ spin slip.

The first two can be thought of as the components of the *(translational) theoretical slip* σ

$$\boldsymbol{\sigma} = \sigma_x \mathbf{i} + \sigma_y \mathbf{j} = \frac{\mathbf{V}_c - \mathbf{V}_r}{V_r} = \frac{\mathbf{V}_s}{V_r} \qquad (2.66)$$

while

$$\varphi = -\frac{\omega_z - \omega_r}{V_r} = -\frac{\omega_{s_z}}{V_r} \qquad (2.67)$$

The longitudinal and lateral slips are dimensionless, whereas the spin slip is not: $[\varphi] = \mathrm{m}^{-1}$.

Quite often tire tests are conducted with $\omega_z = 0$. In that case, $\mathbf{V}_o = \mathbf{V}_c$ and the spin slip simply becomes

$$\varphi = \frac{\omega_r}{V_r} = -\frac{\sin \gamma \, (1 - \varepsilon_r)}{r_r} \qquad (2.68)$$

On the other hand, if only the yaw rate contribution is present (i.e., $\gamma = 0$), it is customary to speak of *turn slip* φ_t

$$\varphi_t = -\frac{\omega_z}{V_r} \qquad (2.69)$$

Summing up, the *pure rolling* conditions (2.47) are therefore equivalent to

$$
\begin{cases}
\sigma_x = 0 \\
\sigma_y = 0 \\
\varphi = 0
\end{cases}
\tag{2.70}
$$

which look simpler, but are useless without the availability of r_r, c_r, and ε_r in (2.49).

2.8.2 The Simple Case (No Camber)

Since in most cars the camber angles are very small, the following simplified expressions can be safely used

$$
\sigma_x = \frac{V_{o_x} - \omega_c\, r_r}{\omega_c\, r_r} = \frac{V_{o_x} - V_r}{V_r} \quad \text{and} \quad \sigma_y = \frac{V_{o_y}}{\omega_c\, r_r} = \frac{V_{o_y}}{V_r}
\tag{2.71}
$$

where the rolling radius r_r is almost constant. They are indeed much simpler than (2.63) and (2.64).

2.8.3 From Slips to Velocities

Inverting (2.63), (2.64), and (2.65), with the realistic assumption $c_r = 0$, we obtain

$$
\begin{aligned}
\frac{V_{o_x}}{\omega_c} &= (1 + \sigma_x)r_r \\
\frac{V_{o_y}}{\omega_c} &= \sigma_y r_r \\
\frac{\omega_z}{\omega_c} &= -\varphi r_r - \sin \gamma (1 - \varepsilon_r)
\end{aligned}
\tag{2.72}
$$

2.8.4 (Not so) Practical Slips

Although, as will be shown, the theoretical slip σ is a better way to describe the tire behavior, it is common practice to use the components of the *practical slip* κ instead

$$\kappa_x = \left(\frac{V_r}{V_{c_x}}\right)\sigma_x = \frac{1}{1+\sigma_x}\sigma_x = \frac{V_{c_x} - V_r}{V_{c_x}} \tag{2.73}$$

$$\kappa_y = \left(\frac{V_r}{V_{c_x}}\right)\sigma_y = \frac{1}{1+\sigma_x}\sigma_y = \frac{V_{c_y}}{V_{c_x}} = -\tan\alpha \simeq -\alpha \tag{2.74}$$

or, conversely

$$\sigma_x = \frac{1}{1-\kappa_x}\kappa_x = \kappa_x(1 + \kappa_x + O(\kappa_x^2)) \tag{2.75}$$

$$\sigma_y = \frac{1}{1-\kappa_x}\kappa_y = \kappa_y(1 + \kappa_x + O(\kappa_x^2)) \tag{2.76}$$

which also shows that practical and theoretical slips are almost equal only when the longitudinal slip is small.

Practical slips are only apparently simpler and their use should be discouraged (for instance, have a look at Fig. 11.32 to appreciate why practical slips are not so practical). The *slip ratio* $\kappa = -\kappa_x$ is also often employed, along with the *slip angle* $\alpha \simeq -\kappa_y$. This approximation is quite good because the slip angle normally does not exceed 15°, that is 0.26 rad.

As discussed in [16, p. 39] and also in [20, p. 597], a number of slip ratio definitions are used worldwide [3, 6–8, 30]. A check, particularly of the sign conventions, is therefore advisable. This can be easily done for some typical conditions like locked wheel ($\omega_c = 0$), or spinning wheel ($\omega_c = \infty$). For instance, with the definitions given here we have $\sigma_x = +\infty$, $\kappa_x = 1$ and $\kappa = -1$ for a travelling locked wheel.

2.8.5 Tire Slips are Rim Slips Indeed

It is worth remarking that *all these tire slip quantities are just a way to describe the motion of the wheel rim, not of the tire.* Therefore they do not provide any direct information on the amount of sliding at any point of the contact patch.

More precisely, *sliding* or *adhesion* are *local* features of any point in the contact patch, whereas *slip* is a *global* property of the *rim* motion as a rigid body. They are completely different concepts. In this regard the name "tire slips" may be misleading. A more appropriate name would have been "rim slips".

This statement is corroborated by the observation that all kinematic quantities introduced in this chapter refer to the rim motion. Actually, to find the kinematics of some points of the tire you have to await till the last chapter.

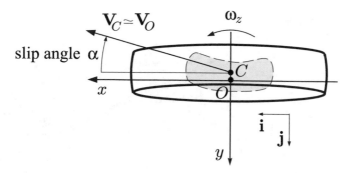

Fig. 2.17 Point O and point C, and slip angle α (top view)

2.8.6 *Slip Angle*

The *slip angle* α is defined as the angle between the rolling velocity $\mathbf{V}_r = V_r\,\mathbf{i}$ and the *travel velocity* $\mathbf{V}_c \simeq \mathbf{V}_o$ (Figs. 2.6 and 2.17)

$$\tan\alpha = -\frac{V_{c_y}}{V_{c_x}} \simeq -\frac{V_{o_y}}{V_{o_x}} \tag{2.77}$$

that is $V_{c_y} = -V_{c_x}\tan\alpha$, basically as in (2.7). For convenience, α is *positive* when measured *clockwise*, that is when it is like in Fig. 2.17.[13]

Of course, a non-sliding rigid wheel has a slip angle constantly equal to zero. On the other hand, a tire may very well exhibit slip angles. However, as will be shown, a wheel with tire can exchange with the road very high longitudinal and lateral forces still with *small* slip angles (as shown in Fig. 11.33). This is one of the reasons why a wheel with tire behaves quite close to a wheel, indeed.

More precisely, (2.77) can be rewritten as

$$\tan\alpha = -\frac{\sigma_y}{1+\sigma_x} = -\frac{\sigma_y}{\sigma_x}\left(\frac{\sigma}{\sigma + \dfrac{\sigma}{\sigma_x}}\right) \tag{2.78}$$

where $\sigma = |\boldsymbol{\sigma}| = \sqrt{\sigma_x^2 + \sigma_y^2}$. As shown in Fig. 2.18, if $\sigma < 0.2$ we have $|\alpha| < 10$deg. This is why real tires are built in such a way to provide the best performances with values of σ below 0.2, as will also be discussed later on with reference to Fig. 11.33.

---

[13] All other angles are positive angles if measured counterclockwise, as usually done in mathematical writing.

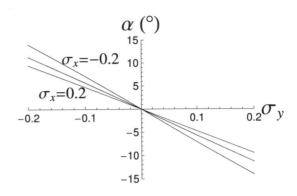

Fig. 2.18 Slip angle α as a function of σ_x and σ_y

2.9 Grip Forces and Tire Slips

In (2.24) it was suggested that the steady-state global mechanical behavior of a wheel with tire could be described by means of forces and moments depending on (h, γ) to identify the rim position, and on other *three* kinematical parameters to determine the rim motion

$$F_x = \widetilde{F}_x \left(h, \gamma, \frac{V_{o_x}}{\omega_c}, \frac{V_{o_y}}{\omega_c}, \frac{\omega_z}{\omega_c} \right)$$

$$F_y = \widetilde{F}_y \left(h, \gamma, \frac{V_{o_x}}{\omega_c}, \frac{V_{o_y}}{\omega_c}, \frac{\omega_z}{\omega_c} \right) \qquad (2.24')$$

$$M_z = \widetilde{M}_z \left(h, \gamma, \frac{V_{o_x}}{\omega_c}, \frac{V_{o_y}}{\omega_c}, \frac{\omega_z}{\omega_c} \right)$$

Moreover, we have shown that the definition of the *pure rolling* conditions ($F_x = F_y = M_z = 0$) leads naturally to the definition of *three tire slips* σ_x, σ_y, and φ.

Inserting (2.72) into (2.24), we end up with these new functions

$$F_x = \widehat{F}_x(h, \gamma, \sigma_x, \sigma_y, \varphi)$$

$$F_y = \widehat{F}_y(h, \gamma, \sigma_x, \sigma_y, \varphi) \qquad (2.79)$$

$$M_z = \widehat{M}_z(h, \gamma, \sigma_x, \sigma_y, \varphi)$$

which provide a better and clearer description of the global mechanical behavior of a tire. Indeed, by definition

$$F_x = \widehat{F}_x(h, \gamma, 0, 0, 0) = 0$$

$$F_y = \widehat{F}_y(h, \gamma, 0, 0, 0) = 0 \qquad (2.80)$$

$$M_z = \widehat{M}_z(h, \gamma, 0, 0, 0) = 0$$

Instead of the vertical height h, it is customary to employ the vertical load F_z as an input variable. This can be safely done since

$$h = h(F_z, \gamma) \tag{2.81}$$

with very little influence by the other parameters (cf. (2.22)). Therefore, the (almost) steady-state global mechanical behavior of a wheel with tire moving not too fast on a flat road is conveniently described by the following functions

$$F_x = F_x(F_z, \gamma, \sigma_x, \sigma_y, \varphi)$$
$$F_y = F_y(F_z, \gamma, \sigma_x, \sigma_y, \varphi) \tag{2.82}$$
$$M_z = M_z(F_z, \gamma, \sigma_x, \sigma_y, \varphi)$$

Similarly, (2.49) can be recast as

$$c_r(F_z, \gamma) \simeq 0, \quad r_r(F_z, \gamma) \simeq r_r(\gamma), \quad \varepsilon_r(F_z, \gamma) \simeq \varepsilon_r \tag{2.83}$$

Unfortunately, it is common practice to employ the following functions, instead of (2.82)

$$F_x = F_x^p(F_z, \gamma, \kappa_x, \alpha, \omega_z)$$
$$F_y = F_y^p(F_z, \gamma, \kappa_x, \alpha, \omega_z) \tag{2.84}$$
$$M_z = M_z^p(F_z, \gamma, \kappa_x, \alpha, \omega_z)$$

They are, in principle, equivalent to (2.82). However, using the longitudinal practical slip κ_x, the slip angle α and the yaw rate ω_z provides a less systematic description of the tire mechanical behavior. It looks simpler, but ultimately it is not.

It is often overlooked that F_x, F_y and M_z (Eqs. (2.79) and (2.82)) depend on *both* the *camber angle* γ and the spin slip φ. In other words, two operating conditions with the same φ, but obtained with different γ's, do not provide the same values of F_x, F_y and M_z, even if F_z, σ_x and σ_y are the same. For instance, the same value of φ can be obtained with no camber γ and positive yaw rate ω_z or with positive γ and no ω_z, as shown in Fig. 2.19. The two contact patches are certainly not equal to each other, and so the forces and moments. The same value of φ means that the rim has the same motion, but not the same position, if γ is different.

We remind that the moment M_z in (2.82) is with respect to a vertical axis passing through a point O chosen in quite an arbitrary way. Therefore, any attempt to attach a physical interpretation to M_z must take care of the position selected for O.

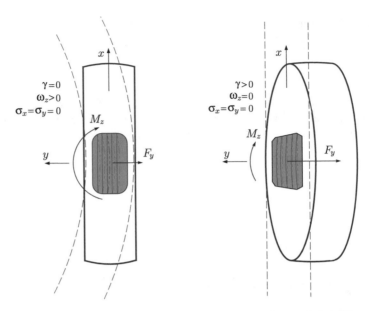

Fig. 2.19 Two different operating conditions, with the same spin slip $\varphi < 0$, but different camber angle γ (see also Fig. 2.13)

2.10   Tire Testing

Tire testing, as in Fig. 2.9, aims to fully identify the three functions (2.82) or (2.84), that is the *relationship* between the *motion* and *position* of the rim and the *force* and *moment* exchanged with the road through the contact patch

$$\text{rim kinematics} \quad \Longleftrightarrow \quad \text{force and moment} \tag{2.85}$$

Actually, this goal had already been stated in Sect. 2.1. The difference is that now we have defined the *tire slips*, that is a precise set of parameters to control the *rim kinematics*.[14]

Indoor tire testing facilities (Fig. 2.9) usually have $\omega_z = 0$ in steady-state tests, and hence lack in generality by imposing a link between γ and φ, as shown in (2.68). However, in most practical applications in road vehicles we have $|\omega_z/\omega_c| < 0,01$ and ω_z can indeed be neglected.[15]

Owing to (2.47) and (2.70), it is meaningful to perform experimental tests for the so-called *pure slip conditions*. Basically it means setting $\gamma = \varphi = 0$ and either $\sigma_y = 0$ or $\sigma_x = 0$. In the first case we have pure longitudinal slip and hence only the

---

[14] Once again, we called *tire slips* what should be called *rim slips*.

[15] In a step steer the steering wheel of a car may reach $\omega_z = 20°/\text{s}=0.35$ rad/s. At a forward speed of 20 m/s, the same wheels have about $\omega_c = 80$ rad/s. The contribution of ω_z to φ is therefore like a camber angle $\gamma \simeq 0,5°$.

longitudinal force $F_x = F_x(F_z, 0, \sigma_x, 0, 0)$, which is a very special case of (2.82). In the second case we have pure lateral slip, which allows for the experimental identification of the functions $F_y = F_y(F_z, 0, 0, \sigma_y, 0)$ and $M_z = M_z(F_z, 0, 0, \sigma_y, 0)$, which are also very special cases.

Unfortunately, the practical longitudinal slip κ_x and the slip angle α usually take the place of σ_x and σ_y, respectively [4].

2.10.1 Tests with Pure Longitudinal Slip

This kind of tests are often called drive/brake tests. Typically, they use longitudinal slip ratio sweeps with constant vertical load, constant forward velocity, and zero lateral velocity (i.e, zero slip angle).

Figure 2.20 shows the typical behavior of the longitudinal force F_x as a function of the practical longitudinal slip κ_x under pure braking conditions, for several values of the vertical load F_z. More precisely, it is the plot of $F_x^p(F_z, 0, \kappa_x, 0, 0)$. It is very important to note that:

- the maximum absolute value of F_x (i.e., the peak value $F_x^{\max}$) was obtained for $\kappa_x \simeq 0, 1$ (i.e., $\sigma_x \simeq 0, 11$);
- F_x grows *less than proportionally* with respect to the vertical load.

Both these aspects of tire behavior have great relevance in vehicle dynamics.

Also quite relevant are the values of the *longitudinal slip stiffness* C_{κ_x}, that is minus the slope of each curve at zero slip

$$C_{\kappa_x}(F_z) = -\left.\frac{\partial F_x^p}{\partial \kappa_x}\right|_{\kappa_x = 0} \tag{2.86}$$

and the *global longitudinal friction coefficient* μ_p^x, that is the ratio between the peak value $F_x^{\max} = \max(|F_x^p|)$ and the corresponding vertical load

$$\mu_p^x(F_z) = \frac{F_x^{\max}}{F_z} \tag{2.87}$$

Typically, as shown in Fig. 2.21, it slightly decreases as the vertical load grows.
On the practical side, it is of some interest to observe that:

- the experimental values are affected by significant errors;
- the tests were carried out till $\kappa_x \simeq 0.3$, to avoid wheel locking and excessive damage to the tire tread;
- the offset of F_x for $\kappa_x = 0$ is due to the rolling resistance: the wheel was (erroneously, but typically) under free rolling conditions, not pure rolling.

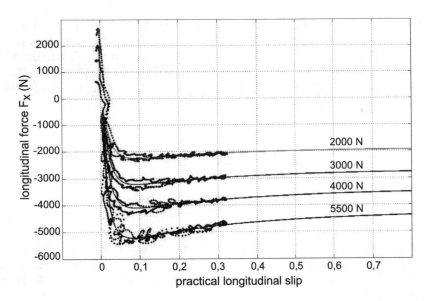

Fig. 2.20 Experimental results for a road tire: longitudinal force F_x vs practical longitudinal slip κ_x for four values of the vertical load F_z

Fig. 2.21 Global longitudinal friction coefficient μ_p^x vs vertical load F_z

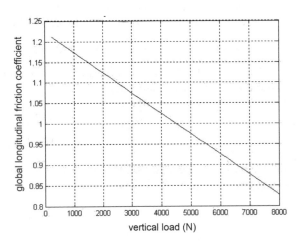

2.10.2 Tests with Pure Lateral Slip

This kind of tests are also called cornering tests. Typically, they use slip angle sweeps with pure rolling, constant vertical load, and constant belt speed V_b (Fig. 2.9). It is worth noting that the wheel forward velocity is $V_{o_x} = V_b \cos \alpha$.

Figure 2.22 shows the typical behavior of the lateral force F_y as a function of the slip angle α, for three values of F_z. More precisely, it is the plot of $F_y^p(F_z, 0, 0, \alpha, 0)$. It is very important to note that:

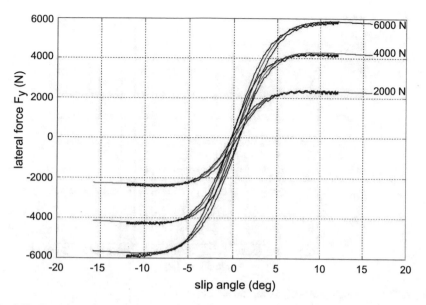

Fig. 2.22 Experimental results for a road tire: lateral force F_y vs slip angle α for three values of the vertical load F_z

- the maximum absolute value of F_y (i.e., the peak value $F_y^{\max}$) was obtained for $\alpha \simeq \pm 8°$ (i.e., $\tan \alpha = -\sigma_y = \pm 0,14$);
- F_y grows *less than proportionally* with respect to the vertical load.

Also quite relevant are the values of the *lateral slip stiffness* C_α, also called *cornering stiffness*

$$C_\alpha(F_z) = \frac{\partial F_y^p}{\partial \alpha}\bigg|_{\alpha=0} \tag{2.88}$$

that is the slope at the origin. As shown in Fig. 2.23, C_α grows less than proportionally with F_z, and actually it can even decrease at exceedingly high values of the vertical load.

Another important quantity is the *global lateral friction coefficient* μ_p^y, that is the ratio between the peak value $F_y^{\max} = \max(|F_y^p|)$ and the vertical load

$$\mu_p^y(F_z) = \frac{F_y^{\max}}{F_z} \tag{2.89}$$

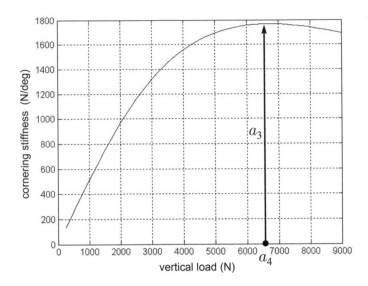

Fig. 2.23 Cornering stiffness C_α vs vertical load F_z

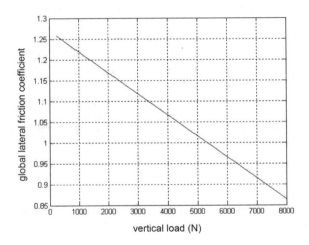

Fig. 2.24 Global lateral friction coefficient μ_p^y vs vertical load F_z

As shown in Fig. 2.24, it slightly decreases with F_z.

Comparing Figs. 2.21 and 2.24 we see that similar peak values for F_x and F_y are obtained for the same vertical load, that is $\mu_p^x \simeq \mu_p^y$. Typically, μ_p^x is slightly greater than μ_p^y.

On the practical side it is to note that:

- the experimental values are affected by small errors;
- the tests were carried out till $\alpha \simeq 12°$, to avoid damaging the tire tread.

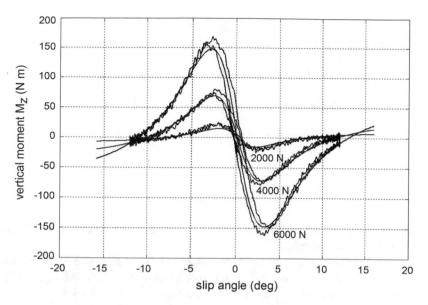

Fig. 2.25 Experimental results: vertical moment M_z vs slip angle α for three values of the vertical load F_z

Figure 2.25 shows an example of the vertical moment M_z as a function of the slip angle α, for three values of F_z, that is the plot of $M_z^p(F_z, 0, 0, \alpha, 0)$. The tests are the same of Fig. 2.22 and similar observations apply.

The behavior of $M_z(\alpha)$ is obviously very much affected by the position of the z-axis, which should be always clearly stated. Therefore, it is hard to speak of "typical behavior" of M_z, unless there is general agreement on where to locate the origin O of the reference system. This aspect could be quite relevant in the comparison and interpretation of tests performed by different institutions, particularly for motorcycle tires at large camber angles.

2.11 Magic Formula

In vehicle dynamics it is useful to have mathematical functions that fit experimental tire response curves, like those in Figs. 2.20 and 2.22. Usually, these curves have similar shapes: they grow less than proportionally, reach a maximum and then tend to a horizontal asymptote. Among the very many functions that share all these features, there is one which is almost exclusively used in vehicle dynamics. It was called *Magic Formula* (MF) by its inventors [1, 2, 23].

Although, over the years, several versions of the Magic Formula have been developed, they are all based on the following anti-symmetric function [20, 24]

$$y(x) = D \sin\{C \arctan[Bx - E(Bx - \arctan(Bx))]\} \qquad (2.90)$$

where the four coefficients are usually referred to as

$$
\begin{array}{ll}
B & \text{stiffness factor} \\
C & \text{shape factor} \\
D & \text{peak value} \\
E & \text{curvature factor}
\end{array}
\qquad (2.91)
$$

Of course, y can be either F_x or F_y, with x being the corresponding practical or theoretical slip component.

The Magic Formula belongs to the so-called *empirical tire models*, in the sense that they mimic some experimental curves, like those in Figs. 2.20 and 2.22, without any modeling of the physical phenomena involved in tire mechanics.

2.11.1 Magic Formula Properties

Let
$$B > 0 \quad E < 1 \quad \text{and} \quad 1 < C < 2 \qquad (2.92)$$

It is quite easy to show that the Magic Formula has the following properties:

- $y(0) = 0$;
- $y'(0) = BCD$ (slope at the origin);
- $y''(0) = 0$;
- $y'''(0) < 0$, if $-(1 + C^2/2) < E$;
- the function is limited: $|y(x)| \leq D$;
- the function has a relative maximum $y_m = y(x_m) = D$, with x_m such that

$$B(1 - E)x_m + E \arctan(Bx_m) = \tan(\pi/(2C)); \qquad (2.93)$$

- the value of the horizontal asymptote is

$$y_a = \lim_{x \to +\infty} y(x) = D \sin(C\pi/2) \qquad (2.94)$$

2.11.2 Fitting of Experimental Data

Probably, the most relevant features of an experimental curve like in Fig. 2.22 are the peak value y_m with the corresponding abscissa x_m, the asymptotic value y_a and the slope at the origin $y'(0)$. Therefore, to determine the four coefficients a possible procedure is as follows. First set the peak value

$$D = y_m \qquad (2.95)$$

then compute the shape factor C employing (2.94)[16]

$$C = 2 - \frac{2}{\pi} \arcsin\left(\frac{y_a}{D}\right) \qquad (2.96)$$

obtain the stiffness factor B as

$$B = \frac{y'(0)}{CD} \qquad (2.97)$$

and, finally, determine the curvature factor E from (2.93), that is by fitting the value of x_m

$$E = \frac{Bx_m - \tan(\pi/(2C))}{Bx_m - \arctan(Bx_m)} \qquad (2.98)$$

It is important that $y_a < y_m$. If they are equal (or almost equal), an unexpected plot may result.

How the four coefficients affect the Magic Formula plot is shown in Figs. 2.26, 2.27, 2.28 and 2.29. In all these plots, the thick line was obtained with $D = 3$, $C = 1.5$, $B = 20$ and $E = 0$.

Fig. 2.26 Changing the peak value D in the Magic Formula

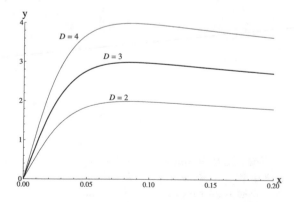

---

[16] $\sin(C\pi/2) = \sin((2 - C)\pi/2)$, since $1 < C < 2$.

Fig. 2.27 Changing the shape factor C in the Magic Formula

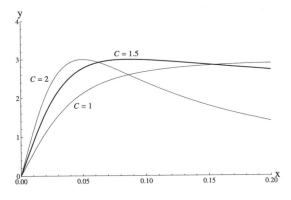

Fig. 2.28 Changing the stiffness factor B in the Magic Formula

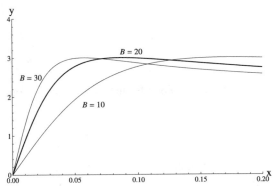

Fig. 2.29 Changing the curvature factor E in the Magic Formula

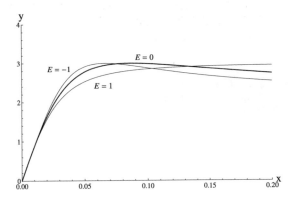

The Magic Formula usually does a good job at approximating experimental curves like in Fig. 2.20 and Fig. 2.22, although, with only four coefficients, the fitting may not be of uniform quality at all points. This aspect will be addressed in Figs. 11.25 and 11.26.

2.11.3 Vertical Load Dependence

Quite often, some coefficients of the Magic Formula are made dependent on the vertical load F_z. According to Figs. 2.21 and 2.24, the global friction coefficient $\mu_p = D/F_z$ decreases almost linearly with F_z, and hence it is quite reasonable to assume

$$D = D(F_z) = \mu_p F_z = (a_1 F_z + a_2) F_z \qquad (2.99)$$

with $a_1 < 0$.

To mimic the pattern shown in Fig. 2.23 for the slope at the origin $y'(0)$, the following formula has been suggested [24]

$$BCD = y'(0) = a_3 \sin(2 \arctan(F_z/a_4)) \qquad (2.100)$$

Actually, the formula to be used is

$$B = B(F_z) = \frac{y'(0)}{CD(F_z)} = \frac{a_3 \sin(2 \arctan(F_z/a_4))}{C(a_1 F_z + a_2) F_z} \qquad (2.101)$$

According to Figs. 2.23 and 2.24, typical values for a road car tire may be $a_1 = -0.05\,\text{kN}^{-1}$, $a_2 = 1.25$, $a_3 = 32\,\text{kN/rad} = 1.8\,\text{kN/deg}$, $a_4 = 6.5\,\text{kN}$. The interpretation of the parameters a_3 and a_4 is shown in Fig. 2.23.

2.11.4 Horizontal and Vertical Shifts

A simple generalization of the MF is by adding a vertical shift y_v and/or a horizontal shift x_h

$$y(x) = y_v + D \sin\left\{C \arctan\left[B(x + x_h) - E\left(B(x + x_h) - \arctan(B(x + x_h))\right)\right]\right\} \qquad (2.102)$$

This version of the MF can cope with rolling resistance and/or tire conicity etc.

2.11.5 Camber Dependence

The camber angle γ has a small, but significant, effect on the lateral force F_y, as will be shown in Figs. 2.37 and 2.38. Therefore, the coefficients of the MF (2.102) should depend on the camber angle as well. In particular, the Pacejka '94 coefficients are

$$C = a_0$$

$$D = (a_1 F_z + a_2) F_z (1 - a_{15} \gamma^2)$$

$$BCD = a_3 \sin(2 \arctan(F_z / a_4))(1 - a_5 |\gamma|)$$

$$B = BCD/(CD) \qquad\qquad (2.103)$$

$$E = (a_6 F_z + a_7)(1 - (a_{16}\gamma + a_{17}) \operatorname{sign}(x + x_h))$$

$$x_h = a_8 F_z + a_9 + a_{10}\gamma$$

$$y_v = a_{11} F_z + a_{12} + (a_{13} F_z + a_{14})\gamma F_z$$

Fitting 18 coefficients may not be an easy task.

An extensive description of the Magic Formula and all its subtleties can be found in [20]. Additional information is available in [11].

2.12 Mechanics of the Wheel with Tire

The main result of this chapter is that to describe the steady-state mechanics of the wheel with tire we need, as a minimum, the functions given in (2.82), that is

$$F_x = F_x(F_z, \gamma, \sigma_x, \sigma_y, \varphi)$$

$$F_y = F_y(F_z, \gamma, \sigma_x, \sigma_y, \varphi) \qquad\qquad (2.82')$$

$$M_z = M_z(F_z, \gamma, \sigma_x, \sigma_y, \varphi)$$

However, taking (2.68) into account, an even simpler formulation for the tire constitutive equations can be adopted in most cases

$$F_x = F_x(F_z, \gamma, \sigma_x, \sigma_y)$$
$$F_y = F_y(F_z, \gamma, \sigma_x, \sigma_y) \qquad\qquad (2.104)$$
$$M_z = M_z(F_z, \gamma, \sigma_x, \sigma_y)$$

Of course, they are not the whole story, and the interested reader will find in Chap. 11 many hints to better understand steady-state and also transient tire behavior.

But let us go back to (2.104). It is very informative to analyze the functions in (2.104) varying only one parameter at the time, while keeping constant (often equal to zero) all the others. These plots are like the filtered (smoothed) version of the experimental plots presented in Sect. 2.10 on tire tests. They are something that any vehicle engineer should always have clear in mind.

The plots hereafter were drawn employing the Magic Formula with the parameters reported below Eq. (2.101). The shape factor C was set equal to 1.65 for the plots of F_x, and equal to 1.3 for the plots of F_y. All forces are in kN.

2.12.1 Braking/Driving

We start with the function $F_x(F_z, 0, \sigma_x, 0) = F_x(\sigma_x)$. Most tires under pure longitudinal slip σ_x behave like in Fig. 2.30. Very near the origin the function is almost linear, but soon becomes strongly nonlinear. Relative maximum/minimum points are attained for $|\sigma_x| \simeq 0.1$. Positive σ_x means braking, negative σ_x means driving.

The effect of changing the vertical load F_z is also shown in Fig. 2.30. Obviously, the higher F_z, the higher $F_x(\sigma_x)$. However, as already mentioned on Sect. 2.10.1 and shown in Fig. 2.20, the growth of F_x with respect to F_z is less than proportional, particularly for low values of $|\sigma_x|$. This is more clearly shown in Fig. 2.31, where we see that the vertical order of the plots of the normalized longitudinal force $F_x^n = F_x(\sigma_x)/F_z$ is reversed with respect to Fig. 2.30. This kind of drawings are often called μ-slip curves.

Fig. 2.30 Longitudinal force F_x due to pure longitudinal slip σ_x, for decreasing vertical loads F_z. More precisely $F_x = F_x(F_z, 0, \sigma_x, 0)$

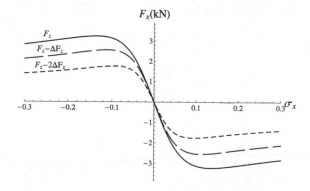

Fig. 2.31 Normalized longitudinal force F_x/F_z due to pure longitudinal slip σ_x, for decreasing vertical loads F_z (line dashing as in Fig. 2.30)

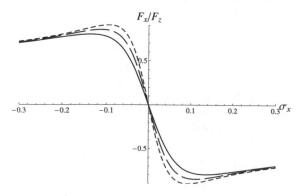

2.12.2 Cornering

Now we consider the function $F_y(F_z, 0, 0, \sigma_y) = F_y(\sigma_y)$. Most tires under pure lateral slip σ_y behave like in Fig. 2.32. Very near the origin the function is almost linear, but it soon becomes strongly nonlinear. Relative maximum/minimum points are attained for $|\sigma_y| \simeq 0.1$. Positive σ_y means negative slip angle α, and viceversa.

Moreover, the effect of changing the vertical load F_z is shown in Fig. 2.32. Again, the growth of F_y with respect to F_z is less than proportional, particularly for low values of $|\sigma_y|$. It is precisely this nonlinearity that is, let us say, activated by anti-roll bars to modify the handling setup of a car. This phenomenon is shown in Fig. 2.33, where we see that the vertical order of the plots of the normalized lateral force $F_y^n = F_y(\sigma_y)/F_z$ is reversed with respect to Fig. 2.32.

It should be noted that functions $F_x(F_z, 0, \sigma_x, 0)$ and $F_y(F_z, 0, 0, \sigma_y)$ behave in a similar way.

The experimental counterpart of Fig. 2.32 was presented in Fig. 2.22.

Fig. 2.32 Lateral force F_y due to pure lateral slip σ_y, for decreasing vertical loads F_z. More precisely $F_y = F_y(F_z, 0, 0, \sigma_y)$

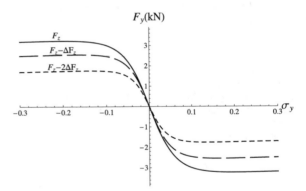

Fig. 2.33 Normalized lateral force F_y/F_z due to pure lateral slip σ_y, for decreasing vertical loads F_z (line dashing as in Fig. 2.32)

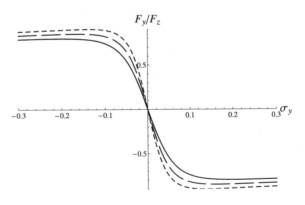

2.12.3 *Combined*

The simultaneous application of σ_x and σ_y affects the grip forces F_x and F_y as shown in Figs. 2.34 and 2.35. Basically, the total force $\mathbf{F}_t$, with components F_x and F_y, is directed like the slip vector $\boldsymbol{\sigma}$, with opposite sign, and has a magnitude almost dependent on $\sigma = |\boldsymbol{\sigma}|$

$$F_x = -\frac{\sigma_x}{\sigma} F_t(\sigma),$$

$$F_y = -\frac{\sigma_y}{\sigma} F_t(\sigma) \tag{2.105}$$

The function $F_t(\sigma)$ can be represented by the Magic Formula.

The tire behavior under combined operating conditions will be thoroughly addressed in Chap 11, where the tire brush model will be developed. At the moment you may have a look at Fig. 11.28, and also at Fig. 11.29.

It is worth noting that the two Figs. 2.34 and 2.35 convey, in different ways, exactly the same information.

Another useful plot is the one shown in Fig. 2.36. For any combination of (σ_x, σ_y), a point in the plane (F_x, F_y) is obtained. All these points fall within a circle of radius $F_t^{\max}$, usually called the *friction circle*. Lines with constant σ_y are also drawn in Fig. 2.36. Lines with constant σ_x are similar, but rotated by 90 degrees around the origin, as shown in Fig. 11.38b.

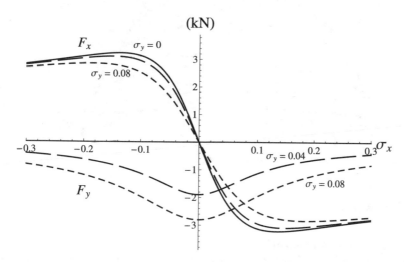

Fig. 2.34 Longitudinal force F_x and lateral force F_y due to combined longitudinal slip σ_x and lateral slip σ_y, for constant vertical load F_z. More precisely $F_x = F_x(F_z, 0, \sigma_x, \sigma_y)$ and $F_y = F_y(F_z, 0, \sigma_x, \sigma_y)$

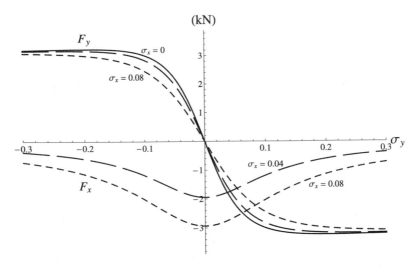

Fig. 2.35 Longitudinal force F_x and lateral force F_y due to combined longitudinal slip σ_x and lateral slip σ_y, for constant vertical load F_z. More precisely $F_x = F_x(F_z, 0, \sigma_x, \sigma_y)$ and $F_y = F_y(F_z, 0, \sigma_x, \sigma_y)$

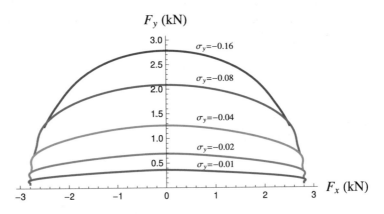

Fig. 2.36 Friction circle with lines at constant σ_y

2.12.4 Camber

Also quite relevant is the effect of the camber angle γ, alone or in combination with σ_y, on the lateral force F_y, as shown in Fig. 2.37 and, for better clarity, also in Fig. 2.38. We see that the camber effects are much stronger at low values of σ_y. However, a right amount of camber can increase a little the maximum lateral force, thus improving the car handling performance.

Fig. 2.37 Lateral force F_y due to lateral slip σ_y, for different values of the camber angle γ and constant vertical load F_z. More precisely $F_y = F_y(F_z, \gamma, 0, \sigma_y)$

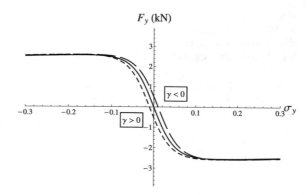

Fig. 2.38 Lateral force F_y due to camber angle γ, for different values of the lateral slip σ_y and constant vertical load F_z. More precisely $F_y = F_y(F_z, \gamma, 0, \sigma_y)$

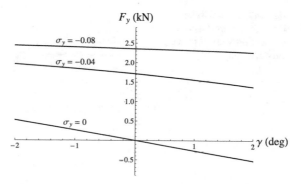

2.12.5 Grip

Finally, the effect of decreasing the grip coefficient μ is investigated. We see in Figs. 2.39 and 2.40 that, as expected, we get lower maximum tangential forces. However, it should also be noted that changing the grip does not affect the slope of the curves at the origin. The reason is that near the origin the tangential force is, by definition, very small, and hence the tire behavior is mainly affected by the tire structure, not by the available amount of grip.

2.12.6 Vertical Moment

The vertical moment M_z as a function of σ_y, with $\sigma_x = 0$, behaves as shown in Fig. 2.41. The reasons for this behavior will be discussed in Chap. 11. Basically, since $M_z = F_t d_t$ (Fig. 2.7), it is the product of a growing force times a decreasing length. It is zero when either of the two is zero.

For much more information on the mechanics of the wheel with tire we suggest to carefully read Chap. 11 on tire models.

Fig. 2.39 Longitudinal force F_x due to pure longitudinal slip σ_x, for constant vertical load F_z and decreasing grip

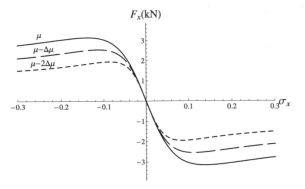

Fig. 2.40 Lateral force F_y due to pure lateral slip σ_y, for constant vertical load F_z and decreasing grip

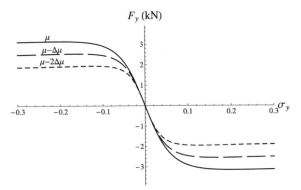

Fig. 2.41 Vertical moment M_z due to pure lateral slip σ_y

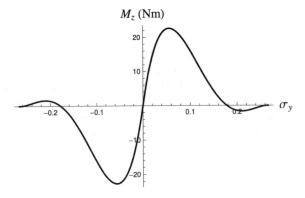

2.13 Rolling Resistance

When a car is driven in a straight line without braking or accelerating, the rolling resistance is mainly caused by the *hysteresis* in the tire due to the deflection of the carcass while rolling. Microslippage in the footprint accounts for less than 5% of total rolling resistance. As shown schematically in Figs. 2.42 and 2.43, the normal pressure p in the leading half of the contact patch is higher than that in the trailing

Fig. 2.42 Torque rolling:
$T = F_z e_x$ and $F_x = 0$

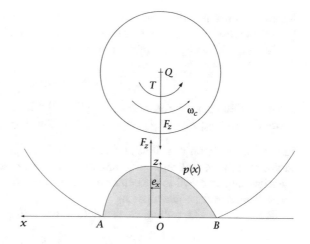

Fig. 2.43 Tractive rolling:
$F_x h = F_z e_x$ and $T = 0$

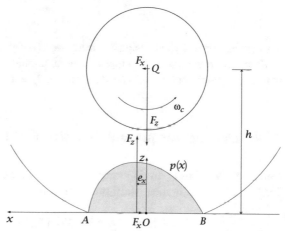

half. Therefore, the vertical resultant $F_z \mathbf{k}$ of the pressure distribution is offset by e_x towards the front of the contact patch, thus generating a *rolling resistance moment*

$$M_y = -F_z e_x \tag{2.106}$$

as already done in (2.10).

The main source of energy dissipation is therefore the visco-elasticity of the materials of which tires are made. Visco-elastic materials lose energy in the form of heat whenever they are deformed. Deformation-induced energy dissipation is the cause of about 90% of rolling resistance [15, 30].

A number of tire operating conditions affect rolling resistance. The most important are load, inflation pressure and temperature. However, as speed increases, tire's internal temperature rises, offsetting some of the increased hysteresis. Therefore, the tire rolling resistance moment is almost constant on a relatively wide range of speeds.

There are basically two different ways to balance the *rolling resistance moment*

$$M_y = -F_z e_x \tag{2.107}$$

- torque rolling: $T = F_z e_x$ and $F_x = 0$ (Fig. 2.42);
- tractive rolling: $F_x = F_z e_x / h = F_z f_r$ and $T = 0$ (Fig. 2.43).

In the first case (torque rolling) we apply a little torque $T \mathbf{j}$ to the rim to balance the moment $M_y \mathbf{j}$, while keeping $F_x = 0$. In the second case (tractive rolling), we apply a horizontal force $F_x \mathbf{i}$ to the center of the rim, which requires an opposite force to be generated in the contact patch.

In case of tractive rolling, we can define the *rolling resistance coefficient f_r*

$$f_r = \frac{e_x}{h} = \frac{F_x}{F_z} \tag{2.108}$$

The values given by tire manufacturers are measured on test drums, usually at 80 km/h in accordance with ISO measurement standards. A typical value of the rolling resistance coefficient f_r for a road car tire is $f_r = 0.006\text{--}0.016$.

2.14 Driving Torque and Tractive Force

Let us apply a (large) driving torque $\mathbf{T} = T \mathbf{j}_c$ to the rim, thus generating a tractive force F_x. Neglecting the (small) moment of inertia of the rim, we have

$$\begin{aligned}
\mathbf{T} = T \mathbf{j}_c &= -((QO \times \mathbf{F} + \mathbf{M}_O) \cdot \mathbf{j}_c) \mathbf{j}_c \\
&= \left(F_x \frac{h}{\cos \gamma} - M_y \cos \gamma - M_z \sin \gamma \right) \mathbf{j}_c
\end{aligned} \tag{2.109}$$

where (2.2) and (2.5) were employed. This expression is fairly simple because the rim axis y_c intersects the z-axis and is perpendicular to the x-axis (Figs. 2.6 and 2.7).

A driving torque $T > 0$ applied to the rim can have a large impact on the offset e_x (Fig. 2.44), and hence on the rolling resistance. As the magnitude of torque applied increases, the rolling resistance first increases mildly. Then, when slippage at the road surface becomes significant, the rolling resistance increases very rapidly. This rapid increase occurs as the maximum torque that the tire can transmit is approached [5, p. 496].

Fig. 2.44 Applying a really
large driving torque

2.14.1   Tractive Force

It is often of interest to evaluate the tractive force F_x generated by a drive torque T.
As shown in Fig. 2.45, if $\gamma = 0$, Eq. (2.109) becomes

$$T = F_x h + F_z e_x = F_x (h + e_z) = F_x a \qquad (2.110)$$

where, obviously

$$e_z = F_z \frac{e_x}{F_x} \qquad (2.111)$$

We see that, ultimately, to know F_x we have to estimate the *drive lever arm a*.

In [32, p. 51] it is shown that this lever arm a can be approximated by the rolling
radius r_r

$$a \simeq r_r \qquad (2.112)$$

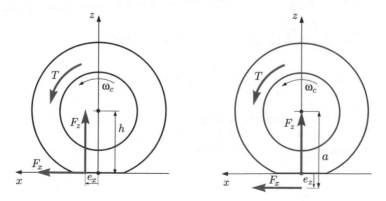

Fig. 2.45 Driving torque T and tractive force F_x: two fully equivalent schemes

Indeed, from power balance and (2.50) we have that

$$T\omega_c = F_x V_r = F_x \omega_c r_r \tag{2.113}$$

It is worth noting that the center of the wheel (and hence the vehicle) moves with speed $V_x = (1 + \sigma_x)\omega_c r_r$, with $\sigma_x < 0$.

From (2.38) and (2.110)–(2.112) we obtain that the ratio e_x/F_x should be almost constant, for given F_z.

2.15 Exercises

2.15.1 Pure Rolling

Explain the difference between torque rolling and trailing rolling in a tire.

Solution

See Sect. 2.13.

2.15.2 Theoretical and Practical Slips

Obtain the relationships between theoretical and practical slip components.

Solution

See (2.73) and (2.74).

2.15.3 Tire Translational Slips and Slip Angle

Find the tire slip angle α in the following cases:

1. $\sigma_y = 0$;
2. $\sigma_y = 0.1$ and $\sigma_x = 0$ *(only cornering)*;
3. $\sigma_y = 0.1$ and $\sigma_x = 0.1$ *(cornering and braking)*;
4. $\sigma_y = 0.1$ and $\sigma_x = -0.1$ *(cornering and driving)*.

Solution

To solve these problems we can use (2.78).

The first case is trivial. Obviously $\alpha = 0$.

The second case is also quite simple, since $\sigma_x = 0$. Therefore, $\alpha = -\arctan(0.1) = -5.7°$.

In the third case we have both lateral and longitudinal slip (cornering and braking). Still according to (2.78), $\alpha = -\arctan(0.1/(1+0.1)) = -5.2°$.

Case number four is similar to case number three, but with negative σ_x (cornering and driving). It provides $\alpha = -\arctan(0.1/(1-0.1)) = -6.3°$.

It is quite interesting to observe how the longitudinal slip affects the slip angle, for given lateral slip. All these results apply to all tires, regardless of their size, type, etc., and are not affected by camber and spin slip. We have simply done kinematics of the rigid rim.

2.15.4 Tire Spin Slip and Camber Angle

Let a tire have a rolling radius $r_r = 0.262$ m and a camber reduction factor $\varepsilon_r = 0.5$. We set the camber angle $\gamma = 3°$. Moreover, suppose the wheel is travelling at $V_r = 10$ m/s, with $\sigma_x = \sigma_y = 0$. Find the spin slip φ in the following cases:

1. *the wheel goes straight ahead;*
2. *the wheel moves clockwise on a circular path of radius $r_p = 10$ m;*
3. *as above, but counterclockwise;*
4. *the wheel moves clockwise on a circular path of radius $r_p = 50$ m.*

Solution

Let ω_c be the angular velocity of the rim around its spindle axis. Since $\sigma_x = 0$, in all cases we have $\omega_c = V_r/r_r = 38.2$ rad/s. Moreover, let ω_z be the yaw rate of the rim.

To answer the first question, which requires $\omega_z = 0$, we can use (2.68). The resulting spin slip is $\varphi = -0.1$ m^{-1}.

To answer question number two (Fig. 2.16) we first compute $\omega_z = -V_r/r_p = -10/10 = -1$ rad/s. Then, we can employ (2.65) to get $\varphi \simeq 0$. Therefore, according to (2.70), the tire is in pure rolling conditions.

In the third question we have $\omega_z = 1$ rad/s. Therefore, again from (2.65), we obtain $\varphi = -0.2$ m^{-1}.

In the last problem we have $\omega_z = -V_r/r_p = -10/50 = -0.2$ rad/s. Applying (2.65) we obtain $\varphi = -0.08$ m^{-1}.

Now we can comment on these results. A camber angle $\gamma = 3°$ is quite high for a car tire. The radius of the path $r_p = 10$ m, which is a rather sharp turn, was chosen to get $\varphi \simeq 0$ in question number two. Notably, it is more or less the kind of radius of the FSAE skid-pad event.

In this exercise we have done kinematics of the rigid rim, but taking also into account two features of the tire. Namely, the rolling radius r_r and the camber reduction factor ε_r.

2.15.5 Motorcycle Tire

Let a tire have a rolling radius $r_r = 0.262$ m and a camber reduction factor $\varepsilon_r = 0$. Moreover, suppose the wheel is travelling at $V_r = 10$ m/s, with $\sigma_x = \sigma_y = 0$ and a camber angle $\gamma = 45°$ Find the spin slip φ in the following cases:

1. *the wheel goes straight ahead;*
2. *the wheel moves clockwise on a circular path of radius $r_p = 10$ m.*

Solution

Let ω_c be the angular velocity of the rim around its spindle axis. Since $\sigma_x = 0$, in all cases we have $\omega_c = V_r/r_r = 38.2$ rad/s. Moreover, let ω_z be the yaw rate of the wheel.

To answer the first question, which requires $\omega_z = 0$, we can use (2.68). The resulting spin slip is $\varphi = -2.7\text{m}^{-1}$. As expected, the spin slip is very high.

To answer question number two (Fig. 2.16) we compute first $\omega_z = -V_r/r_p = -10/10 = -1$ rad/s. Then, we can employ (2.65) to get $\varphi = -2.6\,\text{m}^{-1}$. We see that the turn slip contribution to the spin slip is quite small.

To have pure rolling we should have $\omega_z/\omega_c = -\sin(\gamma) = -r_r/r_p = -0.7$. That is a path with radius $r_p = r_r/0.7 = 0.37$ m.

2.15.6 Finding the Magic Formula Coefficients

The results obtained in a purely lateral test on an FSAE tire are shown in Fig. 2.46. This test was conducted with an almost constant vertical load $F_z = 700$ N on a free rolling wheel with zero camber angle. We want to find a fairly good set of Magic Formula coefficients to fit these data.

Solution

The first step is finding the peak value y_m. We see that the positive and negative peak values are not exactly the same. This is quite typical. Setting $D = -y_m = -1100$ N seems a reasonable choice.

Incidentally, we observe that this tire has a global lateral friction coefficient $\mu_p^y = 1100/700 = 1.57$. Not bad.

The second step looks more tricky. We need the asymptotic value y_a, but this value is not readily available from the plot, as tests are carried out up to about $|\sigma_y| = 0.2$,

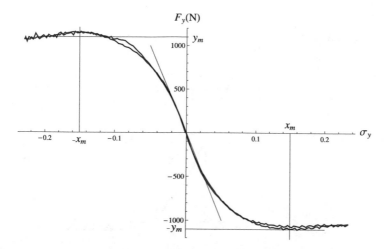

Fig. 2.46 Lateral force F_y due to pure lateral slip σ_y for an FSAE tire

that is $|\alpha| = 12°$, to avoid tire damage. We try $y_a = 800\,\text{N}$, and, according to (2.96), we get the shape factor coefficient $C = 1.48$.

The third step needs also the slope in the origin. From Fig. 2.46 we obtain $F_y'(0) = -20000\,\text{N}$, and hence, according to (2.97), $B = 20000/(CD) = 12.27$.

Finally, we see that the peak values are attained for $\sigma_y = x_m = 0.15$. Therefore, according to (2.98) and employing the just found values of C and B, we get the curvature factor $E = 0.07$.

Now we can check whether the Magic Formula with our set of parameters

$$F_y(\sigma_y) = -1100 \sin\left\{1.48 \arctan\left[12.27\sigma_y - 0.07\left(12.27\sigma_y - \arctan(12.27\sigma_y)\right)\right]\right\} \tag{2.114}$$

provides a good approximation of the experimental data of Fig. 2.46. This is done in Fig. 2.47. We see that, indeed, the smooth curve hits the target.

Actually, we observe that it has been too easy. Indeed, our guess for y_a was not really supported by available data, but nonetheless the final result is very good. Therefore, we repeat the whole procedure, employing the same values of y_m, $F_y'(0)$, and x_m, but with a very different guess about the asymptotic value y_a. For instance $y_a = 550\,\text{N}$. The resulting new set of parameters is $D = -1100\,\text{N}$, $C = 1.67$, $B = 10.91$, and $E = 0.41$. As expected, we got very different values.

Let us do it once more, with an unrealistic low value $y_a = 200\,\text{N}$. After the same steps we get $D = -1100\,\text{N}$, $C = 1.88$, $B = 9.65$, and $E = 0.72$.

You see, we selected three very different asymptotic values for y_a. Which provided very different values of C, B, and E. But what about their corresponding plots? Surprisingly enough, as shown in Fig. 2.48, they are practically indistinguishable in the range of interest, that is $-0.2 < \sigma_y < 0.2$. Therefore, the selection of y_a is not tricky at all, contrary to our first impression. Much more important are the other three conditions on y_m, $F_y'(0)$, and x_m. Indeed, requiring a function to start with a given

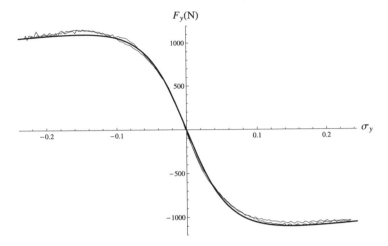

Fig. 2.47 Experimental tire data and Magic Formula fitting

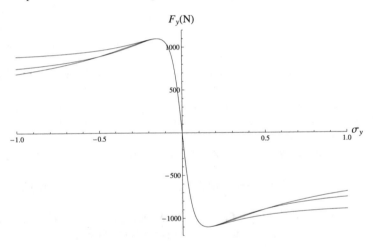

Fig. 2.48 Comparison of three Magic Formula fittings with different y_a

slope at the origin, and then to reach a maximum at a given point does not leave much room.

These results may have relevant practical implications: the very same tire behavior can be associated to very different sets (C, B, E) of three out of four Magic Formula parameters. For instance, in our case, the three sets of the coefficients (C, B, E)

- $(1.48, 12.27, 0.07)$;
- $(1.67, 10.91, 0.41)$;
- $(1.88, \quad 9.65, 0.72)$;

are pretty much equivalent in the range of interest for σ_y (Fig. 2.48). Therefore, looking at the MF parameters may not be a good way to promptly understand the mechanical behavior of a wheel with tire. It is somehow an ill-conditioned problem. More precisely, the final MF fitting plot is almost insensitive to the asymptotic value y_a, at least for $0.2 < y_a/y_m < 0.85$, whereas (C, B, E) change a lot.

From another point of view, in most cases we can select y_a in such a way to have $E = 0$, thus simplifying the Magic Formula, but still with a very good fitting.

2.16   Summary

In this chapter we have first pursued the goal of clearly describing the relevant kinematics of a wheel with tire, mainly under steady-state conditions. This had led to the definitions of slips as a measure of the extent to which the wheel with tire departs from pure rolling conditions. The slip angle has been also defined and discussed. It has been shown that a wheel with tire resembles indeed a rigid wheel because slip angles are quite small. Tire experimental tests show the relationships between the kinematics and the forces/couples the tire exchanges with the road. The Magic Formula provides a convenient way to represent these functions. Finally, the mechanics of the wheel with tire has been summarized with the aid of a number of plots.

2.17   List of Some Relevant Concepts

Section 2.1 a wheel with tire is barely a wheel;
Section 2.4.2 there are two distinct contributions to the spin velocity of the rim;
Section 2.4.2 in a wheel, longitudinal velocities are expected to be much higher than lateral ones;
Section 2.5.1 the name "self-aligning torque" is meaningless and even misleading;
Section 2.6.3.1 rim kinematics depends on six variables, but often (not always) only five may be relevant for the tire;
Section 2.7 a reasonable definition of pure rolling for a wheel with tire is that the grip local actions have no global effect;
Section 2.8 tire slips measure the distance from pure rolling;
Section 2.8.5 tire slips do not provide any direct information on the amount of sliding at any point of the contact patch;
Section 2.10 tire forces and moments depend on both the camber angle and the spin slip;
Section 2.12.6 pure rolling and free rolling are different concepts.

2.18   Key Symbols

a	drive lever arm
B	stiffness factor
C	point of virtual contact
C	shape factor
c_r	distance OC
D	peak value
E	curvature factor
F_x	longitudinal force
F_y	lateral force
F_z	vertical force
h	height above ground of the center of the rim
M_x	overturning moment
M_y	rolling resistance moment
M_z	vertical moment
O	center of the footprint
r_r	rolling radius
$\mathbf{V}_c$	travel velocity
V_{o_x}	longitudinal velocity of O
V_{o_y}	lateral velocity of O
V_r	rolling velocity
$\mathbf{V}_s$	slip velocity
α	slip angle
γ	camber angle
ε_r	camber reduction factor
μ_p^x	longitudinal friction coefficient
μ_p^y	lateral friction coefficient
σ_x	longitudinal slip
σ_y	lateral slip
φ	spin slip
ω_c	angular velocity of the rim around its axis
ω_r	rolling yaw rate of the reference system
ω_z	yaw rate of the reference system

References

1. Bakker E, Pacejka H, Lidner L (1987) Tyre modelling for use in vehicle dynamics studies. SAE Trans 96:190–204
2. Bakker E, Pacejka H, Lidner L (1989) A new tire model with an application in vehicle dynamics studies. SAE Trans 98:101–113

3. Bastow D, Howard G, Whitehead JP (2004) Car suspension and handling, 4th edn. SAE International, Warrendale
4. Bergman W (1977) Critical review of the state-of-the-art in the tire and force measurements. SAE Trans 86:1436–1450
5. Clark SK (2008) The pneumatic tire. NHTSA-DOT HS 810 561
6. Dixon JC (1991) Tyres, suspension and handling. Cambridge University Press, Cambridge
7. Font Mezquita J, Dols Ruiz JF (2006) La Dinámica del Automóvil. Editorial de la UPV, Valencia
8. Gillespie TD (1992) Fundamentals of vehicle dynamics. SAE International, Warrendale
9. Jazar RN (2017) Vehicle dynamics, 3rd edn. Springer, New York
10. Johnson KL (1985) Contact mechanics. Cambridge University Press, Cambridge
11. Leneman F, Schmeitz A (2008) MF-Tyre/MF-Swift 6.1.1. TNO Automotive, Helmond
12. Meirovitch L (1970) Methods of analytical dynamics. McGraw-Hill, New York
13. Michelin (2001) The tyre encyclopaedia. Part 1: grip. Société de Technologie Michelin, Clermont–Ferrand, [CD-ROM]
14. Michelin (2002) The tyre encyclopaedia. Part 2: comfort. Société de Technologie Michelin, Clermont–Ferrand, [CD-ROM]
15. Michelin (2003) The tyre encyclopaedia. Part 3: rolling resistance. Société de Technologie Michelin, Clermont–Ferrand, [CD-ROM]
16. Milliken WF, Milliken DL (1995) Race car vehicle dynamics. SAE International, Warrendale
17. Murray RM, Li Z, Sastry SS (1994) A mathematical introduction to robot manipulation. CRC Press, Boca Raton
18. NHTSA (2009) The effects of varying the levels of nitrogen in the inflation gas of tires on laboratory test performance. Report DOT HS 811 094, National Highway Traffic Safety Administration, Washington
19. Pacejka HB (1996) The tyre as a vehicle component. In: 26th FISITA congress '96: engineering challenge human friendly vehicles, Prague, June 17–21, pp 1–19
20. Pacejka HB (2002) Tyre and vehicle dynamics. Butterworth-Heinemann, Oxford
21. Pacejka HB (2005) Slip: camber and turning. Veh Syst Dyn 43(Supplement):3–17
22. Pacejka HB (2012) Tire and vehicle dynamics, 3rd edn. Butterworth-Heinemann, Oxford
23. Pacejka HB, Bakker E (1992) The magic formula tyre model. Veh Syst Dyn 21:1–18. https://doi.org/10.1080/00423119208969994
24. Pacejka HB, Sharp RS (1991) Shear force development by pneumatic tyres in steady state conditions: a review of modelling aspects. Veh Syst Dyn 20:121–176
25. Popov VL (2010) Contact mechanics and friction. Springer, Berlin
26. Purdy JF (1963) Mathematics underlying the design of pneumatic tires. Edwards Brothers, Ann Arbor
27. Pytel A, Kiusalaas J (1999) Engineering mechanics-statics. Brooks/Cole, Pacific Grove
28. Schramm D, Hiller M, Bardini R (2014) Vehicle dynamics. Springer, Berlin
29. Seward D (2014) Race car design. Palgrave, London
30. Wong JY (2001) Theory of ground vehicles. Wiley, New York
31. Wright C (2013) The contact patch. http://the-contact-patch.com/book/road/c1610-rubber-tyres
32. Zegelaar P (1998) The dynamic response of tyres to brake torque variations and road unevenesses. Delft University of Technology, Delft

Chapter 3
Vehicle Model for Handling and Performance

In Chap. 1 vehicle modeling was approached in general terms. To get quantitative information there is the need to be more specific.

As already stated in Sect. 1.2, in the study of *handling* and *performance* the road is assumed to be perfectly *flat* (no bumps) and with uniform features. Typically a good paved road, either dry or wet [4, 16].

The vehicle model for these operating conditions fulfills all the assumptions listed on Sect. 1.2, with the addition of:

1. small suspension deflections;
2. small tire vertical deformations;
3. small steering angles (otherwise, steering axes passing through the center of the corresponding wheel and perpendicular to the road);
4. perfectly rigid steering system.

Mathematically these additional assumptions amount to having the vehicle always in its reference configuration, as shown in Fig. 1.4, with the exception of the steering angles δ_{ij} of each wheel (δ_{11} being front-left, δ_{12} front-right, etc.). More precisely, a_1, a_2, l, t_1, t_2 and h are all *constant* during the vehicle motion. This is fairly reasonable under most operating conditions.

Typically, the steering axis (pivot line) is something like in Fig. 3.1, with a *caster angle* and a *kingpin inclination angle*. Therefore, there are a *trail* and a *scrub radius*. They are key quantities in the design of the steering system. However, their effects on the dynamics of the whole vehicle may be neglected in some cases, particularly with small steering angles and perfectly rigid steering systems (as assumed here).

The net effect of all these hypotheses is that the vehicle body has a *planar motion* parallel to the road. This is quite a remarkable fact since it greatly simplifies the

---

The original version of this chapter was revised: Figures 3.70 to 3.75 (six figures) has been updated with high resolution. The correction to this chapter is available at https://doi.org/10.1007/978-3-031-06461-6_12

M. Guiggiani, *The Science of Vehicle Dynamics*, https://doi.org/10.1007/978-3-031-06461-6_3

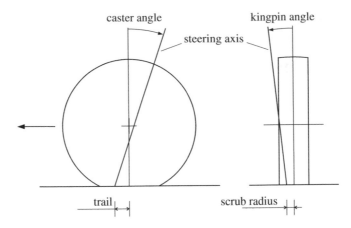

Fig. 3.1 Steering axis

analysis. Moreover, the wheel centers have a *fixed position* with respect to the vehicle body. This also helps a lot.

Notwithstanding its (apparent) simplicity, this vehicle model still exhibits a very rich and interesting dynamic behavior, and has proven to be a valuable tool to capture and understand many aspects of the dynamics of real vehicles. Of course, the underlying hypotheses impose some restrictions on its applicability, which a vehicle engineer should master.

3.1   Mathematical Framework

Basically, a vehicle model (like most physical models) is made of three separate sets of equations:

- congruence (kinematic) equations;
- equilibrium equations;
- constitutive (tire) equations.

It is convenient to consider first the whole vehicle, and then its suspensions.

3.1.1   Vehicle Axis System

As shown in Fig. 1.4, and also in Fig. 3.2, it is useful to define the *vehicle axis system* $S = (x, y, z; G)$, (also called body-fixed reference system) with unit vectors $(\mathbf{i}, \mathbf{j}, \mathbf{k})$. It has origin in the center of mass G of the whole vehicle, and axes fixed relative to the vehicle. Setting the origin in G is customary, but not mandatory.

The x axis marks the forward direction, while the y axis indicates the lateral direction (to the left from the driver's viewpoint). The z axis is vertical, that is perpendicular to the road, with positive direction upward. This *vehicle axis system* is like in [12, p. 631], but not like in [9, p. 8], [16, p. 116] and [25, p. 336] where the SAE axis system is employed. Both reference systems are right-handed, both have the same x-axis, but differ in the directions of the y and z axes.

The drawback of the SAE reference system is that the vertical loads that the road applies to the vehicle, being upward, are negative. This is the main reason for having reversed here the z axis and, consequently, also the y axis.

3.2   Vehicle Congruence (Kinematic) Equations

Kinematic equations are the mathematical relationships between the parameters that describe the vehicle motion. They involve positions, velocities and accelerations, without consideration of the masses nor the forces that caused the motion.

3.2.1   Velocity of G, and Yaw Rate of the Vehicle

The motion of the vehicle body may be completely described by its angular velocity $\boldsymbol{\Omega}$ and by the velocity $\mathbf{V}_G$ of the center of mass G, although any other point would do as well. Owing to the assumed planarity of the vehicle motion, $\mathbf{V}_G$ is horizontal and $\boldsymbol{\Omega}$ is vertical. More precisely

$$\mathbf{V}_G = u\,\mathbf{i} + v\,\mathbf{j} \tag{3.1}$$

and

$$\boldsymbol{\Omega} = r\,\mathbf{k} \tag{3.2}$$

The component u is called *forward velocity* of G, and v is called *lateral velocity* of G. The quantity r is the vehicle *yaw rate* (i.e., angular velocity).

Like in (2.1), the velocity $\mathbf{V}_P$ of any point $P = (x, y)$ of the vehicle body is given by the well-known formula for rigid bodies [14]

$$\begin{aligned}
\mathbf{V}_P &= \mathbf{V}_G + \boldsymbol{\Omega} \times GP \\
&= (u\,\mathbf{i} + v\,\mathbf{j}) + r\,\mathbf{k} \times (x\,\mathbf{i} + y\,\mathbf{j}) \\
&= (u - ry)\,\mathbf{i} + (v + rx)\,\mathbf{j} \\
&= V_{Px}\,\mathbf{i} + V_{Py}\,\mathbf{j}
\end{aligned} \tag{3.3}$$

Therefore, the kinematics of the vehicle body is completely described by, e.g., the three state variables $u(t)$, $v(t)$ and $r(t)$, as shown in Fig. 3.2.

Under normal operating conditions $u > 0$, with

Fig. 3.2 Vehicle axis system
and global kinematics of a
vehicle in planar motion

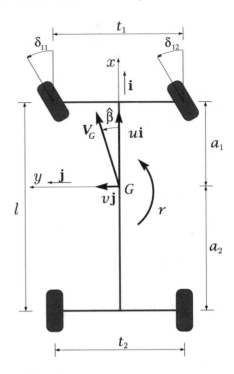

$$u \gg |v| \quad \text{and} \quad u \gg |r|\, l \tag{3.4}$$

that is the vehicle has to behave as shown in Fig. 1.2.

It is very important to realize from the very beginning that the yaw rate r is a property of the vehicle body as a whole. On the other hand, u and v are velocities of G, that is of just one point of the vehicle body. Selecting the center of mass G is customary, but quite arbitrary indeed. Other reasonable choices could be the midpoint between the two axles, or the position of the driver seat.

3.2.2 Yaw Angle of the Vehicle, and Trajectory of G

Let $\mathbf{S}_0 = (x_0, y_0, z_0; O_0)$ be a ground-fixed reference system, with unit vectors $(\mathbf{i}_0, \mathbf{j}_0, \mathbf{k}_0 = \mathbf{k})$, as shown in Fig. 3.3. Therefore

$$\mathbf{i}_0 \cdot \mathbf{i} = \cos \psi \quad \text{and} \quad \mathbf{j}_0 \cdot \mathbf{i} = \sin \psi \tag{3.5}$$

where ψ is the vehicle *yaw angle*. Accordingly, the velocity (3.1) of the center of mass G can also be expressed in the ground-fixed reference frame

$$\mathbf{V}_G = u\,\mathbf{i} + v\,\mathbf{j} = \dot{x}_0^G\,\mathbf{i}_0 + \dot{y}_0^G\,\mathbf{j}_0 \tag{3.6}$$

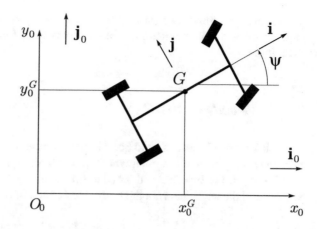

Fig. 3.3 Ground-fixed coordinate system S_0 and vehicle yaw angle ψ

with

$$\dot{x}_0^G = u\cos\psi - v\sin\psi$$
$$\dot{y}_0^G = u\sin\psi + v\cos\psi \qquad (3.7)$$
$$\dot{\psi} = r$$

The yaw angle ψ of the vehicle, at any time $t = \hat{t}$, is given by

$$\psi(\hat{t}) = \psi(0) + \int_0^{\hat{t}} r(t)dt \qquad (3.8)$$

Once the function of time $\psi(t)$ is known, the absolute position of G with respect to a frame S_0 fixed to the road is readily obtained by integrating the first two equations in (3.7)

$$x_0^G(\hat{t}) = x_0^G(0) + \int_0^{\hat{t}} \dot{x}_0 dt = x_0^G(0) + \int_0^{\hat{t}} [u(t)\cos(\psi(t)) - v(t)\sin(\psi(t))]dt$$

$$y_0^G(\hat{t}) = y_0^G(0) + \int_0^{\hat{t}} \dot{y}_0 dt = y_0^G(0) + \int_0^{\hat{t}} [u(t)\sin(\psi(t)) + v(t)\cos(\psi(t))]dt$$

$$(3.9)$$

The two functions $x_0^G(t)$ and $y_0^G(t)$ are the parametric equations of the *trajectory* of G with respect to the fixed reference system S_0. An example of the trajectory of a race car is shown in Fig. 8.58.

Equations (3.7) can be inverted to get

$$u(t) = \cos(\psi(t))\dot{x}_0(t) + \sin(\psi(t))\dot{y}_0(t)$$
$$v(t) = -\sin(\psi(t))\dot{x}_0(t) + \cos(\psi(t))\dot{y}_0(t) \qquad (3.10)$$
$$r(t) = \dot{\psi}(t)$$

These equations show that $u(t)$ and $v(t)$, despite being velocities, cannot be expressed as derivatives of other functions.[1] In other words, a formula like $v = \dot{y}$ is totally meaningless.

3.2.3 Velocity Center C

As well known, if $r \neq 0$, a rigid body in planar motion has an *instantaneous center of zero velocity* C, that is a point whose velocity $\mathbf{V}_C = \mathbf{0}$.

With the aid of Figs. 3.2 and 3.4, and for given u, v and r, it is easy to obtain, at any instant, the coordinates in the body-fixed frame of point C

$$GC = S\mathbf{i} + R\mathbf{j} \tag{3.11}$$

It suffices to observe that

$$\mathbf{V}_G = r\,\mathbf{k} \times CG = r\,\mathbf{k} \times (-R\mathbf{j} - S\mathbf{i}) = rR\mathbf{i} - rS\mathbf{j} = u\,\mathbf{i} + v\,\mathbf{j} \tag{3.12}$$

where

$$R = \frac{u}{r} \tag{3.13}$$

is the distance of C from the vehicle axis, and

$$S = -\frac{v}{r} \tag{3.14}$$

is the longitudinal position of C. Quite surprisingly, while R is very popular, S is hardly mentioned anywhere else. S is called *rotating length*[2] in [18, p. 172].

The instantaneous center of zero velocity C, or *velocity center*, is often misunderstood. Indeed, it is correct to say that the velocity field of the rigid body is like a pure rotation around C, that is the velocity $\mathbf{V}_P$ of any point P is

$$\mathbf{V}_P = r\,\mathbf{k} \times CP \tag{3.15}$$

but it is totally incorrect to think that the same property extends to the acceleration field as well. As a matter of fact, the acceleration $\mathbf{a}_C$ of point C is not zero in general, as shown in (3.48). There is another point, the *acceleration center* K (discussed in Sect. 3.2.9) which has zero acceleration, but nonzero velocity. Therefore, the velocity field is rotational around C, while the acceleration field is rotational around K.

---

[1] The reason is that $df = \cos\psi\,dx_0 + \sin\psi\,dy_0$ is not an exact differential since there does not exist a differentiable function $f(x_0, y_0, \psi)$.

[2] In this book, lengths are usually indicated by a lower case letter. R and S are exceptions.

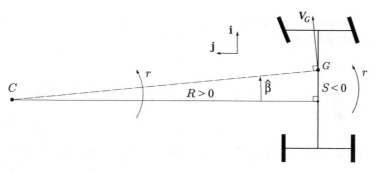

Fig. 3.4 Instantaneous velocity center C and definition of its coordinates S and R in the vehicle axis system

3.2.4 Fundamental Ratios β and ρ

Besides $R = u/r$ and $S = -v/r$, other ratios appear to be relevant in vehicle kinematics. They are

$$\beta = \frac{v}{u} = -\frac{S}{R} \qquad (3.16)$$

and

$$\rho = \frac{r}{u} = \frac{1}{R} \qquad (3.17)$$

The ratio β is closely related to the *vehicle slip angle* $\hat{\beta}$

$$\hat{\beta} = \arctan(\beta) \qquad (3.18)$$

that is the angle between $\mathbf{V}_G$ and $\mathbf{i}$ (Fig. 3.2, and also Fig. 3.4). Actually, in most cases

$$\beta \simeq \hat{\beta} \qquad (3.19)$$

since $|\hat{\beta}| < (0.157 \, \text{rad} = 9°)$, and they can be considered as synonymous $(\tan(0.157) = 0.158)$.

Instead of ρ, it is common practice in vehicle dynamics to employ

$$l\rho = l\frac{r}{u} = \frac{l}{R} \qquad (3.20)$$

where l is the wheelbase of the vehicle. This is the very classical *Ackermann angle* l/R. However, in our opinion, ρ is more fundamental than l/R, as will be shown. For the moment it suffices to note that the wheelbase l is totally irrelevant for the description of the kinematics of the vehicle body. What matters are only u, v and r, or their combinations (ratios), like β and ρ. In this context, the wheelbase l is quite an intruder. And by the way, what is the wheelbase in a three-axle vehicle? A discussion on three-axle vehicles is given in Sect. 3.16.

3.2.5 Acceleration of G and Angular Acceleration of the Vehicle

The acceleration of any point P of the vehicle body is given by the well known formula

$$\mathbf{a}_P = \mathbf{a}_G + \dot{\boldsymbol{\Omega}} \times GP + \boldsymbol{\Omega} \times (\boldsymbol{\Omega} \times GP) \qquad (3.21)$$

which, in case of planar motion, simplifies into

$$\mathbf{a}_P = \mathbf{a}_G + \dot{r}\,\mathbf{k} \times GP - r^2 GP \qquad (3.22)$$

According to (3.2), the angular acceleration is simply given by

$$\dot{\boldsymbol{\Omega}} = \dot{r}\,\mathbf{k} = \ddot{\psi}\,\mathbf{k} = (u\dot{\rho} + \dot{u}\rho)\,\mathbf{k} \qquad (3.23)$$

Typically, $|u\dot{\rho}| \gg |\dot{u}\rho|$ (Fig. 3.5). As a matter of fact, high values of $|\dot{u}|$, i.e. intense braking or acceleration, are possible only when the vehicle trajectory is almost straight, i.e. when $|\rho|$ is very small. It is worth noting that the evaluation of $\dot{r}$ is quite challenging in practice.

A little more involved is the evaluation of the absolute acceleration $\mathbf{a}_G$ of G. Differentiating (3.1) we obtain

$$\begin{aligned}
\mathbf{a}_G = \frac{d\mathbf{V}_G}{dt} &= \dot{u}\,\mathbf{i} + ur\,\mathbf{j} + \dot{v}\,\mathbf{j} - vr\,\mathbf{i} \\
&= (\dot{u} - vr)\,\mathbf{i} + (\dot{v} + ur)\,\mathbf{j} \\
&= a_x\,\mathbf{i} + a_y\,\mathbf{j}
\end{aligned} \qquad (3.24)$$

where

$$\frac{d\mathbf{i}}{dt} = r\,\mathbf{j} \quad \text{and} \quad \frac{d\mathbf{j}}{dt} = -r\,\mathbf{i} \qquad (3.25)$$

since the vehicle reference system S rotates with the vehicle body.

3.2.5.1 Longitudinal and Lateral Components

Equation (3.24) defines the *longitudinal acceleration* a_x of G

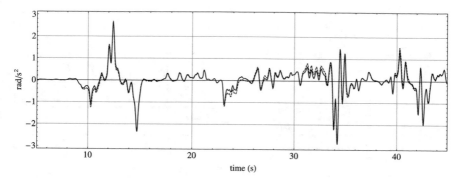

Fig. 3.5 Filtered telemetry data of a Formula car. Comparison between the angular acceleration $\dot{r} = u\dot{\rho} + \dot{u}\rho$ (solid line) and $u\dot{\rho}$ (dashed line)

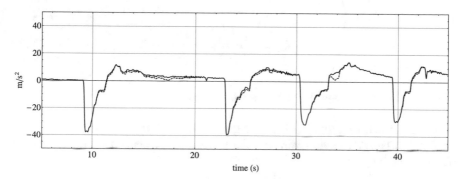

Fig. 3.6 Filtered telemetry data of a Formula car. Comparison between the longitudinal acceleration $a_x = \dot{u} - vr$ (solid line) and $\dot{u}$ (dashed line)

$$a_x = \dot{u} - vr$$
$$= \dot{u} - u^2\rho\beta \tag{3.26}$$

and the *lateral acceleration a_y* of G

$$a_y = \dot{v} + ur$$
$$= u\dot{\beta} + \dot{u}\beta + ur \tag{3.27}$$
$$= u\dot{\beta} + \dot{u}\beta + u^2\rho$$

where longitudinal and lateral refer to the vehicle axis x (defined in Sect. 3.1.1), not to the trajectory of G.

The main contribution to a_x comes from $\dot{u}$, as shown in Fig. 3.6. Therefore, $|\dot{u}| \gg |vr|$. On the other hand, the main contribution to a_y comes from ur, as shown in Fig. 3.7. Therefore, $|ur| \gg |\dot{v}|$. More precisely, $|u^2\rho| \gg |u\dot{\beta}| \gg |\dot{u}\beta|$.

Under *steady-state conditions* ($\dot{u} = \dot{v} = 0$), the lateral acceleration of G becomes

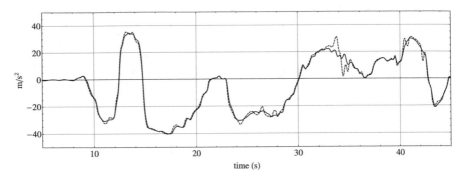

Fig. 3.7 Filtered telemetry data of a Formula car. Comparison between the lateral acceleration $a_y = \dot{v} + ur$ (solid line) and $ur = u^2\rho$ (dashed line)

$$\tilde{a}_y = ur = u^2\rho = \frac{u^2}{R} \tag{3.28}$$

3.2.5.2   Tangential and Centripetal Components

In general, the trajectory of G is not tangent to the vehicle axis x, that is $\hat{\beta} \neq 0$, as in Fig. 3.2. The unit vector $\mathbf{t}$, directed like $\mathbf{V}_G$ (and hence tangent to the trajectory of G), is given by

$$\mathbf{t} = \frac{\mathbf{V}_G}{|\mathbf{V}_G|} = \cos\hat{\beta}\,\mathbf{i} + \sin\hat{\beta}\,\mathbf{j} \tag{3.29}$$

where

$$|\mathbf{V}_G| = V_G = \sqrt{u^2 + v^2} \tag{3.30}$$

and

$$\sin\hat{\beta} = \frac{v}{\sqrt{u^2 + v^2}} \simeq \left(\beta = \frac{v}{u}\right)$$

$$\cos\hat{\beta} = \frac{u}{\sqrt{u^2 + v^2}} \simeq 1 \tag{3.31}$$

Moreover, we can define the normal unit vector $\mathbf{n}$, orthogonal to $\mathbf{V}_G$

$$\mathbf{n} = \mathbf{k} \times \mathbf{t} = -\sin\hat{\beta}\,\mathbf{i} + \cos\hat{\beta}\,\mathbf{j} \tag{3.32}$$

As shown in Fig. 3.8, the acceleration $\mathbf{a}_G$ can be also expressed as

$$\mathbf{a}_G = \mathbf{a}_t + \mathbf{a}_n = a_t\mathbf{t} + a_n\mathbf{n} \tag{3.33}$$

with *tangential* component a_t (directed like $\mathbf{V}_G$)

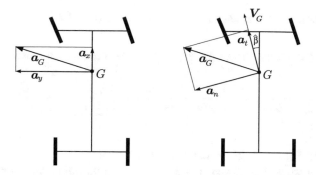

Fig. 3.8 Two noteworthy decompositions of the same acceleration of the center of gravity G

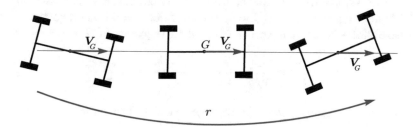

Fig. 3.9 Motion with zero acceleration $\mathbf{a}_G$, although $u, v, r, \dot{u}, \dot{v}$ and $\dot{r}$ are all non zero

$$a_t = \mathbf{a}_G \cdot \mathbf{t} = a_x \cos \hat{\beta} + a_y \sin \hat{\beta} = \frac{a_x u + a_y v}{\sqrt{u^2 + v^2}} = \frac{\dot{u}u + \dot{v}v}{\sqrt{u^2 + v^2}} = \frac{dV_G}{dt} \qquad (3.34)$$

and *centripetal* (or normal) component a_n (orthogonal to $\mathbf{V}_G$)

$$a_n = \mathbf{a}_G \cdot \mathbf{n} = -a_x \sin \hat{\beta} + a_y \cos \hat{\beta} = \frac{-a_x v + a_y u}{\sqrt{u^2 + v^2}} = \frac{r(u^2 + v^2) + \dot{v}u - \dot{u}v}{\sqrt{u^2 + v^2}}$$
$$(3.35)$$

The accelerations $a_x = \dot{u} - vr$ and $a_y = \dot{v} + ur$ are not, in general, the second derivatives of some functions. In other words, a formula like $a_y = \ddot{y}$ is meaningless, and hence wrong (cf. [19, p. 28]). The same remark applies to a_n.

On the other hand, as shown in the last term in (3.34), the tangential acceleration a_t is the first derivative of V_G, and hence the second derivative of the arc length along the trajectory. Indeed, $a_t(t)$ is the only acceleration component that does not depend on the yaw rate $r(t)$.

The lateral acceleration $a_y = ur + \dot{v}$ consists of two terms. Sentences like: "The first is the centrifugal force term and the second is the direct lateral acceleration term." [16, p. 146] may look reasonable, but they are not correct. The (counter)example shown in Fig. 3.9 should clarify the matter: the center of mass G moves with constant velocity $\mathbf{V}_G$ on a straight line, while the vehicle has a yaw rate r. Therefore, even if $a_y = 0$, we have $ur \neq 0$ and $\dot{v} \neq 0$.

3.2.6   Radius of Curvature of the Trajectory of G

The *radius of curvature* R_G of the trajectory of G is readily obtained as

$$R_G = \frac{V_G^2}{a_n} = \frac{(u^2 + v^2)^{\frac{3}{2}}}{r(u^2 + v^2) + \dot{v}u - \dot{u}v} = \frac{V_G}{r + \dfrac{\dot{v}u - \dot{u}v}{V_G^2}} = \frac{V_G^3}{a_y u - a_x v} \tag{3.36}$$

where also (3.30) and (3.35) were taken into account.

It is worth remarking, as shown in Fig. 3.10, that the velocity center C is *not* the center of curvature E_G of the trajectory of G. As will be discussed with reference to Fig. 3.12, to have $E_G = C$, and hence $R_G = R/\cos\hat{\beta}$, it has to be $\dot{v}u - v\dot{u} = 0$, which is more general than $\dot{u} = \dot{v} = 0$. The condition $\dot{r} = 0$ is not required.

Also useful is the curvature $\rho_G = 1/R_G$ of the trajectory of G

$$\rho_G = \frac{1}{R_G} = \frac{a_n}{V_G^2} = \frac{r}{V_G} + \frac{\dot{v}u - v\dot{u}}{V_G^3} = \frac{a_y u - a_x v}{V_G^3} \tag{3.37}$$

Plots of the curvature ρ_G are usually more readable than those of the radius of curvature R_G.

Under normal operating conditions, $V_G \simeq u$, i.e. $|\beta| \ll 1$, and hence

$$\left(\rho_G = \frac{1}{R_G}\right) \simeq \left(\frac{r + \dot{\beta}}{u} = \rho + \frac{\dot{\beta}}{u} = \frac{1}{R} + \frac{\dot{\beta}}{u}\right) \tag{3.38}$$

Quite a compact and interesting formula. It provides physical insights on the relationship between R_G and R.

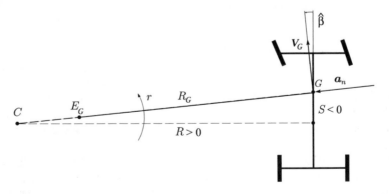

Fig. 3.10 Radius of curvature R_G and center of curvature E_G of the trajectory of G. In general, $E_G \neq C$

3.2.7 Radius of Curvature of the Trajectory of a Generic Point

The velocity $\mathbf{V}_P$ of a generic point $P = (x, y)$ was obtained in (3.3). The acceleration $\mathbf{a}_P$ can be obtained employing (3.22) and (3.24)

$$
\begin{aligned}
\mathbf{a}_P &= \mathbf{a}_G + \dot{r}\,\mathbf{k} \times (x\,\mathbf{i} + y\,\mathbf{j}) - r^2\,(x\,\mathbf{i} + y\,\mathbf{j}) \\
&= (a_x - \dot{r}y - r^2 x)\,\mathbf{i} + (a_y + \dot{r}x - r^2 y)\,\mathbf{j} \qquad (3.39) \\
&= a_{Px}\,\mathbf{i} + a_{Py}\,\mathbf{j}
\end{aligned}
$$

We are interested in the centripetal (normal) component a_{Pn} of $\mathbf{a}_P$ (cf. (3.35))

$$
a_{Pn} = \frac{a_{Py} V_{Px} - a_{Px} V_{Py}}{V_P} \qquad (3.40)
$$

where $V_P = |\mathbf{V}_P|$.

Like in (3.36), the radius of curvature R_P of the trajectory of P is given by

$$
R_P = \frac{V_P^2}{a_{Pn}} \qquad (3.41)
$$

3.2.8 Telemetry Data and Mathematical Channels

Typical *telemetry data* are, among others, $u(t)$, $\hat{\beta}(t)$, $r(t)$, $a_x(t)$, and $a_y(t)$, but not directly the derivatives $\dot{u}(t)$, $\dot{v}(t)$ and $\dot{r}(t)$. It is well known that differentiating experimental signals is a very unreliable process. Therefore, when dealing with experimental data, mathematically equivalent formulas to create the so-called *mathematical channels* [20] may not provide exactly the same results. For instance, in (3.34), $(a_x u + a_y v)$ is more accurate than $\dot{u}u + \dot{v}v$, because it avoids employing derivatives of telemetry data. Similarly, the last formula in (3.37) is the most reliable to compute ρ_G.

As will be discussed, several interesting handling features of a race car depend on $\dot{r}(t)$. Therefore, along with $a_x(t)$ and $a_y(t)$, it would be very useful to measure *directly* also $\dot{r}(t)$, instead of having to differentiate $r(t)$. The results would be much more accurate and reliable. To the best of our knowledge, $\dot{r}(t)$ is never measured directly, probably because it has not been realized how important it is.

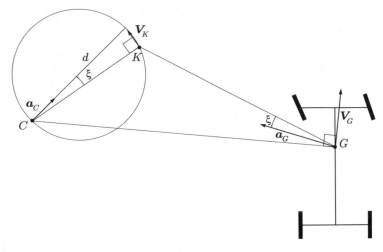

Fig. 3.11 Velocity center C, acceleration center K and inflection circle (with diameter d)

3.2.9   Acceleration Center K

The acceleration field of a rigid body in planar motion is like a pure rotation around the *acceleration center* K, that is a point which has $\mathbf{a}_K = \mathbf{0}$. According to (3.22), with G replaced by K, the acceleration $\mathbf{a}_P$ of any point P must be given by

$$\mathbf{a}_P = \dot{r}\,\mathbf{k} \times KP - r^2 KP \qquad (3.42)$$

Therefore, the angle ξ between $\mathbf{a}_P$ and KP is such that

$$\tan \xi = \frac{\dot{r}}{r^2} \qquad (3.43)$$

By setting $P = G$ in (3.42), as shown in Fig. 3.11

$$\mathbf{a}_G = \dot{r}\,\mathbf{k} \times KG - r^2 KG \qquad (3.44)$$

we obtain that

$$|KG| = \frac{a_G}{\sqrt{\dot{r}^2 + r^4}} \qquad (3.45)$$

or, more precisely (cf. (3.11))

$$GK = \left(\frac{a_x r^2 - a_y \dot{r}}{r^4 + \dot{r}^2} \right) \mathbf{i} + \left(\frac{a_x \dot{r} + a_y r^2}{r^4 + \dot{r}^2} \right) \mathbf{j} \qquad (3.46)$$

3.2.10 Inflection Circle

The *inflection circle* is the set of points of a rigid body in planar motion whose trajectories have an inflection point (i.e., zero curvature). Several formulas concerning the inflection circle, including its diameter d, are given in Sect. 5.2.2.

The acceleration center K lies necessarily on the *inflection circle*. Point K spans the inflection circle depending on the value of the ratio $\dot{r}/r^2$, as shown in Fig. 3.11. This topic will be addressed in detail in Chap. 5, entirely devoted to the kinematics of cornering.

The velocity center C does not belong to the inflection circle, although it looks like (Fig. 3.11). Actually, its trajectory has a cusp, i.e. zero radius of curvature. According to (3.22), the velocity center C has acceleration

$$\mathbf{a}_C = \mathbf{a}_G + \dot{r}\,\mathbf{k} \times GC - r^2 GC \tag{3.47}$$

Then, taking into account (3.11) and (3.24), we obtain (see also (5.7))

$$\mathbf{a}_C = (a_x + vr - \dot{r}u/r)\,\mathbf{i} + (a_y - ur - \dot{r}v/r)\,\mathbf{j}$$
$$= \left(\frac{\dot{u}r - u\dot{r}}{r}\right)\mathbf{i} + \left(\frac{\dot{v}r - v\dot{r}}{r}\right)\mathbf{j} \tag{3.48}$$
$$= r(\dot{R}\,\mathbf{i} - \dot{S}\,\mathbf{j})$$

The first expression in (3.48) is the most accurate as mathematical channel of telemetry data. The second expression is more compact and will be discussed in a while. The third expression provides a clear physical interpretation.

Whenever $\mathbf{a}_C = \mathbf{0}$, the velocity center and the acceleration center coincide, that is $C = K$, and the inflection circle collapses to a point ($d = 0$). This is what typically happens (or should happen) when a car is at the mid-point of a well driven curve. Mathematically, from the second expression in (3.48), $\mathbf{a}_C = \mathbf{0}$ means that

$$\begin{cases} \dot{u}r - u\dot{r} = 0 \\ \dot{v}r - v\dot{r} = 0 \end{cases} \tag{3.49}$$

that can also be read as

$$\frac{\dot{u}}{u} = \frac{\dot{v}}{v} = \frac{\dot{r}}{r} \tag{3.50}$$

which implies (cf. (3.37))

$$\dot{v}u - \dot{u}v = u^2\dot{\beta} = 0 \tag{3.51}$$

but not the other way around. Indeed, there are cases, like in Fig. 3.12, where (3.51) is fulfilled, but not (3.49): the segment GC is tangent to the inflection circle and hence $\dot{\beta} = 0$, albeit $\dot{R} \neq 0$ and/or $\dot{S} \neq 0$.

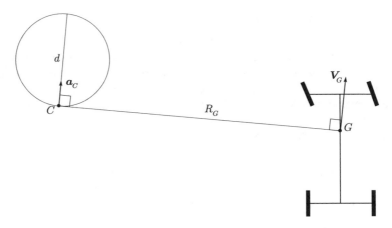

Fig. 3.12 Inflection circle tangent to the segment GC, that is $\dot{v}u - \dot{u}v = u^2\dot{\beta} = 0$

While the condition $d = 0$, i.e. (3.49), is peculiar to the vehicle motion as a whole, the condition $\dot{\beta} = 0$, i.e. (3.51), is peculiar only for (the arbitrarily selected) point G, as shown in (3.38). More precisely, if $d = 0$, *any* point of the rigid body has velocity components that fulfill (3.49), whereas only points on the straight line through C and tangent to the inflection circle have velocity components that fulfill (3.51), if $d \neq 0$.

3.3 Tire Kinematics (Tire Slips)

So far only the kinematics of the vehicle body has been addressed. As a matter of fact, the wheel steer angles δ_{ij} (Fig. 3.2) have not been employed yet. Roughly speaking, the kinematics of the vehicle body is what mostly matters to the driver. However, vehicle engineers are also interested in the kinematics of the wheels, since it strongly affects the forces exerted by the tires, as discussed in Chap. 2.

According to (3.3), the velocity of the center O_{11} of the left front wheel is given by

$$\mathbf{V}_{11} = \mathbf{V}_G + r\,\mathbf{k} \times GO_{11} = (u\,\mathbf{i} + v\,\mathbf{j}) + r\,\mathbf{k} \times \left(a_1\,\mathbf{i} + \frac{t_1}{2}\,\mathbf{j}\right) \qquad (3.52)$$

Performing the same calculation for the centers of all wheels yields

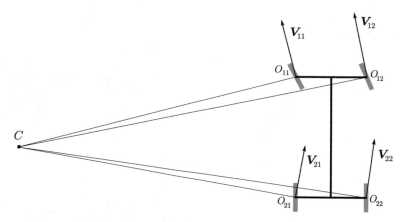

Fig. 3.13 Relationship between the velocities of the centers of the wheels

$$\mathbf{V}_{11} = \left(u - \frac{rt_1}{2}\right)\mathbf{i} + (v + ra_1)\mathbf{j}$$

$$\mathbf{V}_{12} = \left(u + \frac{rt_1}{2}\right)\mathbf{i} + (v + ra_1)\mathbf{j}$$

$$\mathbf{V}_{21} = \left(u - \frac{rt_2}{2}\right)\mathbf{i} + (v - ra_2)\mathbf{j}$$

$$\mathbf{V}_{22} = \left(u + \frac{rt_2}{2}\right)\mathbf{i} + (v - ra_2)\mathbf{j}$$

(3.53)

Of course, these velocities are not independent of each other. They must be like in Fig. 3.13 with respect to the velocity center C, that is

$$\mathbf{V}_{ij} = r\,\mathbf{k} \times C O_{ij} \qquad (3.54)$$

Therefore, the angles $\hat{\beta}_{ij}$ between the vehicle longitudinal axis $\mathbf{i}$ and $\mathbf{V}_{ij}$ can be obtained as (Fig. 3.14)

$$\tan\hat{\beta}_{11} = \frac{v + ra_1}{u - rt_1/2} = \beta_{11} = \tan(\delta_{11} - \alpha_{11})$$

$$\tan\hat{\beta}_{12} = \frac{v + ra_1}{u + rt_1/2} = \beta_{12} = \tan(\delta_{12} - \alpha_{12})$$

$$\tan\hat{\beta}_{21} = \frac{v - ra_2}{u - rt_2/2} = \beta_{21} = \tan(\delta_{21} - \alpha_{21})$$

$$\tan\hat{\beta}_{22} = \frac{v - ra_2}{u + rt_2/2} = \beta_{22} = \tan(\delta_{22} - \alpha_{22})$$

(3.55)

Fig. 3.14 Kinematics of the
centers of the wheels

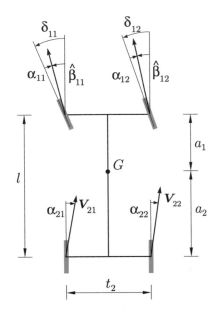

or, equivalently, using (3.13) and (3.14)

$$\beta_{11} = \frac{a_1 - S}{R - t_1/2} \qquad \beta_{12} = \frac{a_1 - S}{R + t_1/2}$$

$$\beta_{21} = \frac{-a_2 - S}{R - t_2/2} \qquad \beta_{22} = \frac{-a_2 - S}{R + t_2/2} \qquad (3.56)$$

where $S \geq -a_2$. Under normal operating conditions we have that $|R| \gg t_i/2$, which
means that

$$\beta_1 = \frac{v + ra_1}{u} \simeq (\beta_{11} \simeq \beta_{12})$$

$$\beta_2 = \frac{v - ra_2}{u} \simeq (\beta_{21} \simeq \beta_{22}) \qquad (3.57)$$

However, particularly in competitions, even a tiny difference can be significant.

As shown in Fig. 3.14, each wheel undergoes a tire *slip angle* α_{ij}. As already
mentioned in Sect. 2.8.6, these angles are taken here as *positive if clockwise*. The tire
slip angles are given by

$$\alpha_{ij} = \delta_{ij} - \hat{\beta}_{ij} \simeq \delta_{ij} - \beta_{ij} \qquad (3.58)$$

The relationship between the steering angles δ_{i1} and δ_{i2} of the two wheels of the same
axle can significantly affect α_{i1} and α_{i2}, and hence the tire friction forces acting on
the vehicle. This aspect is addressed in Sect. 3.4.

As thoroughly discussed in Sect. 2.8, tire (rim) kinematics can, in most cases, be conveniently described by means of the translational slips σ_x and σ_y and the spin slip φ, defined in (2.63), (2.64) and (2.65), respectively.

3.3.1 Translational Slips

According to (2.50), the rolling velocity of each wheel is equal to $\omega_{ij} r_i$, where ω_{ij} is the angular velocity of the rim and r_i is the wheel rolling radius, as defined in (2.36). The travel velocity $\mathbf{V}_{ij}$ of each wheel was obtained in (3.53). Considering also the steering angles δ_{ij}, we obtain for each tire

- longitudinal slips:

$$\sigma_{x_{11}} = \frac{[(u - rt_1/2)\cos(\delta_{11}) + (v + ra_1)\sin(\delta_{11})] - \omega_{11} r_1}{\omega_{11} r_1}$$

$$\sigma_{x_{12}} = \frac{[(u + rt_1/2)\cos(\delta_{12}) + (v + ra_1)\sin(\delta_{12})] - \omega_{12} r_1}{\omega_{12} r_1}$$

$$\sigma_{x_{21}} = \frac{[(u - rt_2/2)\cos(\delta_{21}) + (v - ra_2)\sin(\delta_{21})] - \omega_{21} r_2}{\omega_{21} r_2}$$

$$\sigma_{x_{22}} = \frac{[(u + rt_2/2)\cos(\delta_{22}) + (v - ra_2)\sin(\delta_{22})] - \omega_{22} r_2}{\omega_{22} r_2}$$

(3.59)

- lateral slips:

$$\sigma_{y_{11}} = \frac{(v + ra_1)\cos(\delta_{11}) - (u - rt_1/2)\sin(\delta_{11})}{\omega_{11} r_1}$$

$$\sigma_{y_{12}} = \frac{(v + ra_1)\cos(\delta_{12}) - (u + rt_1/2)\sin(\delta_{12})}{\omega_{12} r_1}$$

$$\sigma_{y_{21}} = \frac{(v - ra_2)\cos(\delta_{21}) - (u - rt_2/2)\sin(\delta_{21})}{\omega_{21} r_2}$$

$$\sigma_{y_{22}} = \frac{(v - ra_2)\cos(\delta_{22}) - (u + rt_2/2)\sin(\delta_{22})}{\omega_{22} r_2}$$

(3.60)

Owing to (3.4), the expressions of the translational slips can be simplified under normal operating conditions and small steering angles

$$\sigma_{x_{11}} \simeq \frac{(u - rt_1/2) - \omega_{11}\,r_1}{\omega_{11}\,r_1} \qquad\qquad \sigma_{y_{11}} \simeq \frac{(v + ra_1) - u\,\delta_{11}}{\omega_{11}\,r_1}$$

$$\sigma_{x_{12}} \simeq \frac{(u + rt_1/2) - \omega_{12}\,r_1}{\omega_{12}\,r_1} \qquad\qquad \sigma_{y_{12}} \simeq \frac{(v + ra_1) - u\,\delta_{12}}{\omega_{12}\,r_1}$$

$$\sigma_{x_{21}} \simeq \frac{(u - rt_2/2) - \omega_{21}\,r_2}{\omega_{21}\,r_2} \qquad\qquad \sigma_{y_{21}} \simeq \frac{(v - ra_2) - u\,\delta_{21}}{\omega_{21}\,r_2} \qquad (3.61)$$

$$\sigma_{x_{22}} \simeq \frac{(u + rt_2/2) - \omega_{22}\,r_2}{\omega_{22}\,r_2} \qquad\qquad \sigma_{y_{22}} \simeq \frac{(v - ra_2) - u\,\delta_{22}}{\omega_{22}\,r_2}$$

3.3.2 Spin Slips

According to (2.65), the evaluation of the spin slips φ_{ij} requires also the knowledge of the wheel yaw rates $\omega_{z_{ij}} = r + \dot{\delta}_{ij}$, of the camber angles γ_{ij}, and of the camber reduction factors ε_i

$$\varphi_{ij} = -\frac{r + \dot{\delta}_{ij} + \omega_{ij}\sin\gamma_{ij}(1 - \varepsilon_i)}{\omega_{ij}\,r_i} \qquad (3.62)$$

The sign conventions are like in Fig. 2.6. Therefore, under static conditions (i.e., vehicle at rest), the two wheels of the same axle have static camber angles of opposite sign

$$\gamma_{i1}^0 = -\gamma_{i2}^0 \qquad (3.63)$$

as will be shown in Fig. 7.5. This is contrary to common practice, but more consistent and more convenient for a systematic treatment.

The kinematic equations for camber variations due to roll motion will be discussed in Sect. 3.10.3.

3.4 Steering Geometry

The main way to turn a vehicle is by steering some (or all) of its wheels. The *amount* of steering is mainly controlled by the driver by turning the steering wheel. Let δ_v be the rotation of the steering wheel.[3]

On the other hand, the *relative steering* of the wheels, that is how much the left-front wheel steers with respect to the right-front wheel, is not under the driver control. It affects the handling behavior and hence it should be carefully selected.

Because of their directional capability, the wheels in a vehicle are arranged such that their heading directions almost "agree", that is they do not conflict too much with each other. However, tires do work pretty well under small slip angles and, as

---

[3] Subscript v was chosen because the Italian translation of steering wheel is *volante*.

Fig. 3.15 Static toe δ_1^0:
toe-in (left) and toe-out
(right) of the front wheels

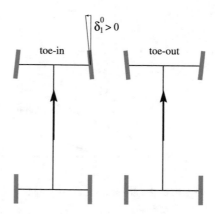

will be discussed, some amount of "disagreement" is not only tolerated, but can be
beneficial.

The mathematical framework is that the two front wheels steer according to[4]

$$\delta_{11} = \delta_{11}(\delta_v) \quad \text{and} \quad \delta_{12} = \delta_{12}(\delta_v) \tag{3.64}$$

such that

$$\delta_{11}(\delta_v) = -\delta_{12}(-\delta_v) \tag{3.65}$$

That is left turning and right turning are alike for the vehicle. Therefore, the Taylor
series expansions up to the second order of these two functions can be written as

$$\delta_{11} \simeq -\delta_1^0 + \tau_1 \delta_v + \varepsilon_1 \frac{t_1}{2l}(\tau_1 \delta_v)^2$$

$$\delta_{12} \simeq \delta_1^0 + \tau_1 \delta_v - \varepsilon_1 \frac{t_1}{2l}(\tau_1 \delta_v)^2 \tag{3.66}$$

These simple formulas cover probably most reasonable options for the steering sys-
tem geometry. The three parameters we can play with are δ_1^0, τ_1, and ε_1. Of course, τ_1
only affects ergonomics, whereas the other two parameters are relevant for vehicle
handling since they affect the relationship between δ_{11} and δ_{12}.

The angle δ_1^0 in (3.66) is called *static toe* setting. As shown in Fig. 3.15, we have
toe-in if $\delta_1^0 > 0$, and *toe-out* if $\delta_1^0 < 0$. In both cases, $\delta_1^0 = \delta_{12}(0)$. Of course, $\delta_1^0 = 0$
is also possible.

The second term $\tau_1 \delta_v$ is the *parallel steering*. It is the same for both wheels and
it is the big one. If there is only this contribution, the two wheels are always parallel
to each other, as shown in Fig. 3.16. Of course, it must be $\tau_1 > 0$. Typically, the

---

[4] The kinematic equations for roll steer, for both front and rear wheels, will be given in (3.210).
Their presentation is delayed till the suspension analysis is completed.

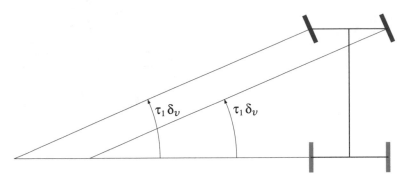

Fig. 3.16 Parallel steering ($\varepsilon_1 = 0$ and $\delta_1^0 = 0$)

gear ratio of the steering system $\tau_1 \simeq 1/20$ in road cars, while race cars have higher values.

The last term is the most intriguing. It is the *dynamic toe* setting, that is a toe that depends on δ_v. Changing the value of ε_1 in (3.66) we can span from 100% Ackermann steering kinematics ($\varepsilon_1 = 1$) to 100% anti-Ackermann ($\varepsilon_1 = -1$), with all possible values in between, including parallel steering ($\varepsilon_1 = 0$). Here we call ε_1 the *Ackermann coefficient*, and the whole last term in (3.66) the *Ackermann correction*.

3.4.1 Ackermann Steering Kinematics

Ackermann steering kinematics was patented in 1818 for horse-drawn carriages, that is for vehicles with rigid wheels,[5] and it is shown in Fig. 3.17. With this arrangement, a vehicle with *rigid* wheels (and no static toe) makes any turn with the wheels that do not fight each other. That is, all wheels can rotate freely with no slip angle.

Incidentally, we point out that C_0 in Fig. 3.17 is not, in general the center of curvature of the trajectories of the front tires, not even if there are no slip angles, as thoroughly discussed in Chap. 5. Therefore, calling it the "geometric center of the vehicle's path of curvature" [22, p. 60] is not correct if the car is not at steady state.

It is a simple calculation (Fig. 3.17) to obtain that Ackermann requires

$$\tan(\delta_{11}) = \frac{l}{R - t_1/2} \qquad \tan(\delta_{12}) = \frac{l}{R + t_1/2} \tag{3.67}$$

which yields

$$\frac{1}{\tan(\delta_{12})} - \frac{1}{\tan(\delta_{11})} = \frac{t_1}{l} \tag{3.68}$$

---

[5] Pneumatic tires were invented about 70 years later.

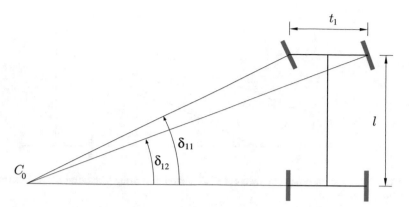

Fig. 3.17 Ackermann steering. More precisely, parallel steering with full positive Ackermann correction ($\varepsilon_1 = 1$)

This relationship can be given in a simpler, approximate form if we perform its Taylor expansion

$$\frac{1}{\delta_{12}} - \frac{1}{\delta_{11}} \simeq \frac{t_1}{l}$$

$$\delta_{11} \simeq \frac{1}{\dfrac{1}{\delta_{12}} - \dfrac{t_1}{l}} = \frac{\delta_{12}}{1 - \dfrac{t_1}{l}\delta_{12}} \tag{3.69}$$

$$\delta_{11} \simeq \delta_{12}\left(1 + \frac{t_1}{l}\delta_{12}\right)$$

and finally

$$\delta_{11} \simeq \delta_{12} + \frac{t_1}{l}\delta_{12}^2 \tag{3.70}$$

which is like (3.66) with $\varepsilon_1 = 1$, and $\delta_1^0 = 0$. As shown in Fig. 3.17, the inner wheel steers more than the outer wheel. We have dynamic toe-out, that is toe-out that grows with δ_v.

Anti-Ackermann, as the name implies, is the other way around, that is $\varepsilon_1 = -1$, as shown in Fig. 3.18. The inner wheel steers less than the outer wheel. We have dynamic toe-in. At first, anti-Ackermann geometry may look as a strange idea, but it is not so strange if we take into account that wheels with tires have slip angles α_{ij}, as shown in Figs. 3.13 and 3.14.

Static toe and dynamic toe look similar, but they are quite different concepts.

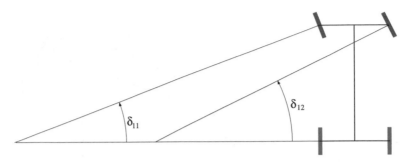

Fig. 3.18 Anti-Ackermann steering. More precisely, parallel steering with full negative Ackermann correction ($\varepsilon_1 = -1$)

3.4.2 Best Steering Geometry

Selecting the best coefficients in (3.66) is not an easy task. As already mentioned, τ_1 only affects the ergonomic features of the vehicle, whereas the other two parameters, that control static and dynamic toe, are relevant for vehicle handling.

Both static toe and dynamic toe have, in different ways, a twofold effect:

1. for a given position of the velocity center C of the vehicle, they affect the tire slips, and hence the values of the tire lateral forces $F_{y_{ij}}$;
2. regardless of the position of C, they affect the directions of the tire lateral forces $F_{y_{ij}}$ of the wheels.

3.4.3 Position of the Velocity Center and Relative Slip Angles

A vehicle with parallel rear wheels and also parallel front wheels is shown in Fig. 3.19. Three cases are presented, with different longitudinal position S of the velocity center C: case (a) has $-a_2 < S < a_1$, case (b) has $S = a_1$, and case (c) has $S > a_1$.

First, let us look at the rear wheels. We see that, for any position of the velocity center C, *the inner rear wheel has always a slip angle a little bigger than that of the outer rear wheel*. This geometric result is obvious if we look at Fig. 3.19, but also quite counterintuitive if we consider the vertical loads: the bigger slip angle belongs to the inner wheel, which has a smaller vertical load than the outer wheel and hence provides a smaller lateral force.

Analysis of the *front slip angles* is a bit more involved, and hence more interesting (Fig. 3.19).

If the vehicle has *parallel steering*, in case (a), it is the front outer wheel to have a bigger slip angle than the front inner wheel. In case (b), both front wheels have the same slip angle. In case (c), which is quite frequent in race cars, it is the front inner wheel to have the bigger slip angle.

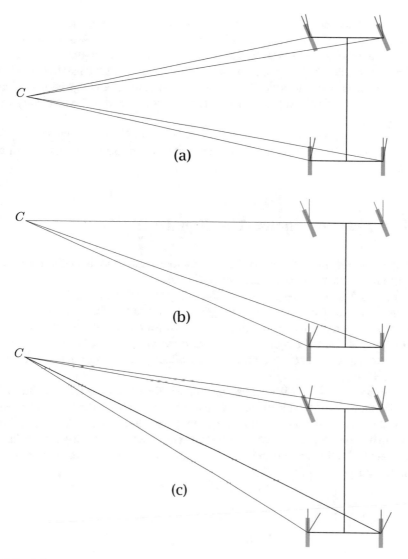

Fig. 3.19 Influence of the coordinate S of the velocity center C on the relative values of the slip angles of the two wheels of the front axle

If the vehicle has *positive Ackermann correction* ($\varepsilon_1 > 0$), that is dynamic toe-out, the inner wheel has the bigger slip angle not only in case (c), but also in case (b) and in a small fraction of case (a).

With *negative Ackermann correction* ($\varepsilon_1 < 0$), that is dynamic toe-in, it is the other way around. The inner wheel has the bigger slip angle only in part of case (c).

Parallel steering is often employed in race cars that operate at high lateral accelerations and hence at high slip angles. Passenger cars have a steering geometry somewhere between Ackermann steering and parallel steering, since they are usually exposed to low lateral accelerations. Anti-Ackermann is a feasible option only in some race cars, but has to be avoided if we often have steer angles $|\delta_{1j}| > 15$ deg.

On the practical side, we have to take into account that "if parallel steering is employed, it is damned difficult to push the car around a sharp corner. If anti-Ackermann geometry is employed, it becomes almost impossible", as correctly stated in [22, p. 61].

On top of everything, if the steer angles are very small, the steering geometry has little relevance. It is crucial in FSAE competitions because of the very small turn radii.

3.5   Vehicle Constitutive (Tire) Equations

In any vehicle model we have to set up equations that relate the vehicle motion to the grip forces each tire exchanges with the road.

Chapters 2 and 11 are devoted to the analysis of the mechanical behavior of wheels with tires. The topic is quite complex. From that analysis, several tire models of increasing complexity can be formulated. However, in all of them the grip forces depend at least on the (theoretical) slips and the vertical loads acting on the tire. These two aspects cannot be omitted. Other effects, like the tire transient behavior can be included if necessary.

As discussed in Sect. 2.9, after having extensively tested a tire the quantities listed in (2.83) should be available to the vehicle dynamicist to properly define the (steady-state) pure rolling conditions. Departing from pure rolling means having grip forces acting in the contact patch. Under steady-state conditions, it is often assumed that, for each wheel with tire, these grip forces and moments obey relations in the following general form

$$F_x = F_x(F_z, \gamma, \sigma_x, \sigma_y, \varphi)$$

$$F_y = F_y(F_z, \gamma, \sigma_x, \sigma_y, \varphi) \tag{3.71}$$

$$M_z = M_z(F_z, \gamma, \sigma_x, \sigma_y, \varphi)$$

where γ is the camber angle, σ_x is the longitudinal theoretical slip, σ_y is the lateral theoretical slip and φ is the spin slip.

As shown in Sect. 3.7.3, and in particular in (3.85), the steering angles δ_{ij} have also to be taken into account to obtain the longitudinal and lateral forces with respect to the vehicle frame.

We recall that using the tire slips simplifies the analysis, but implicitly discards any possible influence of the forward speed on the tire behavior. In race cars, this influence may not be negligible.

3.6 Vehicle Equilibrium Equations

The classical dynamic equilibrium equations for a rigid body are [14]

$$m\,\mathbf{a}_G = \mathbf{F}$$
$$\dot{\mathbf{K}}_G^r = \mathbf{M}_G \tag{3.72}$$

where m is the total mass of the vehicle and $\dot{\mathbf{K}}_G^r$ is the time rate of change of the angular momentum with respect to the center of mass G. The total mass is the sum of the sprung mass m_s and of the unsprung mass m_u

$$m = m_s + m_u \tag{3.73}$$

In a vehicle like an automobile or a motorcycle, particularly when studying ride motions (see Chap. 10), it is useful to distinguish between sprung mass and unsprung mass. The *sprung mass* m_s is the portion of the vehicle's total mass that is supported *above* the suspension, thus including the body, frame, internal components, passengers and cargo. On the other hand, wheels, wheel bearings, brake rotors, calipers belong to the *unsprung mass* m_u, since they are not above the suspension. Of course $m = m_s + m_u$. The sprung mass is usually much bigger than the unsprung mass. Typically, $m_s/m_u = 7 - 10$.

As clearly shown by (3.72), we cannot think of the center of mass G as the point where the entire mass can be considered to be concentrated [21, p. 4], unless $\dot{\mathbf{K}}_G^r = \mathbf{0}$. Unfortunately, it is a very common mistake in vehicle dynamics to believe that the inertia force $-m\mathbf{a}_G$ is always applied at the center of mass G. The inertia force is indeed equal to the product of the mass multiplied by the acceleration of G, but this does not imply that the line of action of the inertia force goes through G.

3.6.1 Inertial Terms

The acceleration $\mathbf{a}_G$ of G has been obtained in (3.24)

$$\mathbf{a}_G = a_x\,\mathbf{i} + a_y\,\mathbf{j}$$
$$= (\dot{u} - vr)\,\mathbf{i} + (\dot{v} + ur)\,\mathbf{j} \tag{3.24'}$$
$$= (\dot{u} - u^2\rho\beta)\,\mathbf{i} + (u\dot{\beta} + \dot{u}\beta + u^2\rho)\,\mathbf{j}$$

The rate of change of the angular momentum $\dot{\mathbf{K}}_G^r$ can be conveniently expressed in terms of the inertia tensor [14, p. 129]

$$\mathbf{J} = \begin{bmatrix} J_x & -J_{xy} & -J_{xz} \\ -J_{yx} & J_y & -J_{yz} \\ -J_{zx} & -J_{zy} & J_z \end{bmatrix} \tag{3.74}$$

in the body-fixed reference frame (see also (9.32))

$$\dot{\mathbf{K}}_G^r = (-J_{zx}\dot{r} + J_{yz}r^2)\,\mathbf{i} + (-J_{zx}r^2 - J_{yz}\dot{r})\,\mathbf{j} + J_z\dot{r}\,\mathbf{k}$$
$$\simeq -J_{zx}(\dot{r}\,\mathbf{i} + r^2\,\mathbf{j}) + J_z\dot{r}\,\mathbf{k} \tag{3.75}$$

since $J_{yz} \simeq 0$. Moreover, it is worth noting that typically $|J_{zx}| \ll J_z$. Therefore, we can often safely assume[6]

$$\dot{\mathbf{K}}_G^r \simeq J_z\dot{r}\,\mathbf{k} \tag{3.76}$$

Strictly speaking, among the inertial terms in (3.75) we are missing the *gyroscopic torques* of the rotating wheels [7, p. 573]. Let J_{w_i} be the axial moment of inertia of a single wheel. The total gyroscopic torque (four wheels) is

$$- 2(J_{w_1}\omega_1 + J_{w_2}\omega_2)r\,\mathbf{i} = -2(J_{w_1}/r_1 + J_{w_2}/r_2)ur\,\mathbf{i} = -L_w\,\mathbf{i} \tag{3.77}$$

where r_i is the rolling radius of the wheels and $\omega_i = u/r_i$ is their angular speed.

Usually, these gyroscopic effects are significant only for race cars (big wheels, high speeds, light vehicles), or in case of solid axle suspensions.

3.6.2   External Force and Moment

The total external force $\mathbf{F}$ and the total external moment $\mathbf{M}_G$ (with respect to G) can be represented in terms of their components in the vehicle (body-fixed) reference system (Fig. 3.2)

$$\mathbf{F} = X\,\mathbf{i} + Y\,\mathbf{j} + Z\,\mathbf{k}$$
$$\mathbf{M}_G = L\,\mathbf{i} + M\,\mathbf{j} + N\,\mathbf{k} \tag{3.78}$$

The components in (3.78) have the following standard names:

- X: longitudinal force;
- Y: lateral or side force;
- Z: vertical or normal force;
- L: rolling moment;
- M: pitching moment;
- N: yawing moment.

---

[6] In a Formula 1 car we have $J_z \simeq 900\,\text{kgm}^2$, $J_y \simeq 800\,\text{kgm}^2$, $J_x \simeq 100\,\text{kgm}^2$, $J_{zx} \simeq 3\,\text{kgm}^2$, and $J_w \simeq 0.8\,\text{kgm}^2$, with $|r| < 1\,\text{rad/s}$ and $|\dot{r}| < 2\,\text{rad/s}^2$.

As already stated, the vehicle body has a planar motion. However, the forces acting on the vehicle do not form a planar system.

3.7 Forces Acting on the Vehicle

There are four different types of external forces acting on a road vehicle:

1. weight (gravitational force);
2. aerodynamic force;
3. road-tire friction forces;
4. road-tire vertical forces.

We discuss each of them separately.

3.7.1 Weight

The *weight* $\mathbf{W}$ is simply given by

$$\mathbf{W} = -W\,\mathbf{k} = -mg\,\mathbf{k} \tag{3.79}$$

where g is the gravitational acceleration. As well known, the weight (only) is applied at G. Therefore, it does not contribute to $\mathbf{M}_G$.

3.7.2 Aerodynamic Force

The *aerodynamic force*

$$\mathbf{F}_a = -X_a\,\mathbf{i} - Y_a\,\mathbf{j} - Z_a\,\mathbf{k} \tag{3.80}$$

depends essentially on the *vehicle shape* and *size*, and on the *relative velocity* $\mathbf{V}_a$ between the vehicle and the air (Fig. 3.20, where β_a is the angle of the relative wind direction). An in-depth discussion on vehicle aerodynamics is beyond the scope of the present work. Here it may suffice to state without proof that

$$X_a = \frac{1}{2}\rho_a S_a C_x V_a^2 \qquad Y_a = \frac{1}{2}\rho_a S_a C_y V_a^2 \qquad Z_a = \frac{1}{2}\rho_a S_a C_z V_a^2 \tag{3.81}$$

where ρ_a is the air density, S_a is the vehicle frontal area (Fig. 3.21), $V_a = |\mathbf{V}_a|$ is the magnitude of the relative velocity, and C_x, C_y, C_z are the dimensionless shape coefficients, all three functions of β_a [1, Chap. 10].

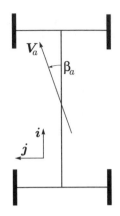

Fig. 3.20 Relative velocity $\mathbf{V}_a$ between the vehicle and the air

Fig. 3.21 Frontal area S_a [15]

Often [1], C_x is called drag coefficient, C_y is called side force coefficient, and C_z is called downforce coefficient. Traditionally, $C_x > 0$, which explains the minus sign in (3.80). According to (3.80), and consistently with its name, $C_z > 0$ means a vertical force directed downward, like weight.[7]

Each term in (3.81) has a well-defined role. The density ρ_a depends on the fluid (air, in this case). The frontal area S_a takes into account the size of the vehicle. The shape of the vehicle, along with the angle β_a, affect the three dimensionless coefficients C_x, C_y and C_z. For instance, if $\mathbf{V}_a$ is directed like the vehicle axis $\mathbf{i}$, that is $\mathbf{V}_a = V_a \mathbf{i}$, the coefficient $C_y = 0$ and hence $Y_a = 0$.

For simplicity, we assume $\beta_a = 0$ hereafter.

In a modern car, the frontal area S_a is about $1.8\,\mathrm{m}^2$ and the *drag coefficient C_x* ranges between 0.20 and 0.35. A Formula 1 car has a frontal area of about $1.3\,\mathrm{m}^2$ and a drag coefficient which ranges between 0.7 and 1.2. It is quite usual to provide directly the product $S_a C_x$ as a more effective way to compare the aerodynamic efficiency of cars. For instance, a Formula 1 car has $S_a C_x$ of about $1.2\,\mathrm{m}^2$, while a commercial one may have it below $0.6\,\mathrm{m}^2$. A Formula 1 car has a *downforce coefficient C_z*

---

[7] In the first edition of this book it was the other way around.

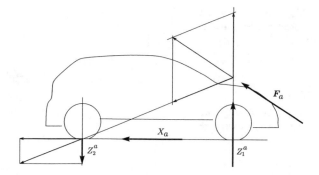

Fig. 3.22 Convenient (and rigorous) decomposition of the total aerodynamic force $\mathbf{F}_a$

with a very high value to achieve a very high aerodynamic downforce. Typically, $S_a C_z = 4 - 5\,\mathrm{m}^2$.

The total aerodynamic force $\mathbf{F}_a$, like all forces, acts along its *line of action*, as shown in Fig. 3.22 and also in Fig. 8.62. Since it is legitimate to move a force along its line of action, the often-used center of pressure concept [1, pp. 212–213] appears to be misleading. In other words, any point on the line of action can be called center of pressure.

In general, the aerodynamic force $\mathbf{F}_a$ does not pass through G (why should it?). Therefore, it contributes to $\mathbf{M}_G$ with an aerodynamic moment $\mathbf{M}_a = M_{a_x}\mathbf{i} + M_{a_y}\mathbf{j} + M_{a_z}\mathbf{k}$, the biggest component being M_{a_y} (pitch moment). However, instead of using the moment components, it is common practice, and indeed very convenient, to do like in Figs. 3.22 and 8.62, thus defining the front and rear aerodynamic vertical forces (positive downward) according to

$$Z_1^a = \frac{1}{2}\rho_a S_a C_{z1} V_a^2 = \zeta_1 V_a^2$$

$$Z_2^a = \frac{1}{2}\rho_a S_a C_{z2} V_a^2 = \zeta_2 V_a^2 \tag{3.82}$$

where the *front and rear downforce coefficients* C_{z1} and C_{z2} have been introduced. In other words, in straight running ($\beta_a = 0$), the aerodynamic force $\mathbf{F}_a$ is perfectly equivalent to two vertical loads Z_1^a and Z_2^a acting *directly on the front and rear axles*, respectively, plus the aerodynamic drag X_a acting *at road level*

$$X_a = \frac{1}{2}\rho_a S_a C_x V_a^2 = \xi V_a^2 \tag{3.83}$$

We remark that this force decomposition is perfectly legitimate, and not arbitrary.

Many race cars have wings and underbody diffuser to create downforces that press the race car against the surface of the track. Therefore, at high speed both Z_1^a and Z_2^a have fairly high positive values, as shown in Figs. 3.23 and 8.62.

For simplicity, here we have tacitly assumed C_x, C_{z1} and C_{z2} to be constant (i.e., speed independent). Actually, this is not strictly true, as shown in Fig. 4.7, because

Fig. 3.23 Aerodynamic forces in a Formula car

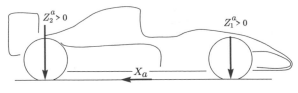

Fig. 3.24 Aerodynamic loads in a road car with $C_z \simeq 0$

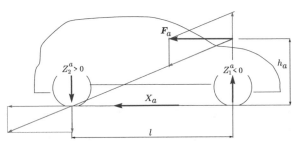

the height from the ground of race cars is typically not constant. Therefore, the assumption of constant aerodynamic shape coefficients should be removed in more advanced analyses.

Road cars usually do not have wings. Therefore, the global coefficient $C_z \simeq 0$, and hence also $Z_a \simeq 0$. However, as shown in Fig. 3.24, since the (almost) horizontal aerodynamic force $\mathbf{F}_a$ is applied at a height h_a from the ground, there are the following (small) vertical loads acting on each axle

$$Z_2^a = -Z_1^a = X_a \frac{h_a}{l} \tag{3.84}$$

In other words, $C_z = 0$ does *not* imply $C_{z1} = C_{z2} = 0$. From (3.81) and (3.84), we obtain in this case $C_{z2} = -C_{z1} = C_x h_a / l$.

3.7.3 Road-Tire Friction Forces

The *road-tire friction forces* $\mathbf{F}_{t_{ij}}$ are the resultant of the tangential stress in each footprint, as shown in (2.14). Typically, for each tire, the tangential force $\mathbf{F}_{t_{ij}}$ is split into a longitudinal component $F_{x_{ij}}$ and a lateral component $F_{y_{ij}}$, as shown in Fig. 3.25. It is very important to note that these two components refer to the wheel reference system shown in Fig. 2.6, not to the vehicle frame.

The driver has more direct control over the tire longitudinal forces $F_{x_{ij}}$, through the brake pedal and the gas pedal, than over the tire lateral forces $F_{y_{ij}}$.

Strictly speaking, the lateral forces $F_{y_{ij}}$ are not applied at the center of the contact patch. In general, there are also the vertical moments $M_{z_{ij}}$. However, these moments have negligible effects on the dynamics of the vehicle as a whole. Indeed, taking

Fig. 3.25 Components of
the road-tire friction forces
in the tire frames

Fig. 3.26 Components of
the road-tire friction forces
in the vehicle frame

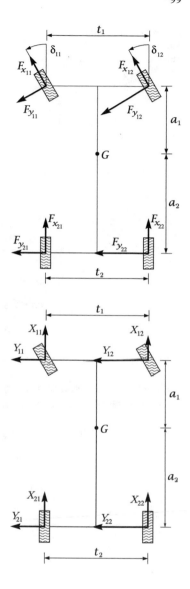

$M_{z_{ij}}$ into account would mean displacing the action lines of $F_{y_{ij}}$ by only a few
centimeters. On the other hand, vertical moments do affect quite a bit the steering
system. In particular, they must be included in vehicle models with compliant steering
system (see Sect. 7.16).

If δ_{ij} is the steering angle of a wheel, the components of the tangential force in
the vehicle frame are given by

Fig. 3.27 Longitudinal and lateral tire forces in the vehicle frame ($X_1 < 0$ in this figure)

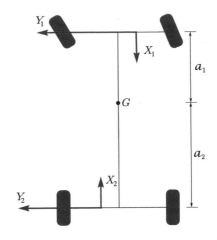

$$\mathbf{F}_{t_{ij}} = X_{ij}\, \mathbf{i} + Y_{ij}\, \mathbf{j}$$

where, as shown in Fig. 3.26

$$X_{ij} = F_{x_{ij}} \cos(\delta_{ij}) - F_{y_{ij}} \sin(\delta_{ij})$$

$$Y_{ij} = F_{x_{ij}} \sin(\delta_{ij}) + F_{y_{ij}} \cos(\delta_{ij})$$

(3.85)

with obvious approximations if δ_{ij} is very small.

To deal with shorter expressions, it is convenient to define (Fig. 3.27)

$$
\begin{aligned}
X_1 &= X_{11} + X_{12} & X_2 &= X_{21} + X_{22} \\
Y_1 &= Y_{11} + Y_{12} & Y_2 &= Y_{21} + Y_{22} \\
\Delta X_1 &= \frac{X_{12} - X_{11}}{2} & \Delta X_2 &= \frac{X_{22} - X_{21}}{2} \\
\Delta Y_1 &= \frac{Y_{12} - Y_{11}}{2} & \Delta Y_2 &= \frac{Y_{22} - Y_{21}}{2}
\end{aligned}
$$

(3.86)

where

$$X_1 = (F_{x_{11}} \cos(\delta_{11}) + F_{x_{12}} \cos(\delta_{12})) - (F_{y_{11}} \sin(\delta_{11}) + F_{y_{12}} \sin(\delta_{12}))$$

$$X_2 = F_{x_{21}} + F_{x_{22}}$$

$$Y_1 = (F_{y_{11}} \cos(\delta_{11}) + F_{y_{12}} \cos(\delta_{12})) + (F_{x_{11}} \sin(\delta_{11}) + F_{x_{12}} \sin(\delta_{12}))$$

$$Y_2 = F_{y_{21}} + F_{y_{22}}$$

$$\Delta X_1 = [(F_{x_{12}} \cos(\delta_{12}) - F_{x_{11}} \cos(\delta_{11})) - (F_{y_{12}} \sin(\delta_{12}) - F_{y_{11}} \sin(\delta_{11}))]/2$$

$$\Delta X_2 = (F_{x_{22}} - F_{x_{21}})/2$$

(3.87)

Fig. 3.28 Components, in
the vehicle frame, of the
lateral forces of the front
wheels ($X_{11} < 0$ and
$X_{12} < 0$ in this figure)

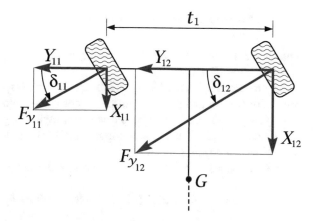

For small steering angles, simpler expressions can be obtained by observing that small errors in the values of the steering angles δ_{ij} have marginal influence on the global equilibrium.[8] More precisely, in the equilibrium equations we can "force" the steering angles of the front wheels δ_{11} and δ_{12} both to be equal to $\delta_1 = (\delta_{11} + \delta_{12})/2$. Similarly, the rear wheels can be set to have the same (often zero) steering, that is $\delta_2 = (\delta_{21} + \delta_{22})/2 \simeq 0$. Therefore, in most practical cases (3.86) becomes

$$
\begin{aligned}
X_1 &= (F_{x_{11}} + F_{x_{12}})\cos(\delta_1) - (F_{y_{11}} + F_{y_{12}})\sin(\delta_1) \\
X_2 &= F_{x_{21}} + F_{x_{22}} \\
Y_1 &= (F_{y_{11}} + F_{y_{12}})\cos(\delta_1) + (F_{x_{11}} + F_{x_{12}})\sin(\delta_1) \\
Y_2 &= F_{y_{21}} + F_{y_{22}} \\
\Delta X_1 &= [(F_{x_{12}} - F_{x_{11}})\cos(\delta_1) - (F_{y_{12}} - F_{y_{11}})\sin(\delta_1)]/2 \\
\Delta X_2 &= (F_{x_{22}} - F_{x_{21}})/2
\end{aligned}
\tag{3.88}
$$

As discussed in Sect. 3.4, the two wheels of the same axle are often intentionally slightly nonparallel. Assuming the two wheels to be parallel is harmless for the global equilibrium of the vehicle, whereas it is quite influential on the tire behavior and hence on the vehicle dynamics.

It is worth noting that we can have $X_1 \neq 0$ and $\Delta X_1 \neq 0$ even in a rear driven car, particularly when the front steer angles are not small. This is mainly due to the fact that $F_{y_{11}} \neq F_{y_{12}}$, as shown in Fig. 3.28.

---

[8] But not on the tire slips.

3.7.4 Road-Tire Vertical Forces

The *road-tire vertical forces* $F_{z_{ij}} \mathbf{k}$ are the resultants of the normal pressure in each footprint, as in (2.12).

As discussed in Sect. 2.13, the displacement with respect to the center of the footprint of the line of action of the vertical forces is the main cause of rolling resistance. This phenomenon can be neglected when studying, e.g., extreme braking or handling, whereas it is of paramount importance for the estimation of fuel consumption or of power losses in general.

It is customary to add the vertical forces of the same axle

$$Z_1 = F_{z_{11}} + F_{z_{12}} \quad \text{and} \quad Z_2 = F_{z_{21}} + F_{z_{22}} \tag{3.89}$$

and to define the differences

$$\Delta Z_1 = \frac{F_{z_{12}} - F_{z_{11}}}{2} \quad \text{and} \quad \Delta Z_2 = \frac{F_{z_{22}} - F_{z_{21}}}{2} \tag{3.90}$$

usually called *lateral load transfers*.

Inverting these equations yields the vertical load on each wheel

$$F_{z_{11}} = \frac{Z_1}{2} - \Delta Z_1 = Z_{11} \qquad F_{z_{12}} = \frac{Z_1}{2} + \Delta Z_1 = Z_{12}$$

$$F_{z_{21}} = \frac{Z_2}{2} - \Delta Z_2 = Z_{21} \qquad F_{z_{22}} = \frac{Z_2}{2} + \Delta Z_2 = Z_{22} \tag{3.91}$$

3.8 Vehicle Equilibrium Equations (More Explicit Form)

The explicit expressions of all the force and moment components in (3.78) are obtained by collecting all the contributions of the external actions. More precisely, we consider:

- Equation (3.81) for the aerodynamic drag;
- Equation (3.82) for the axle vertical aerodynamic loads;
- Equation (3.87) for the axle-road friction forces;
- Equation (3.89) for the axle-road vertical forces;
- Equation (3.90) for the axle-road lateral load transfers.

For a two-axle vehicle we obtain the forces

$$X = X_1 + X_2 - X_a$$

$$Y = Y_1 + Y_2$$

$$Z = Z_1 + Z_2 - (mg + Z_1^a + Z_2^a)$$

and the moments with respect to G

$$L = -\Delta Z_1 t_1 - \Delta Z_2 t_2 + (Y_1 + Y_2)h$$
$$M = -Z_1 a_1 + Z_2 a_2 - (X_1 + X_2 - X_a)h + Z_1^a a_1 - Z_2^a a_2 \qquad (3.92)$$
$$N = Y_1 a_1 - Y_2 a_2 + \Delta X_1 t_1 + \Delta X_2 t_2$$

These expressions can be inserted into (3.78) and then into (3.72) to obtain the six *global equilibrium equations*.

$$m(\dot{u} - vr) = X$$
$$m(\dot{v} + ur) = Y$$
$$0 = Z$$

and   (3.93)

$$-J_{zx}\dot{r} - 2(J_{w_1}/r_1 + J_{w_2}/r_2)ur = L$$
$$-J_{zx}r^2 = M$$
$$J_z\dot{r} = N$$

Actually, it is more convenient to split them into the following *two* sets of equations. A first set of three *in-plane* equilibrium equations, which deal explicitly with the vehicle motion

$$ma_x = m(\dot{u} - vr) = X = X_1 + X_2 - X_a$$
$$ma_y = m(\dot{v} + ur) = Y = Y_1 + Y_2 \qquad (3.94)$$
$$J_z\dot{r} = N = Y_1 a_1 - Y_2 a_2 + \Delta X_1 t_1 + \Delta X_2 t_2$$

and a second set of *out-of-plane* equilibrium equations that involve the constraint forces (vertical loads) to make the vehicle comply with the flatness of the road surface

$$0 = Z = Z_1 + Z_2 - (mg + Z_1^a + Z_2^a)$$
$$-J_{zx}r^2 = M = -(Z_1 - Z_1^a)a_1 + (Z_2 - Z_2^a)a_2 - (X_1 + X_2 - X_a)h$$
$$-J_{zx}\dot{r} - 2(J_{w_1}/r_1 + J_{w_2}/r_2)ur = L = (Y_1 + Y_2)h - \Delta Z_1 t_1 - \Delta Z_2 t_2$$

$$(3.95)$$

Equations (3.94) and (3.95) are really important. If fully understood, they provide a lot of information on vehicle dynamics.

Combining (3.94) and (3.95), the second set can be recast in a form which better highlights the interplay between vertical loads and vehicle motion

$$mg + Z_1^a + Z_2^a = Z_1 + Z_2$$

$$- ma_x h + J_{zx} r^2 + Z_1^a a_1 - Z_2^a a_2 = Z_1 a_1 - Z_2 a_2 \qquad (3.96)$$

$$ma_y h + J_{zx} \dot{r} + 2(J_{w_1}/r_1 + J_{w_2}/r_2) ur = \Delta Z_1 t_1 + \Delta Z_2 t_2$$

where $J_{zx} r^2$ and $J_{zx} \dot{r}$ are usually negligible. The gyroscopic torque may not be negligible in race cars.

We remind that aerodynamic downforces mean $Z_i^a > 0$.

It is convenient to define

$$N_Y = Y_1 a_1 - Y_2 a_2$$
$$\qquad (3.97)$$
$$N_X = \Delta X_1 t_1 + \Delta X_2 t_2$$

to highlight in (3.94) the different origins of the two contributions to the yawing moment

$$N = N_Y + N_X = J_z \dot{r} \qquad (3.98)$$

As a matter of fact, N_Y is due to the lateral forces, while N_X comes from the difference between the longitudinal forces of the same axle.

From

$$Y = Y_1 + Y_2$$
$$\qquad (3.99)$$
$$N_Y = Y_1 a_1 - Y_2 a_2 = J_z \dot{r} - N_X$$

we obtain the lateral (grip) forces exerted by the road on each axle

$$Y_1 = \frac{Y a_2 + N_Y}{l} = \frac{Y a_2^b}{l} \quad \text{and} \quad Y_2 = \frac{Y a_1 - N_Y}{l} = \frac{Y a_1^b}{l} \qquad (3.100)$$

where

$$a_1^b = a_1 - x_N \quad \text{and} \quad a_2^b = a_2 + x_N$$
$$\qquad (3.101)$$
$$\text{with} \quad x_N = \frac{N_Y}{Y} = \frac{J_z \dot{r} - N_X}{Y}$$

Therefore $a_1^b + a_2^b = a_1 + a_2 = l$. In a Formula 1 car, typically $|x_N| < 0.3\,\text{m}$, and often much lower, with a wheelbase l of about 3.5 m.

Whenever $N_Y \neq 0$, the force Y is *not* applied at the center of gravity G. Often the distance x_N between Y and G is very small, but nonetheless it is not identically equal to zero. Therefore, it is not correct that "All accelerative forces acting on a body can be considered to act through the center of gravity of that body" [22, p. 29].

An equivalent, more "dynamic", form of (3.100) is

$$Y_1 = \frac{ma_y a_2}{l} + \frac{N - N_X}{l} \quad \text{and} \quad Y_2 = \frac{ma_y a_1}{l} - \frac{N - N_X}{l} \qquad (3.102)$$

where $N = J_z \dot{r}$.

Most classical vehicle dynamics is strongly based on the single track model, and hence assumes $N_X = 0$; this is correct except when the vehicle:

- operates with high wheel steer angles;
- has a limited-slip (or locked) differential;
- has ESP and it has been activated;
- is braking with locked wheels on a road with nonuniform grip coefficients.

3.9 Vertical Loads and Load Transfers

Load transfers need additional discussion. Indeed, the vertical load acting on a tire does affect very much its behavior. Therefore, it is important to discuss the relationships between vehicle dynamics and vertical loads (3.91).

During vehicle motion, the vertical loads change whenever there are accelerations. In case of substantial aerodynamic vertical loads, the vehicle speed also affects the vertical loads, as shown in (3.82). We remind again that aerodynamic downforces, as in Formula cars, mean $Z_i^a > 0$.

3.9.1 Longitudinal Load Transfer

From the first and second equations in (3.96) it is easy to obtain, for a two-axle vehicle, the vertical loads that the road applies on each axle

$$
\begin{aligned}
Z_1 &= Z_1^0 + Z_1^a + \Delta Z \\
Z_2 &= Z_2^0 + Z_2^a - \Delta Z
\end{aligned}
\tag{3.103}
$$

where

$$
\Delta Z = -\frac{m a_x h}{l} + \frac{J_{zx} r^2}{l} \simeq -\frac{m a_x h}{l}
\tag{3.104}
$$

is the *longitudinal load transfer* due to the longitudinal acceleration a_x, and

$$
Z_1^0 = \frac{m g a_2}{l} \qquad Z_2^0 = \frac{m g a_1}{l}
\tag{3.105}
$$

are the *static loads* on each axle. In a motionless vehicle the vertical loads have to balance only the vehicle weight.

3.9.2 Lateral Load Transfers

Lateral load transfers ΔZ_1 and ΔZ_2 appear explicitly only in the second equation in (3.95), which may be recast as

$$
\begin{aligned}
\Delta Z_1 t_1 + \Delta Z_2 t_2 &= Yh + J_{zx}\dot{r} + 2(J_{w_1}/r_1 + J_{w_2}/r_2)ur \\
&= ma_y h + J_{zx}\dot{r} + 2(J_{w_1}/r_1 + J_{w_2}/r_2)ur \qquad (3.106) \\
&\simeq ma_y h + 2(J_{w_1}/r_1 + J_{w_2}/r_2)a_y
\end{aligned}
$$

where $a_y = \dot{v} + ur$ is the lateral acceleration. Obviously, the gyroscopic torque affects the lateral load transfers, but has no direct influence on the lateral forces.

Of course, one equation is not enough to single out ΔZ_1 and ΔZ_2.

Even under *static conditions*, (3.106) yields, for a two-axle vehicle

$$
\Delta Z_1^0 t_1 + \Delta Z_2^0 t_2 = 0 \qquad (3.107)
$$

which shows that the static lateral load transfers ΔZ_1^0 and ΔZ_2^0 may have in principle any value. However, with the aid of four scales, it is part of the setup procedure to achieve $\Delta Z_1^0 = \Delta Z_2^0 = 0$.

Suspension geometry and compliances influence directly the ratio $\Delta Z_1/\Delta Z_2$. This is a fundamental aspect of vehicle dynamics, as discussed in the next few sections.

3.9.3 Vertical Load on Each Tire

The global amount of lateral load transfer is determined by (3.106), but how much of it goes to the front and how much to the rear cannot be found without looking at the suspensions and at the tires (unless the vehicle is a three-wheeler). This is the motivation for Sect. 3.10, where some of the front and rear suspension features will be exploited.

Summing up, the vertical loads on each tire are

$$
\begin{aligned}
Z_{11} &= 0.5(Z_1^0 + Z_1^a + \Delta Z) - \Delta Z_1 \\
Z_{12} &= 0.5(Z_1^0 + Z_1^a + \Delta Z) + \Delta Z_1 \\
Z_{21} &= 0.5(Z_2^0 + Z_2^a - \Delta Z) - \Delta Z_2 \\
Z_{22} &= 0.5(Z_2^0 + Z_2^a - \Delta Z) + \Delta Z_2
\end{aligned}
\qquad (3.108)
$$

or, more explicitly

$$Z_{11} = \frac{1}{2}\left(\frac{mga_2}{l} + \frac{1}{2}\rho_a S_a C_{z1} u^2 - \frac{ma_x h - J_{zx} r^2}{l} \right) - \Delta Z_1$$

$$Z_{12} = \frac{1}{2}\left(\frac{mga_2}{l} + \frac{1}{2}\rho_a S_a C_{z1} u^2 - \frac{ma_x h - J_{zx} r^2}{l} \right) + \Delta Z_1$$

$$Z_{21} = \frac{1}{2}\left(\frac{mga_1}{l} + \frac{1}{2}\rho_a S_a C_{z2} u^2 + \frac{ma_x h - J_{zx} r^2}{l} \right) - \Delta Z_2$$

$$Z_{22} = \frac{1}{2}\left(\frac{mga_1}{l} + \frac{1}{2}\rho_a S_a C_{z2} u^2 + \frac{ma_x h - J_{zx} r^2}{l} \right) + \Delta Z_2$$

(3.109)

where ΔZ_1 and ΔZ_2 will be obtained after having carried out the suspension analysis (see Sect. 3.10.12).

A general treatment of the gyroscopic torques is quite involved, as it strongly depends on the suspension features. However, in road cars with independent suspensions the gyroscopic torques have very little influence. Therefore, we consider them only in Sect. 3.11 on solid axle suspensions, and in Chap. 8 on race cars.

3.10 Suspension First-Order Analysis

Consistently with the hypotheses listed on this chapter, the suspension mechanics will be analyzed assuming *very small* suspension deflections and tire deformations. This is what a first-order analysis is all about. Of course, it is not the whole story, but it is a good starting point.[9]

More precisely, the following aspects will be addressed:

1. suspension reference configuration (Sect. 3.10.1);
2. suspension internal coordinates (Sect. 3.10.2);
3. camber variation (Sect. 3.10.3);
4. track width variation (Sect. 3.10.4);
5. vehicle internal coordinates (Sect. 3.10.5);
6. suspension and tire stiffnesses (roll and vertical, Sect. 3.10.6);
7. suspension internal equilibrium (Sect. 3.10.7);
8. no-roll centers and no-roll axis (Sect. 3.10.9);
9. roll angles and lateral load transfers (Sect. 3.10.12).

We remark that without the results of this (quite long) section it is not possible to single out the front and rear lateral load transfers. The reader not interested in the detail can jump to Sect. 3.10.13 for the final results.

---

[9] At first it may look paradoxical, but it is not. Actually it is common practice in engineering. Just take the most classical cantilever beam, of length l with a concentrated load F at its end. Strictly speaking, the bending moment at the fixed end is not exactly equal to Fl, since the beam deflection takes the force a little closer to the wall. But this effect is usually neglected.

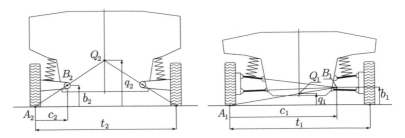

Fig. 3.29 Suspensions in their reference configuration: swing axle (left) and double wishbone suspension (right)

3.10.1   Suspension Reference Configuration

Figure 3.29 shows two possible independent suspensions in their *reference configuration* (vehicle going straight at constant speed). It also serves the purpose of defining some relevant geometric parameters: t_i, b_i, c_i, and consequently q_i.

The reference configuration is supposed to be perfectly *symmetric*. More precisely, the left and right sides are exactly alike (including springs). As usual, t_1 and t_2 are the front and rear track widths.

Points A_i mark the centers of the tire contact patches (i.e., point O in Fig. 2.6). Points B_i are the instantaneous centers of rotation of the wheel hub with respect to the vehicle body. They are often called swing centers [5, p. 150]. Here, for simplicity, the suspension linkage is supposed to be *rigid* and *planar*. In a swing axle suspension, point B_2 is indeed the center of a joint, whereas in a double wishbone suspension (right) point B_1 has to be found by a well-known method. In both cases, the distances c_i and b_i set the position of B_i with respect to A_i (Fig. 3.29).

Also shown in Fig. 3.29 are points Q_1 and Q_2. They are given by the intersection of the straight lines connecting A_i and B_i on both sides. Because of symmetry, they lay on the centerline at heights q_1 and q_2. Points Q_1 and Q_2 are the so-called *roll centers* and their role in vehicle dynamics will be addressed shortly. Moreover, it will be demonstrated that a better name is *no-roll centers*.

3.10.1.1   First-Order Geometric Features

It is worth noting that (Fig. 3.29)

$$q_i = \frac{t_i}{2} \frac{b_i}{c_i} \tag{3.110}$$

which means that only three out of four geometric suspension parameters are really independent. Moreover, in general there are quite a lot of restrictions on the value of

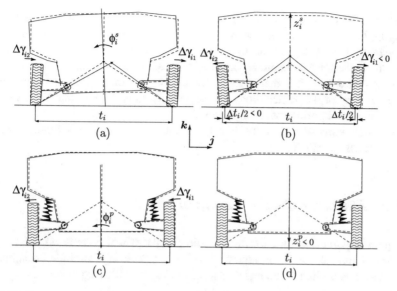

Fig. 3.30 Suggested selection of internal coordinates (front view): **a** roll angle ϕ_i^s due to suspension deflections only, **b** vertical displacement z_i^s due to suspension deflections only, **c** roll angle ϕ_i^p due to tire deformations only, **d** vertical displacement z_i^p due to tire deformations only

the track width t_i. Therefore, a suspension scheme is basically characterized by *two* parameters, say b_i and c_i (in the framework of a first-order analysis).

3.10.2 Suspension Internal Coordinates

For each axle, *four* "internal" coordinates are necessary to monitor the suspension conditions with respect to a reference configuration. A possible selection of coordinates[10] may be as follows (Fig. 3.30)

- body roll angle ϕ_i^s due to suspension deflections only (Fig. 3.30a);
- vertical displacement z_i^s of any point of the body centerline due to suspension deflections only (Fig. 3.30b);
- body roll angle ϕ_i^p due to tire deformations only (Fig. 3.30c);
- vertical displacement z_i^p of any point of the body centerline due to tire deformations only (Fig. 3.30d).

All these coordinates are, by definition, equal to zero in the reference configuration. In Sect. 9.5.1 it is explained why we can take *any* point of the body centerline to define z_i^s and z_i^p.

---

[10] A more precise definition of roll angle is given in Sect. 9.2.

Figure 3.30 shows qualitatively how each single coordinate changes the vehicle configuration for a swing axle suspension. These four coordinates are, obviously, *independent*. It will depend on the vehicle dynamics whether they change or not. In other words, the kinematic schemes of Fig. 3.30 have nothing to do with real operating conditions. It is therefore legitimate, but not mandatory at all, to define, e.g., the roll ϕ_i^s of the vehicle body keeping the vertical displacement z_i^s fixed and without any tire deformation, as in Fig. 3.30.

Any other kinematic quantity, like the camber variations, is a function of the selected set of coordinates $(\phi_i^s, z_i^s, \phi_i^p, z_i^p)$.

3.10.3 Kinematic Camber Variation

It is quite important to monitor the variation of the wheel *camber angles* γ_{ij} as a function of the selected coordinates $(\phi_i^s, z_i^s, \phi_i^p, z_i^p)$. In a *first-order* analysis, the investigation is limited to the linear term of the Taylor series expansion

$$\gamma_{ij} = \gamma_{ij}^0 + \Delta\gamma_{ij} \simeq \gamma_{ij}^0 + \frac{\partial\gamma_{ij}}{\partial\phi_i^s}\,\phi_i^s + \frac{\partial\gamma_{ij}}{\partial z_i^s}\,z_i^s + \frac{\partial\gamma_{ij}}{\partial\phi_i^p}\,\phi_i^p + \frac{\partial\gamma_{ij}}{\partial z_i^p}\,z_i^p \qquad (3.111)$$

where the static camber angles γ_{ij}^0 were defined in (3.63), and all derivatives are evaluated at the reference configuration (therefore they are numbers, not functions).

From Fig. 3.29, and also with the aid of Fig. 3.30, we obtain the following *general* results for *any symmetric* planar suspension (cf. Fig. 9.5)

$$
\begin{aligned}
\frac{\partial\gamma_{i1}}{\partial\phi_i^s} &= \frac{\partial\gamma_{i2}}{\partial\phi_i^s} = -\frac{q_i - b_i}{b_i} = -\frac{t_i/2 - c_i}{c_i} \\[2mm]
\frac{\partial\gamma_{i1}}{\partial z_i^s} &= -\frac{\partial\gamma_{i2}}{\partial z_i^s} = -\frac{1}{c_i} \\[2mm]
\frac{\partial\gamma_{i1}}{\partial\phi_i^p} &= \frac{\partial\gamma_{i2}}{\partial\phi_i^p} = 1 \\[2mm]
\frac{\partial\gamma_{i1}}{\partial z_i^p} &= \frac{\partial\gamma_{i2}}{\partial z_i^p} = 0
\end{aligned}
\qquad (3.112)
$$

Combining (3.111) and (3.112), we obtain the following formulas for camber variations (Fig. 3.30)

$$
\begin{aligned}
\Delta\gamma_{i1} &\simeq -\left(\frac{t_i/2 - c_i}{c_i}\right)\phi_i^s - \frac{1}{c_i}z_i^s + \phi_i^p \\[2mm]
\Delta\gamma_{i2} &\simeq -\left(\frac{t_i/2 - c_i}{c_i}\right)\phi_i^s + \frac{1}{c_i}z_i^s + \phi_i^p
\end{aligned}
\qquad (3.113)
$$

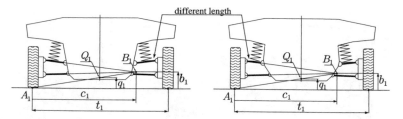

Fig. 3.31 Different double wishbone suspensions with the same first-order behavior

The sign convention for the camber variations $\Delta\gamma_{ij}$ is like in Fig. 2.6 (see also Figs. 7.6 and 9.10). In the suspension shown in Fig. 3.30a we have $t_i/2 > c_i$, whereas in the suspension shown in Fig. 3.30b we have $t_i/2 < c_i$. For an intuitive illustration of what the suspension contributions really are you may have a look at Fig. 3.32.

Equation (3.113) is quite remarkable. It is simple, yet profound. The two suspension schemes of Fig. 3.29, which look so different, do have indeed very different values of the first two partial derivatives in (3.112). On the other hand, it should not be forgotten that (3.111) is just a first-order approximation. As shown in Fig. 3.31, different suspensions can exhibit the same first-order behavior.

Equation (3.113) is merely a *kinematic* relationship. There is no dynamics in it. Therefore, we must be careful not to attempt to extract from it information it cannot provide at all.

Another common mistake is to state, e.g., that a suspension scheme has a typical value of the partial derivative $\partial\gamma_{ij}/\partial\phi_i^s$, without specifying which are the other three internal coordinates. The value of the partial derivative is very much affected by which other coordinates are kept constant.

3.10.4 Kinematic Track Width Variation

Also relevant is the variation of the track widths t_i as a function of the selected coordinates $(\phi_i^s, z_i^s, \phi_i^p, z_i^p)$.

The first-order relationship between the vertical suspension displacement z_i^s and the track length variation Δt_i is given by (Figs. 3.29 and 3.30b)

$$\Delta t_i \simeq -\frac{2b_i}{c_i} z_i^s = -\frac{4q_i}{t_i} z_i^s \qquad (3.114)$$

It does not depend on ϕ_i^s and ϕ_i^p because of left-right *symmetry* of the suspension system typical of most cars.

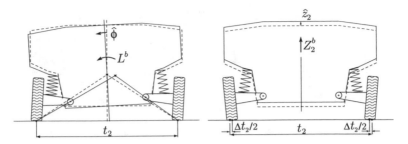

Fig. 3.32 Fictitious moment L^b and fictitious vertical force Z_2^b to measure roll and vertical stiffnesses

3.10.5 Vehicle Internal Coordinates

Three internal coordinates are necessary to monitor the vehicle body condition with respect to a reference (often static) configuration. A suitable choice may be to take as coordinates the *vehicle body roll angle* ϕ and the front and rear vertical displacements z_1, z_2 of the vehicle centerline (Fig. 3.32).

These three coordinates are, of course, independent. Whether they change or not will depend on the vehicle dynamics.

The total roll angle ϕ of the vehicle body is given by

$$\phi = \phi_1^s + \phi_1^p = \phi_2^s + \phi_2^p \tag{3.115}$$

that is, by the roll angle ϕ_i^s due to the suspension deflections plus the roll angle ϕ_i^p due to the tire deformations [5, p. 90].

Similarly, the front and rear vertical displacements z_1, z_2 of the vehicle centerline are

$$z_1 = z_1^s + z_1^p \quad \text{and} \quad z_2 = z_2^s + z_2^p \tag{3.116}$$

where z_i^s are the vertical displacements of the vehicle centerline due to suspension deflections only and z_i^p are the vertical displacements due to the tire deformations only.

Equations (3.115) and (3.116) precisely relate the eight suspension internal coordinates to the three vehicle internal coordinates.

3.10.6 Definition of Roll and Vertical Stiffnesses

The goal of this section is to define the stiffness associated with each internal coordinate.

It is important to realize that the *symmetric* behavior of the two suspensions of the same axle plays a key role here. If, for some reason, the two suspensions were different, then we should also have to consider the cross-coupled stiffnesses.

3.10.6.1 Roll Stiffnesses

To evaluate the roll stiffnesses, we assume to apply first a (small) pure rolling moment $L^b \mathbf{i}$ to the vehicle body.

As shown in Fig. 3.32, application of a (small) pure rolling moment $L^b \mathbf{i}$ to the vehicle body results in a (small) measurable pure roll rotation $\hat{\phi} \, \mathbf{i}$ such that[11]

$$k_\phi \hat{\phi} = (k_{\phi_1} + k_{\phi_2})\hat{\phi} = L^b \tag{3.117}$$

where k_ϕ is, by definition, the *global roll stiffness* of the vehicle. Therefore

$$k_\phi = \frac{L^b}{\hat{\phi}} \tag{3.118}$$

is now known.

Moreover, by measuring the corresponding load transfers (Fig. 3.33)

$$L_1^b = \Delta Z_1^L t_1 = k_{\phi_1}\hat{\phi} \quad \text{and} \quad L_2^b = \Delta Z_2^L t_2 = k_{\phi_2}\hat{\phi} \tag{3.119}$$

also the *roll stiffnesses* k_{ϕ_1} and k_{ϕ_2} of the *front and rear axles*, respectively, can be obtained

$$k_{\phi_1} = \frac{\Delta Z_1^L t_1}{\hat{\phi}} \quad \text{and} \quad k_{\phi_2} = \frac{\Delta Z_2^L t_2}{\hat{\phi}} \tag{3.120}$$

Of course, as a check, it has to be $L_1^b + L_2^b = L^b$ and (springs in parallel)

$$k_\phi = k_{\phi_1} + k_{\phi_2} \tag{3.121}$$

The load transfers ΔZ_1^L and ΔZ_2^L depend on the combined deflections of suspensions and tires. Of course, $z_1 = z_2 = 0$, since they are not affected by L^b. This is true only if the left and right suspensions have a perfectly *symmetric* behavior. For instance, the so-called *contractive suspensions* do not behave the same way and, therefore, a pure rolling moment also yields some vertical displacement.

For further developments, it is necessary to determine how much of $\hat{\phi}$ is due to the suspension springs and, possibly, anti-roll bar (Fig. 3.34), and how much to the tire vertical deflections. More precisely, it is necessary to single out the *suspension roll stiffnesses* $k_{\phi_1}^s$ and $k_{\phi_2}^s$ from the *tire roll stiffnesses* $k_{\phi_1}^p$ and $k_{\phi_2}^p$.

---

[11] The symbol $\hat{\phi}$ (instead of just ϕ) is used to stress that this is not the roll angle under operating conditions, but the roll angle due to a pure rolling moment.

Fig. 3.33 Load transfers ΔZ_i^L due to a pure rolling moment L_i^b (no need of lateral forces)

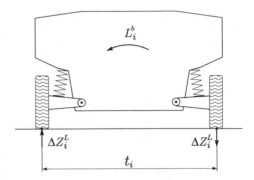

Fig. 3.34 Increasing the suspension roll stiffness by means of the anti-roll bar [13]

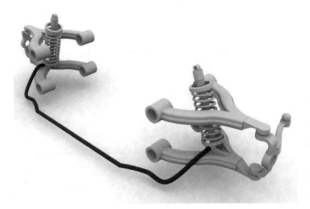

Under the same global pure rolling moment L^b **i**, the tires and the suspensions of the same axle behave like *springs in series*. Therefore

$$k_{\phi_i} = \frac{k_{\phi_i}^s k_{\phi_i}^p}{k_{\phi_i}^s + k_{\phi_i}^p} \qquad (3.122)$$

which means that, for each axle

$$\Delta Z_i^L t_i = k_{\phi_i}\hat\phi = k_{\phi_i}^s \hat\phi_i^s = k_{\phi_i}^p \hat\phi_i^p = L_i^b \quad \text{with} \quad \hat\phi = \hat\phi_i^s + \hat\phi_i^p \qquad (3.123)$$

where $\hat\phi_i^s$ and $\hat\phi_i^p$ are the roll angles due, respectively, to the suspension and tire deflections that the vehicle body undergoes under the action of a pure rolling moment L^b **i**.

If p_1 and p_2 are the measured vertical stiffnesses of a single front and rear tire, respectively (in a first-order analysis, a linear behavior can be consistently assumed), the *tire roll stiffnesses* are given by

$$k_{\phi_i}^P = \frac{p_i t_i^2}{2} \tag{3.124}$$

Once $k_{\phi_i}^P$ are known from (3.124), the *suspension roll stiffness* $k_{\phi_i}^s$ for each axle can be obtained from (3.122)

$$k_{\phi_i}^s = \frac{k_{\phi_i}^P k_{\phi_i}}{k_{\phi_i}^P - k_{\phi_i}} \tag{3.125}$$

Summing up, by applying L^b and measuring $\hat{\phi}$, $\Delta Z_1^L t_1$ and $\Delta Z_2^L t_2$ we get k_{ϕ_1} and k_{ϕ_2}. Then, from the measurement of $k_{\phi_1}^P$ and $k_{\phi_2}^P$, we can compute $k_{\phi_1}^s$ and $k_{\phi_2}^s$. Therefore, all relevant roll stiffnesses have been defined. Moreover, a possible experimental procedure has been outlined.

3.10.6.2 Vertical Stiffnesses

Similarly, to obtain the vertical stiffnesses, small vertical loads Z_i^b are assumed to be applied over each axle.

As shown in Fig. 3.32, application to the vehicle body centerline, exactly over the front axle, of an upward (small) vertical load $Z_1^b \mathbf{k}$ results only in a (small) vertical displacement $\hat{z}_1$ such that[12]

$$Z_1^b = k_{z_1} \hat{z}_1 \tag{3.126}$$

which defines the *global front vertical stiffness* k_{z_1}. Doing the same on the rear axle provides

$$Z_2^b = k_{z_2} \hat{z}_2 \tag{3.127}$$

which defines the *global rear vertical stiffness* k_{z_2}.

Again, to single out the suspension and tire contributions, first observe that together the two tires of each axle have a vertical stiffness

$$k_{z_i}^P = 2p_i \tag{3.128}$$

Therefore, the corresponding *suspension* vertical stiffness $k_{z_i}^s$ can be obtained from

$$k_{z_i} = \frac{k_{z_i}^s k_{z_i}^P}{k_{z_i}^s + k_{z_i}^P} \tag{3.129}$$

which means that for each axle

$$k_{z_i}\hat{z}_i = k_{z_i}^s \hat{z}_i^s = k_{z_i}^P \hat{z}_i^P = Z_i^b \quad \text{with} \quad \hat{z}_i = \hat{z}_i^s + \hat{z}_i^P \tag{3.130}$$

---

[12] The symbols $\hat{z}_1$ and $\hat{z}_2$ (instead of just z_1 and z_2) are used to stress that these are not the vertical displacements under operating conditions.

where $\hat{z}_i^s$ and $\hat{z}_i^p$ are the vertical displacements of the centerline due, respectively, to the suspension and tire deflections.

3.10.6.3 First-Order Elastic Features

The four numbers $k_{\phi_1}^s$, $k_{z_1}^s$, $k_{\phi_2}^s$ and $k_{z_2}^s$ completely characterize the first-order *elastic* features of the front and rear *suspensions*. Similarly, the four numbers $k_{\phi_1}^p$, $k_{z_1}^p$, $k_{\phi_2}^p$ and $k_{z_2}^p$ completely characterize the first-order *elastic* features of the front and rear *tires*.

3.10.7 Suspension Internal Equilibrium

The forces exerted by the *road* on each *tire* are transferred to the vehicle body by the *suspensions*.

It is important to find out how much of these loads goes *through the suspension linkages*, and how much *through the suspension springs*, thus requiring suspension deflections.

As already discussed in Sect. 3.7, each tire is subject to a force $X_{ij}\,\mathbf{i} + Y_{ij}\,\mathbf{j} + Z_{ij}\,\mathbf{k}$. Here, for simplicity, it is assumed that this force is applied at the center of the contact patch.

3.10.8 Effects of a Lateral Force

So far the suspension geometry has played no role (except in Sect. 3.10.3 on the camber variations), at least not explicitly. This was done purposely to highlight which vehicle features are not directly related to the suspension kinematics.

The fundamental reason that makes the suspension geometry so relevant is that vehicle bodies are subject to *horizontal (lateral) forces* (inertial and aerodynamic forces).

Starting from the reference configuration, and according to the equilibrium equation (3.94), let us apply to the vehicle body an *inertia lateral force* $-Y\,\mathbf{j}$, with $Y = ma_y$. As shown in Fig. 3.35, be this force located at height h above the road and at distances a_1^b and a_2^b from the front and rear axles, respectively. As shown in (3.100), a_1^b and a_2^b differ from a_1 and a_2 whenever the yaw moment $N_Y \neq 0$.

Exactly like in (3.100), in a two-axle vehicle the lateral forces exerted by the road on each axle to balance Y are given by (Fig. 3.36)

$$Y_1 = \frac{Y a_2^b}{l} \quad \text{and} \quad Y_2 = \frac{Y a_1^b}{l} \tag{3.131}$$

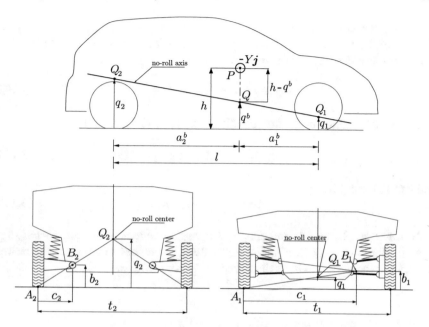

Fig. 3.35 No-roll centers and no-roll axis for a swing arm suspension (left) and a double wishbone suspension (right)

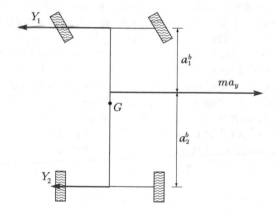

Fig. 3.36 Lateral forces for dynamic equilibrium

It is very important to remind that these two forces can be obtained from the global equilibrium equations alone. Therefore, they are not affected by the suspensions, by the type of tires, by the amount of grip, etc. Moreover, like in (3.106),

$$Yh = \Delta Z_1 t_1 + \Delta Z_2 t_2 \tag{3.132}$$

This is all that can be achieved from global equilibrium.

Among the effects of the inertia force $-Y\mathbf{j}$ there is, in general, a (small) *roll angle* ϕ of the vehicle body. This angle ϕ is the sum of ϕ_i^s due to the suspension deformations and ϕ_i^p due to the tire deflections

$$\phi = \phi_1^s + \phi_1^p = \phi_2^s + \phi_2^p \tag{3.133}$$

From the definition of the tire roll stiffnesses (3.124), it arises that

$$\Delta Z_1 t_1 = k_{\phi_1}^p \phi_1^p \quad \text{and} \quad \Delta Z_2 t_2 = k_{\phi_2}^p \phi_2^p \tag{3.134}$$

and hence, from (3.132)

$$Yh = k_{\phi_1}^p \phi_1^p + k_{\phi_2}^p \phi_2^p \tag{3.135}$$

However, to obtain ΔZ_1 and ΔZ_2, it is necessary to look at the *suspension kinematics*. More precisely, in a first-order analysis, it suffices to consider the no-roll centers and the no-roll axis, as discussed in the next section.

3.10.9 No-Roll Centers and No-Roll Axis

The commonly called roll centers are renamed here *no-roll centers*. Similarly, the roll axis is renamed here *no-roll axis*. The reasons for departing from the traditional naming are explained in this section and in Chap. 9.

However, we state from the very beginning what the outcome of our analysis will be: the roll axis, as that axis about which the vehicle rolls, does not exist. There is no such thing as an axis about which the vehicle rolls, albeit the vehicle rolls indeed. A similar conclusion was obtained also in [3, 11] and in [2, p. 400].

Let us have a closer look at the suspension linkages. In case of purely transversal independent suspensions, like those shown, e.g., in Fig. 3.35, it is easy to obtain the instantaneous center of rotation B_i of each wheel hub with respect to the vehicle body. Another useful point is the center A_i of each contact patch (i.e., point O in Fig. 2.6).

The same procedure can be applied also to the MacPherson strut. The kinematic scheme is shown in Fig. 3.37, while a possible practical design is shown in Fig. 3.38. The MacPherson strut is the most widely used front suspension system, especially in cars of European origin. It is the only suspension to employ a slider, marked by number 2 in Fig. 3.37. Usually, the slider is the damper, which is then part of the suspension linkage. To obtain the instantaneous center of rotation B_i of each wheel hub with respect to the vehicle body it suffices to draw two lines, one along joints 3 and 4, and the other through joint 1 and perpendicular to the slider (not to the steering axis, which goes from joint 1 and 3, as also shown in Fig. 3.37).

In all suspension schemes (Figs. 3.35 and 3.37), the intersection of lines connecting A_i and B_i on both side of the same axle provides, for each axle, the *no-roll*

Fig. 3.37 No-roll center for
a MacPherson strut

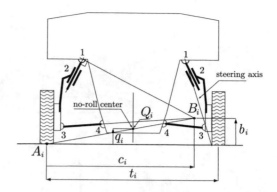

Fig. 3.38 Example of front
MacPherson strut [13]

center Q_i. The signed distance of Q_i from the road is named q_i in Fig. 3.35. A no-roll
center below the road level would have $q_i < 0$.

Therefore, a two-axle vehicle has two no-roll centers Q_1 and Q_2. The unique
straight line connecting Q_1 and Q_2 is called the *no-roll axis* (Fig. 3.35).

Some comments are in order here:

- the procedure just described to obtain the no-roll centers Q_i is not ambiguous,
  provided the motion of the wheel hub with respect to the vehicle body is planar
  and has one degree of freedom;
- points A_i are well defined and are not affected by the tire vertical compliance;

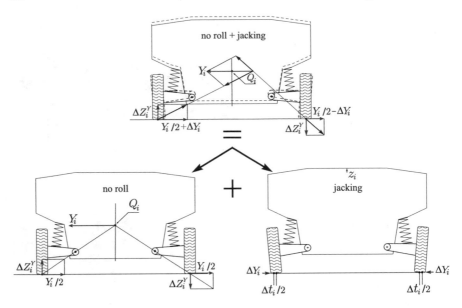

Fig. 3.39 Lateral load transfer without suspension roll, but with vehicle raising due to suspension jacking

- a three-axle vehicle has three points Q_i. Therefore, in general there is *not* a straight line connecting Q_1, Q_2 and Q_3. How to define, if possible, something like a no-roll axis for a three-axle vehicle will be addressed in Sect. 3.16.

What is the motivation for having defined the no-roll centers, and afterwards the no-roll axis?

Figure 3.39 shows how a lateral force Y_i, if *applied at Q_i*, is transferred to the ground *by the suspension linkage*, with no intervention of the springs. Therefore, *a force applied at the roll center does not produce any suspension roll*, although it produces a load transfer ΔZ_i^Y

$$Y_i q_i = \Delta Z_i^Y t_i \tag{3.136}$$

This is the key feature of the roll centers Q_i, which are renamed here *no-roll centers*.

The no-roll axis is useful because the two lateral forces Y_1 and Y_2 must be like in (3.131) for the global equilibrium to be fulfilled. A lateral force Y applied at any point of the line connecting Q_1 and Q_2 is indeed equivalent to a force Y_1 applied at Q_1 and a force Y_2 applied at Q_2, both that obey (3.131). This is the motivation for defining the roll axis. Again, a better name is *no-roll axis*.

Summing up, *application of a force to the vehicle body at any point of the no-roll axis does not produce suspension roll*. More precisely, a force (of any direction) applied to the vehicle body and whose line of action goes through the no-roll axis may affect the vehicle roll angle, but only because of tire deflections, with no contribution from the suspensions. In addition, there may be variations of z_1 and z_2, as discussed in Sect. 3.10.10.

3.10.10 Suspension Jacking

However, no suspension roll does not mean no other effect at all.

Indeed, there are always lateral load transfers $\Delta Z_i^Y = Y_i q_i / t_i$, and hence also some rolling of the vehicle body related to the tire vertical deflections.

Moreover, since the lateral forces exerted by the road on the left and right tires are not equal to each other (they will be equal to $Y_i/2 \pm \Delta Y_i$, where ΔY_i depends on the tire behavior), there is also a small *rising* z_i^s of the vehicle body (Fig. 3.39)

$$z_i^s = \frac{2b_i}{c_i} \frac{\Delta Y_i}{k_{z_i}^s} = \frac{4q_i}{t_i} \frac{\Delta Y_i}{k_{z_i}^s} \tag{3.137}$$

associated, as in (3.114), with a small *track variation* Δt_i

$$\Delta t_i = -\frac{2b_i}{c_i} z_i^s = -\frac{4q_i}{t_i} z_i^s = -\left(\frac{4q_i}{t_i}\right)^2 \frac{\Delta Y_i}{k_{z_i}^s} \tag{3.138}$$

and *suspension jacking*, as in Fig. 3.30b (see also [6, p. 121]). The stiffness of the tires does not appear in (3.137) and (3.138).

The lateral forces exerted by the road on the left and right tires are not equal to each other because of the different vertical loads and different tire slips. This aspect is thoroughly discussed in Sect. 7.5.3.

It is worth noting that the lower the absolute value $|q_i|$ of the no-roll center height, the lower the suspension jacking. This is one of the reasons for avoiding suspensions like in Fig. 3.39 [21, p. 67].

Of course, suspension jacking is also induced by variations in the vertical loads due to longitudinal acceleration and/or aerodynamic effects.

3.10.11 Roll Moment

Let us go back to a purely lateral force $-Y\mathbf{j}$ applied at P (not necessarily the center of mass G), as shown in Fig. 3.35. Since the global equilibrium dictates the values of Y_1 and Y_2 in (3.131), we conveniently decompose the lateral force $-Y\mathbf{j}$ into a force $-Y_1\mathbf{j}$ applied at the front no-roll center Q_1 and a force $-Y_2\mathbf{j}$ applied at the other no-roll center Q_2, plus a suitable moment $L^b\mathbf{i}$.

There is a simple two-step procedure to obtain this result. First, consider that $-Y\mathbf{j}$ at P is equivalent to the same force $-Y\mathbf{j}$ applied at point Q on the no-roll axis, right

Fig. 3.40 How to evaluate
the roll moment L^b

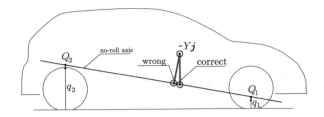

below P, plus a pure (horizontal) roll moment

$$L^b\mathbf{i} = Y(h - q^b)\mathbf{i} \tag{3.139}$$

where

$$q^b = \frac{a_2^b q_1 + a_1^b q_2}{a_1^b + a_2^b} \tag{3.140}$$

Then, it is obvious that the force $-Y\mathbf{j}$ applied at Q is exactly equivalent to a force $-Y_1\mathbf{j}$ applied at the front no-roll center Q_1 and a force $-Y_2\mathbf{j}$ applied at the other no-roll center Q_2. Indeed

$$Yq^b = Y_1 q_1 + Y_2 q_2 \tag{3.141}$$

with $Y_1 a_1^b = Y_2 a_2^b$, and hence

$$Yh = Y(h - q^b) + Y_1 q_1 + Y_2 q_2 \tag{3.142}$$

This way we have decomposed the lateral force Y into two forces Y_1 and Y_2 at the two no-roll centers, each one of the magnitude imposed by the equilibrium equations (3.131), plus a *horizontal roll moment* $L^b = Y(h - q^b)$.

It is important to note that it would be wrong to take the shortest distance from P to the roll axis to compute the roll moment L^b (cf. [21, p. 67]). It is precisely the *vertical* distance $(h - q^b)$ that has to be taken as the force moment arm, as shown in Figs. 3.35 and 3.40.

Summing up, a lateral force $-Y\mathbf{j}$ at P is totally equivalent to a lateral force $-Y_1\mathbf{j}$ at Q_1 and another lateral force $-Y_2\mathbf{j}$ at Q_2, plus the *roll moment* $L^b\mathbf{i} = Y(h - q^b)\mathbf{i}$ applied to the vehicle body (Figs. 3.35, 3.39 and 3.41).

Figure 3.39 shows how each force Y_i, applied at Q_i, is transferred to the ground by the suspension linkage, *without producing any suspension roll*. This is the key feature of the no-roll center Q_i.

Quite remarkably, this is true whichever the direction of the force there applied, and hence it is correct to speak of a no-roll center point (at first, Fig. 3.39 might suggest the idea of a roll center height q_i).

The moment $L^b = Y(h - q^b)$ is the sole responsible for *suspension roll*. More precisely

Fig. 3.41 Lateral load transfers and suspension internal force distribution (left turn, front view)

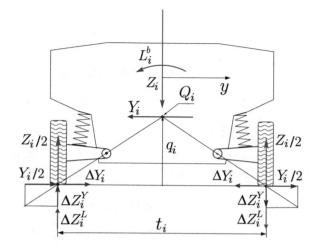

$$L^b = Y(h - q^b) = L_1^b + L_2^b = k_{\phi_1}^s \phi_1^s + k_{\phi_2}^s \phi_2^s = \Delta Z_1^L t_1 + \Delta Z_2^L t_2 \qquad (3.143)$$

exactly like in (3.123).

The total lateral load transfer ΔZ_i on each axle is therefore given by (Fig. 3.41)

$$\Delta Z_i t_i = (\Delta Z_i^Y + \Delta Z_i^L) t_i = Y_i q_i + k_{\phi_i}^s \phi_i^s = k_{\phi_i}^p \phi_i^p \qquad (3.144)$$

that is by the sum of the part due to the suspension linkage ΔZ_i^Y and the part due to the suspension springs ΔZ_i^L (Eqs. (3.123) and (3.136)). This is the last of the very many equations obtained for the first-order suspension analysis. They are solved in the next section.

Incidentally, we note that L^b defined in (3.139) is not equal to the moment L defined in (3.92). Similar symbols, but different meaning.

3.10.12 Roll Angles and Lateral Load Transfers

All relevant equations for the first-order suspension analysis have been obtained. Solving them provides the relationship between Y and the total roll angle ϕ and, more importantly, the relationship between the front and rear load transfers ΔZ_1 and ΔZ_2.

The main equations are gathered here to have them available at a glance:

$$Y = Y_1 + Y_2 \qquad (3.94')$$

$$Yh = \Delta Z_1 t_1 + \Delta Z_2 t_2 \qquad (3.132')$$

$$\phi = \phi_i^s + \phi_i^p \qquad (3.133')$$

$$\Delta Z_i t_i = k^P_{\phi_i} \phi^P_i \tag{3.134'}$$

$$Y h = k^P_{\phi_1} \phi^P_1 + k^P_{\phi_2} \phi^P_2 \tag{3.135'}$$

$$Y(h - q^b) = k^s_{\phi_1} \phi^s_1 + k^s_{\phi_2} \phi^s_2 \tag{3.143'}$$

$$\Delta Z_i t_i = (\Delta Z^Y_i + \Delta Z^L_i) t_i = Y_i q_i + k^s_{\phi_i} \phi^s_i \tag{3.144'}$$

These equations are really of great relevance in vehicle dynamics.

The front and rear roll angles due to the suspension and tire deflections can be obtained by solving the following system of equations

$$\begin{aligned} \phi^s_1 + \phi^P_1 &= \phi^s_2 + \phi^P_2 \\ Y(h - q^b) &= k^s_{\phi_1} \phi^s_1 + k^s_{\phi_2} \phi^s_2 \\ Y_1 q_1 + k^s_{\phi_1} \phi^s_1 &= k^P_{\phi_1} \phi^P_1 \\ Y_2 q_2 + k^s_{\phi_2} \phi^s_2 &= k^P_{\phi_2} \phi^P_2 \end{aligned} \tag{3.145}$$

The roll angles due to tire deflections are

$$\phi^P_1 = \frac{1}{k^P_{\phi_1}} \frac{k_{\phi_1} k_{\phi_2}}{k_\phi} \left[\frac{Y(h - q^b)}{k_{\phi_2}} + \frac{Y_1 q_1}{k^s_{\phi_1}} + \frac{Y_1 q_1}{k^s_{\phi_2}} + \frac{Y_1 q_1 + Y_2 q_2}{k^P_{\phi_2}} \right] = \phi^P_1(Y_1, Y_2)$$

$$\phi^P_2 = \frac{1}{k^P_{\phi_2}} \frac{k_{\phi_1} k_{\phi_2}}{k_\phi} \left[\frac{Y(h - q^b)}{k_{\phi_1}} + \frac{Y_2 q_2}{k^s_{\phi_1}} + \frac{Y_2 q_2}{k^s_{\phi_2}} + \frac{Y_1 q_1 + Y_2 q_2}{k^P_{\phi_1}} \right] = \phi^P_2(Y_1, Y_2) \tag{3.146}$$

and the roll angles due to suspension (spring) deflections are

$$\phi^s_1 = \frac{1}{k^s_{\phi_1}} \frac{k_{\phi_1} k_{\phi_2}}{k_\phi} \left[\frac{Y(h - q^b)}{k_{\phi_2}} - \frac{Y_1 q_1}{k^P_{\phi_1}} + \frac{Y_2 q_2}{k^P_{\phi_2}} \right] = \phi^s_1(Y_1, Y_2)$$

$$\phi^s_2 = \frac{1}{k^s_{\phi_2}} \frac{k_{\phi_1} k_{\phi_2}}{k_\phi} \left[\frac{Y(h - q^b)}{k_{\phi_1}} - \frac{Y_2 q_2}{k^P_{\phi_2}} + \frac{Y_1 q_1}{k^P_{\phi_1}} \right] = \phi^s_2(Y_1, Y_2) \tag{3.147}$$

where

$$k_\phi = k_{\phi_1} + k_{\phi_2} = \frac{k^s_{\phi_1} k^P_{\phi_1}}{k^s_{\phi_1} + k^P_{\phi_1}} + \frac{k^s_{\phi_2} k^P_{\phi_2}}{k^s_{\phi_2} + k^P_{\phi_2}} \tag{3.148}$$

is the total roll stiffness, like in (3.117).

Equations (3.146) and (3.147) show how the tire and suspension stiffnesses interact with each other and with the first-order suspension geometry (i.e., the no-roll axis position). According to them, the total roll angle ϕ produced by a lateral force $-Y\,\mathbf{j}$

applied at P (Fig. 3.35) is given by

$$k_\phi \phi = Y(h - q^b) + Y_1 q_1 \frac{k_{\phi_1}}{k_{\phi_1}^P} + Y_2 q_2 \frac{k_{\phi_2}}{k_{\phi_2}^P} \tag{3.149}$$

3.10.13 Explicit Expressions of the Lateral Load Transfers

Lateral load transfers ΔZ_i are among the most influential quantities in vehicle dynamics. They can be obtained, e.g., combining (3.134) and (3.146)

$$\Delta Z_1 t_1 = \frac{k_{\phi_1} k_{\phi_2}}{k_\phi} \left[\frac{Y(h - q^b)}{k_{\phi_2}} + \frac{Y_1 q_1}{k_{\phi_1}^s} + \frac{Y_1 q_1}{k_{\phi_2}^s} + \frac{Y_1 q_1 + Y_2 q_2}{k_{\phi_2}^P} \right]$$

$$\tag{3.150}$$

$$\Delta Z_2 t_2 = \frac{k_{\phi_1} k_{\phi_2}}{k_\phi} \left[\frac{Y(h - q^b)}{k_{\phi_1}} + \frac{Y_2 q_2}{k_{\phi_1}^s} + \frac{Y_2 q_2}{k_{\phi_2}^s} + \frac{Y_1 q_1 + Y_2 q_2}{k_{\phi_1}^P} \right]$$

which can also be recast as

$$\Delta Z_1 = \frac{1}{t_1} \left[\frac{k_{\phi_1}}{k_\phi} Y\left(h - q^b\right) + Y_1 q_1 + \frac{k_{\phi_1} k_{\phi_2}}{k_\phi} \left(\frac{Y_2 q_2}{k_{\phi_2}^P} - \frac{Y_1 q_1}{k_{\phi_1}^P} \right) \right] = \frac{k_{\phi_1}^P \phi_1^P}{t_1}$$

$$\Delta Z_2 = \frac{1}{t_2} \left[\frac{k_{\phi_2}}{k_\phi} Y\left(h - q^b\right) + Y_2 q_2 + \frac{k_{\phi_1} k_{\phi_2}}{k_\phi} \left(\frac{Y_1 q_1}{k_{\phi_1}^P} - \frac{Y_2 q_2}{k_{\phi_2}^P} \right) \right] = \frac{k_{\phi_2}^P \phi_2^P}{t_2}$$

$$\tag{3.151}$$

In a first-order analysis, the ratio $\Delta Z_1 / \Delta Z_2$ does not depend on the roll angle ϕ.

In (3.151) the interplay between stiffnesses and (first-order) suspension geometry looks quite tricky. First observe that

$$\Delta Z_1^Y = Y_1 q_1 / t_1$$
$$\Delta Z_2^Y = Y_2 q_2 / t_2 \tag{3.152}$$

are the load transfers through the suspension links. They are often called *kinematic load transfer components*. A key role is played by the no-roll center heights q_1 and q_2.

Moreover, there are the load transfers ΔZ_i^L due to the roll moment $Y(h - q^b)$

$$\Delta Z_1^L t_1 = \frac{k_{\phi_1}}{k_\phi} Y\left(h - q^b\right) + \frac{k_{\phi_1} k_{\phi_2}}{k_\phi} \left(\frac{Y_2 q_2}{k_{\phi_2}^P} - \frac{Y_1 q_1}{k_{\phi_1}^P} \right) = k_{\phi_1}^s \phi_1^s$$

$$\tag{3.153}$$

$$\Delta Z_2^L t_2 = \frac{k_{\phi_2}}{k_\phi} Y\left(h - q^b\right) + \frac{k_{\phi_1} k_{\phi_2}}{k_\phi} \left(\frac{Y_1 q_1}{k_{\phi_1}^P} - \frac{Y_2 q_2}{k_{\phi_2}^P} \right) = k_{\phi_2}^s \phi_2^s$$

They are often called *elastic load transfer components*. In these equations there is a first fairly intuitive term, followed by what can be called the "flexible tire correction". Indeed, the last terms would disappear when assuming tires to be rigid.

The ratio λ, often called *roll balance*

$$\lambda = \frac{\Delta Z_1}{\Delta Z_1 + \Delta Z_2} = \frac{\eta_1}{\eta_1 + \eta_2} \tag{3.154}$$

is of paramount importance since it strongly affects the handling behavior. This aspect will be thoroughly discussed in Sect. 7.5.3. Typical values of λ in a F1 car are in the range 0.54–0.62.

Summing up, in a first-order vehicle analysis the lateral load transfers are linear functions of Y and N, and hence of Y_1 and Y_2, that is

$$\Delta Z_1 = \xi_{11} Y_1 + \xi_{12} Y_2$$
$$\Delta Z_2 = \xi_{21} Y_1 + \xi_{22} Y_2 \tag{3.155}$$

Since we are *neglecting the inertial effects of roll motion*, the lateral forces are simply given by

$$Y = ma_y = Y_1 + Y_2$$
$$Y_1 = \frac{ma_y a_2}{l} + \frac{J_z \dot{r} - (\Delta X_1 t_1 + \Delta X_2 t_2)}{l} = \frac{ma_y a_2}{l} + \frac{N_Y}{l} \tag{3.156}$$
$$Y_2 = \frac{ma_y a_1}{l} - \frac{J_z \dot{r} - (\Delta X_1 t_1 + \Delta X_2 t_2)}{l} = \frac{ma_y a_1}{l} - \frac{N_Y}{l}$$

Therefore, ultimately, the lateral load transfers ΔZ_i are (linear) functions of the lateral acceleration a_y, and, just a little, of the angular acceleration $\dot{r}$. Moreover, in vehicle with limited-slip differential or ESP, the contribution due to ΔX_i can be rather relevant.

Incidentally, we observe that we have avoided to split the mass into front and rear masses [21, p. 133], or to introduce such concepts like the mass centroid axis [22, p. 29].

3.10.14 Lateral Load Transfers with Rigid Tires

If the tire vertical deflections are neglected (i.e., $k_{\phi_i}^p \to \infty$ and $k_{\phi_i}^s \to k_{\phi_i}$), all expressions simplify considerably. For instance, Eq. (3.149) becomes

$$k_\phi \phi = Y(h - q^b) \tag{3.157}$$

This is a most classical result, which deserves to be analyzed.

With rigid tires the roll angle $\phi = \phi_1^s = \phi_2^s$ is basically a function of $Y = Y_1 + Y_2$ only, whereas with flexible tires it depends on Y_1 and Y_2. Actually, q^b is still affected by Y_1/Y_2, but very little, since the no-roll axis is usually almost horizontal.

Also for the lateral load transfers, a much simpler expression than (3.144) or (3.150) is obtained

$$\Delta Z_i t_i = \frac{k_{\phi_i}}{k_\phi} Y(h - q^b) + Y_i q_i = k_{\phi_i}\phi + Y_i q_i \qquad (3.158)$$

However, particularly in Formula cars, it may be not so safe to assume the tires to be perfectly rigid in the vertical direction (they are not at all!). Inclusion of tire compliance should be done according to (3.150), not by simply softening the suspension stiffness. Indeed, loosely speaking, the tires counteract the rolling moment Yh, whereas the suspension springs have to deal with $Y(h - q^b)$. This point should not be overlooked.

3.11   Dependent Suspensions (Solid Axle)

In a dependent suspension the two wheels of the same axle are rigidly connected together. Nowadays very few cars are equipped with dependent suspensions. Nonetheless, it is still a type of suspension which is widely employed in commercial vehicles or the like, that is on vehicles that need to carry large loads compared to the vehicle weight.

Perhaps, the most classical lateral location linkage for dependent suspensions is the *Panhard rod* (also called Panhard bar or track bar), schematically shown in Fig. 3.42. A rendering of a complete dependent suspension with Panhard rod is shown in Fig. 3.43. The Panhard rod is a rigid bar running sideways in the same plane as the axle, connecting one end of the axle to the car body on the opposite side of the vehicle. The bar is attached at both ends with pivots that allow it to swivel upwards and downwards only, so that the axle can move in the vertical plane only. However, to effectively locate the axle longitudinally, it is usually used in conjunction with trailing arms. Obviously, the rigid axle has two degrees of freedom with respect to the vehicle body, exactly like two independent suspensions.

Fig. 3.42 Planar scheme of a dependent suspension with Panhard rod

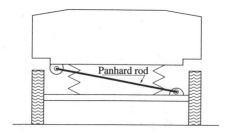

Fig. 3.43 Dependent
suspension with Panhard rod
[13]

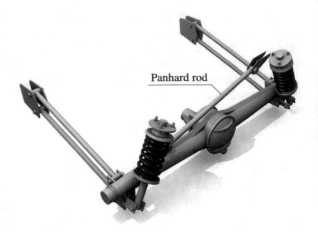

Most of the analysis developed for independent suspensions is applicable to dependent suspensions as well. For instance, the suspension internal coordinates listed on Sect. 3.10.4 are still meaningful (except track variation, which is obviously zero in the present case). Vertical stiffness and roll stiffness are also well defined. The only thing that needs to be addressed is the determination of the no-roll center Q_i.

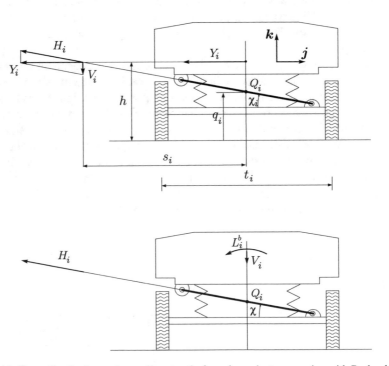

Fig. 3.44 Force distribution and no-roll center Q_i for a dependent suspension with Panhard rod

Following the method explained in Sect. 3.10.9, we apply a lateral force Y_i, like in Fig. 3.44 (top). This force can be decomposed into a force H_i, which is counteracted by the Panhard rod, and a vertical force V_i, which must be counteracted by the springs, and whose line of action is located at a distance s_i from the vehicle centerline (Fig. 3.44 (bottom)). It is easy to obtain

$$H_i = \frac{Y_i}{\cos \chi_i}$$

$$V_i = Y_i \tan \chi_i$$

$$s_i = \frac{h - q_i}{\tan \chi_i}$$

(3.159)

and hence $\quad L_i^b = V_i s_i = Y_i (h - q_i)$

where χ_i is the inclination of the Panhard rod. The lower χ_i, the better.

The moment $L_i^b = V_i s_i = Y_i (h - q_i)$ is the sole responsible of the vehicle body roll, as shown in Fig. 3.45 (top), and the force V_i is the only responsible for the body vertical displacement, as shown in Fig. 3.45 (bottom).

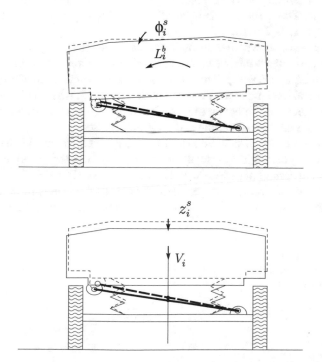

Fig. 3.45 Roll angle and vertical displacement of a dependent suspension with Panhard rod

To have zero suspension roll we need zero moment, and this is possible if and only if $h = q_i$. Therefore, the *no-roll center* is point Q_i in Fig. 3.44.

In any case, we have a small body vertical displacement, either upward or downward, depending on if we are turning left or right. The vertical displacement would be zero if and only if $\chi_i = 0$, which is clearly impossible in practice. Therefore, the Panhard rod is a simple linkage, with the disadvantage of a certain degree of asymmetry.

Of course, dependent suspensions do not exhibit suspension jacking, nor camber variations.

3.11.1 *Unsprung Masses and Lateral Load Transfers*

In a vehicle like an automobile or a motorcycle, particularly when studying ride motions (see Chap. 10), it is useful to distinguish between sprung mass and unsprung mass. The *sprung mass m_s* is the portion of the vehicle's total mass that is supported *above* the suspension, thus including the body, frame, internal components, passengers and cargo. On the other hand, wheels, wheel bearings, brake rotors, calipers belong to the *unsprung mass m_u*, since they are not above the suspension. Of course $m = m_s + m_u$. The sprung mass is usually much bigger than the unsprung mass. Typically, $m_s/m_u = 7 - 10$.

Here we investigate whether the notion of unsprung mass may be relevant also in handling, in particular when we want to evaluate lateral load transfers.

In vehicles equipped with independent suspensions we can immediately apply the results for lateral load transfers presented in Sect. 3.10.13, where m is the total mass of the vehicle and h is the height of its global center of mass G. In other words, we do not have to bother about sprung and unsprung masses.

In vehicles equipped with solid axles things are a bit different. We have a sprung mass m_s with G_s at height h_s (and distances a_{1s}, a_{2s} from the axles), and, for each solid axle, an unsprung mass m_{u_i}, with G_{u_i} at height h_{u_i} (Fig. 3.46).

The load distribution due to the inertial effects of the unsprung mass m_{u_i} is schematically shown in Fig. 3.46. Basically, a centrifugal force $Y_{u_i} = m_{u_i}a_y$ acts

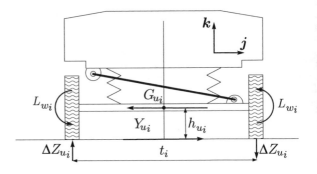

Fig. 3.46 Load distribution due to the inertial effects of a solid axle (front view, left turn)

on each solid axle. Moreover, there are two gyroscopic torques $L_{w_i} = (J_w/r_i)ur$, often not quite negligible in solid axles. The equilibrium of each suspension requires

$$0 = Y_{u_i} h_{u_i} - \Delta Z_{u_i} t_i + 2L_{w_i} \qquad (3.160)$$

which yields the load transfer ΔZ_{u_i}

$$\Delta Z_{u_i} = \frac{Y_{u_i} h_{u_i} + 2L_{w_i}}{t_i} \qquad (3.161)$$

Each solid axle contribute to the lateral load transfer independently of the other one.

In practical terms, in (3.151) it suffices to set $Y = m_s a_y$ at height h_s, to modify Y_1 and Y_2 accordingly, and to add on the r.h.s. the terms ΔZ_{u_1} and ΔZ_{u_2}, one for each equation.

How to modify (3.146) and (3.147), that is the equations for the roll angles, is left to the reader.

3.12 Linked Suspensions

So far we have considered independent suspensions and dependent suspensions. Basically, an independent suspension is a one degree of freedom mechanism. Therefore, the two wheels of an axle have two degrees of freedom. Similarly, a dependent suspension is a two-degree-of-freedom mechanism for a rigid axle (Fig. 3.42).

This path of reasoning leads to a more general class of suspensions: a two degree of freedom mechanism for the axle, without the requirement of rigid axle. An example of what can be called *linked suspension* is given in Fig. 3.47. Black hinges are connected to the vehicle body. We see there is a kinematic link between the left and right wheel. Therefore the two wheels are not independent, neither they are connected by a rigid beam.

Apparently, this kind of suspension has been employed only in some model racing cars. With suitable geometry, it can perform very well.

Fig. 3.47 Schematic of a linked suspension

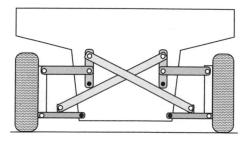

3.13 Differential Mechanisms

The two wheels of a driven axle must be linked together so that a *single* engine and transmission can turn *both* wheels. The mechanism that links the two driven wheels of the same axle is called the *differential* (Fig. 3.48). The distinguishing feature of all differential mechanisms is that they allow each driven wheel to rotate at *different angular speeds* (as the name of the mechanism implies).[13]

However, this is only half the story (or maybe, one third). As a matter of fact, a differential also applies *torques* to the driven wheels. The possible combinations of torque split depend on the type of limited-slip differential (LSD).

Ultimately, the actual combination of angular speeds and torques comes from the interaction between the differential, the tires and the vehicle motion.

In this section the *equations* governing (almost) any type of *passive* differential mechanism are discussed in detail (see also [8, Chap. 13] and [16, Chap. 20]). Qualitative descriptions are interesting, but only apparently simpler. Without equations we cannot achieve an in-depth understanding of what happens in a limited-slip differential.

3.13.1 Relative Angular Speeds

Regardless of the specific mechanical design, a differential is essentially a housing ("a" in Fig. 3.48) with two *aligned* shafts ("d" in Fig. 3.48). The alignment of the two shafts is necessary to have an *epicycloidable* mechanism, that is a mechanism with, possibly, a rotating housing. Needless to say, in a car the housing (also called cage) is connected to the gearbox and each shaft is connected to a wheel. For instance, in Fig. 3.48, the housing is connected to the gearbox by means of a hypoid gear set.

Fig. 3.48 Typical differential mechanism

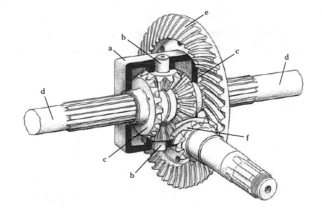

---

[13] Let us state from the beginning that it is not always true that, when rounding a corner, the inner wheel rotates at a slower angular speed than the outer wheel.

Fig. 3.49 Possible absolute
(top) and relative (bottom)
angular speeds in a
differential mechanism

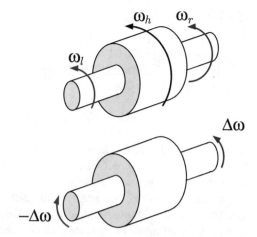

Let ω_l, ω_h and ω_r be the *absolute angular speeds* of the left shaft, of the housing and of the right shaft, respectively (Fig. 3.49-top). They are all positive.

All differentials must fulfill one very specific *kinematic* requirement: *the two shafts must have, with respect to the housing, opposite angular speeds* (Fig. 3.49-bottom). This is necessary to have a vehicle that behaves the same way in left turns and right turns. Therefore, employing the well known Willis formula, differential mechanisms must be such that

$$\frac{\omega_l - \omega_h}{\omega_r - \omega_h} = -1 \qquad \text{(Willis formula)} \qquad (3.162)$$

This is the fundamental *kinematic relationship* of differentials.

It can be rewritten as

$$\omega_l + \omega_r = 2\omega_h \qquad (3.163)$$

which confirms the already mentioned key *kinematic* feature of the differential: if one wheel rotates *faster* than the housing by $\Delta\omega$, the other wheel must rotate *slower* than the housing by the same $\Delta\omega$

$$\Delta\omega = |\omega_l - \omega_h| = |\omega_r - \omega_h| = \frac{|\omega_r - \omega_l|}{2} \qquad (3.164)$$

In other words, the two shafts have opposite angular speeds $\pm\Delta\omega$ with respect to the housing (Fig. 3.49-bottom). Since the differential has two degrees of freedom, Willis formula alone cannot say anything about the value of $\Delta\omega$ and whether $\omega_l \lessgtr \omega_r$.

Of course, in a locked differential we have, by definition, $\omega_l = \omega_r = \omega_h$, that is $\Delta\omega = 0$. A locked differential is not a differential.

3.13.2 Torque Balance

As shown in Fig. 3.50, let M_l, M_h and M_r be the moments (torques) applied to the left shaft, to the housing, and to the right shaft, respectively. Neglecting the (small) inertial effects, these three moments must be such that

$$M_l + M_h + M_r = 0 \quad \text{(torque balance)} \tag{3.165}$$

This is the fundamental *equilibrium relationship*: the three torques applied to the differential must sum to zero. Of course, this statement holds true for locked differentials as well, that is when $\omega_l = \omega_r = \omega_h$.

It should be noted that negative torques M_l and M_r for the differential mean positive (driving) longitudinal forces applied to the wheel from the road (car in power-on conditions). For instance, as discussed in Sect. 2.14.1, in a rear-driven car we have (Fig. 2.45)

$$X_l = F_{x_{21}} = -M_l/r_2 = \widetilde{M}_l/r_2 \quad \text{and} \quad X_r = F_{x_{22}} = -M_r/r_2 = \widetilde{M}_r/r_2 \tag{3.166}$$

where r_2 is (more or less) the rolling radius of the rear tires.

Equation (3.165) can be rewritten as

$$\widetilde{M}_l + \widetilde{M}_r = M_h \tag{3.167}$$

We often prefer to use $\widetilde{M}_l$ and $\widetilde{M}_r$, because they are the torques applied by the differential to the wheels.

The vehicle is under power-on conditions if $M_h > 0$ (accelerating), and under power-off conditions if $M_h < 0$

$$M_h = (F_{x_{21}} + F_{x_{22}})r_2 \tag{3.168}$$

According to (3.167), given M_h we know the sum $\widetilde{M}_l + \widetilde{M}_r$, but we cannot say anything about $\Delta\widetilde{M}$, that is the difference between the two torques $\widetilde{M}_r$ and $\widetilde{M}_l$

$$\Delta\widetilde{M} = \widetilde{M}_l - \widetilde{M}_r = (-M_l) - (-M_r) \tag{3.169}$$

Fig. 3.50 Possible moments acting on the differential during power-on conditions ($M_h > 0$, while $M_l < 0$ and $M_r < 0$)

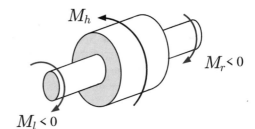

which is the key feature of limited-slip differentials (LSD). We need additional information (i.e., equations), as discussed in the next sections.

Always remember that the features of the differential strongly affect, among other things, the vehicle yawing moment ΔX_2, defined in (3.86) and (3.87).

3.13.3 Global Power Balance

Speed differentiation is activated, by definition, whenever $\omega_l \neq \omega_r$ (wheels rotating at different absolute angular speeds). The subsequent relative motions *inside* the housing may yield a small friction power loss $W_d \geq 0$. Therefore, the global power balance for the differential is (Fig. 3.51)

$$M_h \omega_h + M_l \omega_l + M_r \omega_r = W_d \quad \text{(global power balance)} \qquad (3.170)$$

This is the fundamental *global power balance relationship*. As we will see shortly, it is precisely the amount of power loss W_d, when $\omega_l \neq \omega_r$, that characterizes the behavior of limited-slip differentials in vehicle dynamics.

Fig. 3.51 Global power balance of the differential mechanism during power-on conditions

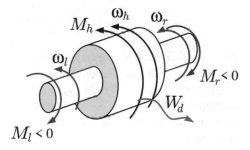

In an open differential, we have $W_d = 0$, even when $\omega_l \neq \omega_r$. Also in a locked differential $W_d = 0$, but because

$$(\omega_l = \omega_r) = \omega_h \qquad (3.171)$$

that is speed differentiation is not activated.

3.13.4 Internal Power Balance

Now we can combine the three fundamental relationships just obtained. More precisely, taking (3.162) and (3.167) into account, statement (3.170) can be given a more interesting, and also less obvious, form

$$M_l(\omega_l - \omega_h) + M_r(\omega_r - \omega_h) = W_d \tag{3.172}$$

that is

$$(\widetilde{M}_l - \widetilde{M}_r)\left(\frac{\omega_r - \omega_l}{2}\right) = W_d \tag{3.173}$$

that can be read as the *internal* power balance of the housing of the differential. More compactly, we can rewrite (3.173) as

$$\Delta\widetilde{M}\,\Delta\widetilde{\omega} = W_d \tag{3.174}$$

where

$$\Delta\widetilde{M} = \widetilde{M}_l - \widetilde{M}_r \quad \text{and} \quad \Delta\widetilde{\omega} = \frac{\omega_r - \omega_l}{2} \tag{3.175}$$

The physical meaning of (3.173) is clear: if the differential action is activated and there is energy dissipation inside the housing, then we have different torques applied to the wheels. Moreover, $\omega_r > \omega_l$ if and only if $\widetilde{M}_l > \widetilde{M}_r$, and viceversa

$$\begin{aligned}
\widetilde{M}_l &= \frac{M_h + \Delta\widetilde{M}}{2} = \frac{M_h}{2} + \frac{W_d}{\omega_r - \omega_l} \\
\widetilde{M}_r &= \frac{M_h - \Delta\widetilde{M}}{2} = \frac{M_h}{2} - \frac{W_d}{\omega_r - \omega_l}
\end{aligned} \tag{3.176}$$

These expressions look very nice, but to make them useful, we have to supply additional information either about W_d or about $\Delta\widetilde{M}$.

One observation is in order here. If there is only dry friction inside the housing, the torque difference $\Delta\widetilde{M}$ is a function of M_h and of the sign of the relative angular velocity $\Delta\widetilde{\omega}$. According to the Coulomb's Law of Friction, the amount of $\Delta\widetilde{\omega}$ does not matter.

3.13.5   Internal Efficiency

Whenever $\omega_r \neq \omega_l$, one shaft provides the *input power* $W_i > 0$ for the internal mechanism of the housing, while the other shaft receives the *output power* W_o, the difference being the power $W_d > 0$ dissipated inside the housing

$$W_i - W_o = W_d \tag{3.177}$$

Of course $W_i > W_o$. Typically, $W_o > 0$, but not necessarily.

We do not know a priori whether W_i comes from the left or right wheel. In any case, the ratio between the output internal power and the input internal power is, by definition, the *internal efficiency* η_h of the differential mechanism

$$\eta_h(M_h) = \frac{W_o}{W_i} = \frac{W_i - W_d}{W_i} \tag{3.178}$$

The internal efficiency is a fundamental concept in the theory and analysis of differential mechanisms. In general, it is a function of M_h.

As expected, $\eta_h \leq 1$. Less obvious is the fact that, in some cases, we may have $\eta_h < 0$ (be patient, it will be explained sooner or later).

In general, the function $\eta_h(M_h)$ is assumed to be known, in the sense that differentials are designed to have η_h varying within a well defined range (see Figs. 3.55, 3.59, 3.60, 3.62, 3.65).

It is worth noting that, even if the differential has a very low internal efficiency η_h, the power loss W_d is a very small fraction of the power $|M_h \omega_h|$ flowing through the drivetrain. This is clear from (3.173), since $|\omega_r - \omega_l| \ll \omega_h$, and $|M_r - M_l| \ll |M_h|$.

3.13.6 Slow Wheel and Fast Wheel

We have to find out which wheel provides, with respect to the housing, the input power W_i and which wheel gets the output power W_o in (3.177). Instead of left and right, now we look at the angular speed ω_f of the *faster*-rotating wheel and at the angular speed ω_s of the *slower*-rotating wheel. Of course, both are positive ($\omega_f > \omega_s > 0$).

We remark that fast and slow do not refer to the forward speeds of the wheels. In other words, in the same curve, the slower-rotating wheel can switch from the inner to the outer wheel, and viceversa.

Let M_f and M_s be the torques applied by the road to the faster-rotating wheel and to the slower-rotating wheel, respectively. It is also convenient to define $\tilde{M}_f = -M_f$ and $\tilde{M}_s = -M_s$. With this new notation, we can rewrite (3.173) as

$$(\tilde{M}_s - \tilde{M}_f)\left(\frac{\omega_f - \omega_s}{2}\right) = W_d \tag{3.179}$$

that is

$$\Delta M \, \Delta \omega = W_d \tag{3.180}$$

Since, by definition

$$\begin{aligned} \Delta \omega &= \omega_f - \omega_h \\ &= \omega_h - \omega_s \\ &= \frac{\omega_f - \omega_s}{2} > 0 \end{aligned} \tag{3.181}$$

we have that also the *differential torque*, ΔM must always be positive

$$\Delta M = M_f - M_s = \tilde{M}_s - \tilde{M}_f > 0 \tag{3.182}$$

In general, it is a function of M_h. It could not be otherwise after (3.169).

From (3.173) and (3.179), we obtain

$$\Delta \widetilde{M} = \Delta M \ \mathrm{sign}\,(\omega_r - \omega_l) \tag{3.183}$$

which goes straight to the behavior of all passive differentials. Indeed, the differential torque ΔM is, along with the internal efficiency η_h, another (equivalent) key concept to understand differentials. Clutch-pack differentials are better characterized using ΔM, while η_h is more suited for geared differentials.

Of course, there is direct proportionality between $\Delta \widetilde{M}$ and ΔX_2

$$\Delta X_2 = \frac{\Delta \widetilde{M}}{2r_2} \tag{3.184}$$

where r_2 is the rear wheel rolling radius.

3.13.7 Torque Split Relationship

As already stated, there are two possible working conditions for a limited-slip differential:

1. *power-on*: positive (driving) torque M_h from the engine. This means $M_s < 0$ for the differential (Fig. 3.52). Therefore, to have $W_i > 0$ and $W_i > W_o$ in (3.177), taking (3.179) into account, it has to be

$$W_i = -M_s \, \Delta\omega \quad \text{and} \quad W_o = -M_f \, \Delta\omega \tag{3.185}$$

   In some cases, we can have $M_f > 0$ and hence $W_o < 0$ (Fig. 3.63).
2. *power-off*: negative (braking) torque M_h from the engine. This means that $0 < M_s < M_f$ for the differential (Fig. 3.53). In this case to have $W_i > 0$ and $W_i > W_o$ in (3.177), it has to be

$$W_i = M_f \, \Delta\omega \quad \text{and} \quad W_o = M_s \, \Delta\omega \tag{3.186}$$

Fig. 3.52 Internal power balance: torques and relative angular speeds under *power-on* working conditions ($M_h > 0$, to be balanced typically by $M_s < M_f < 0$, although in some cases it can be $0 < M_f$)

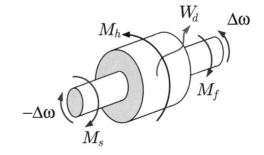

Fig. 3.53 Internal power
balance: torques and relative
angular speeds under
power-off working
conditions ($M_h < 0$, to be
balanced by $0 < M_s < M_f$)

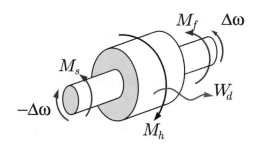

Now we can obtain the value of the ratio M_s/M_f by taking into account the
internal efficiency η_h of the differential mechanism, defined in (3.178). Combining
(3.178) with (3.185) for power-on, and with (3.186) for power-off, we obtain

$$\text{power-on:} \quad \frac{-M_f \, \Delta\omega}{-M_s \, \Delta\omega} = \frac{M_f}{M_s} = \eta_h(M_h)$$

$$\text{power-off:} \quad \frac{M_s \, \Delta\omega}{M_f \, \Delta\omega} = \frac{M_s}{M_f} = \eta_h(M_h) \tag{3.187}$$

These are the *torque split relationships*. Perhaps, a look at Figs. 3.56 and 3.57 may
be useful to better understand the physical phenomenon.

Also useful is this expression of the internal efficiency

$$\eta_h = \left(\frac{M_h - \Delta M}{M_h + \Delta M} \right)^s \quad \text{where} \quad s = \text{sign}(M_h) \tag{3.188}$$

which is another, more compact, way to write (3.187). We see that the internal
efficiency dictates the ratio between the torques applied to the wheels.

The differential torque $\Delta M \geq 0$ was defined in (3.182). It can be obtained from
(3.188)

$$\Delta M = \frac{1 - (\eta_h)^s}{1 + (\eta_h)^s} \, M_h = \frac{1 - \eta_h}{1 + \eta_h} |M_h| \tag{3.189}$$

where $\eta_h = \eta_h(M_h)$ (beware, not $\eta_h(|M_h|)$).

These are results of paramount importance. However, we remark once again that
(3.187) and (3.189) hold true if and only if $\Delta\omega > 0$, that is only if the *differential
action is active*. Therefore, these results cannot be applied to a locked differential.

Usually, M_s and M_f have the *same sign*. During power-on (i.e., $M_h > 0$), they
are both negative, as in Fig. 3.52, while during power-off (i.e., $M_h < 0$) they are
both positive, as in Fig. 3.53. However, power-on conditions like in Fig. 3.63 (i.e.,
$\eta_h < 0$) are possible with a preloaded clutch-pack differential.

It is often erroneously stated that geared differentials have $\eta_h < 0$.[14] Actually, for a geared differential to behave like a differential, it must be $\eta_h > 0$.

3.13.7.1 Torque Bias Ratio (TBR)

Instead of the internal efficiency η_h, it is common practice to use the Torque Bias Ratio (TBR),[15] which is exactly equal to $1/\eta_h$

$$\text{TBR} = \frac{1}{\eta_h} \tag{3.190}$$

Of course, TBR ≥ 1. It is a simple exercise to rewrite (3.187) using the TBR

$$\text{power-on:} \quad \frac{M_s}{M_f} = \text{TBR}$$

$$\text{power-off:} \quad \frac{M_f}{M_s} = \text{TBR} \tag{3.191}$$

3.13.8 Locking Coefficient

Another common and useful way to convey the same information is by means of the *locking coefficient* ε_h

$$\varepsilon_h = \frac{1 - \eta_h}{1 + \eta_h} = \frac{\text{TBR} - 1}{\text{TBR} + 1} \tag{3.192}$$

that is

$$\eta_h = \frac{1 - \varepsilon_h}{1 + \varepsilon_h} \tag{3.193}$$

The locking coefficient ε_h arises naturally from (3.189)

$$\Delta M = \frac{1 - \eta_h}{1 + \eta_h} |M_h| = \varepsilon_h |M_h| \tag{3.194}$$

where, in general, $\varepsilon_h = \varepsilon_h(M_h)$ is a function of M_h (not of $|M_h|$). An open differential has $\varepsilon_h = 0$.

---

[14] Here is an example of such wrong sentences: "A Torsen works on the principle that a spinning worm gear can rotate the wheel, but the rotating wheel cannot spin the worm gear".

[15] Incidentally [10], we mention that a symbol like TBR could be interpreted as the product of three quantities if written using a mathematical font like in TBR.

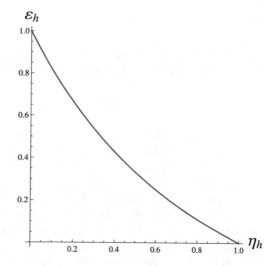

Fig. 3.54 Relationship between internal efficiency η_h and locking coefficient ε_h

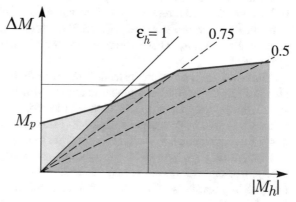

Fig. 3.55 Generic example of all possible working conditions for a differential

Equation (3.194) shows why the locking coefficient ε_h is called locking coefficient, indeed. For instance, if $\eta_h = 1/2$, we obtain $\varepsilon_h = 1/3$ (Fig. 3.54). Therefore, we know that, under power-on conditions, $M_f = M_s/2$ and $\Delta M = |M_h|/3$.

A compact way to characterize the behavior of a differential is by means of a plot like in Fig. 3.55. The border (red) line is the plot of ΔM in (3.194), that is of all working conditions with $\Delta \omega \neq 0$. Whenever ΔM along the red line is a function of $|M_h|$ we say that the differential is *torque sensitive*.

Points in the shaded area below the red line correspond to working conditions with $\Delta \omega = 0$ (locked differential). That is, the value of ΔM is not sufficient to overcome the internal friction. Points above the red line cannot be reached by that differential.

3.13.9   Rule of Thumb

From (3.179), we can extract a simple rule of thumb: *it is always the slower wheel
that receives from the road the higher longitudinal force*

$$(\tilde{M}_s - \tilde{M}_f)(\omega_f - \omega_s) > 0 \quad \text{that is} \quad \tilde{M}_s > \tilde{M}_f \tag{3.195}$$

where "slower" means "with lower absolute angular speed". The relationship
between torque and longitudinal force is discussed in Sect. 2.14.

As shown in Fig. 3.56 for, e.g., a left turn, the outcome of a *power-on* condition
strongly depends on the vertical loads acting on the two wheels. At low lateral
acceleration, that is at low lateral load transfers, the inner (left) wheel is slower and
receives from the road the higher longitudinal force, that is $X_l = X_r/\eta_h$ (Fig. 3.56a).
On the contrary, at high lateral acceleration the inner wheel is barely touching the
ground because of the high lateral load transfer, and it is the outer wheel which
is slower and gets the higher longitudinal force, that is $X_l = X_r\eta_h$ (Fig. 3.56c).
In between there is a range of medium values of lateral acceleration in which the
two wheels are *locked together*, (Fig. 3.56b). More precisely, we have $\Delta\omega = 0$ and
$X_r\eta_h < X_l < X_r/\eta_h$. That is, there is ΔM but it is not strong enough to unlock the
differential.

The *power-off* condition, instead, is more predictable, as shown in Fig. 3.57 (and
also in Fig. 3.74). The inner wheel is always slower than the outer wheel. At first
sight it may look that Fig. 3.57 violates the aforementioned rule of thumb, but it
is not so. Both wheels receive a negative (braking) force such that $X_l = \eta_h X_r$, and
hence it is correct that $X_l > X_r$. Never forget to take negative signs into account.

Comparing Figs. 3.56 and 3.57, you may ask why power-off is more predictable
than power-on. The reason is that the wheels have always positive absolute angular speeds, regardless of the sign of M_h. Think about it, maybe with the aid of
Sect. 3.13.17.

3.13.10   A Simple Mathematical Model

We consider here only limited-slip differentials with *constant* internal efficiency
$\eta_h < 1$. In this case it is convenient to use a mathematical trick to model (3.187)

$$\eta_h^\zeta X_l = X_r \tag{3.196}$$

where

$$\zeta = \frac{\arctan(\chi \, \Delta\tilde{\omega} \, \text{sign}(M_h))}{\pi/2} \tag{3.197}$$

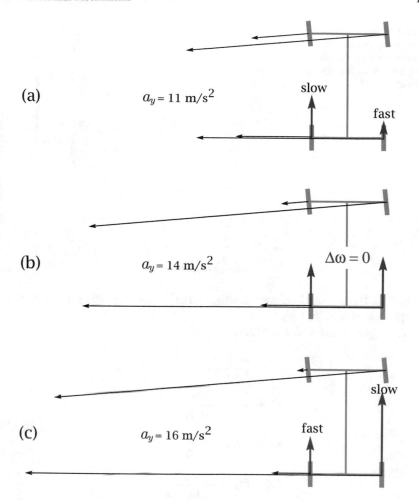

(a) $a_y = 11 \text{ m/s}^2$

slow

fast

(b) $a_y = 14 \text{ m/s}^2$

$\Delta\omega = 0$

(c) $a_y = 16 \text{ m/s}^2$

slow

fast

Fig. 3.56 Longitudinal (and lateral) forces during *power-on* in a vehicle equipped with a limited-slip differential with constant $\eta_h = 0.5$: **a** low lateral acceleration, **b** medium lateral acceleration, **c** high lateral acceleration

with χ a positive *big number*, something around 1000 s/rad. This way, the limited-slip differential action is activated whenever $\Delta\tilde{\omega}$ has significant values, with a smooth transition through the locked state of the differential ($\Delta\tilde{\omega} \simeq 0$).

By setting $\eta_h = 1$ in (3.196) we obtain the behavior of the open differential.

3.13.11 Alternative Governing Equations

Equation (3.178) that defines the internal efficiency η_h can be safely replaced by

Fig. 3.57 Longitudinal forces during *power-off* (coasting mode) in a vehicle equipped with a limited-slip differential with $\eta_h = 0.5$

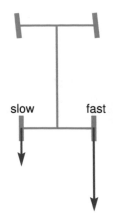

slow   fast

$$M_l \varepsilon_l = M_r \varepsilon_r \qquad (3.198)$$

where $\varepsilon_l > 0$ and $\varepsilon_r > 0$. Of course, what matters is only the ratio $\varepsilon_l/\varepsilon_r$, which must be equal either to η_h or $1/\eta_h$.

We suggest to the reader to obtain

$$\omega_h = \frac{\omega_l + \omega_r}{2}$$

$$M_l = -M_h \frac{\varepsilon_r}{\varepsilon_l + \varepsilon_r}$$

$$M_r = -M_h \frac{\varepsilon_l}{\varepsilon_l + \varepsilon_r} \qquad (3.199)$$

$$W_d = M_h \frac{(\varepsilon_r - \varepsilon_l)(\omega_r - \omega_l)}{2(\varepsilon_l + \varepsilon_r)}$$

This is an alternative, perhaps clearer, way to describe the behavior of the differential. We invite the reader to investigate this aspect.

3.13.12 Open Differential

Most road cars are equipped with an *open differential* (Fig. 3.48), which is characterized by having $W_d \simeq 0$, that is $\eta_h \simeq 1$, and hence $\varepsilon_h = 0$. Indeed a better name would be friction-free differential. From (3.187) it immediately follows that, in a vehicle with an open differential, the two driven wheels always receive the same torque, regardless of their angular speeds

$$(M_l = M_r) = -M_h/2 \quad \text{that is} \quad X_l = X_r \qquad (3.200)$$

Actually, the phenomenon is the other way around. To fulfill (3.200), the tires of the inner and of the outer wheels need specific longitudinal slips. The angular speed of each wheel is then automatically selected according to (3.205). An example will be shown in Fig. 3.68.

The open differential works very well unless one wheel can exchange with the road a very small longitudinal force, because, e.g., it is on a slippery surface (off-road vehicles, Fig. 3.68), or it is barely touching the ground (inner wheel in race cars). This is the main motivation for using limited-slip differentials.

3.13.13 Limited-Slip Differentials (LSD)

Differentials that do not obey (3.200) are called *limited-slip differentials*, or self-locking differentials. They come in many different types, but most of them rely on significant mechanical dry *friction* inside the housing, which means $W_d > 0$ in (3.180), provided $\Delta\omega \neq 0$ and $\Delta M \neq 0$.

More precisely, they have low *internal efficiency* η_h, that is

$$\eta_h \ll 1 \tag{3.201}$$

and hence high TBR. In general, η_h may be a function of M_h.

Limited-slip differentials are employed very often in race cars. It is customary to define two categories of limited-slip differentials, *geared* and *clutch-pack*, depending on the way friction is generated inside the housing.

3.13.14 Geared Differentials

The distinguishing feature of *geared differentials*, such as the Torsen differential (Fig. 3.58) and the Quaife differential, is to employ low-efficiency gear trains. The result is an *almost constant* internal efficiency η_h, and hence an almost constant locking coefficient ε_h. Therefore, according to (3.194), the difference ΔM between the left and right driving torques is a *linear* function of the applied driving torque $|M_h| = |M_s + M_f|$

$$\Delta M = \varepsilon_h |M_h| \tag{3.202}$$

as shown in Fig. 3.59. In other words, the differential is *linearly torque sensitive*. Of course, this is equivalent to a *constant ratio* between the torques applied to the wheels, as stated in (3.187).

Figure 3.59 shows the linear relationship (red line) between the total torque $|M_h|$ and the differential torque ΔM to activate the differential action, that is to have $\Delta\omega \neq 0$. The locking coefficient ε_h depends on the design of the gear set, but it is

Fig. 3.58 Cutaway view of
the Type-A Torsen
differential (courtesy of
JTEKT Torsen)

Fig. 3.59 Constant locking
coefficient $\varepsilon_h = 0.5$, and
hence linear torque
sensitivity

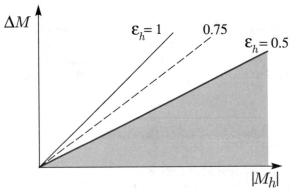

basically constant for a given differential. Points inside the shaded area correspond to
locked conditions, that is $\Delta\omega = 0$. Points outside the shaded area cannot be reached.

3.13.15 Clutch-Pack Differentials

All *clutch-pack differentials* exhibit a clutch torque preload M_p, also called break-
away torque. That is, the clutches have some spring preload. Therefore, to activate
the differential action, even when no torque is applied to the differential housing

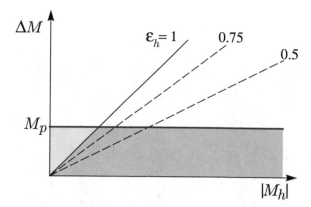

Fig. 3.60 Constant clutch torque preload M_p (differential with no wedging action)

($M_h = 0$), there has to be a difference $\Delta M = M_p$ between the left and right driving torques, as shown by the red line in Fig. 3.60.

However, clutch-pack differentials do not behave all the same, the main difference being on how the clutches are further loaded by the wedging action of the input torque M_h.

One type, often referred to as the Salisbury differential, is shown in Fig. 3.61. It behaves as shown in Fig. 3.62. The differential action is activated along the red line.

In the Salisbury differential the torque sensitivity is linear only for fairly high values of $|M_h|$. In this linear part, the locking coefficient ε_h depends on the shape of the so-called ramps (numbered 3 in Fig. 3.61). Usually, the constant value of ε_h during power-on is typically different from the constant value of ε_h during power-off. This is achieved using different ramp angles on driving and braking sides. Mathematically, it means that we have different pictures like Fig. 3.62 for $M_h > 0$ and $M_h < 0$.

For small values of $|M_h|$, we have a constant difference $\Delta M = M_p$ between the left and right driving torques, and hence a varying ε_h. Points inside the shaded area correspond to locked conditions, that is $\Delta \omega = 0$. Points outside the shaded area cannot be reached.

Points in the gray region have $\varepsilon_h > 1$, which means longitudinal forces like in Fig. 3.63. The case shown in Fig. 3.63 is possible only in a spool axle or in a clutch-pack differential with clutch preload. In a geared differential, since $M_p = 0$, it is not possible to have the two longitudinal forces pointing in opposite directions, even when it is locked. Indeed, in Fig. 3.59 there is no gray region.

The comparison of Fig. 3.59 with Fig. 3.62 clearly shows that the Torsen differential operates in a manner very similar to a Salisbury differential, but with zero preload, that is $M_p = 0$.

There is another type of clutch-pack differentials, as the one shown in Fig. 3.64, that behaves like in Fig. 3.65, that is according to the following relationship, with constant k_h

$$\Delta M = M_p + k_h |M_h| \tag{3.203}$$

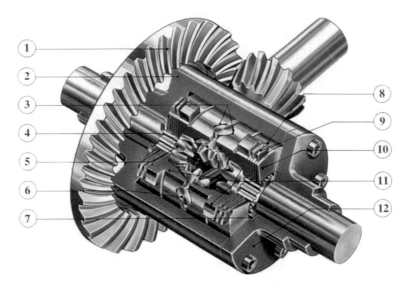

Fig. 3.61 Cutaway view of a Salisbury *clutch-pack* differential: 1—crown gear, 2—differential housing, 3—ramps, 4—spider gear, 5—side gear, 6—cross shaft, 7—lugs, 8—inner clutch disc, 9—outer clutch disc, 10—preload spring

Fig. 3.62 Torque sensitivity in a Salisbury differential

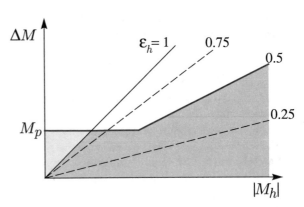

In this case, ε_h and η_h are never constant along the red line, unless $M_p = 0$. Therefore, also these differentials are torque sensitive, but not linearly.

The meaning of lines and areas in Fig. 3.65 are like in Fig. 3.62. The different behavior depends on the different location of the preload springs. Think about it and try to figure out why.

Figures 3.59, 3.62 and 3.65 characterize these limited-slip differentials. However, the real behavior of any kind of differential depends on the interaction with tire mechanics, as will be discussed shortly.

Fig. 3.63 Possible longitudinal forces in a vehicle equipped with a preloaded clutch-pack differential

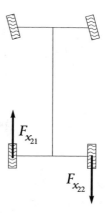

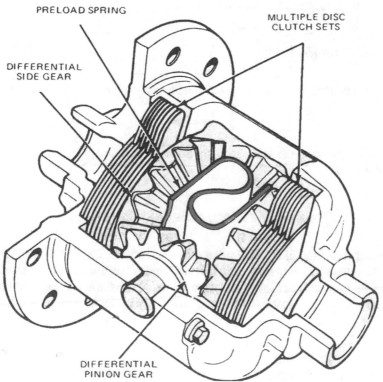

Fig. 3.64 Cutaway view of a Trac-loc *clutch-pack* differential

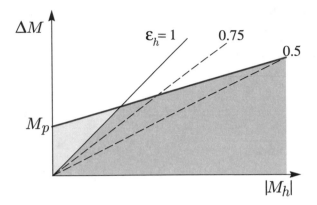

Fig. 3.65 Torque sensitivity in a Trac-loc differential

3.13.16 Spindle Axle

In this case there is no differential at all, and hence we have

$$\omega_l = \omega_r \tag{3.204}$$

The two wheels have the same angular velocity. Of course, we do not have direct control on ΔM.

3.13.17 Differential-Tire Interaction

Something very important is still missing in our analysis of the differential mechanism: the interaction between the differential mechanism and the tire mechanics.

Let us assume that we control the torque M_h applied to the housing and the housing angular speed ω_h, and that our differential has the behavior as, e.g., in Fig. 3.55. We only know:

- $\omega_l + \omega_r = 2\omega_h$ (Eq. (3.163));
- $\widetilde{M}_l + \widetilde{M}_r = M_h$ (Eq. (3.167));
- the maximum achievable value of $\Delta M(M_h)$ (Fig. 3.55).

Therefore, at the moment, we cannot obtain the values of $\widetilde{M}_l$, $\widetilde{M}_r$, ω_l, and ω_r. Nor we know whether the differential is locked or not.

To truly understand how a differential works it is necessary to investigate how it interacts with the mechanics of the wheel with tire. It is the *interaction with tires* that provides the missing additional information $\widetilde{M}_l$, $\widetilde{M}_r$, ω_l, and ω_r. This interaction often leads to not-so-obvious results.

For instance, it is often erroneously stated that in a curve the inner wheel has necessarily a lower angular velocity than the outer wheel. In a curve, the inner wheel has necessarily a *lower forward velocity* than the outer wheel, but to obtain the angular velocities we must also take into account the *longitudinal slips* $\sigma_{x_{2j}}$ of both tires.

More precisely, owing to (2.63), we have the following kinematic relationships for a rear-driven axle

$$\omega_l = \omega_{21} = \frac{V_{x_{21}}}{(1 + \sigma_{x_{21}})r_2} \quad \text{and} \quad \omega_r = \omega_{22} = \frac{V_{x_{22}}}{(1 + \sigma_{x_{22}})r_2} \qquad (3.205)$$

where r_2 is the rolling radius and $V_{x_{2j}}$ are the forward velocities of the wheels. These equations clearly show that $V_{x_{21}} < V_{x_{22}}$ does *not* necessarily imply $\omega_{21} < \omega_{22}$. In fact, one of the main motivations for using a limited-slip differential, particularly in race cars, is to mitigate the effects of the inner wheel spinning faster than the outer during power-on. We recall that in power-on conditions $\sigma_{x_{2j}} < 0$.

3.13.17.1 Virtual Test Rig

To investigate the differential-tire interaction, it is convenient to use the (virtual) test rig shown in Fig. 3.66. It is a driven axle with limited-slip differential.

We assume that we can set directly and independently the values of the following quantities:

1. the vertical loads Z_l and Z_r acting on each wheel;
2. the grip coefficients μ_l and μ_r of each wheel with the road;
3. the forward speeds V_l and V_r of each wheel;
4. the theoretical lateral slip σ_y of both wheels;
5. the theoretical longitudinal slip σ_x of, say, the left wheel. This is a way to control the torque M_h and the angular speed ω_h;
6. the mechanical efficiency η_h of the differential (not necessarily constant).

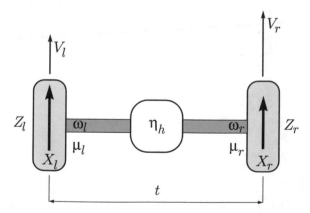

Fig. 3.66 Virtual rig for testing a driven axle with limited-slip differential

Of course we also know the value of the tire rolling radius r_r.

With this virtual test rig we can simulate and analyze some working conditions for the differential-tire system, without having to perform full vehicle dynamics computations. The goal is to monitor $X_l = \widetilde{M}_l/r_r$, $X_r = \widetilde{M}_r/r_r$, ω_l, and ω_r.

For the tire behavior we refer to Chap. 2. Therefore, we assume that each longitudinal force X_l and X_r, for fixed and given vertical loads and grip coefficients, is a known function (Magic Formula) of its theoretical slip σ. More precisely, see (at least) Sect. 2.11 for the Magic Formula, and Sect. 2.12 for a short description of the tire behavior.

3.13.17.2 Same Speeds and Loads, but Different Grip (μ-Split)

We start with a very typical case (Fig. 3.67). The car goes straight ($V_l = V_r$), in power-on, with equal vertical loads ($Z_l = Z_r$), but the left wheel has a lower grip coefficient than the right wheel ($(\mu_l = 0.4) < (\mu_r = 1)$). Both lateral slips are equal to zero.

The two solid curves in Figs. 3.67 and 3.68 are examples of the tire behavior, already described in Fig. 2.39, for tires with the same vertical load and different grip. Since $V_l = V_r$ we can use $-\sigma_x$ to monitor the wheel kinematics: the higher $-\sigma_x$, the higher the angular velocity of the wheel.

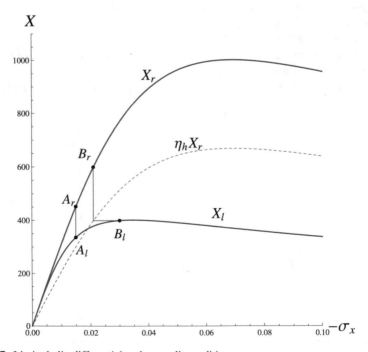

Fig. 3.67 Limited-slip differential under μ-split conditions

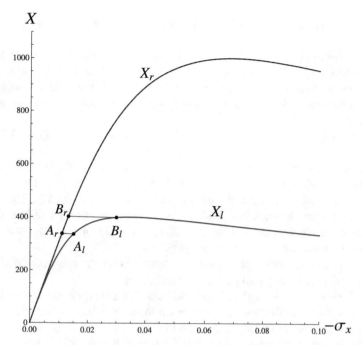

Fig. 3.68 Open differential under μ-split conditions

First, consider a limited-slip differential with constant $\eta_h = 0.67$, that is that behaves like in Fig. 3.59 with locking coefficient $\varepsilon_h = 0.2$.

For small values of $-\sigma_x$, that is for small longitudinal forces, the differential is locked: the two wheels have the same longitudinal slip and hence the same angular velocity (points A_r and A_l in Fig. 3.67). The differential action is inhibited because none of the wheels can provide a longitudinal force sufficiently high to overcome the other force plus the internal friction, while having a lower $-\sigma_x$.

After the intersection between the curves X_l and $\eta_h X_r$, the differential action starts: the left wheel, that is the wheel with less grip, rotates faster than the right wheel and the longitudinal forces are such that $X_l = \eta_h X_r$ (points B_l and B_r in Fig. 3.67), as required by the first equation in (3.187), that is by the friction inside the housing.

On the contrary, as shown in Fig. 3.68, with an open differential ($\eta_h = 1$), we always have $X_l = X_r$. Therefore, points A_r and A_l (which now are like B_r and B_l) must be at the same force level. Of course, because of the different grip, the two wheels must always operate with different longitudinal slips, and hence different angular velocities, to provide equal longitudinal forces.

Comparing Figs. 3.67 and 3.68 we clearly see that, in general, a limited-slip differential provides a higher maximum global longitudinal force than an open differential. In the present case, 1000 N instead of 800 N.

3.13.17.3   Same Grip, but Different Forward Speeds and Vertical Loads

This second example is also paradigmatic. In the test rig of Fig. 3.66, we set different forward speeds $V_l < V_r$ of the wheels. For instance, $V_r = 1.025 V_l$, with the left (inner) forward speed $V_l = 20$ m/s. Therefore, $V_r = 20.5$ m/s. If we supplement these data with the vehicle track $t = 1.5$ m, we have the inner wheel running on a circle of radius 60 m.

The (geared) differential is assumed to have internal efficiency $\eta_h = 0.5$, that is TBR $= 2$.

Since we are simulating a left turn, we must have $Z_l < Z_r$. We set the vertical loads to be $Z_l = 1500$ N and $Z_r = 4500$ N. Both wheels have the same grip coefficient $\mu_l = \mu_r = \mu = 1$ and the same rolling radius $r_r = 0.25$ m. The tire longitudinal forces are shown in Fig. 3.69. For simplicity, the influence of lateral forces is neglected. Here we are investigating the differential-tire interaction, not doing full vehicle dynamics simulations (well, not yet).

All these quantities are kept *constant* while we change the longitudinal slip σ_x of the left wheel to have different power-on conditions.

Since here we have to deal with two wheels that have different forward speeds ($V_l \neq V_r$), it turns out that a clearer approach to monitor their kinematics is to use the angular velocities ω_l and ω_r, according to (3.205), instead of the tire longitudinal slips. This makes it much easier to apply the rule of thumb (3.195) that the slower wheel has always the higher longitudinal force.

Four possible power-on working conditions, for increasing values of $-\sigma_x$ (and hence of M_h and ω_h), are shown in Figs. 3.70, 3.71, 3.72 and 3.73. In all these figures there are the same tire plots $X_l(\omega_l, \mu, Z_l)$ and $X_r(\omega_r, \mu, Z_r)$ (solid curves), along with the plots $\eta_h X_l$ and $\eta_h X_r$ (dashed curves). Under pure rolling conditions, the inner wheel has an angular velocity of 80 rad/s, while the outer wheel has an angular velocity of 82 rad/s. Honestly, for negative longitudinal forces, the dashed curves are the plots of X_l/η_h and X_r/η_h. Think about it.

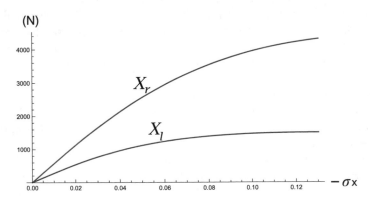

Fig. 3.69 Tire longitudinal forces X_l and X_r

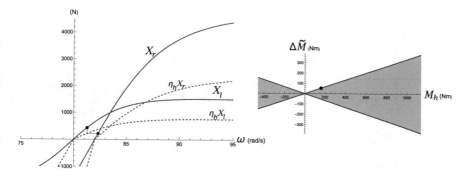

Fig. 3.70 Differential action activated under power-on, with slower left wheel. Therefore, $X_r = \eta_h X_l$

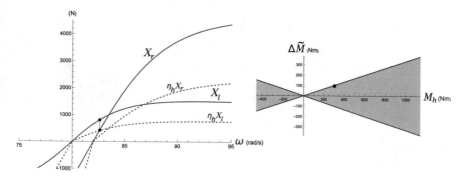

Fig. 3.71 Locked limited-slip differential, with force X_l higher than X_r

In Fig. 3.70 the differential action is activated, with the left wheel being slower ($\omega_l < \omega_r$) and hence providing the higher force according to the rule $\eta_h X_l = X_r$. On the right of Fig. 3.70 we see that the black point lies indeed on the border of the shaded region.

In Fig. 3.71, the differential action has just stopped, that is the differential has locked and hence $\omega_l = \omega_r$. However, the left wheel is still providing the higher force, i.e. $X_l > X_r$, albeit being subject to a lower vertical load.

In general, we have a locked differential, that is $\omega_l = \omega_r = \omega$, in the range of angular speeds whose boundaries fulfill the following equations

$$\eta_h X_l(\omega, \mu, Z_l) = X_r(\omega, \mu, Z_r) \quad \text{(lower bound for } \omega\text{)}$$
$$X_l(\omega, \mu, Z_l) = \eta_h X_r(\omega, \mu, Z_r) \quad \text{(upper bound for } \omega\text{)}$$

(3.206)

In practice, just look at the intersections between solid and dashed curves.

In Fig. 3.72 we still have a locked differential ($\omega_l = \omega_r$), but now with $X_l < X_r$. This apparently paradoxical situation is due to the difference between the vertical loads ($Z_l \ll Z_r$), along with the left tire approaching its maximum longitudinal force.

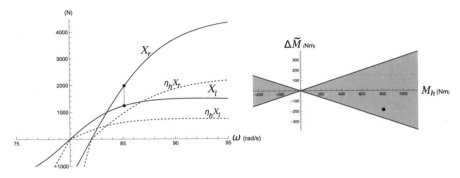

Fig. 3.72 Locked limited-slip differential, with force X_r higher than X_l. The black point lies inside the shaded region

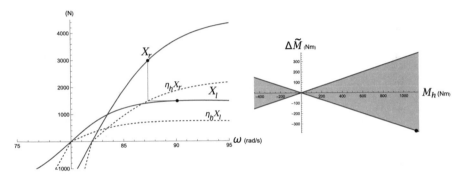

Fig. 3.73 Differential action activated, with slower right wheel. Therefore, $X_l = \eta_h X_r$

In Fig. 3.73, we see that the differential action has activated again, now with the right (outer) wheel being slower ($\omega_l > \omega_r$), and hence providing the higher longitudinal force $X_r = X_l/\eta_h$. This last case is typical of race cars exiting a curve.

In Fig. 3.74 a possible power-off condition (i.e., $M_h < 0$) is shown. Both longitudinal forces are negative, like in Fig. 3.57.

In all cases, it is the *interaction between the limited-slip differential and the tire mechanics to rule the outcome*. More precisely, in Figs. 3.70, 3.71, 3.72, 3.73 and 3.74 the type of differential sets η_h, the tire behavior sets the solid curves, and together they provide the dashed curves. Then it is up to the driver to set the total longitudinal force $X_l + X_r$.

3.13.17.4 Power-Off with a Preloaded Clutch-Pack LSD

Working conditions like in Fig. 3.63 are possible in a preloaded clutch-pack differential, as shown in Fig. 3.75. If this happens, we have $\eta_h < 0$.

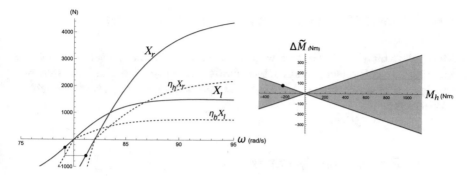

Fig. 3.74 Differential action activated under power-off, with slower left wheel. Therefore, $X_r = X_l/\eta_h$

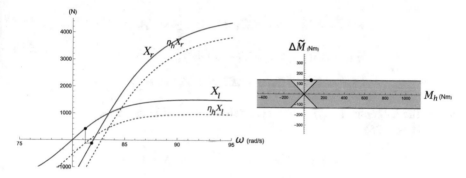

Fig. 3.75 Possible working condition with a simply preloaded clutch-pack differential

3.13.18 Informal Summary About the Differential Behavior

The behavior of a limited-slip differential looks at first far from obvious. And strong is the temptation to rush to apparently reasonable explanations.

To get a sound knowledge we need to write down all the relevant mathematical equations and instruct our intuition to read the physical phenomena behind them. It is hard, but feasible. The main point is to realize that we cannot ultimately look at the differential alone. We need to take into account the tire mechanical behavior as well.

The virtual test rig of Fig. 3.66 can be a tool to somehow simplify the problem because we have many parameters under direct control. Plots like those presented in Sect. 3.13.17 allow a simple, yet quantitative, analysis of the differential-tire interaction. After a little effort, everything becomes clear and predictable.

3.14   Vehicle Model for Handling and Performance

After quite a bit of work, we are now ready to set up our first-order vehicle model for
handling and performance analysis. Essentially, setting up a model means collecting
all relevant equations, their order being not important. Of course, a two-axle vehicle
is considered.

3.14.1   Equilibrium Equations

We have three *in-plane* equilibrium equations for the whole vehicle

$$ma_x = m(\dot{u} - vr) = X_1 + X_2 - \frac{1}{2}\rho S C_x u^2 = X$$

$$ma_y = m(\dot{v} + ur) = Y_1 + Y_2 = Y \tag{3.94'}$$

$$J_z \dot{r} = Y_1 a_1 - Y_2 a_2 + \Delta X_1 t_1 + \Delta X_2 t_2 = N$$

where the tangential (grip) forces were defined in (3.86) with respect to the vehicle
frame (Fig. 3.25)

$$X_1 = X_{11} + X_{12} \qquad\qquad X_2 = X_{21} + X_{22}$$

$$Y_1 = Y_{11} + Y_{12} \qquad\qquad Y_2 = Y_{21} + Y_{22} \tag{3.86'}$$

$$\Delta X_1 = \frac{X_{12} - X_{11}}{2} \qquad\qquad \Delta X_2 = \frac{X_{22} - X_{21}}{2}$$

and in (3.87) to exploit the contribution of each single tire ($\delta_2 \simeq 0$)

$$X_1 = (F_{x_{11}}\cos(\delta_{11}) + F_{x_{12}}\cos(\delta_{12})) - (F_{y_{11}}\sin(\delta_{11}) + F_{y_{12}}\sin(\delta_{12}))$$
$$X_2 = F_{x_{21}} + F_{x_{22}}$$
$$Y_1 = (F_{y_{11}}\cos(\delta_{11}) + F_{y_{12}}\cos(\delta_{12})) + (F_{x_{11}}\sin(\delta_{11}) + F_{x_{12}}\sin(\delta_{12}))$$
$$Y_2 = F_{y_{21}} + F_{y_{22}}$$
$$\Delta X_1 = [(F_{x_{12}}\cos(\delta_{12}) - F_{x_{11}}\cos(\delta_{11})) - (F_{y_{12}}\sin(\delta_{12}) - F_{y_{11}}\sin(\delta_{11}))]/2$$
$$\Delta X_2 = (F_{x_{22}} - F_{x_{21}})/2 = \Delta\widetilde{M}/(2r_r)$$

$$\tag{3.87'}$$

Moreover, there are three *out-of-plane* equilibrium equations

$$mg + Z_1^a + Z_2^a = Z_1 + Z_2$$

$$-ma_x h + (Z_1^a a_1 - Z_2^a a_2) \simeq Z_1 a_1 - Z_2 a_2 \tag{3.96'}$$

$$ma_y h \simeq \Delta Z_1 t_1 + \Delta Z_2 t_2$$

The first two equations link the vertical loads Z_1 and Z_2 acting on each axle to the vehicle features, to the vehicle forward velocity u, and to the vehicle longitudinal acceleration a_x

$$
\begin{aligned}
Z_1 &= Z_1^0 + Z_1^a + \Delta Z \\
&= \frac{mga_2}{l} + \frac{1}{2}\rho_a S_a C_{z1} u^2 - \frac{ma_x h}{l} \\
Z_2 &= Z_2^0 + Z_2^a - \Delta Z \\
&= \frac{mga_1}{l} + \frac{1}{2}\rho_a S_a C_{z2} u^2 + \frac{ma_x h}{l}
\end{aligned}
\tag{3.103'}
$$

The third equation in (3.96) shows how the sum of the lateral load transfers ΔZ_1 and ΔZ_2 is related to the lateral acceleration a_y.

3.14.2   Roll Angles

Six *internal* constitutive equations involve the two roll angles due to the pneumatic tires

$$
\begin{aligned}
\phi_1^p &= \frac{1}{k_{\phi_1}^p}\frac{k_{\phi_1}k_{\phi_2}}{k_\phi}\left[\frac{Y(h-q^b)}{k_{\phi_2}} + \frac{Y_1 q_1}{k_{\phi_1}^s} + \frac{Y_1 q_1}{k_{\phi_2}^s} + \frac{Y_1 q_1 + Y_2 q_2}{k_{\phi_2}^p}\right] = \phi_1^p(Y_1, Y_2) \\
\phi_2^p &= \frac{1}{k_{\phi_2}^p}\frac{k_{\phi_1}k_{\phi_2}}{k_\phi}\left[\frac{Y(h-q^b)}{k_{\phi_1}} + \frac{Y_2 q_2}{k_{\phi_1}^s} + \frac{Y_2 q_2}{k_{\phi_2}^s} + \frac{Y_1 q_1 + Y_2 q_2}{k_{\phi_1}^p}\right] = \phi_2^p(Y_1, Y_2)
\end{aligned}
\tag{3.146'}
$$

the two roll angles due to the suspensions

$$
\begin{aligned}
\phi_1^s &= \frac{1}{k_{\phi_1}^s}\frac{k_{\phi_1}k_{\phi_2}}{k_\phi}\left[\frac{Y(h-q^b)}{k_{\phi_2}} - \frac{Y_1 q_1}{k_{\phi_1}^p} + \frac{Y_2 q_2}{k_{\phi_2}^p}\right] = \phi_1^s(Y_1, Y_2) \\
\phi_2^s &= \frac{1}{k_{\phi_2}^s}\frac{k_{\phi_1}k_{\phi_2}}{k_\phi}\left[\frac{Y(h-q^b)}{k_{\phi_1}} - \frac{Y_2 q_2}{k_{\phi_2}^p} + \frac{Y_1 q_1}{k_{\phi_1}^p}\right] = \phi_2^p(Y_1, Y_2)
\end{aligned}
\tag{3.147'}
$$

and the two vertical displacements z_i^s due to suspension deflections only

$$
z_i^s = \frac{4q_i}{t_i}\frac{\Delta Y_i}{k_{z_i}^s} = z_i^s(\Delta Y_i)
\tag{3.137'}
$$

3.14.3 Lateral Load Transfers

To single out ΔZ_1 and ΔZ_2, we have to take into account the features (i.e., stiffnesses and kinematics) of the vehicle suspensions

$$\Delta Z_1 = \frac{1}{t_1}\left[\frac{k_{\phi_1}}{k_\phi}Y\left(h - q^b\right) + Y_1 q_1 + \frac{k_{\phi_1}k_{\phi_2}}{k_\phi}\left(\frac{Y_2 q_2}{k_{\phi_2}^P} - \frac{Y_1 q_1}{k_{\phi_1}^P}\right)\right] = \Delta Z_1(Y_1, Y_2)$$

$$\Delta Z_2 = \frac{1}{t_2}\left[\frac{k_{\phi_2}}{k_\phi}Y\left(h - q^b\right) + Y_2 q_2 + \frac{k_{\phi_1}k_{\phi_2}}{k_\phi}\left(\frac{Y_1 q_1}{k_{\phi_1}^P} - \frac{Y_2 q_2}{k_{\phi_2}^P}\right)\right] = \Delta Z_2(Y_1, Y_2)$$

$$(3.151')$$

where

$$q^b = \frac{a_2^b q_1 + a_1^b q_2}{a_1^b + a_2^b} \qquad\qquad (3.140')$$

with a_1^b and a_2^b defined in (3.100).

3.14.4 Total Vertical Loads

Therefore, the total vertical loads Z_{ij} acting on each wheel are as follows

$$Z_{11} = F_{z_{11}} = \frac{1}{2}\left(Z_1^0 + Z_1^a + \Delta Z\right) - \Delta Z_1$$

$$Z_{12} = F_{z_{12}} = \frac{1}{2}\left(Z_1^0 + Z_1^a + \Delta Z\right) + \Delta Z_1$$

$$(3.108')$$

$$Z_{21} = F_{z_{21}} = \frac{1}{2}\left(Z_2^0 + Z_2^a - \Delta Z\right) - \Delta Z_2$$

$$Z_{22} = F_{z_{22}} = \frac{1}{2}\left(Z_2^0 + Z_2^a - \Delta Z\right) + \Delta Z_2$$

Of course, we also have the following inequalities

$$\sqrt{X_{ij}^2 + Y_{ij}^2} \leq \mu_p Z_{ij} \qquad \text{and} \qquad Z_{ij} \geq 0 \qquad\qquad (3.207)$$

that is grip limitation and unilateral contact.

3.14.5 Static Camber and Camber Variations

Very important is the kinematic link between the suspension internal coordinates and the (first-order) camber variations, given in (3.113)

$$\Delta\gamma_{i1} \simeq -\left(\frac{t_i/2 - c_i}{c_i}\right)\phi_i^s + \phi_i^p - \frac{1}{c_i}z_i^s$$

$$\Delta\gamma_{i2} \simeq -\left(\frac{t_i/2 - c_i}{c_i}\right)\phi_i^s + \phi_i^p + \frac{1}{c_i}z_i^s \tag{3.113'}$$

Different suspensions with the same no-roll centers share only the same value of q_i (Figs. 3.29 and 3.30). Therefore they behave differently.

According to (3.111), the total camber angles are

$$\gamma_{ij} = \gamma_{ij}^0 + \Delta\gamma_{ij} \tag{3.208}$$

where $\gamma_{i2}^0 = -\gamma_{i1}^0 = \gamma_i^0$.

3.14.6 Steer Angles

Suspension roll angles ϕ_i^s may affect the steering angles δ_{ij} of the wheels. It is the so-called *roll steer*. This feature has some relevance in vehicle handling and can be modelled by modifying (3.66) in the following way

$$\delta_{ij} = \delta_{ij}(\delta_v, \phi_i^s) \tag{3.209}$$

For left turning and right turning to be alike for the vehicle, the two functions of the same axle must be such that $\delta_{i1}(\delta_v, \phi_i^s) = -\delta_{i2}(-\delta_v, -\phi_i^s)$. Therefore, the Taylor series expansions of the two functions for the steering angle of the wheels of the same axle can be written as

$$\delta_{i1} = -\delta_i^0 + \tau_i\delta_v + \varepsilon_i\frac{t_i}{2l}(\tau_i\delta_v)^2 + \Upsilon_i\phi_i^s$$

$$\delta_{i2} = \delta_i^0 + \tau_i\delta_v - \varepsilon_i\frac{t_i}{2l}(\tau_i\delta_v)^2 + \Upsilon_i\phi_i^s \tag{3.210}$$

where δ_i^0 is the static toe, τ_i is the gear ratio of the whole steering system, ε_i is the Ackermann coefficient for dynamic toe, and Υ_i is the roll steer coefficient. Most cars have $\tau_2 = \varepsilon_2 = 0$, that is no direct steering of the rear wheels.

3.14.7   Tire Slips

Congruence equations are, by definition, a link between kinematic quantities. In a vehicle they relate the vehicle motion to the tire kinematics (translational slips, spin slips, camber angles, steering angles).

The longitudinal and lateral slips were defined for a single wheel with tire in (2.63) and (2.64), respectively. The slips for the four wheels of a vehicle were given in (3.59) and (3.60):

- longitudinal slips:

$$\sigma_{x_{11}} = \frac{[(u - rt_1/2)\cos(\delta_{11}) + (v + ra_1)\sin(\delta_{11})] - \omega_{11} r_1}{\omega_{11} r_1}$$

$$\sigma_{x_{12}} = \frac{[(u + rt_1/2)\cos(\delta_{12}) + (v + ra_1)\sin(\delta_{12})] - \omega_{12} r_1}{\omega_{12} r_1}$$

$$\sigma_{x_{21}} = \frac{[(u - rt_2/2)\cos(\delta_{21}) + (v - ra_2)\sin(\delta_{21})] - \omega_{21} r_2}{\omega_{21} r_2} \tag{3.59'}$$

$$\sigma_{x_{22}} = \frac{[(u + rt_2/2)\cos(\delta_{22}) + (v - ra_2)\sin(\delta_{22})] - \omega_{22} r_2}{\omega_{22} r_2}$$

- lateral slips:

$$\sigma_{y_{11}} = \frac{(v + ra_1)\cos(\delta_{11}) - (u - rt_1/2)\sin(\delta_{11})}{\omega_{11} r_1}$$

$$\sigma_{y_{12}} = \frac{(v + ra_1)\cos(\delta_{12}) - (u + rt_1/2)\sin(\delta_{12})}{\omega_{12} r_1}$$

$$\sigma_{y_{21}} = \frac{(v - ra_2)\cos(\delta_{21}) - (u - rt_2/2)\sin(\delta_{21})}{\omega_{21} r_2} \tag{3.60'}$$

$$\sigma_{y_{22}} = \frac{(v - ra_2)\cos(\delta_{22}) - (u + rt_2/2)\sin(\delta_{22})}{\omega_{22} r_2}$$

Owing to (3.4), the expressions of the translational slips can be simplified under normal operating conditions and small steering angles

$$\sigma_{x_{11}} \simeq \frac{(u - rt_1/2) - \omega_{11} r_1}{\omega_{11} r_1} \qquad \sigma_{y_{11}} \simeq \frac{(v + ra_1) - u\,\delta_{11}}{\omega_{11} r_1}$$

$$\sigma_{x_{12}} \simeq \frac{(u + rt_1/2) - \omega_{12} r_1}{\omega_{12} r_1} \qquad \sigma_{y_{12}} \simeq \frac{(v + ra_1) - u\,\delta_{12}}{\omega_{12} r_1}$$

$$\sigma_{x_{21}} \simeq \frac{(u - rt_2/2) - \omega_{21} r_2}{\omega_{21} r_2} \qquad \sigma_{y_{21}} \simeq \frac{(v - ra_2) - u\,\delta_{21}}{\omega_{21} r_2} \qquad (3.61')$$

$$\sigma_{x_{22}} \simeq \frac{(u + rt_2/2) - \omega_{22} r_2}{\omega_{22} r_2} \qquad \sigma_{y_{22}} \simeq \frac{(v - ra_2) - u\,\delta_{22}}{\omega_{22} r_2}$$

According to (2.83), the rolling radii r_i should depend on the vertical load and the camber angle. However, such dependence is so weak in a car that they can be safely assumed as constant.

Compactly, we have that the translational slips are as follows

$$\sigma_{x_{ij}} = \sigma_{x_{ij}}(v, r, u, \omega_{ij}, \delta_{ij})$$
$$\sigma_{y_{ij}} = \sigma_{y_{ij}}(v, r, u, \omega_{ij}, \delta_{ij}) \qquad (3.211)$$

The spin slip was defined for a single wheel with tire in (2.65). For the four wheels of a car they are

$$\varphi_{ij} = -\frac{r + \dot{\delta}_{ij} + \omega_{ij}\sin\gamma_{ij}(1 - \varepsilon_i)}{\omega_{ij} r_i} \qquad (3.62')$$

However, even in a Formula 1 car, the yaw rate $|r|$ is lower than 1 rad/s, that is $60°/s$, and $|\dot{\delta}_{ij}|$ is about four times smaller. The bigger contribution comes from the last term, which ranges between 1 and 5 rad/s. Therefore, as in (2.68)

$$\varphi_{ij} \simeq -\frac{\sin\gamma_{ij}(1 - \varepsilon_i)}{r_i} \qquad (3.212)$$

which shows that, approximately, the spin slips affect the tire behavior like the camber angle. The camber reduction factor ε_i can be assumed as constant (cf. (2.83)).

3.14.8 Tire Constitutive Equations

Each tire behaves according to its *constitutive equations* (2.82), as shown in Sect. 2.12. Both the longitudinal force $F_{x_{ij}}$ and the lateral force $F_{y_{ij}}$ depend on the vertical load $F_{z_{ij}}$, the camber angle γ_{ij}, the translational slips $\sigma_{x_{ij}}$ and $\sigma_{y_{ij}}$, and the spin slip φ_{ij}

$$F_{x_{ij}} = F_{x_{ij}}(F_{z_{ij}}, \gamma_{ij}, \sigma_{x_{ij}}, \sigma_{y_{ij}}, \varphi_{ij})$$
$$F_{y_{ij}} = F_{y_{ij}}(F_{z_{ij}}, \gamma_{ij}, \sigma_{x_{ij}}, \sigma_{y_{ij}}, \varphi_{ij}) \qquad (2.82')$$

However, as shown in (2.104), the dependence on the spin slip is often omitted.

The Magic Formula [17], discussed in Sect. 2.11, can be conveniently used to mathematically represent these functions.

Of course, in this analysis we assume the grip to be a known parameter. This is quite a strong assumption, as the grip depends on a lot of other parameters, like tire temperature, tire pressure, road conditions, to mention but a few.

3.14.9 Equations Governing the Differential Mechanisms

A differential with internal efficiency η_h (possibly equal to one in an open differential), provides a link between the longitudinal forces due to the engine power. For a rear driven vehicle we have

$$\eta_h^{\zeta(t)} F_{x_{21}} = F_{x_{22}} \tag{3.196'}$$

where, as discussed in Sect. 3.13.10, a simple, yet effective, model can be

$$\zeta(t) = \frac{\arctan\left(\chi\, \widetilde{\Delta\omega}(t)\, \mathrm{sign}(M_h(t))\right)}{\pi/2} \tag{3.197'}$$

with χ a big positive number and

$$\widetilde{\Delta\omega} = \frac{\omega_{22} - \omega_{21}}{2} \tag{3.213}$$

Moreover, the differential also provides a link between the angular velocities of the wheels

$$\omega_{21} = \omega_h - \widetilde{\Delta\omega} \quad \text{and} \quad \omega_{22} = \omega_h + \widetilde{\Delta\omega} \tag{3.214}$$

The torque M_h applied to the differential housing is given by

$$M_h = (F_{x_{21}} + F_{x_{22}})r_r \tag{3.168'}$$

3.14.10 Summary

The equations listed in this section may look at first a bit complicated. However, there are only three differential equations, namely the equilibrium equations (3.94). All other equations are algebraic. This means that, ultimately, the model is governed by three equations of motion, all first-order differential equations, in the unknown functions $u(t)$, $v(t)$ and $r(t)$. Given (i.e., input) functions may be the angular speeds $\omega_h(t)$ of the housing of the differential along with the steering wheel angle $\delta_v(t)$. Both are controlled by the driver.

However, it is very common to assume the forward velocity $u(t)$ to be a given function, instead of $\omega_h(t)$, thus having only two differential equations.

To investigate the steady-state behavior, just set to zero all time derivatives.

3.15 The Structure of This Vehicle Model

It is advisable to extract simplified models tailored for specific vehicles and/or operating conditions. The goal is to obtain models simple enough for training human beings in learning and understanding vehicle dynamics. Of course, we will pay attention to state the additional assumptions needed for the simplified model to be meaningful.

A list of possible options can be:

1. braking on a straight road with uniform grip;
2. accelerating on a straight road with uniform grip;
3. handling at constant and given forward velocity u:

   a. vehicle with open differential;
   b. vehicle with limited-slip differential;
   c. vehicle without wings (no downforce);
   d. vehicle with wings.

In the next chapters, most of these options will be elaborated in detail. In addition, we will develop vehicle models for studying ride and road holding, and we will extend the handling model to take into account roll motion. A final chapter will address what happens in the contact patch between tire and road.

But there is one more topic to be discussed here.

3.16 Three-Axle Vehicles

Most vehicles have two axles, but many have three (or more) [23, 24]. Just consider trucks. As we have already discussed, each axle has a no-roll center. So there are three no-roll centers, each at a different height, in general. But what about the no-roll axis? As a matter of fact, a straight line is defined by two points, not three!

It is quite amazing that such an (apparently) fundamental concept like the classical roll axis turns out to be totally meaningless for an important class of vehicles. And it is even more surprising that vehicle dynamics often employs such a weak concept.

Having said that, let us address the problem with open mind. First of all, consider that the vehicle knows nothing about no-roll axis and the like. It behaves according to the fundamental laws of dynamics. And for sure, the vehicle body (assumed rigid) has an instantaneous screw axis, but it has nothing to do with the roll axis as commonly defined.

Actually, the really big difference between a vehicle with two and a vehicle with three (or more) axles is that with two axles we have in many respects a statically

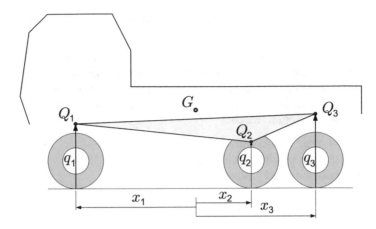

Fig. 3.76 Triangle of possible no-roll points ($x_2 < 0$ and $x_3 < 0$)

determinate (or isostatic) structure, whereas with three or more axles we have always to deal with a statically indeterminate (or hyperstatic) structure.

For instance, the *static* vertical loads in a two-axle vehicle can be obtained by the equilibrium equations only and are not affected by the suspension stiffnesses

$$Z_1^0 = \frac{mga_2}{l} \qquad Z_2^0 = \frac{mga_1}{l} \tag{3.105'}$$

On the other hand, with three axles the static vertical loads cannot be obtained by the equilibrium equations only, that is without taking into account the suspension and tire vertical stiffnesses.

Let, $x_1 = a_1$, $x_2 = -a_2$ and x_3 be the longitudinal coordinate of each axle in the vehicle reference frame (Fig. 3.76). The three static vertical loads on each axle must obey to the following equilibrium equations (in case of negligible aerodynamic loads)

$$0 = Z_1^0 + Z_2^0 + Z_3^0 - mg$$
$$0 = Z_1^0 x_1 + Z_2^0 x_2 + Z_3^0 x_3 \tag{3.215}$$

We have two equations with three unknowns. Therefore, there are infinitely many solutions. For instance, we could set $X_3 = 0$ by raising the two wheels of the third axle, thus restoring the common two-axle architecture.

Exactly the same observation applies to the lateral forces Y_i: there are infinitely many possible combinations of lateral forces Y_1, Y_2 and Y_3 to balance Y.

Notwithstanding these difficulties, the analysis to *define* (and measure) the roll and vertical stiffnesses $k_{\phi_i}^s$, $k_{\phi_i}^p$, $k_{z_i}^s$, $k_{z_i}^p$ still applies entirely. We can then proceed to collect all relevant equations, like in Sect. 3.10.12.

$$Y = Y_1 + Y_2 + Y_3 \tag{3.216}$$

$$Y x_N = Y_1 x_1 + Y_2 x_2 + Y_3 x_3 \tag{3.217}$$

$$Y q^b = Y_1 q_1 + Y_2 q_2 + Y_3 q_3 \tag{3.218}$$

$$Y h = \Delta Z_1 t_1 + \Delta Z_2 t_2 + \Delta Z_3 t_3 \tag{3.219}$$

$$mg = Z_1 + Z_2 + Z_3 \tag{3.220}$$

$$X h = -(Z_1 x_1 + Z_2 x_2 + Z_3 x_3) \tag{3.221}$$

$$\Delta Z_i t_i = k_{\phi_i}^P \phi_i^P \tag{3.222}$$

$$\Delta Z_i t_i = Y_i q_i + k_{\phi_i}^s \phi_i^s \tag{3.223}$$

$$Z_i^0 - Z_i = k_{z_i}^P z_i^P \tag{3.224}$$

$$Z_i^0 - Z_i = k_{z_i}^s z_i^s - \Delta Y_i \frac{4 q_i}{t_i} \tag{3.225}$$

$$\phi = \phi_i^s + \phi_i^P \tag{3.226}$$

$$\frac{z_1 - z_3}{x_1 - x_3} = \frac{z_1 - z_2}{x_1 - x_2} \tag{3.227}$$

which imply

$$Y h = k_{\phi_1}^P \phi_1^P + k_{\phi_2}^P \phi_2^P + k_{\phi_3}^P \phi_3^P \tag{3.228}$$

$$Y(h - q^b) = k_{\phi_1}^s \phi_1^s + k_{\phi_2}^s \phi_2^s + k_{\phi_3}^s \phi_3^s \tag{3.229}$$

It is worth noting that the suspension jacking, as in (3.137), affects the vertical loads because the system is hyperstatic. Therefore it also interacts with lateral load transfers, with the lateral forces and, ultimately, with roll motion. In other words, in a three-axle vehicle there is interaction between suspension jacking and roll angles. It was not so in a two-axle vehicle.

Moreover, under a given lateral force Y, the roll angles are also affected by the amount of grip of each axle, and vice versa.

But maybe the most interesting and, somehow, surprising result is that, as shown in Fig. 3.76, in a three-axle vehicle the no-roll axis must be replaced by a *triangle of possible no-roll points*. The three no-roll centers Q_i are the vertices of this triangle of possible no-roll points. The actual height q^b depends not only on the heights q_i of the no-roll centers, but also on the value of each lateral force Y_i.

This result generalizes the concept of no-roll axis and confirms that sentences like "The vehicle has two roll centers about which it rolls when cornering" are incorrect.

3.17 Exercises

3.17.1 Center of Curvature E_G of the Trajectory of G

With the aid of Fig. 3.10, obtain the coordinates of E_G in the vehicle reference system.

Solution

The sought coordinates are

$$- R_G \sin \hat{\beta} \quad \text{and} \quad R_G \cos \hat{\beta} \tag{3.230}$$

with R_G given in (3.36).

3.17.2 Track Variation

A car has a rear track width $t_2 = 1.65$ m and independent suspensions. Employing the four internal coordinates defined in Sect. 3.10.2, compute the track variation Δt_2 due to a body vertical displacement $z_2^s = 5$ cm for three possible heights q_2 of the no-roll center: 0, 5, 10 cm (Fig. 3.29).

Solution

We can employ (3.114) to obtain Δt_2 equal to 0 cm, -0.6 cm and -1.2 cm, respectively (Fig. 3.30b). Of course, these results are first-order approximations. It is worth noting that, in this framework, two different suspensions, but with the same t_2 and q_2, provide the same Δt_2.

3.17.3 Camber Variation

As in the former exercise, a car has independent suspensions. Compute the camber variations $\Delta \gamma_{21}$ and $\Delta \gamma_{22}$ due to a body vertical displacement $z_2^s = 5$ cm and/or a body roll angle $\phi = 5$ deg for three possible values of c_2: 160, 80, 40 cm (Fig. 3.29). Assume rigid tires.

Solution

The equation to be employed is (3.113).

If we apply the 5 cm vertical displacement alone, we obtain for $\Delta \gamma_{22} = - \Delta \gamma_{21}$ (opposite sign) the following values: 1.8 deg, 3.6 deg, 7.2 deg. This is called suspension jacking (Figs. 3.39 and 9.7).

If we apply the 5 deg roll angle alone, we obtain for $\Delta\gamma_{22} = \Delta\gamma_{21}$ (same sign) the following values: 2.4 deg, -0.2 deg, -5.3 deg. Quite interestingly, in the first case the wheels lean like the vehicle body (Fig. 9.5-bottom), while in the last case they lean in the opposite direction (Fig. 9.5-top).

If we apply both the vertical displacement and the roll angle, we obtain for $\Delta\gamma_{22}$ the following values: 4.2 deg, 3.4 deg, 1.9 deg. At the same time, for $\Delta\gamma_{21}$ we obtain 0.6 deg, -3.8 deg, -12.5 deg.

Actually, a roll angle of 5 deg and/or a vertical displacement of 5 cm are quite large for a first-order analysis to be very accurate. Repeat the exercise with smaller values.

3.17.4 Power Loss in a Limited-Slip Differential

Estimate the power loss W_d inside the differential housing in the four cases shown in Figs. 3.70, 3.71, 3.72 and 3.73. Assume a rolling radius $r_r = 0.25$ m.

Solution

We know from Sect. 3.13.17.3 that $\eta_h = 0.67$. Obviously, $W_d = 0$ in Figs. 3.71 and 3.72, because the differential is locked.

In Fig. 3.70 we have, approximately, $\omega_l = 81.7\,\text{rad/s}$, $\omega_r = 82.5\,\text{rad/s}$, $X_l = 222\,\text{N}$ and $X_r = 148\,\text{N}$.

First, we can compute the internal efficiency $\eta_h = (148 \times 82.5)/(222 \times 81.7) = 0.67$. It is ok.

Then, according to (3.179), we have $W_d = 0.25(222 - 148)(82.5 - 81.7)/2000 = 0.005\,\text{kW}$, while the total power flowing through the differential is about $W_h\omega_h = 0.25(148 \times 82.5 + 222 \times 81.7)/1000 + 0.005 = 7.6\,\text{kW}$.

In Fig. 3.73 we have, approximately, $\omega_l = 86.0\,\text{rad/s}$, $\omega_r = 83.6\,\text{rad/s}$, $X_l = 350\,\text{N}$ and $X_r = 530\,\text{N}$.

As done above, we can compute the internal efficiency $\eta_h = (350 \times 86.0)/(530 \times 83.6) = 0.68$. It is almost ok.

Then we have $W_d = 0.25(530 - 350)(86.0 - 83.6)/2000 = 0.34\,\text{kW}$, while the total power flowing through the differential is about $W_h\omega_h = 0.25(350 \times 86.0 + 530 \times 83.6)/1000 + 0.34 = 19.0\,\text{kW}$. In this case W_d is not negligible because the differential has to cope with a very critical situation.

3.17.5 Differential-Tires Interaction

The notation is like in Sect. 3.13.17. In particular, we are using the virtual test rig of Fig. 3.66. Here are five plots without captions (Figs. 3.77, 3.78, 3.79, 3.80 and 3.81), and the following five captions without plots:

Fig. 3.77 See exercise
3.17.5

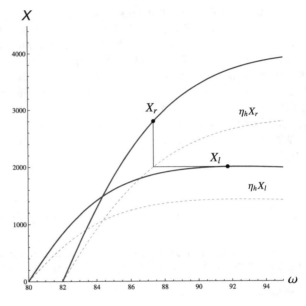

Fig. 3.78 See exercise
3.17.5

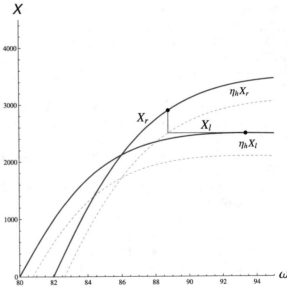

Fig. 3.79 See exercise 3.17.5

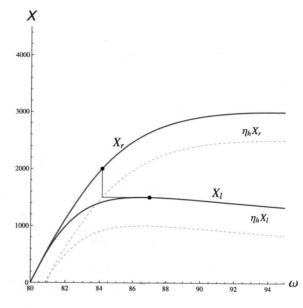

Fig. 3.80 See exercise 3.17.5

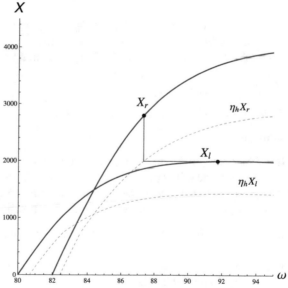

Fig. 3.81 See exercise
3.17.5

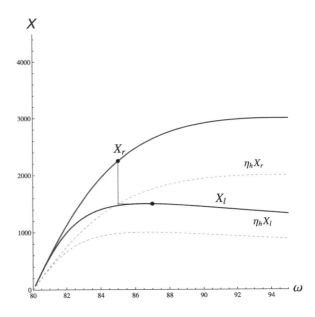

1. clutch-pack differential with no wedging action under μ-split conditions;
2. geared differential under μ-split conditions;
3. clutch-pack differential with no wedging action in a left turn;
4. geared differential in a left turn;
5. Salisbury differential in a left turn.

Try to match plot and caption, and explain why.

Solution

To solve this problem we can look at some relevant details. Solid curves having the same slope at the initial point are typical of μ-split cases. Dashed curve and the corresponding solid curve with the same initial point are typical of geared differentials. Therefore, we have the following plot-caption matches:

1. clutch-pack differential with no wedging action under μ-split conditions: Fig. 3.79;
2. geared differential under μ-split conditions: Fig. 3.81;
3. clutch-pack differential with no wedging action in a left turn: Fig. 3.78;
4. geared differential in a left turn: Fig. 3.77;
5. Salisbury differential in a left turn: Fig. 3.80.

3.18 Summary

This is the main chapter of this book, the core of it. Therefore it covers a lot of topics.

At the beginning, the simplifying assumptions to formulate a simple, yet signifi-cant, vehicle model have been listed. Then the kinematics of the vehicle as a whole has been described in detail, followed by the kinematics of each wheel with tire. For-mulation of the constitutive (tire) equations and of the global equilibrium equations has been the next step.

A lot of work has been devoted to lateral load transfers, which has required an in-depth suspension analysis. This has led to the definition of suspension and vehicle internal coordinates, of no-roll centers and no-roll axis, for both independent and dependent suspensions. The case of three-axle vehicles has been also considered.

In the end, the vehicle model for handling and performance has been formulated in a synthetic, yet precise way. A general description of the mechanics of both open and limited-slip differential mechanisms has been included.

3.19 List of Some Relevant Concepts

Section 3.2.1—a vehicle can have a lateral velocity v, although it is normally much lower than the forward velocity u;

Section 3.2.2—the lateral and forward velocities cannot be expressed as derivatives of other functions;

Section 3.2.3—the velocity center C is not, in general, the center of curvature;

Section 3.10.2—for each axle, four internal coordinates are necessary to monitor the suspension conditions with respect to a reference configuration;

Section 3.10.9—no suspension roll does not mean no other effect at all. There can be suspension jacking and tire roll;

Section 3.10.11—application of a force at any point of the no-roll axis does not produce suspension roll;

Section 3.13.5—a fundamental parameter in a differential mechanism is its internal efficiency η_h;

Section 3.16—in a three-axle vehicle, the no-roll axis must be replaced by a triangle of possible no-roll points.

3.20 Key Symbols

a_1	distance of G from the front axle
a_2	distance of G from the rear axle
a_n	centripetal acceleration
a_t	tangential acceleration
a_x	longitudinal acceleration
a_y	lateral acceleration
C	velocity center
C_x, C_y, C_z	aerodynamic coefficients

d	diameter of the inflection circle
$F_{x_{ij}}$	tire longitudinal forces
$F_{y_{ij}}$	tire lateral forces
$F_{z_{ij}}$	tire vertical forces
g	gravitational acceleration
G	center of mass
h	height of G
J_x, J_y, J_z	moments of inertia
J_w	moment of inertia of a wheel
K	acceleration center
k_ϕ	total roll stiffness
k_{ϕ_i}	global roll stiffness of ith axle
$k_{\phi_i}^p$	tire roll stiffness of ith axle
$k_{\phi_i}^s$	suspension roll stiffness of ith axle
l	wheelbase
L_{w_i}	gyroscopic torque
m	mass
M_f	torque applied to the differential by the faster spinning wheel
M_h	torque applied to the differential housing
$\widetilde{M}_l = -M_l$	torque applied to the left wheel
$\widetilde{M}_r = -M_r$	torque applied to the right wheel
M_s	torque applied to the differential by the slower spinning wheel
N	yaw moment
q_1	height of the front no-roll center
Q_1	front no-roll center
q_2	height of the rear no-roll center
Q_2	rear no-roll center
r	yaw rate
R	lateral coordinate of C
r_i	rolling radii
r_r	rolling radius
S	longitudinal coordinate of C
S_a	frontal area
t_1	front track
t_2	rear track
u	longitudinal velocity
v	lateral velocity
W_d	friction power loss inside the differential
W_i	input power
W_o	output power
X	longitudinal force
X_a	aerodynamic drag
X_l	longitudinal forces applied to the left wheel from the road
X_r	longitudinal forces applied to the right wheel from the road

Y	lateral force
Z	vertical force
Z_i	vertical load on ith axle
Z_i^0	static vertical load on ith axle
Z_i^a	aerodynamic vertical load on ith axle
ΔZ	longitudinal load transfer
ΔZ_i	lateral load transfer on ith axle
α_{ij}	tire slip angles
β	ratio v/u
$\hat{\beta}$	vehicle slip angle
γ_{ij}	camber angles
δ_{ij}	steer angle of the wheels
δ_v	steering wheel angle of rotation
ε_1	Ackermann coefficient
ε_h	locking coefficient
ζ	exponent
η_h	internal efficiency of the differential housing
ρ	ratio r/u
ρ_a	air density
$\sigma_{x_{ij}}$	tire longitudinal slips
$\sigma_{y_{ij}}$	tire lateral slips
τ	steer gear ratio
ϕ	roll angle
φ_{ij}	spin slips
ψ	yaw angle
ω_f	angular speed of the faster spinning wheel
ω_h	angular velocity of the differential housing
ω_{ij}	angular velocities of the wheels
ω_l	angular velocity of the left wheel
ω_r	angular velocity of the right wheel
ω_s	angular speed of the slower spinning wheel

References

1. Barnard RH (2001) Road vehicle aerodynamic design, 2nd edn. Mechaero Publishing, Hertfordshire
2. Blundell M, Harty D (2004) The multibody systems approach to vehicle dynamics. Butterworth-Heinemann, Oxford
3. Dixon JC (1987) The roll-centre concept in vehicle handling dynamics. Proc Inst Mech Eng Part D: Transp Eng 201(1):69–78
4. Dixon JC (1991) Tyres, suspension and handling. Cambridge University Press, Cambridge

5. Dixon JC (2009) Suspension geometry and computation. Wiley, Chichester
6. Ellis JR (1994) Vehicle handling dynamics. Mechanical Engineering Publications, London
7. Genta G, Morello L (2009) The automotive chassis, vol 2. Springer, Berlin
8. Genta G, Morello L (2009) The automotive chassis, vol 1. Springer, Berlin
9. Gillespie TD (1992) Fundamentals of vehicle dynamics. SAE International, Warrendale
10. Guiggiani M, Mori LF (2008) Suggestions on how not to mishandle mathematical formulæ. TUGboat 29:255–263
11. Innocenti C (2007) Questioning the notions of roll center and roll axis for car suspensions. In: Deuxieme Congres International Conception et Modelisation des Systemes Mecaniques, Monastir
12. Jazar RN (2014) Vehicle dynamics, 2nd edn. Springer, New York
13. Longhurst C (2013). https://www.carbibles.com/guide-to-car-suspension/
14. Meirovitch L (1970) Methods of analytical dynamics. McGraw-Hill, New York
15. Michelin (2003) The tyre encyclopedia. Part 3: rolling resistance. Société de Technologie Michelin, Clermont–Ferrand [CD-ROM]
16. Milliken WF, Milliken DL (1995) Race car vehicle dynamics. SAE International, Warrendale
17. Pacejka HB (2002) Tyre and vehicle dynamics. Butterworth-Heinemann, Oxford
18. Pauwelussen JP (2015) Essentials of vehicle dynamics. Butterworth-Heinemann, Oxford
19. Rajamani R (2012) Vehicle dynamics and control, 2nd edn. Springer, New York
20. Segers J (2008) Analysis techniques for racecar data acquisition. SAE International, Warrendale
21. Seward D (2014) Race car design. Palgrave, London
22. Smith C (1978) Tune to win. Aero Publishers, Fallbrook
23. Williams DE (2011) On the equivalent wheelbase of a three-axle vehicle. Veh Syst Dyn 49(9):1521–1532
24. Williams DE (2012) Generalised multi-axle vehicle handling. Veh Syst Dyn 50(1):149–166
25. Wong JY (2001) Theory of ground vehicles. Wiley, New York

Chapter 4
Braking Performance

Driving a vehicle involves, among other things, braking [1]. Fortunately, most of the times, we brake very softly, far from the braking performance limit. Most drivers, perhaps, never need to experience the limit braking performance of their car in everyday traffic. However, engineers must know very well the mechanics of braking a vehicle, to allow it to stop as soon as possible in case of emergency. Actually, this problem has been somehow mitigated by the advent of ABS systems [2], which now equip every road car. However, many race cars do not have ABS and hence brake design and balance is still a relevant topic in vehicle dynamics.

By *brake balance* or bias, we mean how much to brake the front wheels with respect to the rear wheels. The goal is to stop the vehicle as soon as possible, but avoiding wheel locking. Cars have only one pedal to brake all wheels and brake balance is left to the car. By the way, wheel locking should be avoided because, in order of importance:

1. the steering/directional capability is totally impaired (most important);
2. the grip is lower;
3. energy dissipation switches from the brakes to the contact patches and tires get damaged.

On the other hand, almost all motorcycles and bicycles have independent brake commands for the front wheel and for the rear wheel, thus leaving the duty of brake balance to the rider. Many bicyclists fear using the front brake because they believe it might cause the bicycle to overturn. Actually, overturning a bicycle with the front brake is much harder than it seems. Not using the front brake is a bad habit, since it drastically impairs the braking performance.

© The Author(s), under exclusive license to Springer Nature Switzerland AG 2023 177
M. Guiggiani, *The Science of Vehicle Dynamics*,
https://doi.org/10.1007/978-3-031-06461-6_4

4.1 Pure Braking

As anticipated, we extract tailored models from the fairly general vehicle model developed in Chap. 3.

When braking on a *flat*, *straight* road, with *uniform grip*, we know beforehand that

$$
\begin{aligned}
Y &= 0 \\
N &= 0 \\
\Delta X_i &= 0 \\
\Delta Z_i &= 0
\end{aligned}
\tag{4.1}
$$

that is, there are no lateral forces, no yaw moment and no lateral load transfers. Accordingly, the vehicle goes straight, with no lateral acceleration and yaw rate (and also no lateral velocity)

$$
\begin{aligned}
a_y &= 0 \\
\dot{r} &= 0 \\
v &= 0 \\
r &= 0
\end{aligned}
\tag{4.2}
$$

Other quantities are usually very small. In particular, if the wheels of the same axle have a bit of convergence (also called toe-in), that means that there are small steering angles and, accordingly, very small lateral slips. Similarly, if the wheels of the same axle have some camber, the tires are subject to a small spin slip:

$$
\begin{aligned}
\delta_{ij} &\simeq 0 \\
\sigma_{y_{ij}} &\simeq 0 \\
\varphi_{ij} &\simeq 0
\end{aligned}
\tag{4.3}
$$

At first, all these quantities can be set equal to zero.

4.2 Vehicle Model for Braking Performance

A simple, yet significant, model to study the limit braking performance of a road vehicle is shown in Fig. 4.1. We are dealing here with road vehicles, without significant aerodynamic downforces (however, have a look at Fig. 3.24). Formula cars are dealt with in Sect. 4.12.

We suppose to brake on a flat and straight road, with uniform grip. Therefore, the vehicle goes straight. Moreover, we assume to apply a constant force to the brake pedal. Therefore, pitch oscillations are negligible.

Summing up, we can employ the two-dimensional model shown in Fig. 4.1. The vehicle is just a single rigid body with mass m, moving horizontally with forward

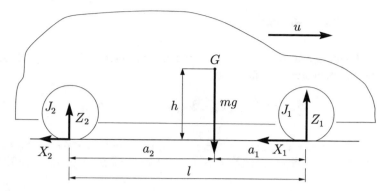

Fig. 4.1 Model for braking performance analysis of road cars

speed u and forward acceleration $\dot{u} < 0$. Beside its own weight mg, it receives two vertical forces Z_1 and Z_2 from the road, one per axle, and two longitudinal (braking) forces X_1 and X_2, again one per axle.

In this chapter only we assume X_1 and X_2 to be positive if directed like in Fig. 4.1. It is more convenient to deal with positive quantities.

4.3 Equilibrium Equations

The three equilibrium equations are readily obtained from Fig. 4.1

$$m\dot{u} = -(X_1 + X_2)$$
$$0 = Z_1 + Z_2 - mg \qquad (4.4)$$
$$0 = (X_1 + X_2)\,h - Z_1\,a_1 + Z_2\,a_2$$

which must be supplemented by the following inequalities

$$|X_i| \le \mu_p^x Z_i \quad \text{and} \quad Z_i \ge 0 \qquad (4.5)$$

where μ_p^x is the global longitudinal friction coefficient defined in (2.87). It is quite obvious that the braking forces cannot exceed the traction limit, nor the vertical forces be negative. For brevity, we will use the symbol μ for μ_p^x in this chapter.

The aerodynamic drag X_a has not been included because in road cars it is really small compared to the braking forces.

The rolling resistance is also very small. The braking forces X_i already include this small contribution.

4.3.1 Rigorous Moment Equation

In the third equation in (4.4) we omitted the rotating inertia J_1 and J_2 of the two axles. The complete equation is

$$(J_1 + J_2)\frac{\dot{u}}{r_r} = (X_1 + X_2)\,h - Z_1\,a_1 + Z_2\,a_2 \tag{4.6}$$

where r_r is the wheel rolling radius. However, the contribution of the rotating inertia is usually negligible. Typically, $(J_1 + J_2)/r_r \simeq 1\,\mathrm{kgm}$, while, e.g., $mh \simeq 600\,\mathrm{kgm}$.

4.4 Longitudinal Load Transfer

When going at constant speed, that is with $\dot{u} = 0$, we have from (4.4) (or, directly, from (3.105)) that the static vertical loads on each axle are

$$Z_1^0 = \frac{mga_2}{l} \qquad Z_2^0 = \frac{mga_1}{l} \tag{4.7}$$

During braking with $\dot{u} < 0$, the two loads change, although their sum must be constantly equal to the vehicle weight mg. We have the so-called longitudinal load transfer ΔZ

$$Z_1 = Z_1^0 + \Delta Z \quad \text{and} \quad Z_2 = Z_2^0 - \Delta Z \tag{4.8}$$

where (cf. (3.104))

$$\Delta Z = -\frac{mh}{l}\dot{u} \tag{4.9}$$

with $\dot{u} < 0$. The front axle is subject to a higher load ($Z_1 > Z_1^0$), while the rear axle to a lower load ($Z_2 < Z_2^0$). It is worth noting that the load transfer does not depend on the type of suspensions.

We have *overturning* of the vehicle if $Z_2 = 0$, that is if

$$|\dot{u}| = a_1 g / h \tag{4.10}$$

This condition is never met in cars, whereas it may limit the brake performance in some motorcycles.

4.5 Maximum Deceleration

The best braking performance $|\dot{u}|_{\max}$ is obtained if both axles brake at their traction limit, that is if

$$X_1 = \mu\, Z_1 \quad \text{and} \quad X_2 = \mu\, Z_2 \tag{4.11}$$

From the equilibrium equations (4.4), it is straightforward to obtain the *limit deceleration*

$$|\dot{u}| = \mu g \tag{4.12}$$

Of course, the maximum deceleration is the minimum between (4.10) and (4.12)

$$|\dot{u}|_{max} = \min(\mu g, \, a_1 g/h) \tag{4.13}$$

Cars have $\mu < a_1/h$, whereas in some motorcycles it can be the other way around. Here we are mainly dealing with cars, and therefore we have

$$|\dot{u}|_{max} = \mu g \tag{4.14}$$

4.6 Brake Balance

When braking at the best braking performance, that is with $\dot{u} = -\mu g$, the longitudinal forces are

$$X_{1_P} = \mu \, Z_{1_P} = \mu \left(Z_1^0 + \frac{mh}{l}\mu g \right) = \mu \frac{mg}{l}(a_2 + \mu h)$$

$$X_{2_P} = \mu \, Z_{2_P} = \mu \left(Z_2^0 - \frac{mh}{l}\mu g \right) = \mu \frac{mg}{l}(a_1 - \mu h) \tag{4.15}$$

The *optimal brake balance* (or brake bias) β_P to have the *best* braking performance is promptly obtained as

$$\beta_P = \frac{X_{1_P}}{X_{2_P}} = \frac{Z_{1_P}}{Z_{2_P}} = \frac{a_2 + \mu h}{a_1 - \mu h} \tag{4.16}$$

Typical values in road cars are $\beta_P \simeq 2$ on dry asphalt ($\mu \simeq 0.8$) and $\beta_P \simeq 1.5$ on wet asphalt ($\mu \simeq 0.4$). More commonly, the same concepts would be expressed as front/rear $= 66/33$ and front/rear $= 60/40$, respectively.

Let μ_1 be the coefficient of friction of the front axle, and μ_2 be the coefficient of friction of the rear axle. Then, the optimal brake balance β_P is given by

$$\beta_P = \frac{X_{1_P}}{X_{2_P}} = \frac{\mu_1(a_2 + \mu_2 h)}{\mu_2(a_1 - \mu_1 h)} \tag{4.17}$$

which generalizes (4.16) when μ_1 (front) $\neq \mu_2$ (rear).

4.7 All Possible Braking Combinations

If the best braking performance is our ultimate goal, we should also look around to see what happens if we employ a brake balance not equal to β_P. All possible braking combinations can be visualized in a simple, yet very useful, figure.

First solve the equilibrium equations (4.4) with $X_1 = \mu Z_1$, thus getting

$$Z_1 = \frac{X_1}{\mu} = Z_1^0 + \frac{h}{l}(X_1 + X_2) \tag{4.18}$$

and hence

$$X_1 = \mu \left(\frac{Z_1^0 + \frac{h}{l} X_2}{1 - \mu \frac{h}{l}} \right) \tag{4.19}$$

This is the relationship between X_1 and X_2 to have limit (threshold) braking at the front wheels.

Similarly, solve the equilibrium equations (4.4) with $X_2 = \mu Z_2$, thus getting

$$X_2 = \mu \left(\frac{Z_2^0 - \frac{h}{l} X_1}{1 + \mu \frac{h}{l}} \right) \tag{4.20}$$

This is the relationship between X_1 and X_2 to have limit (threshold) braking at the rear wheels.

In the plane (X_2, X_1) we can now draw the two straight lines (4.19) and (4.20), as shown in Fig. 4.2. The region inside the two lines contains all possible (admissible) braking combinations. Trying to trespass the upper line means front wheels lock-up. Trying to trespass the right line means rear wheels lock-up. Point P is the condition of best braking performance. It requires the combination of the braking forces X_{1_P} and X_{2_P}, which were obtained in (4.15).

Points with the same level of deceleration all belong to straight lines with slope $45°$, that is lines with constant $X_1 + X_2 = -m\dot{u}$. The maximum deceleration corresponds to the line passing through point P. Braking with balance β_P means moving along the line OP.

Some other relevant cases are shown in Fig. 4.3. Region 1 corresponds to low decelerations. So small that they can be obtained with any balance between front and rear braking forces, or even with only a rear braking force X_{2_0}

$$X_{2_0} = \frac{\mu Z_2^0}{1 + \mu \frac{h}{l}} \tag{4.21}$$

Fig. 4.2 Region of all
admissible braking
combinations

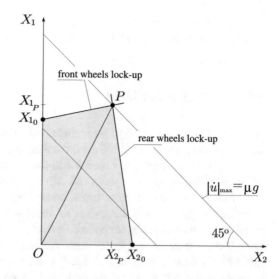

Fig. 4.3 Region of all
admissible braking
combinations with indication
of some particular cases

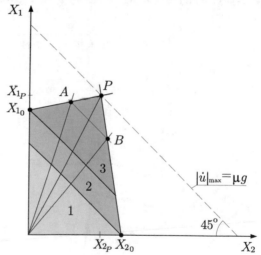

Region 2 needs necessarily some braking force at the front wheels, but even the front
wheels alone, with a braking force X_{1_0}, would do (front/rear = 100/0)

$$X_{1_0} = \frac{\mu Z_1^0}{1 - \mu \dfrac{h}{l}} \tag{4.22}$$

Region 3, that is high decelerations, require intervention of both axles. The higher
the deceleration, the narrower the range A–B.

To complete our discussion we have to address the effects of changing the grip coefficient μ and/or the position of G, that is a_1/a_2, and maybe h.

4.8 Changing the Grip

The formulation developed so far includes the grip coefficient as a parameter. Therefore, we have already obtained all formulas to deal with different values of μ. To understand what happens it is helpful to draw the admissible region for, say, three different values $\mu_l < \mu_o < \mu_h$ of the grip coefficient,[1] as shown in Fig. 4.4.

Let us assume that our car has a brake balance that follows line OP_2, that is optimized for $\mu = \mu_o$. If the grip is lower, that is $\mu_l < \mu_o$, there will be less load transfer ΔZ and a lower brake balance would be optimal. If we still follow line OP_2, we exit the admissible region at point A, that is for a deceleration lower than $\mu_l g$ and with the front wheels at lock-up. It can be shown that the deceleration is equal to $\varepsilon_l \mu_l g$, with the braking efficiency $\varepsilon_l < 1$ given by

$$\varepsilon_l = \frac{a_2}{a_2 + h(\mu_o - \mu_l)}, \qquad \text{if } \mu_l < \mu_o \tag{4.23}$$

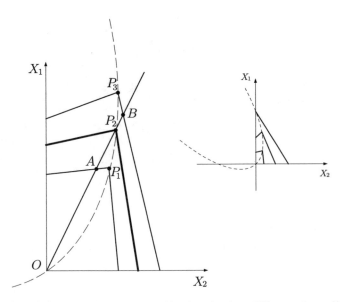

Fig. 4.4 Region of all admissible braking combinations for three different grip coefficients (left) and parabola of limit points (right)

---

[1] In this section we assume to have the same grip in both axles.

Braking efficiency $\varepsilon_h < 1$ is also obtained when the out-of-balance is due to a higher value $\mu_h > \mu_o$ of the grip coefficient. As shown in Fig. 4.4, we exit the admissible region at point B, which is not optimal. Rear wheels are about to lock up and the deceleration is equal to $\varepsilon_h \mu_h g$, with the braking efficiency $\varepsilon_h < 1$ given by

$$\varepsilon_h = \frac{a_1}{a_1 + h(\mu_h - \mu_o)}, \qquad \text{if } \mu_h > \mu_o \qquad (4.24)$$

Also shown in Fig. 4.4 is the parabola that collects all vertices P when varying the coefficient μ. Point P located on the X_1 axis means that maximum deceleration is limited by overturning.

4.9 Changing the Weight Distribution

The longitudinal position of G affects the static load distribution. Therefore, it affects the brake balance, but not the maximum deceleration μg. Accordingly, we get an admissible region like in Fig. 4.5, with a new vertex $\hat{P}$ still on the same line at 45°, and with sides parallel to those of the original region.

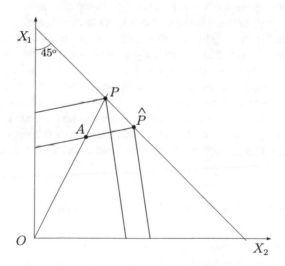

Fig. 4.5 Region of admissible braking combinations for two different weight distributions

4.10 A Numerical Example

A numerical example may be useful to understand better the braking performance of a road car. We take a small car with the following features: mass $m = 1000\,\text{kg}$, wheelbase $l = 2.4\,\text{m}$, $a_1 = a_2 = l/2$, height of the center of mass $h = 0.5\,\text{m}$.

Assuming a grip coefficient $\mu = 0.8$, the maximum deceleration is vehicle independent and it is equal to $|\dot{u}|_{max} = \mu g = 7.84\,\text{m/s}^2$, with $g = 9.81\,\text{m/s}^2$.

According to (4.7), the static vertical loads for both axles are $Z_1^0 = Z_2^0 = 4900\,\text{N}$. The load transfer at maximum deceleration is $\Delta Z = 1633\,\text{N}$. Therefore, the vertical loads acting on each axle are $Z_{1_P} = 6533\,\text{N}$ and $Z_{2_P} = 3267\,\text{N}$, which means a brake balance $\beta_P = 2$. This is the optimal value for that car if $\mu = 0.8$.

Should the grip coefficient drop to 0.4 because, e.g., of rain, we would end up with a braking efficiency $\varepsilon_1 = 0.86$. An increase of the grip coefficient up to 1.2 would still bring a reduced braking efficiency $\varepsilon_2 = 0.86$.

4.11 Braking, Stopping, and Safe Distances

The braking distance refers to the distance a vehicle will travel from the point when its brakes are fully applied to when it comes to a complete stop.

The total stopping distance is the sum of the perception-reaction distance and the braking distance. The perception-reaction time ranges from 0.75 to 1.5 s.

In everyday traffic, the driver must keep a safe distance between his/her vehicle and the vehicle in front in order to avoid collision if the car in front brakes or stops. The safe distance corresponds to the distance covered by the vehicles in the perception-reaction time. This is the rationale for the *three-second rule*, by which a driver can easily maintain a safe trailing distance at any speed. The rule is that a driver should ideally stay at least three seconds behind any vehicle that is directly in front. Of course, it can be applied at any speed and with any weather condition.

4.12 Braking Performance of Formula Cars

Formula cars have aerodynamic devices that provide very high downforces at high speed, as briefly explained in Sect. 3.7.2. These loads affect braking pretty much. The first, obvious, effect is that the maximum longitudinal deceleration is speed dependent. In a Formula 1 car it can be up to 5 g at 350 km/h, although the physical grip μ rarely exceeds 1.6. The second, perhaps less obvious, effect is that also the optimal brake balance β_P is speed dependent.

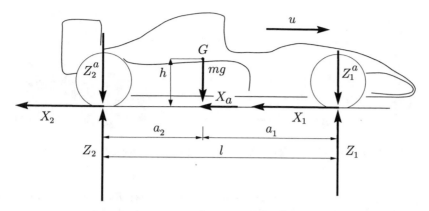

Fig. 4.6 Vehicle model for braking performance of a Formula car (all forces are positive)

4.12.1 Equilibrium Equations

The equilibrium equations (4.4) must be supplemented by the aerodynamic loads. According to Sect. 3.7.2 and as shown in Figs. 3.22 and 4.6, the total aerodynamic force $\mathbf{F}_a$ is equivalent to three forces: a drag force X_a at road level and two vertical forces Z_1^a and Z_2^a acting directly on the front and rear axles, respectively. Therefore, the equilibrium equations become

$$m\dot{u} = -(X_1 + X_2) - X_a$$
$$0 = Z_1 + Z_2 - mg - Z_1^a - Z_2^a \qquad (4.25)$$
$$0 = (X_1 + X_2 + X_a)h - (Z_1 - Z_1^a)a_1 + (Z_2 - Z_2^a)a_2$$

Unlike in (3.94) and (3.95), here we assume X_1 and X_2 to be positive if directed like in Fig. 4.6, that is to be indeed braking forces. As shown in Fig. 2.43, the tire rolling resistance is part of the braking (grip) forces X_i.

We recall that (cf. (3.81) and (3.82))

$$X_a = \frac{1}{2}\rho_a S_a C_x u^2 = \xi u^2$$
$$Z_1^a = \frac{1}{2}\rho_a S_a C_{z1} u^2 = \zeta_1 u^2 \qquad (4.26)$$
$$Z_2^a = \frac{1}{2}\rho_a S_a C_{z2} u^2 = \zeta_2 u^2$$

where, as it is common practice among race engineers, $C_x > 0$ and $C_{zi} > 0$.

For simplicity, here we assume $S_a C_x$, $S_a C_{z1}$ and $S_a C_{z2}$ to be constant (i.e., speed independent). Actually, this is not strictly true, as shown in Fig. 4.7, because the

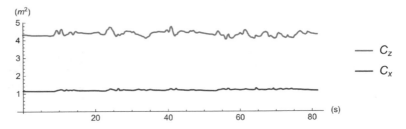

Fig. 4.7 Measured values of $S_a C_z$ and $S_a C_x$ in one lap of an F1 race car

height from the ground of the car is not constant. Therefore, the assumption of constant coefficients should be removed in more advanced analyses.

4.12.2   Vertical Loads

The vertical loads on each axle are given by the static loads (4.7) (zero speed), plus the aerodynamic (speed dependent) loads (4.26), plus or minus the inertial longitudinal load transfer (cf. (4.8))

$$Z_1 = Z_1^0 + \zeta_1 u^2 + \Delta Z$$
$$Z_2 = Z_2^0 + \zeta_2 u^2 - \Delta Z$$
(4.27)

Like in (4.9), the *inertial* longitudinal load transfer ΔZ is given by

$$\Delta Z = -\frac{mh}{l}\dot{u}$$
(4.28)

with $\dot{u} < 0$. When braking, the front axle is subject to a higher load, while the rear axle to a lower load, with respect to the static loads. It is a purely inertial effect. However, at high speed the drag force X_a is not negligible and can significantly affect the vertical loads, even if $C_z = 0$ as shown in Fig. 3.24.

4.12.3   Maximum Deceleration

The maximum deceleration is promptly obtained by assuming that both axles are at their limit braking conditions, that is $X_1 = \mu Z_1$ and $X_2 = \mu Z_2$

$$|\dot{u}|_{\max} = \mu\left(g + \frac{\zeta_1 + \zeta_2}{m}u^2\right) + \frac{\xi}{m}u^2$$
(4.29)

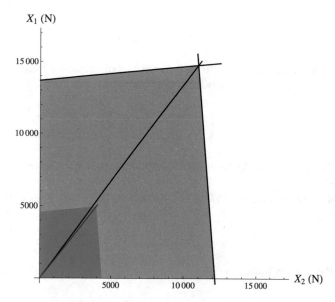

Fig. 4.8 Regions of all admissible braking combinations of a Formula car at 200 km/h, with and without aerodynamic downforces

This formula generalizes (4.14). Of course $|\dot{u}|_{\max}$ is very speed dependent, as also shown in Fig. 4.8.

More detailed information can be obtained by looking at (4.29) as a differential equation in the unknown function $u(t)$

$$\dot{u} = -\mu g - \frac{u^2}{k} \qquad (4.30)$$

where

$$k = \frac{m}{\mu(\zeta_1 + \zeta_2) + \xi} \qquad (4.31)$$

Setting the initial speed $u(0) = u_0$, the analytical solution is as follows

$$u(t) = u_d \tan\left(\arctan\left(\frac{u_0}{u_d}\right) - \frac{t}{t_d}\right) \qquad (4.32)$$

where

$$u_d = \sqrt{\mu g k} \quad \text{and} \quad t_d = \sqrt{\frac{k}{\mu g}} \qquad (4.33)$$

Moreover, to have $u(t) > 0$, it must be $t < t_0$, with

$$t_0 = t_d \arctan \left(\frac{u_0}{u_d} \right) \tag{4.34}$$

Therefore, t_0 is the shortest time to stop the car from speed u_0. Integrating (4.32), we can obtain the distance s_0 travelled by the car to come to a stop

$$s_b = \frac{1}{2} t_d u_d \ln \left(1 + \frac{u_0^2}{u_d^2} \right) = \frac{k}{2} \ln \left(1 + \frac{u_0^2}{\mu g k} \right) \tag{4.35}$$

It is worth comparing this equation with its counterpart (4.52) for cars without aerodynamic devices.

A plot of (4.35) is available in Fig. 4.12 (lower curve), along with a plot of (4.52) (upper curve).

4.12.4   Brake Balance

To brake at the best braking performance, that is with $\dot{u} = -|\dot{u}|_{\max}$, the longitudinal forces must be

$$
\begin{aligned}
X_{1_P} &= \mu \left(Z_1^0 + \zeta_1 u^2 + \frac{mh}{l} |\dot{u}|_{\max} \right) \\
&= \mu \left(Z_1^0 + \zeta_1 u^2 + \frac{h}{l} [\mu g m + \mu (\zeta_1 + \zeta_2) u^2 + \xi u^2] \right) \\
X_{2_P} &= \mu \left(Z_2^0 + \zeta_2 u^2 - \frac{mh}{l} |\dot{u}|_{\max} \right) \\
&= \mu \left(Z_2^0 + \zeta_2 u^2 - \frac{h}{l} [\mu g m + \mu (\zeta_1 + \zeta_2) u^2 + \xi u^2] \right)
\end{aligned} \tag{4.36}
$$

Having the right brake balance is very important for lap performance. The *optimal brake balance* (or brake bias) β_P to have the *best* braking performance is promptly obtained as

$$\beta_P(u) = \frac{X_{1_P}}{X_{2_P}} = \frac{(a_2 + h\mu)gm + u^2[(a_1 + a_2)\zeta_1 + h\xi + h(\zeta_1 + \zeta_2)\mu]}{(a_1 - h\mu)gm + u^2[(a_1 + a_2)\zeta_2 - h\xi - h(\zeta_1 + \zeta_2)\mu]} \tag{4.37}$$

which generalizes (4.16). As expected, in general now β_P is explicitly speed dependent.

Fig. 4.9 The line of action
of the global aerodynamic
force must pass through G to
have speed independent
brake balance

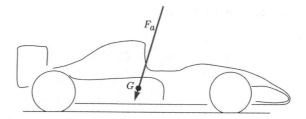

4.12.5   Speed Independent Brake Balance

To avoid explicit speed dependence of β_P, and hence to enhance the lap performance,
it must be in (4.37)

$$
\beta_P = \frac{a_2 + h\mu}{a_1 - h\mu} = \frac{(a_1 + a_2)\zeta_1 + h\xi + h(\zeta_1 + \zeta_2)\mu}{(a_1 + a_2)\zeta_2 - h\xi - h(\zeta_1 + \zeta_2)\mu} \tag{4.38}
$$

that is, with $C_z = C_{z1} + C_{z2}$

$$
\beta_P = \frac{a_2 + h\mu}{a_1 - h\mu} = \frac{(a_1 + a_2)C_{z1} + h(C_x + C_z\mu)}{(a_1 + a_2)C_{z2} - h(C_x + C_z\mu)} \tag{4.39}
$$

This condition should be taken into account during setup (see also the next two
sections).

Interestingly enough, there is a simple physical interpretation of (4.38) and (4.39):
the line of action of the global aerodynamic force $\mathbf{F}_a = -\xi u^2\, \mathbf{i} - (\zeta_1 + \zeta_2)u^2\, \mathbf{k}$
must pass through the center of mass G, as shown in Fig. 4.9 and as discussed
in Sect. 4.13.4.

4.12.6   Practical Brake Balance

In addition to the brake balance $\beta = X_1/X_2$, we define, as commonly done by race
engineers, the *practical brake balance* η

$$
\eta = \frac{X_1}{X_1 + X_2} \tag{4.40}
$$

along with the *weight distribution* ω

$$
\omega = \frac{Z_1^0}{Z_1^0 + Z_2^0} = \frac{a_2}{a_1 + a_2} \tag{4.41}
$$

and the *aero balance* α

$$\alpha = \frac{C_{z_1}}{C_{z_1} + C_{z_2}} = \frac{\zeta_1}{\zeta_1 + \zeta_2} = \frac{\zeta_1}{\zeta} \tag{4.42}$$

where

$$\zeta = \zeta_1 + \zeta_2 \tag{4.43}$$

We obtain the following result[2] for the *optimal* practical brake balance η_P

$$\eta_P = \frac{mg(\mu\frac{h}{l} + \omega) + \zeta u^2(\mu\frac{h}{l} + \alpha) + \xi u^2 \frac{h}{l}}{mg + \zeta u^2} \tag{4.44}$$

which is the counterpart of (4.37). Using η_P or β_P is just a matter of taste.
   In case of different front to rear grip ($\mu_1 \neq \mu_2$), we have[3]

$$\eta_P = \frac{\mu_1 \left[mg(\mu_2\frac{h}{l} + \omega) + \zeta u^2(\mu_2\frac{h}{l} + \alpha) + \xi u^2\frac{h}{l} \right]}{mg[\mu_1\omega + \mu_2(1 - \omega)] + \zeta u^2[\mu_1\alpha + \mu_2(1 - \alpha)] + \xi u^2\frac{h}{l}(\mu_1 - \mu_2)}$$
$$\tag{4.45}$$

4.12.7   Speed Independent Practical Brake Balance

Of course, in general, η_P is speed dependent. To avoid speed dependence of η_P, it must be $\partial \eta_p / \partial u = 0$, that means (see also Fig. 4.9)

$$(\omega - \alpha)\zeta - \frac{h}{l}\xi = 0 \tag{4.46}$$

or, equivalently

$$(\omega - \alpha)C_z - \frac{h}{l}C_x = 0 \tag{4.47}$$

which can be rewritten as

$$\omega = \alpha + \frac{C_x}{C_z}\frac{h}{l} \tag{4.48}$$

Quite a compact and interesting formula. We see that, to avoid speed dependence of η_p, it is necessary that $\omega > \alpha$, but just a little. For instance, in a Formula car, it results in $\omega - \alpha \simeq 0.02$.
   The last three equations are the counterpart of (4.39). However, they look simpler to be kept in mind.

---

[2] Ernesto Desiderio, personal communication, 6 May 2020.
[3] Federico Sánchez Motellón, personal communication, 22 June 2020.

4.12.8 Sensitivities

From (4.44) we can easily compute the sensitivities of η_P.

The sensitivity of the optimal practical brake balance η_p with respect to the aero balance α is

$$\frac{\partial \eta_P}{\partial \alpha} = \frac{\zeta u^2}{mg + \zeta u^2} \tag{4.49}$$

Very simple formula. No μ, no h/l, no C_x. Strong speed dependence at high speed.

Similarly, the sensitivity of the optimal practical brake balance η_p with respect to the weight distribution ω is

$$\frac{\partial \eta_P}{\partial \omega} = \frac{mg}{mg + \zeta u^2} \tag{4.50}$$

Again, a very simple formula. No μ, no h/l, no C_x. Strong speed dependence at low speed.

It may be interesting to observe that

$$\frac{\partial \eta_P}{\partial \alpha} + \frac{\partial \eta_P}{\partial \omega} = 1 \tag{4.51}$$

4.12.9 Typical F1 Braking Performance

A typical braking performance of an F1 car is shown in Fig. 4.10. The deceleration grows suddenly up to about 38 m/s^2. Then, as the speed u (m/s) decreases, also the aerodynamic load decreases, thus requiring the driver to gradually release the brake pedal. Meanwhile, the car is already negotiating the curve, as shown by the lateral acceleration and wheel steer angle (deg). Also shown in Fig. 4.10 is the acceleration ($a_x > 0$) when the car exits the curve.

It is interesting to compare the total acceleration $\sqrt{a_x^2 + a_y^2}$ (lower line in Fig. 4.11) with the potential maximum deceleration (4.29) (upper line in Fig. 4.11). Whenever possible, the driver tries to stay as close as possible to the limit. This can be done in all curves that are grip-limited. Of course, not in those curves that are speed-limited (like, e.g., curve 3 in the Barcelona circuit).

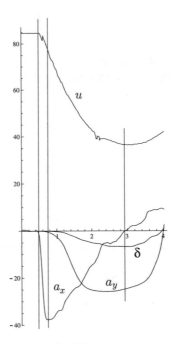

Fig. 4.10 Typical braking performance of an F1 car

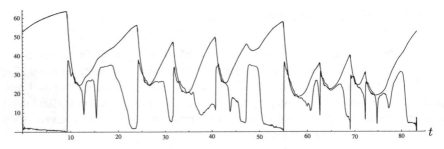

Fig. 4.11 Comparison between the total acceleration (lower line) and the potential maximum acceleration (upper line) of an F1 car

4.13   Exercises

4.13.1   Minimum Braking Distance

For the road car described in Sect. 4.10, *compute the minimum braking distance assuming the following data:*

- *grip coefficient $\mu = 0.8$ (dry asphalt);*
- *initial speed $u_0 = 100$ km/h;*
- *braking efficiency $\varepsilon = 1$.*

Then, for the sake of comparison, repeat the same calculation in case the car has only the front brakes, and then in case the car has only the rear brakes.

Solution

First we convert the initial speed in SI units: $u_0 = 100/3.6 = 27.8\,\text{m/s}$.

In our model, the maximum deceleration is equal to $\mu g = 7.85\,\text{m/s}^2$. Therefore, it is not affected by the position of G and by the mass m.

We know by elementary physics that the speed decreases from u_0 to zero according to $u(t) = u_0 - \mu g t$. Therefore, we obtain the braking time $t_b = u_0/(\mu g) = 3.54\,\text{s}$, which is a linear function of the initial speed.

We can now compute the distance covered by the car to come to a stop

$$s_b = \frac{1}{2}\mu g t_b^2 = \frac{u_0^2}{2\mu g} = 49\,\text{m} \tag{4.52}$$

Of course, to get this minimum distance, the brake balance β_P must be set according to (4.16). In this case $\beta_P = 2$, as shown in Sect. 4.10.

It can be of some interest to compare this expression of s_b for road cars with (4.35) for Formula cars.

The braking distance can also be found by determining the work required to dissipate the vehicle kinetic energy, that is $0.5mu_0^2 = m\mu g s_b$. Of course, the result is the same, but without the byproduct of the braking time t_b.

As well known, the braking distance of a road car is a quadratic function of the initial speed u_0. Doubling the speed makes the braking distance four times longer.

If braking with the front wheels only, we can at most get the braking force X_{1_0} given by (4.22). We see that now the position of G becomes relevant. With a bit of algebra, we obtain that in this case the maximum deceleration is

$$a_f = \mu g \frac{a_2}{l}\left(\frac{1}{1-\mu\dfrac{h}{l}}\right) = \mu g \frac{a_2}{l-\mu h} = 4.71\,\text{m/s}^2 \tag{4.53}$$

The braking distance is therefore given by $7.85/4.71 \times 49 = 82\,\text{m}$, that is

$$s_f = \frac{u_0^2}{2a_f} \tag{4.54}$$

Similarly, but employing (4.21), we obtain that in case of rear braking only the deceleration is

$$a_r = \mu g \frac{a_1}{l}\left(\frac{1}{1+\mu\dfrac{h}{l}}\right) = \mu g \frac{a_1}{l+\mu h} = 3.36\,\text{m/s}^2 \tag{4.55}$$

and the braking distance is $7.85/3.36 \times 49 = 114$ m. As expected, front braking only is more efficient than rear braking only. This is particularly true in bicycles and motorcycles. Try to guess why.

4.13.2 Braking with Aerodynamic Downforces

A GP2 race car has the following features (notation as in Sect. 3.7.2):

- $m = 680$ kg;
- $l = a_1 + a_2 = 3.025$ m;
- $a_1/a_2 = 1.27$, *that is weight distribution front/rear of 0.44/0.56;*
- $S_a C_{z1} = 1.5$ m^2;
- $S_a C_{z2} = 2.1$ m^2;
- $S_a C_x = 1.1$ m^2;
- $\mu = 1.35$;
- $h = 0.27$ m;
- *air density* 1.25 kg/m^3.

Compute the minimum braking distance and the minimum braking time when it is running straight at 150 km/h *and at 300* km/h.

Solution

In this race car we have (see (4.26))

- $\zeta_1 = 0.5 \times 1.25 \times 1.5 = 0.9375$ kg/m
- $\zeta_2 = 0.5 \times 1.25 \times 2.1 = 1.3125$ kg/m
- $\xi = 0.5 \times 1.25 \times 1.1 = 0.6875$ kg/m

Therefore, in (4.32) we obtain $u_d = 49.17$ m/s and $t_d = 3.71$ s.

If the initial speed is $150/3.6 = 41.67$ m/s, we obtain from (4.34) the minimum braking time $t_0 = 2.61$ s. According to (4.29), the highest deceleration is 22.75 m/s^2. The braking distance is $s_b = 49.4$ m. It is obtained integrating numerically (7.237) from 0 to 2.61 s.

If the initial speed is $300/3.6 = 83.33$ m/s, we obtain from (4.34) the minimum braking time $t_0 = 3.85$ s. According to (4.29), the highest deceleration is 51.28 m/s^2. The braking distance is $s_b = 123.6$ m. It is obtained integrating numerically (4.32) from 0 to 3.85 s.

Just out of curiosity, this car would stop in about 25 m if running at 100 km/h. The time would be less than 2 s.

From Fig. 4.12 we can appreciate how important the aerodynamic loads are in Formula cars. The braking distances with all aerodynamic forces, with drag but no downforces, with no aerodynamics at all, are quite far apart at high speeds.

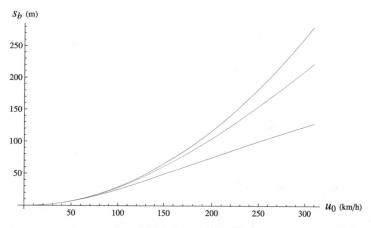

Fig. 4.12 Comparison between the braking distance of a GP2 car (lower curve), the braking distance for the same car, but without any aerodynamic effect (upper curve), the braking distance with drag, but no downforces (intermediate curve)

4.13.3 GP2 Brake Balance

The brake distances computed in the former exercise need a perfect brake balance β_P at any speed. Compute the value of the perfect brake balance for the same GP2 car at 100, 150 and 300 km/h, and comment on it.
Solution

First of all, let us test whether this car fulfills (4.39), which would make β_P speed insensitive. The l.h.s. term makes 1.275, while the r.h.s. term makes 1.295. Very good.

Indeed, we have $\beta_P = 1.280$ at 100 km/h, $\beta_P = 1.283$ at 150 km/h, and $\beta_P = 1.290$ at 300 km/h. We see that this car has a brake balance which is almost speed independent.

4.13.4 Speed Independent Brake Balance

Check the physical interpretation of (4.38).

Solution

The physical interpretation of (4.38) requires the global aerodynamic force $\mathbf{F}_a = -(X_a \mathbf{i} + Z_a \mathbf{k})$ to pass through the center of mass G (Fig. 4.9). Therefore, we have to solve the following system of equations (cf. (4.25))

$$m\dot{u} = -(X_1 + X_2) - X_a$$

$$0 = Z_1 + Z_2 - mg - Z_a$$

$$0 = (X_1 + X_2)h - Z_1 a_1 + Z_2 a_2 \tag{4.56}$$

$$X_1 = \mu_1 Z_1$$

$$X_2 = \mu_2 Z_2$$

where, for greater generality, the front grip μ_1 is not necessarily equal to the rear grip μ_2.

The resulting brake balance is

$$\beta_P = \frac{X_1}{X_2} = \frac{\mu_1(a_2 + h\mu_2)}{\mu_2(a_1 - h\mu_1)} \tag{4.57}$$

which is, indeed, speed independent, and generalizes (4.38). Quite a useful result to optimize the braking performance of a Formula car.

4.14 Summary

The goal of this chapter has been to understand how to stop a vehicle as soon as possible, avoiding wheel locking. This result can be achieved only if the vehicle has the right brake balance. Unfortunately, brake balance is affected by the value of the grip and by the position of the center of mass. This topic has been addressed in detail, both analytically and graphically, through the region of all possible braking conditions. The peculiarity of the braking performance of a Formula car has been also discussed.

4.15 List of Some Relevant Concepts

Section 4.4—the longitudinal load transfer does not depend on the type of suspensions;
Section 4.5—maximum deceleration is limited by either grip or overturning (supposing brakes are powerful enough);
Section 4.6—brake balance depends on grip and weight distribution;
Section 4.7—all possible braking combinations can be represented by a simple figure;
Section 4.12.2—wings do not affect load transfer directly;
Section 4.12.4—brake balance is affected by wings;
Section 4.12.5—the line of action of the global aerodynamic force must pass through G to have speed independent brake balance.

4.16 Key Symbols

a_1	distance of G from the front axle
a_2	distance of G from the rear axle
C_x, C_{z_i}	aerodynamic coefficients
g	gravitational acceleration
G	center of mass
h	height of G
J_y	moment of inertia
l	wheelbase
m	mass
S_a	frontal area
u	longitudinal velocity
$\dot{u}$	longitudinal acceleration
X_i	braking force acting on the i-th axle
Z_i	vertical load on i-th axle
Z_i^0	static vertical load on i-th axle
Z_i^a	aerodynamic vertical load on i-th axle
ΔZ	longitudinal load transfer

α	aero balance
β	brake balance
β_P	optimal brake balance
ε_i	braking efficiency
η	practical brake balance
η_P	optimal practical brake balance
$\mu = \mu_p^x$	coefficient of grip
ρ_a	air density
ω	weight distribution

References

1. Heißing B, Ersoy M (eds) (2011) Chassis handbook. Springer, Wiesbaden
2. Savaresi SM, Tanelli M (2010) Active braking control systems design for vehicles. Springer, London

Chapter 5
The Kinematics of Cornering

Cars have to negotiate corners. Everybody knows that. But not all cars do that the same way [4]. This is particularly evident in race cars, where the ability to negotiate a corner is a crucial aspect to minimize lap time.

In this chapter we will exploit the kinematics of a vehicle while taking a corner. At first sight, taking a corner looks quite a trivial task. But designing a vehicle that does it properly is one of the main challenges faced by a vehicle engineer [2]. Therefore, there is the need to investigate what really happens during the cornering process. It will be shown that some very significant kinematical quantities must follow precise patterns for the car to get around corners in a way that makes the driver happy. In some sense, the geometric features of the trajectory must adhere to some pretty neat criteria.

Before digging into the somehow mysterious kinematics of cornering, we will recall some kinematical concepts. Strangely enough, it appears that they have never been employed before in vehicle dynamics, although all of them date back to Euler or so.

5.1 Planar Kinematics of a Rigid Body

As discussed at the beginning of Chap. 3, in many cases a vehicle can be seen as a rigid body in planar motion. Basically, we need a flat road and small roll angles. The congruence (kinematic) equations for this case were given in Sect. 3.2. We will extensively use the symbols defined therein.

Here we recall some fundamental concepts of planar kinematics of a rigid body [1, 3, 6]. They will turn out to be very useful to understand how a car takes a corner.

The original version of this chapter was revised. Legends of figures 5.21 and 5.22 were updated. The correction to this chapter is available at https://doi.org/10.1007/978-3-031-06461-6_13

M. Guiggiani, *The Science of Vehicle Dynamics*, https://doi.org/10.1007/978-3-031-06461-6_5

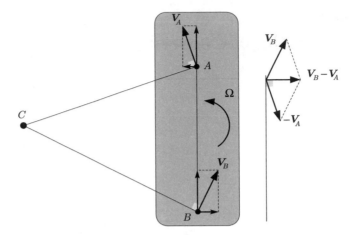

Fig. 5.1 Relationship between the velocities of two points of the same rigid body in planar motion

5.1.1 Velocity Field and Velocity Center

In a rigid body, by definition, the distance between any two points is constant. Accordingly, taken two such points, say A and B, their velocities must have the same component along the direction AB, as shown in Fig. 5.1. More precisely, the two velocities are related by the following equation

$$\mathbf{V}_B = \mathbf{V}_A + \mathbf{\Omega} \times AB = \mathbf{V}_A + \mathbf{V}_{BA} \tag{5.1}$$

where $\mathbf{\Omega}$ is the angular velocity. This is the fundamental equation of the kinematics of rigid bodies, planar or three-dimensional. It had been already given in (2.1) and (3.3).

It is worth noting that $\mathbf{\Omega}$ is the same for all points. It is a kinematic feature of the rigid body as a whole.

Another way to state the fundamental equation (5.1) is saying that the relative velocity $\mathbf{V}_{BA} = \mathbf{V}_B - \mathbf{V}_A$ is orthogonal to the segment AB and proportional to the length of AB, that is $|\mathbf{V}_{BA}| = |\mathbf{\Omega}||AB|$ (Fig. 5.1).

It can be shown [1, 3, 6] that in case of planar motion, that is $\mathbf{\Omega} = r\,\mathbf{k}$, and with $r \neq 0$, at any instant there is one point C of the (extended) rigid body that has zero velocity. Therefore, applying (5.1) to A and C, and then to B and C we have

$$\mathbf{V}_A = r\,\mathbf{k} \times CA \quad \text{and} \quad \mathbf{V}_B = r\,\mathbf{k} \times CB \tag{5.2}$$

as shown in Fig. 5.1.

Several different names are commonly in use to refer to point C:

- instantaneous center of velocity;
- velocity center;

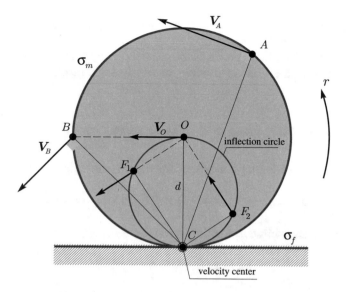

Fig. 5.2   Velocity field of a rigid wheel σ_m rolling on a flat road σ_f

- instantaneous center of zero velocity;
- instantaneous center of rotation.

5.1.2   *Fixed and Moving Centrodes*

As the body changes its position, the point of the rigid body with zero velocity changes as well. If we follow the positions of this sequence of points we obtain a curve σ_f in the fixed plane, called the *fixed centrode* or space centrode, and another curve σ_m on the moving plane, called the *moving centrode* or the body centrode. It can be shown that the moving centrode rolls without slipping on the fixed centrode, the point of rolling contact being C.

A simple example should help clarify the matter. Just consider a rigid circle rolling without slipping on a straight line, as shown in Fig. 5.2. It is exactly like a rigid wheel rolling on a flat road. The two centrodes are the circle σ_m and the straight line σ_f. Point C as a point of the circle has zero velocity. However, the geometric point[1] $\hat{C}$ that at each instant coincides with C moves on the road with a speed

$$V_{\hat{C}} = rd \tag{5.3}$$

---

[1] By geometric point we mean a point not belonging to the rigid body.

where d is the diameter of the inflection circle (already defined in Sect. 3.2.10 and to be discussed in detail in Sect. 5.1.4).

The velocity field is like a pure rotation around C (Fig. 5.1). But the acceleration field is not! In fact, the wheel is travelling on the road, not turning around C.

5.1.3   Acceleration Field and Acceleration Center

The counterpart of (5.1) for the accelerations of points of a rigid body is

$$\mathbf{a}_B = \mathbf{a}_A + \dot{\boldsymbol{\Omega}} \times AB + \boldsymbol{\Omega} \times (\boldsymbol{\Omega} \times AB) = \mathbf{a}_A + \mathbf{a}_{BA} \qquad (5.4)$$

In case of planar motion it simplifies into (Fig. 5.3)

$$\mathbf{a}_B = \mathbf{a}_A + \dot{r}\,\mathbf{k} \times AB - r^2 AB \qquad (5.5)$$

The relative acceleration $\mathbf{a}_{BA} = \mathbf{a}_B - \mathbf{a}_A$ between any two points is proportional to the length $|AB|$ and forms an angle ξ with the segment AB (Fig. 5.3)

$$\tan \xi = \frac{\dot{r}}{r^2} \qquad (5.6)$$

As discussed in Sect. 3.2.9, it can be shown that in case of planar motion, that is $\boldsymbol{\Omega} = r\,\mathbf{k}$, and with $r \neq 0$, at any instant there is one point K of the (extended) rigid body that has zero acceleration. In general, $K \neq C$. The absolute acceleration of any point A forms an angle ξ with the segment KA, as shown in Fig. 5.3. Therefore, the acceleration field is like a pure rotation around K.

Several different names are commonly in use to refer to point K:

• instantaneous center of acceleration;
• acceleration center;
• instantaneous center of zero acceleration.

The velocity and acceleration fields are superimposed in Fig. 5.4.

5.1.4   Inflection Circle and Radii of Curvature

Let us consider again, as an example, the rigid wheel rolling on a flat road. For the moment let us also assume that it rolls at constant speed. The center O of the wheel has zero acceleration, and hence it is the acceleration center K, as shown in Fig. 5.5. The acceleration field is centripetal towards $O = K$. It is worth noting that the acceleration of C is not zero

$$\mathbf{a}_C = \mathbf{n} r^2 d \qquad (5.7)$$

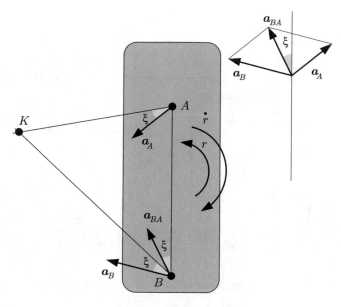

Fig. 5.3 Relationship between the acceleration of two points of the same rigid body in planar motion

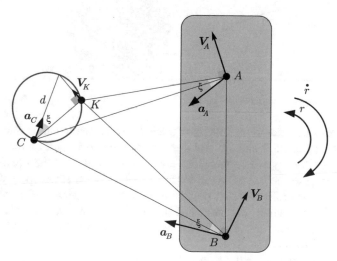

Fig. 5.4 Velocity center, acceleration center, and inflection circle

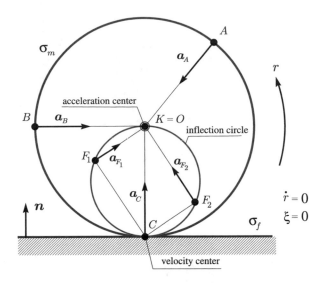

Fig. 5.5 Acceleration field of a rigid wheel rolling at *constant speed* on a flat road

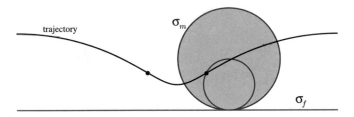

Fig. 5.6 Trajectory with two inflection points

where d is the diameter of the inflection circle (already mentioned in Sect. 3.2.10).

Comparing Figs. 5.2 and 5.5, we see that at a given instant of time there are points, like F_1 and F_2, whose velocities and accelerations have the same direction. They all belong to the inflection circle [5, Sect. 4.5]. Even if we apply an angular acceleration $\dot{r}$, as in Fig. 5.7, the points on the inflection circle still have collinear velocity and acceleration. The points of the rigid body on the inflection circle, as the name implies, have a trajectory with an *inflection point*, that is a point with zero curvature, as shown in Fig. 5.6.

Point C has a nice property: its acceleration is not affected by $\dot{r}$. In other words, Eq. (5.7) holds true even if $\dot{r} \neq 0$. Therefore, it is possible to obtain the diameter d of the inflection circle from the knowledge of $\mathbf{a}_C$ and r.

The inflection circle turns out to be very useful to evaluate the *radius of curvature* of the trajectory of *any* point of the rigid body. The rule is very simple, and it is exemplified in Fig. 5.8. Let us take, for instance, point A. The center of curvature E_A of its trajectory must fulfill the following relationship

$$|AC|^2 = |AE_A||AF_A| \quad \text{or, more compactly} \quad a^2 = ef \qquad (5.8)$$

Fig. 5.7 Acceleration field of a rigid wheel rolling at *non-constant speed* on a flat road

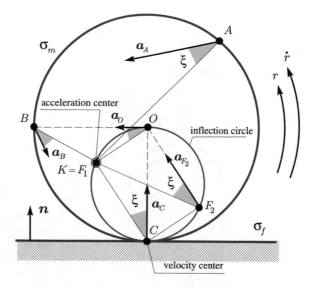

where E_A and F_A are always on the same side with respect to A (this is why point A is always first in the three terms in (5.8)). As a consequence, points E_A and F_A are always on opposite sides with respect to C. The distance $R_A = |AE_A|$ is the radius of curvature of the trajectory of A.

Exactly the same rules apply to any other point, like point B in Fig. 5.8.

Quite interestingly, we can obtain the following formula for the centripetal (normal) component of the acceleration of A

$$a_A^n = \frac{V_A^2}{|AE_A|} = \frac{(ra)^2}{e} = r^2 f = r^2|AF_A| \tag{5.9}$$

The same kind of formula applies to any other point of the rigid body.

5.2 The Kinematics of a Turning Vehicle

Driving a vehicle to make a turn amounts, roughly speaking, to forcing it to follow a path with *variable radius of curvature*. The traditional approach looks only at the kinematics for a given instant of time, as shown in Fig. 5.9. This is a good starting point, but not the whole story. For instance, from Fig. 5.9 we cannot know the radius of curvature of the trajectory of G (which, of course, is not equal to CG, in general). But let us make the reasoning more precise.

A vehicle has infinitely many points and hence infinitely many trajectories. However, as a rigid body, these trajectories are not independent of each other. It suffices to look at the trajectory (path) of two points. It is perhaps advisable to select the

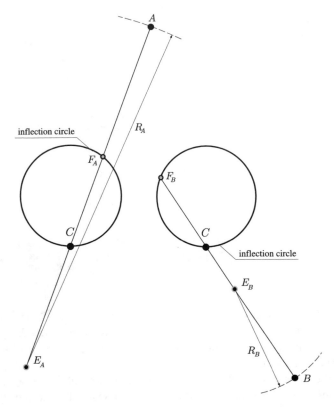

Fig. 5.8 How the inflection circle relates to the centers of curvature of the trajectories of the points of a rigid body

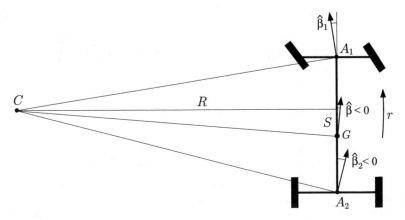

Fig. 5.9 Definition of front slip angle $\hat{\beta}_1$ and rear slip angle $\hat{\beta}_2$ for a turning vehicle

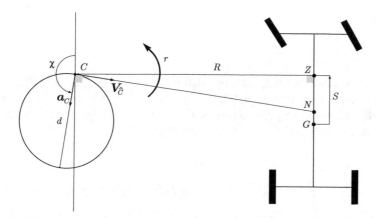

Fig. 5.10 Inflection circle and definition of some relevant quantities

midpoint A_1 of the front axle and the midpoint A_2 of the rear axle (Fig. 5.9). It is not mandatory at all, but maybe convenient.

Looking at the trajectories also implies monitoring the radii of curvature and how they relate to each other.

To monitor whether a vehicle is performing well, or not so well, we can consider also the *fixed* and *moving centrodes*, along with the *inflection circle* (Fig. 5.10). Indeed, we should have clear in mind that the position of the velocity center C changes continuously in time, thus tracing the two centrodes. Therefore, the two centrodes "contain" all the geometric features of the kinematics of the turning vehicle.

However, the centrodes of a vehicle are "built" by the vehicle itself, under the driver actions. In some sense, a vehicle can be seen as a *centrode builder*. It is not like in Fig. 5.2, or in single-degree-of-freedom planar linkages, where the centrodes are completely determined and cannot be modified.

5.2.1 Moving and Fixed Centrodes of a Turning Vehicle

By definition, the centrodes are generated by the successive positions of the velocity center C.

The *moving centrode* σ_m is given by the successive positions of C in the body-fixed reference system $\mathsf{S} = (x, y, z; G)$, that is with respect to the vehicle. As already obtained in (3.11), the position of C with respect to the vehicle is given by (Fig. 5.9)

$$\mathbf{D} = GC = S\,\mathbf{i} + R\,\mathbf{j} \tag{5.10}$$

where, as usual, $S = -v/r$ and $R = u/r$. More explicitly, the parametric equations of the moving centrode in S are

$$x_m(\hat{t}) = S(\hat{t})$$
$$y_m(\hat{t}) = R(\hat{t})$$

(5.11)

where we use $\hat{t}$, instead of t, to remark that it is a parameter (like in (3.9)).

The parametric equations $(x_f(\hat{t}), y_f(\hat{t}))$ of the *fixed centrode* σ_f in the ground-fixed reference system S_0 can be obtained from the knowledge of the absolute coordinates of G, given in (3.9), and of the yaw angle (3.8)

$$x_f(\hat{t}) = x_0^G(\hat{t}) + S(\hat{t}) \cos \psi(\hat{t}) - R(\hat{t}) \sin \psi(\hat{t})$$
$$y_f(\hat{t}) = y_0^G(\hat{t}) + S(\hat{t}) \sin \psi(\hat{t}) + R(\hat{t}) \cos \psi(\hat{t})$$

(5.12)

By definition, the vehicle belongs precisely to the same rigid plane of the moving centrode. They move together. Therefore, the parametric equations of the *moving centrode*, at time t, in the ground-fixed reference system are

$$x_m^f(t, \hat{t}) = x_0^G(t) + S(\hat{t}) \cos \psi(t) - R(\hat{t}) \sin \psi(t)$$
$$y_m^f(t, \hat{t}) = y_0^G(t) + S(\hat{t}) \sin \psi(t) + R(\hat{t}) \cos \psi(t)$$

(5.13)

Again, the parameter to draw the moving centrode is $\hat{t}$, while t sets the instant of time.

The typical shape of the fixed and moving centrodes of a vehicle making a turn are shown in Figs. 5.11 and 5.12. We see that the moving centrode σ_m is pretty much a straight line, while the fixed centrode σ_f is made of two distinct parts, as is the kinematics of turning: entering the curve and exiting the curve. The velocity center C is the point of rolling contact of the two centrodes.

Actually, the centrodes shown in Figs. 5.11 and 5.12 are typical of a vehicle making a curve the good way. The centrodes changes abruptly if the vehicle does not make the curve properly. This may happen, e.g., if the speed is too high. An example of "bad" centrodes, and hence of bad performance, is shown in Fig. 5.13. We see that the centrodes for the exiting phase (Fig. 5.13c) are totally different from those in Fig. 5.12. The vehicle spins out.

Quite interestingly, as shown in Fig. 5.13b, the two centrodes start having a bad shape although the vehicle still has an apparent good behavior. Therefore, the two centrodes could be used as a warning of handling misbehavior. They depart from the proper shape a little before the vehicle shows unwanted behavior.

To confirm that this is real stuff, we show in Fig. 5.14 the centrodes of a Formula car making Turn 5 of the Barcelona circuit. In this case everything was fine, as confirmed by the "good" shape of both centrodes. Also shown are the trajectory of G and the inflection circle.

But not all laps are the same. Figure 5.15 shows the centrodes for the same curve in a case in which the Formula car did not perform well.

Fig. 5.11 Vehicle *entering a curve*: moving centrode rolling on the fixed centrode

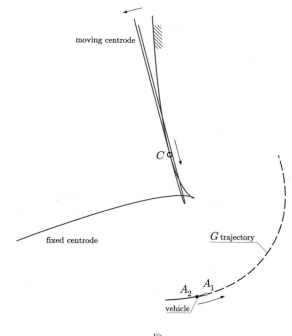

Fig. 5.12 Vehicle *exiting a curve*: moving centrode rolling on the fixed centrode

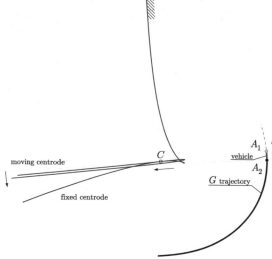

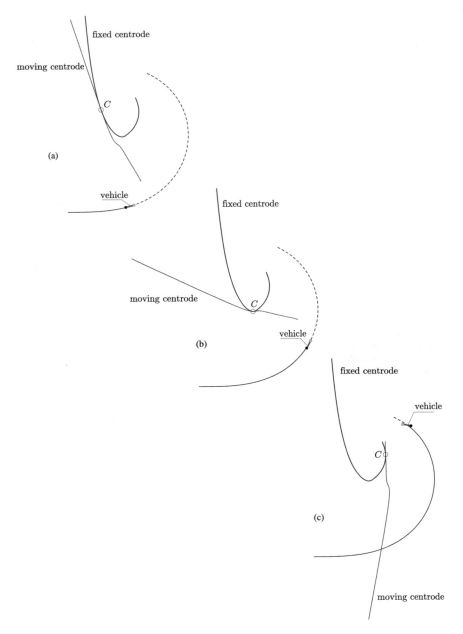

Fig. 5.13 Centrodes of a turning vehicle with handling misbehavior in the final part of the curve (the car spins out)

Fig. 5.14 Centrodes of a Formula car making Turn 5 of the Barcelona circuit (the inflection circle is also shown)

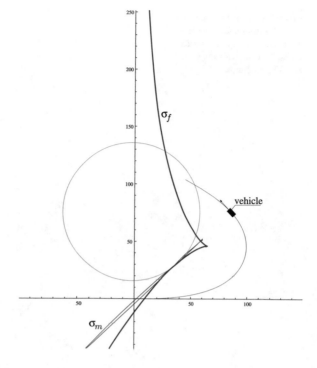

Fig. 5.15 Centrodes of a Formula car *badly* making Turn 5 of the Barcelona circuit (the inflection circle is also shown)

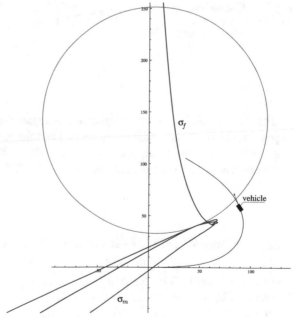

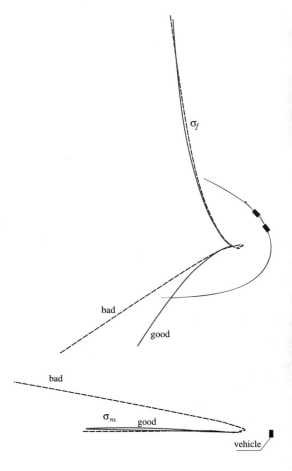

Fig. 5.16 Comparison of the fixed centrodes and of the trajectories of a Formula car making Turn 5 of the Barcelona circuit

Fig. 5.17 Comparison of the moving centrodes of a Formula car making Turn 5 of the Barcelona circuit

The fixed centrodes for the two cases are compared in Fig. 5.16. The entering part is pretty much the same, whereas the central and the exiting parts are very different. It is worth noting that the trajectories of G are almost the same.

The moving centrodes are compared in Fig. 5.17. Again, they differ markedly in the exiting part.

5.2.2 Inflection Circle of a Turning Vehicle

The inflection circle (Fig. 5.10), that is all those points whose trajectory have an inflection point, can be obtained at any instant of time from telemetry data. Perhaps, the main formula is (5.7), that links the diameter d of the inflection circle to the acceleration of the velocity center C. The acceleration $\mathbf{a}_C$ was given in (3.48), which is repeated here for ease of reading (see also (5.19))

$$\mathbf{a}_C = (a_x + vr - \dot{r}u/r)\,\mathbf{i} + (a_y - ur - \dot{r}v/r)\,\mathbf{j}$$

$$= \left(\frac{\dot{u}r - u\dot{r}}{r}\right)\mathbf{i} + \left(\frac{\dot{v}r - v\dot{r}}{r}\right)\mathbf{j} \tag{3.48'}$$

$$= r(\dot{R}\,\mathbf{i} - \dot{S}\,\mathbf{j})$$

We see that we also need $\dot{r}$, which is not commonly measured directly, although it should be.

Here we list, with reference to Fig. 5.10, some relevant formulas

$$\sigma = \chi - \pi, \qquad \sin\chi = -\sin\sigma, \qquad \cos\chi = -\cos\sigma \tag{5.14}$$

$$d = \frac{1}{r^2}\sqrt{\left(\frac{\dot{v}r - v\dot{r}}{r}\right)^2 + \left(\frac{\dot{u}r - u\dot{r}}{r}\right)^2} = \sqrt{\frac{\dot{R}^2 + \dot{S}^2}{r^2}} \tag{5.15}$$

$$d\cos\chi = \left(\frac{\dot{u}r - u\dot{r}}{r}\right)\frac{1}{r^2} = \frac{\dot{R}}{r} \tag{5.16}$$

$$d\sin\chi = \left(\frac{\dot{v}r - v\dot{r}}{r}\right)\frac{1}{r^2} = -\frac{\dot{S}}{r} \tag{5.17}$$

$$\mathbf{d} = d\cos\chi\,\mathbf{i} + d\sin\chi\,\mathbf{j} = \frac{\dot{R}\,\mathbf{i} - \dot{S}\,\mathbf{j}}{r} \tag{5.18}$$

$$\mathbf{a}_C = r^2\mathbf{d} = r(\dot{R}\,\mathbf{i} - \dot{S}\,\mathbf{j}) = r^2 d(\cos\chi\,\mathbf{i} + \sin\chi\,\mathbf{j}) \tag{5.19}$$

$$\mathbf{V}_{\hat{C}} = \dot{S}\,\mathbf{i} + \dot{R}\,\mathbf{j} = rd(-\sin\chi\,\mathbf{i} + \cos\chi\,\mathbf{j}) \tag{5.20}$$

$$\mathbf{D} = S\,\mathbf{i} + R\,\mathbf{j} \tag{5.21}$$

$$r\mathbf{d}\cdot\mathbf{D} = \dot{R}S - R\dot{S} \tag{5.22}$$

$$\dot{\mathbf{D}} = \dot{S}\,\mathbf{i} + \dot{R}\,\mathbf{j} - S\dot{\chi}\,\mathbf{j} + R\dot{\chi}\,\mathbf{i} = (\dot{S} + R\dot{\chi})\,\mathbf{i} + (\dot{R} - S\dot{\chi})\,\mathbf{j} \tag{5.23}$$

$$\dot{d} = \frac{1}{r^3 d}\left[r(\dot{R}\ddot{R} + \dot{S}\ddot{S}) - \dot{r}(\dot{R}^2 + \dot{S}^2)\right] \tag{5.24}$$

$$\frac{\mathrm{d}}{\mathrm{d}t}\left(\frac{\mathbf{D}}{d}\right) = \frac{\dot{\mathbf{D}}d - \mathbf{D}\dot{d}}{d^2} = \frac{1}{d^2}\left\{[(\dot{S} + R\dot{\chi})d - S\dot{d}]\mathbf{i} + [(\dot{R} - S\dot{\chi})d - R\dot{d}]\mathbf{j}\right\} \tag{5.25}$$

These equations cover many aspects (Fig. 5.10):

• the diameter d of the inflection circle;
• the orientation χ of $\mathbf{a}_C$, and hence also of the inflection circle, with respect to the vehicle longitudinal axis;
• the acceleration $\mathbf{a}_C$ of the velocity center C;
• the speed $\mathbf{V}_{\hat{C}}$ of the geometric point $\hat{C}$;
• the rate of change of d;
• the rate of change of the vector $\mathbf{D} = GC$.

It is worth noting that almost all quantities depend on r, $\dot{R}$ and $\dot{S}$, that is on u, v, r, $\dot{u}$, $\dot{v}$, and $\dot{r}$

$$\dot{R} = \frac{\dot{u}r - u\dot{r}}{r^2} = \frac{a_x - r^2 S - \dot{r}R}{r}$$

$$-\dot{S} = \frac{\dot{v}r - v\dot{r}}{r^2} = \frac{a_y - r^2 R + \dot{r}S}{r}$$

(5.26)

As already discussed in Sect. 3.2.8, mathematically equivalent formulas may not be equivalent at all when dealing with experimental data. Probably, it is better avoiding $\dot{u}$ and $\dot{v}$, and use a_x and a_y instead. It would be also very beneficial to measure directly $\dot{r}$, instead of differentiating the yaw rate r.

It is worth noting that, although S is the longitudinal coordinate of the velocity center C with respect to center of mass G, the quantity $\dot{S}$ is not related to G. It is a global quantity, like u, r, $\dot{r}$, R, $\dot{R}$. Quantities strictly related to G, and hence less general and less reliable, are v, $\dot{v}$, β, $\dot{\beta}$.

As shown in Fig. 5.10, along the axis of the vehicle there are, at any instant of time, some special points. Point Z has zero slip angle, that is, $\beta_Z = 0$, or equivalently $\mathbf{V}_Z = u\,\mathbf{i}$. Point N has $\dot{\beta}_N = 0$. Good handling requires these two points Z and N to be fairly close to each other and not too far from the front axle. Therefore, good handling behavior, like in Fig. 5.14, maybe requires small values of $|\dot{S}|$ or $|r\dot{S}|$. This is a topic that deserves further investigation. See also Sect. 5.2.3.

Also interesting is to observe that

$$|\mathbf{a}_C| = r^2 d \quad \text{and} \quad |\mathbf{V}_{\hat{C}}| = |rd|$$

(5.27)

They are strictly related.

5.2.3 Tracking the Curvatures of Front and Rear Midpoints

To better understand the kinematics of a turning vehicle, we also consider the curvature of the trajectories and how they change in time under the driver action on the steering wheel. In particular, we monitor the trajectories of the midpoints $A_1 = (a_1, 0)$ and $A_2 = (-a_2, 0)$ of both axles (Fig. 5.9), and their centers of curvature E_1 and E_2, respectively. There is a nice interplay between radii of curvature, the velocity center and the inflection circle.

We know from (3.3) that the velocities $\mathbf{V}_1$ and $\mathbf{V}_2$ of A_1 and A_2 are

$$\mathbf{V}_1 = u\,\mathbf{i} + (v + ra_1)\,\mathbf{j}$$
$$\mathbf{V}_2 = u\,\mathbf{i} + (v - ra_2)\,\mathbf{j}$$

(5.28)

and hence $V_1 = |\mathbf{V}_1| = \sqrt{u^2 + (v + ra_1)^2}$ and $V_2 = |\mathbf{V}_2| = \sqrt{u^2 + (v - ra_2)^2}$. The corresponding front and rear vehicle slip angles $\hat{\beta}_1$ and $\hat{\beta}_2$ (Fig. 5.9), respectively, are such that

$$\tan(\hat{\beta}_1) = \frac{v + ra_1}{u} = \beta_1$$

$$\tan(\hat{\beta}_2) = \frac{v - ra_2}{u} = \beta_2$$

(5.29)

From (3.39) we obtain the accelerations of A_1 and A_2

$$\mathbf{a}_1 = (a_x - r^2 a_1)\mathbf{i} + (a_y + \dot{r}a_1)\mathbf{j}$$

$$\mathbf{a}_2 = (a_x + r^2 a_2)\mathbf{i} + (a_y - \dot{r}a_2)\mathbf{j}$$

(5.30)

Through the knowledge of velocity and acceleration we can compute the centripetal (normal) component of the two accelerations, as in (3.40)

$$a_{1n} = \frac{-(a_x - r^2 a_1)(v + ra_1) + (a_y + \dot{r}a_1)u}{V_1}$$

$$a_{2n} = \frac{-(a_x + r^2 a_2)(v - ra_2) + (a_y - \dot{r}a_2)u}{V_2}$$

(5.31)

or, more explicitly (but numerically less reliably)

$$a_{1n} = \frac{-(\dot{u} - vr - r^2 a_1)(v + ra_1) + (\dot{v} + ur + \dot{r}a_1)u}{V_1}$$

$$a_{2n} = \frac{-(\dot{u} - vr + r^2 a_2)(v - ra_2) + (\dot{v} + ur - \dot{r}a_2)u}{V_2}$$

(5.32)

The curvatures ρ_1 and ρ_2 of the trajectories of A_1 and A_2 are now promptly obtained (cf. (3.37))

$$\rho_1 = \frac{a_{1n}}{V_1^2} = \frac{r}{V_1} + \frac{(\dot{v} + \dot{r}a_1)u - (v + ra_1)\dot{u}}{V_1^3} = \frac{r + d\hat{\beta}_1/dt}{V_1} \simeq \frac{r + \dot{\beta}_1}{u}$$

$$\rho_2 = \frac{a_{2n}}{V_2^2} = \frac{r}{V_2} + \frac{(\dot{v} - \dot{r}a_2)u - (v - ra_2)\dot{u}}{V_2^3} = \frac{r + d\hat{\beta}_2/dt}{V_2} \simeq \frac{r + \dot{\beta}_2}{u}$$

(5.33)

The *entering* phase of making a left turn is characterized by increasing steer angles and diminishing radii of curvature. Moreover, as we have already seen, the velocity center C gets closer and closer to the vehicle. The corresponding transient kinematics is shown in Fig. 5.18. It is worth noting that, according to (5.8), the radius of curvature of point A_1 is equal to $E_1 A_1$, and hence it is shorter than $C A_1$. On the contrary, the radius of curvature of point A_2 is equal to $E_2 A_2$, which is longer than $C A_2$. This happens because the vehicle slip angle $\hat{\beta}_1$ at point A_1 is increasing, while the vehicle slip angle $\hat{\beta}_2$ at point A_2 is diminishing (in the sense that it gets bigger, but it is negative), as shown in Fig. 5.18 and according to (5.33).

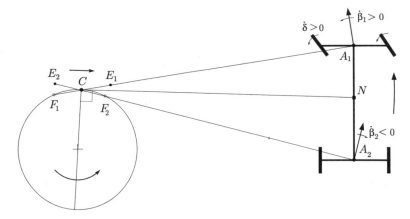

Fig. 5.18 Radii of curvature of a vehicle *entering* a turn properly

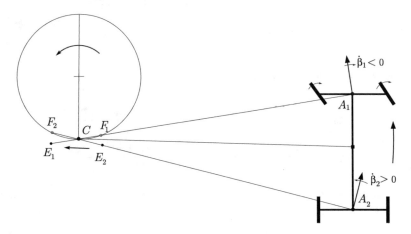

Fig. 5.19 Radii of curvature of a vehicle *exiting* a turn properly

The kinematics of a vehicle *exiting* properly a turn is shown in Fig. 5.19. We see that many things go the other way around with respect to entering.

In both cases, the knowledge of the inflection circle immediately makes clear the relationship between the position of the velocity center C and the centers of curvature E_1 and E_2.

But things may go wrong. Bad kinematic behaviors are shown in Fig. 5.20. We see that the time derivatives of $\hat{\beta}_1$ and $\hat{\beta}_2$ are not as they should be. Indeed, point C is travelling also longitudinally. Again, the positions and orientations of the inflection circle immediately convey the information about the unwanted kinematics of the vehicle.

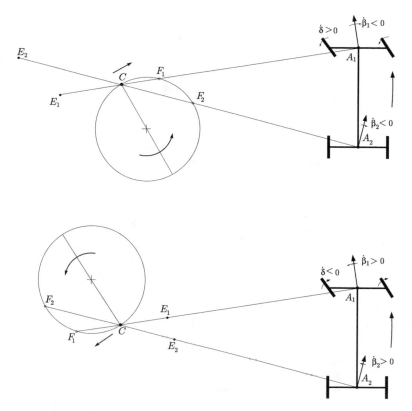

Fig. 5.20 Examples of undesirable kinematics in a turn

5.2.4 *Evolutes*

Let us go back to good turning behavior. The *evolute* of a curve is the locus of all its centers of curvature. The evolutes of the trajectories of points A_1 and A_2, that is the midpoints of each axle, are shown in the lower part of Figs. 5.22 and 5.21. Also shown are the centers of curvature E_1 and E_2 at a given instant of time, along with the corresponding inflection circle (this one drawn in the upper part with the centrodes). We see that the two evolutes are almost coincident. The relative positions of E_1 and E_2 are consistent with Figs. 5.18 and 5.19.

At the onset of bad turning behavior, the two evolutes depart abruptly from each other. Therefore, monitoring the evolutes of A_1 and A_2 can be another objective way to investigate the handling features of a vehicle.

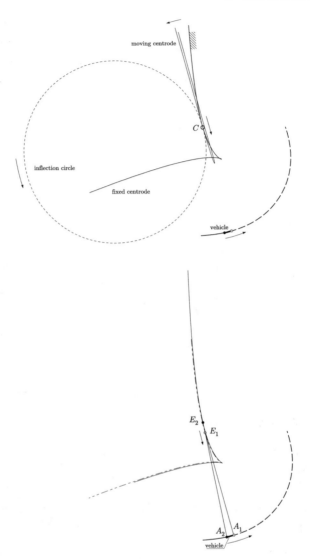

Fig. 5.21 Vehicle *entering* a curve: inflection circle (top) and centers of curvatures with the corresponding evolutes (bottom)

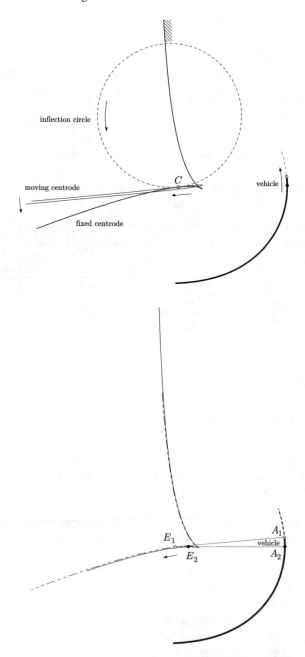

Fig. 5.22 Vehicle *exiting* a curve: inflection circle (top) and centers of curvatures with the corresponding evolutes (bottom)

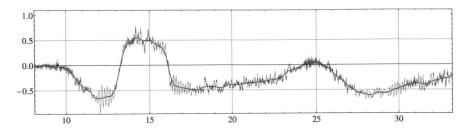

Fig. 5.23 Yaw rate $r(t)$, in rad/s: comparison between raw and filtered data

Table 5.1 Radii of curvature of the trajectories of the axle midpoints

	t (s)	Turn No	R_1 (m)	R_2 (m)
1	10.87	1	-135.01	-148.94
2	12.66	1	-52.54	-50.58
3	30.98	4	-104.20	-102.97
4	61.84	10	75.55	125.94
5	63.87	10	26.19	25.23
6	64.87	10	73.65	59.86

5.3 Exercises

5.3.1 Front and Rear Radii of Curvature

With the data of Tables 8.1 and 8.2 and assuming $a_1 = 1.68$ m and $a_2 = 1.32$ m, compute the radii of curvature R_1 and R_2 of the trajectories of the axle midpoints A_1 and A_2 (Fig. 5.18). Discuss the results.

Solution

According to (5.33), to compute $R_i = 1/\rho_i$ we need the centripetal component a_{in} of the acceleration and the speed V_i. The most reliable formula for a_{in} should be (5.31), because it avoids the computation of $\dot{u}$ and $\dot{v}$. Unfortunately, with the usual telemetry sensors, we cannot avoid the computation of $\dot{r}$. Therefore, the results will strongly depend upon the filter applied to the raw yaw rate before differentiating it, as obvious from Fig. 5.23.

The computed radii of curvature are shown in Table 5.1. The reader is invited to check whether they are consistent with Figs. 5.18 and 5.19, and also to compare them with R_G, already computed in Table 8.2.

5.3.2 Drawing Centrodes

Select the parametric equations to be employed to draw the moving centrode in Fig. 5.17. Then, do the same for Fig. 5.13.

Solution

In Fig. 5.17 the moving centrodes are plotted in the vehicle reference plane. Therefore, we must use (5.10).

On the other hand, in Fig. 5.13 there are three different positions of the moving centrodes. That means that we must use (5.13), for three different instants of time t.

5.4 List of Some Relevant Concepts

Section 5.1.2—the moving centrode rolls without slipping on the fixed centrode, the point of rolling contact being C;
Section 5.1.2—the velocity field is like a pure rotation around C, but the acceleration field is not;
Section 5.1.3—the acceleration field is like a pure rotation around K;
Section 5.1.4—the inflection circle makes it possible to easily evaluate the radius of curvature of the trajectory of any point of a rigid body;
Section 5.2.2—handling misbehavior strongly affects the shape of centrodes;
Section 5.2.4—monitoring the evolutes can be another objective way to investigate the handling features of a vehicle.

5.5 Key Symbols

$\mathbf{a}_C$	acceleration of C
a_x	longitudinal acceleration of G
a_y	lateral acceleration of G
C	velocity center
d	diameter of the inflection circle
G	center of mass;
K	acceleration center
r	yaw rate
R	lateral coordinate of G
S	longitudinal coordinate of G
u	longitudinal velocity
v	lateral velocity of G
χ	orientation of $\mathbf{a}_C$
ψ	yaw angle

References

1. Bottema O (1979) Theoretical kinematics. Dover Publications, New York
2. Heißing B, Ersoy M (eds) (2011) Chassis handbook. Springer, Wiesbaden
3. Pytel A, Kiusalaas J (1999) Engineering mechanics—statics. Brooks/Cole, Pacific Grove
4. Rossa FD, Mastinu G, Piccardi C (2012) Bifurcation analysis of an automobile model negotiating a curve. Veh Syst Dyn 50(10):1539–1562
5. Sandor GN, Erdman AG (1984) Advanced mechanism design: analysis and synthesis, vol 2. Prentice-Hall, Englewood Cliffs
6. Waldron KJ, Kinzel GL (2004) Kinematics, dynamics, and design of machinery. Wiley, New York

Chapter 6
Map of Achievable Performance (MAP)

In this chapter we present a new approach, a *global* one, called Map of Achievable Performance (MAP), which provides a way to analyze the overall steady-state handling features of road/race cars. It is completely general, in the sense that is can be employed for any real car, and for any mathematical model as well.

A vehicle is in steady-state conditions when all time derivatives are zero, that is $\dot{u} = \dot{v} = \dot{r} = 0$. In practical terms, that means having the vehicle going round along a circular path of constant radius, at constant forward speed.

Two concepts play a central role in MAP: the achievable region, that is the totality of the achievable trim conditions for a given vehicle, and the *level (handling) curves* inside the achievable region, to highlight the vehicle peculiar features.

The envelope of level curves is often a good practical way to obtain the achievable regions, as will be shown shortly.

6.1 MAP Fundamental Idea

The physics behind any MAP is, in principle, fairly simple: the driver sets the angular position δ_v of the steering wheel, and the forward speed u, and, at steady state, the vehicle reacts with a constant lateral velocity v and a constant yaw rate r.

Mathematically, it means that the steady-state behavior of any vehicle is completely characterized by two maps[1]

$$v = \hat{v}(\delta_v, u) \quad \text{and} \quad r = \hat{r}(\delta_v, u) \tag{6.1}$$

The use of v and r to monitor the vehicle behavior is not mandatory. In fact, to have a more geometric description of the vehicle motion, we prefer to use $\beta = v/u$ and $\rho = r/u$

---

[1] In this chapter some subscripts are dropped to make equations more readable.

M. Guiggiani, *The Science of Vehicle Dynamics*,
https://doi.org/10.1007/978-3-031-06461-6_6

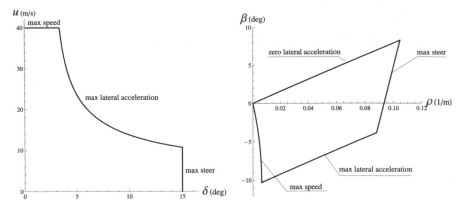

Fig. 6.1 Achievable *input* region (left) and achievable *output* region (right) for an understeer vehicle

$$\beta = \hat{\beta}(\delta_v, u) = \frac{\hat{v}(\delta_v, u)}{u} \qquad \text{and} \qquad \rho = \hat{\rho}(\delta_v, u) = \frac{\hat{r}(\delta_v, u)}{u} \tag{6.2}$$

as shown in Fig. 6.1.

Obviously, other kinematic quantities can be mapped as well. They are all functions of (δ_v, u).

6.2   Input Achievable Regions

As already stated, the driver controls the steering wheel angle δ_v and the forward speed u. In some cases it is useful to consider also the lateral acceleration $\tilde{a}_y = ur = u^2\rho$ as a possible input quantity.

Let us start with the achievable region in the plane (δ, u) (Fig. 6.2), where $\delta = \tau\delta_v$, as in (3.66).

First, we observe that input quantities are subject to obvious practical limitations: maximum steer angle $\delta_{\max}$ (black line in Fig. 6.2), and maximum vehicle speed $u_{\max}$ (red line in Fig. 6.2). However, this achievable region cannot be a rectangle. Owing to limited grip, the achievable forward speed depend on δ, as marked by the (green) curved boundary $u_{\lim}(\delta)$, connecting the two straight sides in Fig. 6.2.

Perhaps, the effect of grip is more evident if we consider the achievable region in the plane (u, a_y), shown in Fig. 6.3. We see that at low speeds the lateral acceleration a_y is limited by the maximum steer angle (black curve), while at medium to high speeds, a_y is grip limited (green curve). The achievable region in Fig. 6.3 has a speed dependent green boundary $a_{y,\lim}(u)$, clearly showing that we are dealing with a Formula car with relevant aerodynamic downforces.

Also interesting is the achievable region in the plane (δ, a_y), shown in Fig. 6.4 for the same Formula car. At small steer angles, the boundary is due to $u_{\max}$ (red curve),

Fig. 6.2 Achievable region in the input space (δ, u), with boundary $u_{\lim}(\delta)$

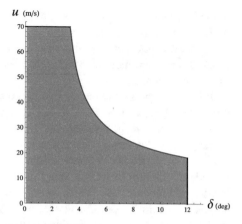

Fig. 6.3 Achievable region in the space (u, a_y), with boundary $a_{y,\lim}(u)$

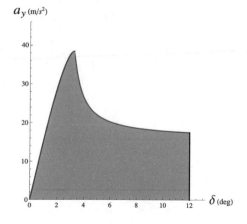

Fig. 6.4 Achievable region in the space (δ, a_y), with boundary $a_{y,\lim}(\delta)$

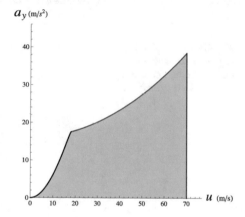

while for medium to high values of δ, the lateral acceleration is limited (green curve) by the grip and by the aerodynamic downforces.

All these figures, unless otherwise specified, are for a simple mathematical model of an understeer vehicle with wings and open differential (basically a Formula car at the center of a bend).

It is very important to realize that all these input achievable regions cover exactly the same set of steady-state working conditions of the vehicle. However, to really make these region fully equivalent it is necessary to draw the level curves for the corresponding "missing" quantity, as done in Figs. 6.5, 6.6 and 6.7. We invite the reader to check that these three MAPs are indeed equivalent, in the sense that they provide the same information about the vehicle steady-state behavior. Moreover, it is interesting to compare Fig. 6.5 with Fig. 7.31.

Aerodynamic devices affect pretty much the shape of achievable regions, as shown in Fig. 6.8. Each region highlights different effects on the vehicle behavior.

Fig. 6.5 Achievable input region (δ, u), with level curves of constant lateral acceleration a_y (m/s^2)

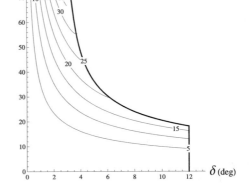

Fig. 6.6 Achievable input region (u, a_y), with level curves of constant steer angle δ (deg)

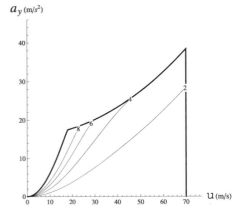

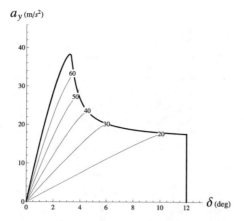

Fig. 6.7 Achievable input region (δ, a_y), with level curves of constant forward speed u (m/s)

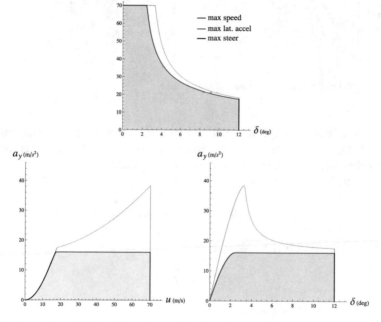

Fig. 6.8 Comparison of achievable regions with and without aerodynamic downloads (thin curves and thick curves, respectively)

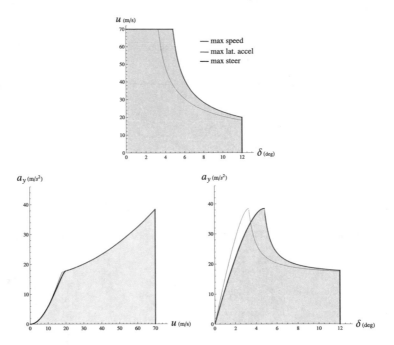

Fig. 6.9 Thick lines are for a more understeer race car

Also interesting is comparing vehicles with different level of understeer, as done in Fig. 6.9. Again, the three regions change in different ways. Of course, it takes time to get used to this approach to analyze the handling behavior.

More formally, we have from Fig. 6.4 that

$$\delta_{\lim}(u) = \delta(a_{y,\lim}(u), u) \qquad (6.3)$$

which can be inverted to get (Fig. 6.2)

$$u_{\lim}(\delta) \qquad (6.4)$$

and hence, as shown in Fig. 6.4

$$a_{y,\lim}(\delta) = a_{y,\lim}(u_{\lim}(\delta)) \qquad (6.5)$$

These functions define the boundary of input achievable regions. Moreover, they are also useful for obtaining the boundaries of output achievable regions.

6.3   Achievable Performances on Input Regions

Level curves of any measurable or computable quantity can be drawn inside an input achievable region. For instance, lines at constant β and constant ρ are shown in Fig. 6.10. The same kind of lines are drawn in Fig. 6.11 and also in Fig. 6.12. These plots look at first very different, but provide, if combined with Figs. 6.5, 6.6 and 6.7, the same information on the steady-state behavior of the vehicle.

Incidentally, we observe that $\beta \simeq -4$ deg characterizes part of the boundary.

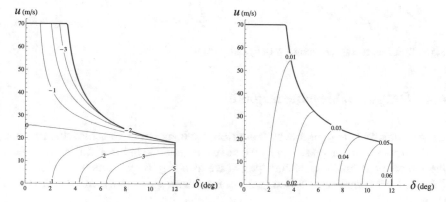

Fig. 6.10   Level curves for constant β (left) and ρ (right)

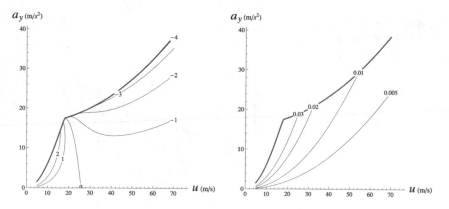

Fig. 6.11   Level curves for constant β (left) and ρ (right)

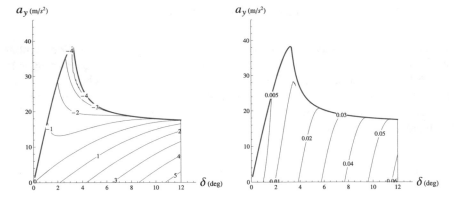

Fig. 6.12 Level curves for constant β (left) and ρ (right)

6.4   Output Achievable Regions

Output achievable regions are, by definition, the image of input achievable regions.

For instance, in the plane (ρ, β) we have a region like in Fig. 6.13. As expected, the boundary is made up of four parts (see also Fig. 6.1): maximum steer angle (black), limit lateral acceleration (green), maximum speed (red), and (almost) zero speed (black).

Fig. 6.13 Output achievable region in the plane (ρ, β)

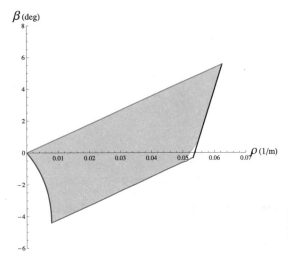

Fig. 6.14 Level curves for constant u (red lines) and constant a_y (green lines) in the output plane (ρ, β)

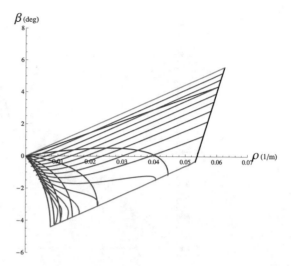

6.5 Achievable Performances on Output Regions

On the output region (ρ, β) it is useful and fairly easy to draw parametric curves keeping constant one input parameter. Therefore, we have curve with constant forward speed u, constant steer angle δ_v, and constant lateral acceleration a_y (Fig. 6.14). It is much more difficult to keep constant some other output quantity. The same observations hold true also for mixed I/O regions.

It is left to the reader to realize why Fig. 6.14 is typical of a race car with high aerodynamic downloads. It can be useful to compare Fig. 6.14 with Fig. 7.31.

6.6 Mixed Input/Output Achievable Regions

Also mixed input/output achievable regions are possible. For instance, in the plane (δ, ρ) we have a region like in Fig. 6.15. Again, the boundary is made up of four parts: maximum steer angle (black), limit lateral acceleration (green), maximum speed (red), and (almost) zero spee d (see also Fig. 7.42).

6.7 Achievable Performances on Mixed I/O Regions

The plane (δ, ρ) is perhaps the most intuitive MAP (Fig. 6.16), particularly if we draw level curves for constant speed u. It immediately shows how a driver can operate on δ and on u to negotiate a bend with curvature ρ.

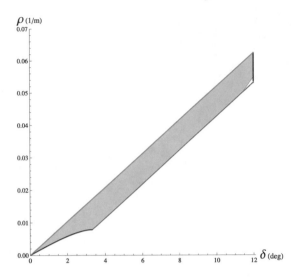

Fig. 6.15 Mixed I/O achievable region in the plane (δ, ρ)

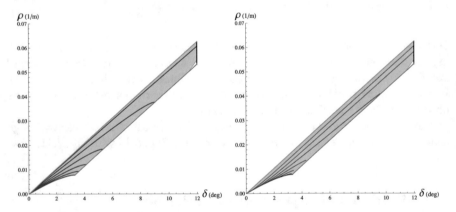

Fig. 6.16 Level curves for constant speed u (left) and constant lateral acceleration a_y (right) in the plane (δ, ρ)

Also interesting is the closeup of Fig. 6.16 at high speeds, shown in Fig. 6.17, which clearly shows the interplay between u and a_y in a car with aerodynamic downloads (see also Fig. 6.14).

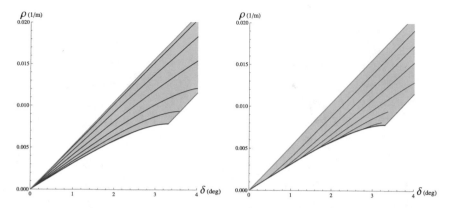

Fig. 6.17   Closeup of level curves for constant speed u (left) and constant lateral acceleration a_y (right) in the plane (δ, ρ)

6.8   MAP from Constant Speed Tests

All the MAPs presented so far in this chapter were obtained with a single track model with aerodynamic downloads. Now, it is time to employ a more realistic double track model, still for a Formula car, as it were a real vehicle to be tested on a proving ground.

The so-called Slowly Increasing Steer (SIS) test is a good way to collect (pseudo) steady-state data. To span several working conditions, we perform the test at several different constant forward speeds u. In all cases, we increase the front wheel steer angle δ, starting from zero, with a constant rate of 0.5 deg/s.

Three possible MAPs with level curves for constant forward speed u are shown in Figs. 6.18, 6.19 and 6.20. We can easily infer the shape of the corresponding achievable regions.

6.9   MAP from Constant Steer Tests

Useful MAPs can also be obtained from tests with constant steer angle and slowly increasing speed. Actually, this is the most robust test procedure: keeping constant the steering wheel position is easy, and certainly easier than keeping constant the forward speed, not to mention how hard it is to keep constant the turning radius.

Again, in Fig. 6.21 the shape of the achievable region is clearly defined. Moreover, we observe that at sufficiently high speed, the boundary of the achievable region can be reached with values of δ as low as about 4 degrees.

Also interesting is adding the constant steer lines to Fig. 6.20, thus obtaining Fig. 6.22. Maybe, this is the MAP that better highlights the different handling features of this race car when "visiting" different points inside the achievable region.

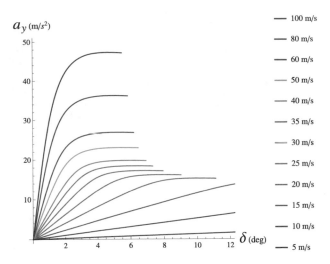

Fig. 6.18 MAP (δ, a_y) with level curves for constant u

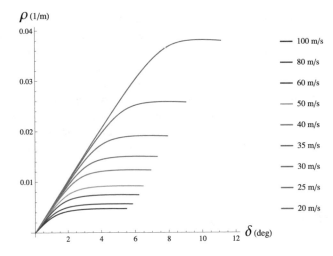

Fig. 6.19 MAP (δ, ρ) with level curves for constant u

6.10 Concluding Remarks

The MAP approach looks promising, but it still needs to be fully developed to become a way to really compare different setups. We have to understand how to read these plots. We have to learn what to look at and why. This is to say that we should refrain from discarding the MAP approach just because it has not provided very good answers in a while.

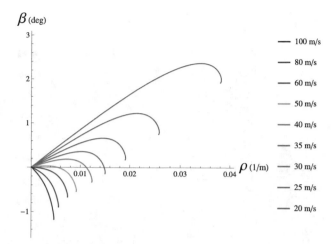

Fig. 6.20 MAP (ρ, β) with level curves for constant u

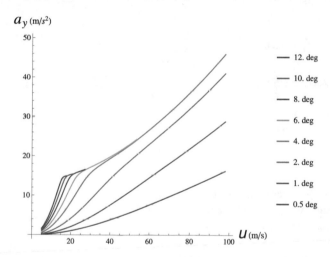

Fig. 6.21 MAP (u, a_y) with level curves for constant δ

More detailed applications of MAPs are provided in the next two chapters.

6.11 List of Some Relevant Concepts

Section 6.2—aerodynamic downloads strongly affect the shape of achievable regions;
Section 6.8—aerodynamic downloads strongly affect the lines at constant lateral acceleration;
Section 6.10—MAPs highlight understeer to oversteer transitions.

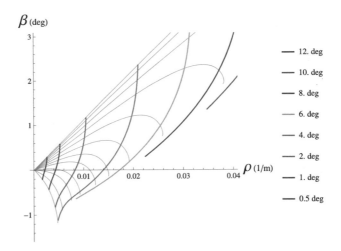

Fig. 6.22 MAP (ρ, β) with also level curves for constant δ

6.12   Key Symbols

a_y	lateral acceleration
G	center of mass;
r	yaw rate
u	longitudinal velocity
v	lateral velocity of G
β	v/u
δ	$\tau \delta_v$ (net steer angle of the wheels)
δ_v	angular position of the steering wheel
ζ	damping ratio
ρ	r/u
τ	gear ratio of the steering system
ω_s	damped natural frequency

Chapter 7
Handling of Road Cars

Ordinary road cars are by far the most common type of motor vehicle. Almost all of them share the following features relevant to handling:

1. four wheels (two axles);
2. two-wheel drive;
3. open differential;
4. no wings (and hence, no significant aerodynamic downforces);
5. no intervention by electronic active safety systems like ABS or ESP under ordinary operating conditions.

Moreover, in the mathematical models it is also typically assumed that the vehicle moves on a *flat* road at almost *constant* forward speed u, thus requiring small longitudinal forces by the tires.

The handling analysis of this kind of vehicles is somehow the simplest that can be envisaged.[1] That does not mean that it is simple at all.

The vehicle model developed in Chap. 3 is employed. However, owing to the above listed features of road cars, several additional simplifications can be made, which first lead to the *double track model* and eventually to the celebrated *single track model*. All the steps that lead to the single track model are thoroughly discussed to clarify when it is a suitable model for vehicle dynamics.

The original version of this chapter was revised: Figures 7.12, to 7.19 has been updated with high resolution. The correction to this chapter is available at https://doi.org/10.1007/978-3-031-06461-6_12

[1] Some sports cars and all race cars have a limited-slip differential. Several race cars also have wings that provide fairly high aerodynamic downforces at high speed. The handling of these vehicles is somehow more involved than that of ordinary road cars and will be addressed in Chap. 8.

© The Author(s), under exclusive license to Springer Nature Switzerland AG 2023,   239
corrected publication 2023
M. Guiggiani, *The Science of Vehicle Dynamics*,
https://doi.org/10.1007/978-3-031-06461-6_7

7.1   Additional Simplifying Assumptions for Road Car Modeling

The vehicle model introduced in Chap. 3, and whose equations were collected in Sect. 3.14, is simplified hereafter, taking into account the distinguishing features relevant to handling of road cars (Fig. 7.1).

7.1.1   Negligible Vertical Aerodynamic Loads

Aerodynamics of road cars is mostly concerned with attaining low drag, because of its impact on fuel consumption. Therefore, road cars normally do not have aerodynamic devices to generate significant vertical loads, that is $Z_1^a \simeq 0$ and $Z_2^a \simeq 0$. Basically, this means that the handling features of a road car are (almost) speed insensitive.

7.1.2   Open Differential

The main simplification is that the vehicle is equipped with an *open differential*. Since there is almost no friction inside an open differential mechanism, in (3.178) we have that its internal efficiency $\eta_h \simeq 1$, and hence $M_l \simeq M_r$. In other words, both driving wheels receive always the same torque from the engine. Therefore, in the global equilibrium equations (3.94), the tire longitudinal forces $F_{x_{ij}}$ (Fig. 7.1) are such that $F_{x_{11}} = F_{x_{12}}$ and $F_{x_{21}} = F_{x_{22}}$, and hence do not contribute to the yaw moment N applied to the vehicle. Summing up, in (3.87)

$$\Delta X_1 = -[F_{y_{12}} \sin(\delta_{12}) - F_{y_{11}} \sin(\delta_{11})]/2$$
$$\Delta X_2 = 0 \tag{7.1}$$

Basically, this means that the handling features of a road car are (almost) insensitive to the radius of curvature of its trajectory, provided the radius is not too small.

A look at Fig. 3.28 can be useful to better understand $\Delta X_1 = (X_{12} - X_{11})/2$.

7.1.3   Almost Constant Forward Speed

If the forward speed u is almost constant ($\dot{u} \simeq 0$, and hence $a_x \simeq 0$), and the aerodynamic drag is not very high (like in ordinary cars, but not in a Formula 1 car, which,

Fig. 7.1 Vehicle basic
scheme (double track model)

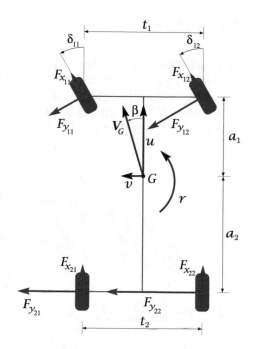

however, does not have an open differential),[2] the tire longitudinal forces are quite
small (Fig. 7.1). That means that also the *longitudinal slips are small and can be
neglected*. Therefore,

$$F_{x_{ij}} \simeq 0$$
$$\sigma_{x_{ij}} \simeq 0 \tag{7.2}$$

which means that all wheels are almost under longitudinal pure rolling conditions.
  As a consequence, we have

$$u \simeq \omega_h r_2 \tag{7.3}$$

where ω_h is the angular velocity of the differential housing, and r_2 is the rolling
radius of the rear wheels.

---

[2] The left and right wheels of the same axle are normally equipped with the same kind of brake.
Therefore, the braking torque is pretty much the same under ordinary operating conditions, and,
again, (7.1) holds true. However, there are important exceptions. The left and right braking forces
can be different if: (a) the grip is different and at least one wheel is locked; (b) the friction coefficients
inside the two brakes is different (for instance, because of different temperatures, which is often the
case in racing cars); (c) some electronic stability system, like ESP or ABS, has been activated.

7.2   Mathematical Model for Road Car Handling

The equations collected in Sect. 3.14 for the fairly general vehicle model described in Chap. 3 are now tailored to the case of road cars with open differential, no wings, and almost constant forward speed.

The general definitions (3.87) for the horizontal (in-plane) forces acting on the vehicle now become (Figs. 7.1 and 7.2)

$$
\begin{aligned}
X_1 &\simeq -[F_{y_{11}} \sin(\delta_{11}) + F_{y_{12}} \sin(\delta_{12})] \\
X_2 &\simeq 0 \\
Y_1 &\simeq F_{y_{11}} \cos(\delta_{11}) + F_{y_{12}} \cos(\delta_{12}) \\
Y_2 &= F_{y_{21}} + F_{y_{22}} \\
\Delta X_1 &\simeq -[F_{y_{12}} \sin(\delta_{12}) - F_{y_{11}} \sin(\delta_{11})]/2 \\
\Delta X_2 &\simeq 0
\end{aligned}
\tag{7.4}
$$

7.2.1   Global Equilibrium

Since the forward speed u in (3.94) is given, the vehicle has basically only *lateral* and *yaw* dynamics (often simply called lateral dynamics), described by the following system of two differential equations (Fig. 7.2)

$$
\begin{aligned}
ma_y &= m(\dot{v} + ur) = Y = Y_1 + Y_2 \\
J_z \dot{r} &= N = Y_1 a_1 - Y_2 a_2 + \Delta X_1 t_1
\end{aligned}
\tag{7.5}
$$

while

$$
X_2 = m(\dot{u} - vr) + [F_{y_{11}} \sin(\delta_{11}) + F_{y_{12}} \sin(\delta_{12})] + \frac{1}{2}\rho_a S_a C_x u^2 \tag{7.6}
$$

is now an algebraic equation, the unknown being the tire longitudinal force X_2 (see (7.4)).

It looks like we are playing a dirty game. First we say $X_2 \simeq 0$, and now we are computing it. This is indeed to check whether X_2 is actually very small.

We recall that u is the vehicle longitudinal velocity, r is the vehicle yaw rate, v is the lateral velocity of G, a_x is the longitudinal acceleration of G, and a_y is the lateral acceleration of G. The vehicle has mass m and moment of inertia J_z with respect to a vertical axis located at G. It is worth noting that u and r are not affected by the position of G.

Fig. 7.2 Global dynamics of the double track model

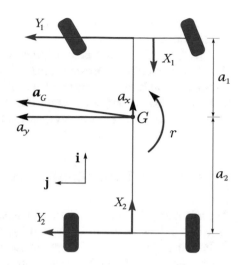

7.2.2 Approximate Lateral Forces

In all two-axle vehicles with an open differential, it is possible to solve (7.5) with respect to the front and rear lateral forces (cf. (3.100))

$$Y_1 = \frac{ma_2}{l} a_y + \frac{J_z \dot{r} - \Delta X_1 t_1}{l} \simeq \frac{ma_2}{l} a_y$$

$$Y_2 = \frac{ma_1}{l} a_y - \frac{J_z \dot{r} - \Delta X_1 t_1}{l} \simeq \frac{ma_1}{l} a_y$$

(7.7)

where, in the last terms, we took into account that $|J_z \dot{r}| \ll |ma_y a_i|$, since in a car $J_z < ma_1 a_2$ and $|\dot{r} a_i| \ll |a_y|$. The other term $\Delta X_1 t_1$ becomes relevant if the wheel steer angle is at least 15 degrees. It is common practice to ignore this contribution. In most cases it is hardly mentioned, and almost always neglected, although it can be far from negligible. The main reason for this "ostracism" is that the analysis is much simpler if $\Delta X_1 t_1$ is set to zero. Well, not quite a reasonable reason...

Moreover, under ordinary operating conditions $|\dot{v}| \ll |ur|$ (Fig. 3.7), and we can use

$$\tilde{a}_y = ur = u^2 \rho$$

(7.8)

already defined in (3.28), instead of the full expression $a_y = \dot{v} + ur$ of the lateral acceleration, to approximately evaluate the axle lateral forces (cf. (3.100))

$$Y_1 \simeq \frac{ma_2}{l} \tilde{a}_y \quad \text{and} \quad Y_2 \simeq \frac{ma_1}{l} \tilde{a}_y$$

(7.9)

Therefore, in a *two-axle* vehicle with *open differential*, the axle lateral forces are approximately *linear* functions of the lateral acceleration $\tilde{a}_y$. This is a simple, yet fundamental result in vehicle dynamics of road cars, which greatly impacts on the whole vehicle model.

Equations (7.9) hold true only when $\dot{v} = \dot{r} = 0$, that is when the vehicle is in steady-state conditions. However, they are sufficiently accurate when employed to estimate lateral load transfers and roll angles, as will be shown. Actually, we should never forget that in the present analysis there is no roll dynamics (except in Chap. 9). Therefore, the roll angle is always assumed to be the angle at steady state.

7.2.3 Lateral Load Transfers and Vertical Loads

According to (3.151) and (3.155), both lateral load transfers ΔZ_1 and ΔZ_2 are *linear* functions of both lateral forces Y_1 and Y_2.

Inserting (7.9) into (3.151), we obtain the following simplified equations for the lateral load transfers in vehicles with open differential and linear springs

$$\Delta Z_1 \simeq \frac{k_{\phi_1}k_{\phi_2}}{t_1 k_\phi}\left(\frac{h-q}{k_{\phi_2}} + \frac{a_2 q_1}{lk_{\phi_1}^s} + \frac{a_2 q_1}{lk_{\phi_2}^s} + \frac{a_2 q_1 + a_1 q_2}{lk_{\phi_2}^p}\right)m\tilde{a}_y = \eta_1 m\tilde{a}_y$$

$$\Delta Z_2 \simeq \frac{k_{\phi_1}k_{\phi_2}}{t_2 k_\phi}\left(\frac{h-q}{k_{\phi_1}} + \frac{a_1 q_2}{lk_{\phi_1}^s} + \frac{a_1 q_2}{lk_{\phi_2}^s} + \frac{a_2 q_1 + a_1 q_2}{lk_{\phi_1}^p}\right)m\tilde{a}_y = \eta_2 m\tilde{a}_y$$

$$(7.10)$$

or, equivalently

$$\Delta Z_1 \simeq \frac{1}{t_1}\left[\frac{k_{\phi_1}}{k_\phi}(h-q) + \frac{a_2 q_1}{l} + \frac{k_{\phi_1}k_{\phi_2}}{k_\phi l}\left(\frac{a_1 q_2}{k_{\phi_2}^p} - \frac{a_2 q_1}{k_{\phi_1}^p}\right)\right]m\tilde{a}_y = \eta_1 m\tilde{a}_y$$

$$\Delta Z_2 \simeq \frac{1}{t_2}\left[\frac{k_{\phi_2}}{k_\phi}(h-q) + \frac{a_1 q_2}{l} + \frac{k_{\phi_1}k_{\phi_2}}{k_\phi l}\left(\frac{a_2 q_1}{k_{\phi_1}^p} - \frac{a_1 q_2}{k_{\phi_2}^p}\right)\right]m\tilde{a}_y = \eta_2 m\tilde{a}_y$$

$$(7.11)$$

where $l = a_1 + a_2$ is the wheelbase, $q = (a_2 q_1 + a_1 q_2)/l$, and

$$k_\phi = k_{\phi_1} + k_{\phi_2} = \frac{k_{\phi_1}^s k_{\phi_1}^p}{k_{\phi_1}^s + k_{\phi_1}^p} + \frac{k_{\phi_2}^s k_{\phi_2}^p}{k_{\phi_2}^s + k_{\phi_2}^p} \qquad (3.148')$$

is the global roll stiffness.

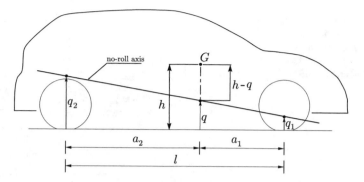

Fig. 7.3 Geometric parameters that affect lateral load transfers

The two quantities η_1 and η_2, and hence the ratio $\Delta Z_1/\Delta Z_2 = \eta_1/\eta_2$, depend in a peculiar way on the track widths t_i, on the roll stiffnesses of the suspensions $k_{\phi_i}^s$, on the roll stiffnesses of the tires $k_{\phi_i}^P$, on the heights q_i of the no-roll centers Q_i,[3] on the longitudinal position (a_1, a_2) and height h of the center of gravity G (Fig. 7.3). The roll stiffnesses are defined in Sect. 3.10.6, and in particular in (3.148). The no-roll centers are defined in Sect. 3.10.9.

If the tires are supposed to be perfectly rigid, that is $k_{\phi_i}^P \to \infty$ and $k_{\phi_i}^s = k_{\phi_i}$, the expressions of the lateral load transfers (7.11) become much simpler

$$\Delta Z_1 \simeq \frac{1}{t_1}\left[\frac{k_{\phi_1}(h-q)}{k_\phi} + \frac{a_2 q_1}{l}\right]m\tilde{a}_y = \eta_1 m\tilde{a}_y$$

$$\Delta Z_2 \simeq \frac{1}{t_2}\left[\frac{k_{\phi_2}(h-q)}{k_\phi} + \frac{a_1 q_2}{l}\right]m\tilde{a}_y = \eta_2 m\tilde{a}_y \tag{7.12}$$

as in (3.158).

Taking (7.9) into account we also obtain that

$$\Delta Z_1 = \eta_1 \frac{l}{a_2} Y_1 \quad \text{and} \quad \Delta Z_2 = \eta_2 \frac{l}{a_1} Y_2 \tag{7.13}$$

The total vertical loads (3.109) on each tire can be further simplified because, in the present case, the longitudinal load transfer ΔZ is negligible. Moreover, cars with an open differential are not so sporty to have significant aerodynamic vertical loads. Therefore, combining (3.109) and (7.10), we obtain

---

[3] We call no-roll center what is commonly called roll center.

Fig. 7.4 Roll angle ϕ_i^s due to suspension deflections only, roll angle ϕ_i^p due to tire deformations only, and total vehicle roll angle ϕ (front view)

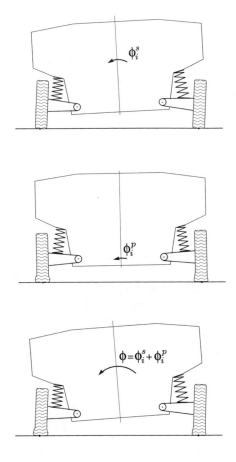

$$Z_{11} = F_{z_{11}} = \frac{Z_1^0}{2} - \Delta Z_1(\tilde{a}_y) = \frac{mga_2}{2l} - \eta_1 m\tilde{a}_y$$

$$Z_{12} = F_{z_{12}} = \frac{Z_1^0}{2} + \Delta Z_1(\tilde{a}_y) = \frac{mga_2}{2l} + \eta_1 m\tilde{a}_y$$

$$Z_{21} = F_{z_{21}} = \frac{Z_2^0}{2} - \Delta Z_2(\tilde{a}_y) = \frac{mga_1}{2l} - \eta_2 m\tilde{a}_y \qquad (7.14)$$

$$Z_{22} = F_{z_{22}} = \frac{Z_2^0}{2} + \Delta Z_2(\tilde{a}_y) = \frac{mga_1}{2l} + \eta_2 m\tilde{a}_y$$

which shows that the variations of vertical loads are (linear) functions of the lateral acceleration $\tilde{a}_y = ur$.

7.2.4 Roll Angles

Also the (steady-state) roll angles due to suspension deflections (3.147) depend upon Y_1 and Y_2, and hence, according to (7.9), can be set as functions of the lateral acceleration only[4]

$$\phi_1^s = \frac{1}{k_{\phi_1}^s} \frac{k_{\phi_1} k_{\phi_2}}{k_\phi} \left[\frac{h-q}{k_{\phi_2}} - \frac{a_2 q_1}{lk_{\phi_1}^P} + \frac{a_1 q_2}{lk_{\phi_2}^P} \right] m\tilde{a}_y = \rho_1^s m\tilde{a}_y$$

$$\phi_2^s = \frac{1}{k_{\phi_2}^s} \frac{k_{\phi_1} k_{\phi_2}}{k_\phi} \left[\frac{h-q}{k_{\phi_1}} - \frac{a_1 q_2}{lk_{\phi_2}^P} + \frac{a_2 q_1}{lk_{\phi_1}^P} \right] m\tilde{a}_y = \rho_2^s m\tilde{a}_y$$

$$(7.15)$$

The same applies to roll angles ϕ_i^P due to tire deformations. According to (3.144) and (7.10) we obtain

$$\phi_1^P = \frac{\Delta Z_1 t_1}{k_{\phi_1}^P} = \frac{\eta_1 t_1}{k_{\phi_1}^P} m\tilde{a}_y = \rho_1^P m\tilde{a}_y$$

$$\phi_2^P = \frac{\Delta Z_2 t_2}{k_{\phi_2}^P} = \frac{\eta_2 t_2}{k_{\phi_2}^P} m\tilde{a}_y = \rho_2^P m\tilde{a}_y$$

$$(7.16)$$

If the tires are supposed to be rigid, we have $\rho_1^P = \rho_2^P = 0$, and $\rho_1^s = \rho_2^s = (h - q)/k_\phi$.

The roll angles (Fig. 7.4) are important because they affect camber angles and steer angles of the wheels, as shown hereafter.

7.2.5 Camber Angle Variations

Let, $\gamma_{i2}^0 = -\gamma_{i1}^0 = \gamma_i^0$ be the camber angles under static conditions (Fig. 7.5), and let $\Delta \gamma_{i1} = \Delta \gamma_{i2} = \Delta \gamma_i$ be the camber variations due to vehicle roll motion (Fig. 7.6). The camber angles of the two wheels of the same axle are thus given by

$$\gamma_{i1} = -\gamma_i^0 + \Delta \gamma_i \qquad \gamma_{i2} = \gamma_i^0 + \Delta \gamma_i \qquad (7.17)$$

where the camber variation $\Delta \gamma_i$, according to (3.113), (7.15) and (7.16), depends on the roll angles, and hence on the lateral acceleration $\tilde{a}_y$

$$\Delta \gamma_i \simeq \left[-\left(\frac{t_i/2 - c_i}{c_i} \right) \rho_i^s + \rho_i^P \right] m\tilde{a}_y = \chi_i m\tilde{a}_y \qquad (7.18)$$

---

[4] In this model the roll inertial effects are totally disregarded.

Fig. 7.5 Positive static
camber γ_i^0 (front view)

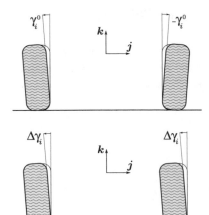

Fig. 7.6 Positive camber
variations $\Delta \gamma_i$ due to roll
motion (front view, left turn)

since the term $\pm z_i^s / c_i$ is usually negligible in road cars on flat roads.

Three suspensions with the same t_i, but with different values of c_i, are shown in
Fig. 7.7. We see that, as expected, the same amount of vehicle roll angle ϕ_i^s yields
different camber variations (tire roll angle ϕ_i^p not considered). In all cases the roll
angle is due to a left turn. Camber variations are negative in the first case.

7.2.6 Steer Angles

According to (3.210) and taking into account (7.15), we obtain the following (approximate, but very good) expressions for the steering angles of the two wheels of the
same axle

$$\delta_{i1} = -\delta_i^0 + \tau_i \delta_v + \varepsilon_i \frac{t_i}{2l} (\tau_i \delta_v)^2 + \Upsilon_i \rho_i^s m \tilde{a}_y = \delta_{i1}(\delta_v, \tilde{a}_y)$$

$$\delta_{i2} = \delta_i^0 + \tau_i \delta_v - \varepsilon_i \frac{t_i}{2l} (\tau_i \delta_v)^2 + \Upsilon_i \rho_i^s m \tilde{a}_y = \delta_{i2}(\delta_v, \tilde{a}_y) \tag{7.19}$$

which are, obviously, functions of the steering wheel rotation δ_v imposed by the
driver and, possibly, of the lateral acceleration $\tilde{a}_y = ur$.

In (7.19), as discussed in Sect. 3.4, δ_i^0 is the static toe angle, τ_i is the first-order
gear ratio of the whole steering system, ε_i is the Ackermann coefficient for dynamic
toe, Υ_i is the roll steer coefficient and $\rho_i^s m \tilde{a}_y$ is the suspension roll angle ϕ_i^s. If
the tires are supposed to be rigid, we have $\rho_1^s = \rho_2^s = (h - q)/k_\phi$. The analysis is
considerably simpler if $\Upsilon_i = 0$, that is if there is no roll steer. Most cars have $\tau_2 = 0$,
that is no direct steering of the rear wheels.

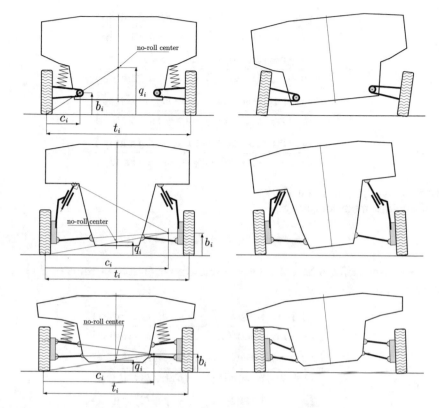

Fig. 7.7 Front view of three different suspensions (right), and their camber variations (left) due to the same positive vehicle roll angle ϕ_i^s (tire roll angle ϕ_i^p not considered)

7.2.7   Tire Slips

As already stated in Sect. 7.1.3, in the model under investigation all wheels are almost under longitudinal pure rolling conditions, that is $\sigma_{x_{ij}} \simeq 0$. Therefore, according to (3.59)

$$\omega_{11}r_1 = (u - rt_1/2)\cos(\delta_{11}) + (v + ra_1)\sin(\delta_{11})$$

$$\omega_{12}r_1 = (u + rt_1/2)\cos(\delta_{12}) + (v + ra_1)\sin(\delta_{12})$$

$$\omega_{21}r_2 = (u - rt_2/2)\cos(\delta_{21}) + (v - ra_2)\sin(\delta_{21})$$

$$\omega_{22}r_2 = (u + rt_2/2)\cos(\delta_{22}) + (v - ra_2)\sin(\delta_{22})$$

$$(7.20)$$

where ω_{ij} is the angular velocity of the corresponding rim and r_i is the wheel rolling radius, as defined in (2.38).

Under these assumed operating conditions, the tire *lateral* slips (3.60) become

$$\sigma_{y_{11}} = \frac{(v + ra_1) \cos(\delta_{11}) - (u - rt_1/2) \sin(\delta_{11})}{(u - rt_1/2) \cos(\delta_{11}) + (v + ra_1) \sin(\delta_{11})}$$

$$\sigma_{y_{12}} = \frac{(v + ra_1) \cos(\delta_{12}) - (u + rt_1/2) \sin(\delta_{12})}{(u + rt_1/2) \cos(\delta_{12}) + (v + ra_1) \sin(\delta_{12})}$$

$$\sigma_{y_{21}} = \frac{(v - ra_2) \cos(\delta_{21}) - (u - rt_2/2) \sin(\delta_{21})}{(u - rt_2/2) \cos(\delta_{21}) + (v - ra_2) \sin(\delta_{21})} \qquad (7.21)$$

$$\sigma_{y_{22}} = \frac{(v - ra_2) \cos(\delta_{22}) - (u + rt_2/2) \sin(\delta_{22})}{(u + rt_2/2) \cos(\delta_{22}) + (v - ra_2) \sin(\delta_{22})}$$

where $\delta_{ij} = \delta_{ij}(\delta_v, ur)$ as in (7.19).

Therefore, more compactly

$$\sigma_{y_{ij}} = \sigma_{y_{ij}}\left(v, r; u, \delta_{ij}(\delta_v, ur)\right) \qquad (7.22)$$

It will turn useful to have these very same slips expressed in terms of $\beta = v/u$ and $\rho = r/u$

$$\sigma_{y_{11}} = \frac{(\beta + \rho a_1) \cos(\delta_{11}) - (1 - \rho t_1/2) \sin(\delta_{11})}{(1 - \rho t_1/2) \cos(\delta_{11}) + (\beta + \rho a_1) \sin(\delta_{11})}$$

$$\sigma_{y_{12}} = \frac{(\beta + \rho a_1) \cos(\delta_{12}) - (1 + \rho t_1/2) \sin(\delta_{12})}{(1 + \rho t_1/2) \cos(\delta_{12}) + (\beta + \rho a_1) \sin(\delta_{12})}$$

$$\sigma_{y_{21}} = \frac{(\beta - \rho a_2) \cos(\delta_{21}) - (1 - \rho t_2/2) \sin(\delta_{21})}{(1 - \rho t_2/2) \cos(\delta_{21}) + (\beta - \rho a_2) \sin(\delta_{21})} \qquad (7.23)$$

$$\sigma_{y_{22}} = \frac{(\beta - \rho a_2) \cos(\delta_{22}) - (1 + \rho t_2/2) \sin(\delta_{22})}{(1 + \rho t_2/2) \cos(\delta_{22}) + (\beta - \rho a_2) \sin(\delta_{22})}$$

and, more compactly

$$\sigma_{y_{ij}} = \sigma_{y_{ij}}\left(\beta, \rho; \delta_{ij}(\delta_v, ur)\right) \qquad (7.24)$$

We see that the "main" dependence on u has disappeared.

7.2.8 Simplified Tire Slips

Equations (7.21) can be simplified without impairing their accuracy too much. More precisely, taking into account that $u \gg |v|$, $u \gg |rt_i|$, $|\delta_{ij}| \ll 1$, and $\omega_{ij} r_i \simeq u$, we obtain (see (3.55) and (3.57))

$$\sigma_{y_{11}} \simeq \frac{v + ra_1}{u} - \delta_{11} = \beta_1 - \delta_{11}$$

$$\sigma_{y_{12}} \simeq \frac{v + ra_1}{u} - \delta_{12} = \beta_1 - \delta_{12}$$

$$\sigma_{y_{21}} \simeq \frac{v - ra_2}{u} - \delta_{21} = \beta_2 - \delta_{21}$$ (7.25)

$$\sigma_{y_{22}} \simeq \frac{v - ra_2}{u} - \delta_{22} = \beta_2 - \delta_{22}$$

More explicitly, according to (7.19), we have

$$\sigma_{y_{11}} \simeq \frac{v + ra_1}{u} - \left(\tau_1 \delta_v - \delta_1^0 + \varepsilon_1 \frac{t_1}{2l}(\tau_1 \delta_v)^2 + \Upsilon_1 \rho_1^s mur \right)$$

$$\sigma_{y_{12}} \simeq \frac{v + ra_1}{u} - \left(\tau_1 \delta_v + \delta_1^0 - \varepsilon_1 \frac{t_1}{2l}(\tau_1 \delta_v)^2 + \Upsilon_1 \rho_1^s mur \right)$$

$$\sigma_{y_{21}} \simeq \frac{v - ra_2}{u} - \left(\tau_2 \delta_v - \delta_2^0 + \varepsilon_2 \frac{t_2}{2l}(\tau_2 \delta_v)^2 + \Upsilon_2 \rho_2^s mur \right)$$ (7.26)

$$\sigma_{y_{22}} \simeq \frac{v - ra_2}{u} - \left(\tau_2 \delta_v + \delta_2^0 - \varepsilon_2 \frac{t_2}{2l}(\tau_2 \delta_v)^2 + \Upsilon_2 \rho_2^s mur \right)$$

Most cars have $\tau_2 = 0$, that is no direct steering of the rear wheels.

Equations (7.21)–(7.26) show how the lateral tire slips σ_{ij} are related to the global vehicle motion, to the kinematic steer angles, to the toe-in/out angles, and to the roll steer angle. None of these contributions can be neglected, in general.

We can also look at the *actual* tire slip angles α_{ij} (Fig. 7.8), defined in (3.55) and (3.58). In this model (not to be taken as a general rule) we have

$$\sigma_{y_{ij}} \simeq -\alpha_{ij}$$ (7.27)

Just compare (3.58) with (7.25).

7.2.9 Tire Lateral Forces

The lateral force exerted by each tire on the vehicle depends on many quantities, as shown in the second equation in (2.82). For sure, there is a strong dependence on the vertical loads Z_{ij} and on the lateral slips $\sigma_{y_{ij}}$, while, in this vehicle model, the longitudinal slips $\sigma_{x_{ij}}$ are negligible. The camber angles γ_{ij} need to be considered as well, since they are quite influential, even if small. According to (3.212), the spin slips φ_{ij} are directly related to γ_{ij}. Therefore, a suitable model for the lateral force of each wheel with tire is (Fig. 7.1)

Fig. 7.8 Actual slip angles α_{ij} in the double track model (see also Fig. 3.13)

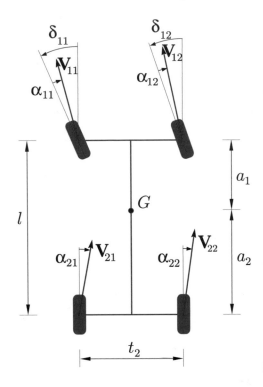

$$F_{y_{ij}} = F_{y_{ij}}\left(Z_{ij}, \gamma_{ij}, \sigma_{y_{ij}}\right) \tag{7.28}$$

Of course, extensive tire testing is required to make these functions available.

Needless to say, many other parameters affect the tire performance: road surface, temperature, inflation pressure, etc.

The lateral force Y_i for each axle of the vehicle is obtained by adding the lateral forces of the left wheel and of the right wheel (cf. (3.85) and (7.4), with $F_{x_{ij}} \simeq 0$)

$$Y_1 = F_{y_{11}} \cos(\delta_{11}) + F_{y_{12}} \cos(\delta_{12})$$

$$Y_2 = F_{y_{21}} + F_{y_{22}} \tag{7.29}$$

$$\Delta X_1 = [F_{y_{11}} \sin(\delta_{11}) - F_{y_{12}} \sin(\delta_{12})]/2$$

In general, the two wheels of the same axle undergo different vertical loads, different camber angles, and different lateral slips. Therefore, the two lateral forces are very different, as shown, e.g., in Fig. 3.28 and also in Fig 7.12. Equations (7.14) and (7.18), when inserted into (7.28), allow to take all these aspects into account. Therefore (Fig. 7.1)

$$Y_1 = F_{y_{11}}\Big(Z_{11}(ur), \gamma_{11}(ur), \sigma_{y_{11}}\Big)\cos\big(\delta_{11}(\delta_v, ur)\big)$$

$$+ F_{y_{12}}\Big(Z_{12}(ur), \gamma_{12}(ur), \sigma_{y_{12}}\Big)\cos\big(\delta_{12}(\delta_v, ur)\big)$$

$$= F_{y_1}(\sigma_{y_{11}}, \sigma_{y_{12}}, \delta_v, ur),$$

$$Y_2 = F_{y_{21}}\Big(Z_{21}(ur), \gamma_{21}(ur), \sigma_{y_{21}}\Big) + F_{y_{22}}\Big(Z_{22}(ur), \gamma_{22}(ur), \sigma_{y_{22}}\Big)$$

$$= F_{y_2}(\sigma_{y_{21}}, \sigma_{y_{22}}, ur),$$

$$\Delta X_1 = F_{y_{11}}\Big(Z_{11}(ur), \gamma_{11}(ur), \sigma_{y_{11}}\Big)\sin\big(\delta_{11}(\delta_v, ur)\big)$$

$$- F_{y_{12}}\Big(Z_{12}(ur), \gamma_{12}(ur), \sigma_{y_{12}}\Big)\sin\big(\delta_{12}(\delta_v, ur)\big)$$

$$= \Delta X_1(\sigma_{y_{11}}, \sigma_{y_{12}}, \delta_v, ur)$$

$$(7.30)$$

It should be clearly understood that the functions in (7.30) are *known* algebraic functions.

A general comment on this vehicle model is in order here: some quantities depend (linearly) *only* on the lateral acceleration $\tilde{a}_y = ur$. However, it must be remarked that this peculiarity needs an open differential, no aerodynamic forces, almost constant forward speed.

7.3 Double Track Model

7.3.1 Governing Equations of the Double Track Model

Summing up, the double track vehicle model for studying the handling of road cars is governed by the following three sets of equations:

- two *equilibrium* equations (lateral and yaw), as in (7.5)

$$m(\dot{v} + ur) = Y_1 + Y_2 = Y$$
$$J_z\dot{r} = Y_1a_1 - Y_2a_2 + \Delta X_1t_1 = N$$
$$(7.31)$$

- three *constitutive* equations, as in (7.30), which are affected by several setup parameters and by the vertical loads

$$Y_1 = F_{y_1}(\sigma_{y_{11}}, \sigma_{y_{12}}, \delta_v, ur)$$
$$Y_2 = F_{y_2}(\sigma_{y_{21}}, \sigma_{y_{22}}, ur)$$
$$\Delta X_1 = \Delta X_1(\sigma_{y_{11}}, \sigma_{y_{12}}, \delta_v, ur)$$
$$(7.32)$$

- four *congruence* equations (tire lateral slips), as in (7.21), which take care, among other things, of the Ackermann coefficient

$$\sigma_{y_{11}} = \sigma_{y_{11}}(v, r; u, \delta_{11}(\delta_v, ur))$$

$$\sigma_{y_{12}} = \sigma_{y_{12}}(v, r; u, \delta_{12}(\delta_v, ur))$$

$$\sigma_{y_{21}} = \sigma_{y_{21}}(v, r; u, \delta_{21}(\delta_v, ur)) \tag{7.33}$$

$$\sigma_{y_{22}} = \sigma_{y_{22}}(v, r; u, \delta_{22}(\delta_v, ur))$$

We have simply $\delta_{ij} = \delta_{ij}(\delta_v)$ if there is no roll steer.

This vehicle model for road vehicle handling is fairly general, and it is usually called *double track model*. A more classical formulation of the same model is obtained taking (7.27) into account. However, using σ_{ij} instead of α_{ij} is conceptually clearer.

7.3.2 Dynamical Equations of the Double Track Model

The dynamical equations for road vehicle handling are now promptly obtained. As a final step, it suffices to insert (7.32) and (7.33) into (7.31)

$$m(\dot{v} + ur) = Y(v, r; u, \delta_v)$$

$$J_z \dot{r} = N(v, r; u, \delta_v) \tag{7.34}$$

This is a dynamical system with two state variables, namely, but not necessarily, $v(t)$ and $r(t)$, as discussed in Sect. 7.3.3. The driver controls the steering wheel angle $\delta_v(t)$ and the forward speed u.

The double track model can be used to simulate and investigate the vehicle handling behavior under steady-state or transient conditions (i.e., nonconstant $\delta_v(t)$).

Unfortunately, the *double track model* is not as popular as the *single track model* (often and mistakenly also named "bicycle model"). The effort required to build a computer program and to run simulations with the double track model is comparable to the effort required by the less accurate single track model (introduced and discussed in Sect. 7.5).

7.3.3 Alternative State Variables (β and ρ)

The use of $v(t)$ and $r(t)$ as state variables is not mandatory, and other options can be envisaged. Other state variables may provide a better insight into vehicle handling, if properly handled.

The state variables $\beta(t)$ and $\rho(t)$ have been already introduced in (3.16) and (3.17). They are repeated here for ease of reading

$$\beta = \frac{v}{u} = -\frac{S}{R} \tag{3.16'}$$

and

$$\rho = \frac{r}{u} = \frac{1}{R} \tag{3.17'}$$

They are just v and r normalized with respect to u.

The corresponding three sets of equations of the double track model become:

- equilibrium equations (cf. (3.23), (3.27) and (7.31))

$$m(\dot{\beta}u + \beta\dot{u} + u^2\rho) = Y = Y_1 + Y_2$$
$$J_z(\dot{\rho}u + \rho\dot{u}) = N = Y_1a_1 - Y_2a_2 + \Delta X_1 t_1 \tag{7.35}$$

- constitutive equations (as in (7.32), with $\tilde{a}_y = ur = u^2\rho$)

$$Y_1 = F_{y_1}(\sigma_{y_{11}}, \sigma_{y_{12}}, \delta_v, u^2\rho)$$
$$Y_2 = F_{y_2}(\sigma_{y_{21}}, \sigma_{y_{22}}, u^2\rho) \tag{7.36}$$
$$\Delta X_1 = \Delta X_1(\sigma_{y_{11}}, \sigma_{y_{12}}, \delta_v, u^2\rho)$$

- congruence equations (cf. (7.33), with $\tilde{a}_y = ur = u^2\rho$)

$$\sigma_{y_{11}} = \sigma_{y_{11}}(\beta, \rho; \delta_{11}(\delta_v, u^2\rho))$$
$$\sigma_{y_{12}} = \sigma_{y_{12}}(\beta, \rho; \delta_{12}(\delta_v, u^2\rho))$$
$$\sigma_{y_{21}} = \sigma_{y_{21}}(\beta, \rho; \delta_{21}(\delta_v, u^2\rho)) \tag{7.37}$$
$$\sigma_{y_{22}} = \sigma_{y_{22}}(\beta, \rho; \delta_{22}(\delta_v, u^2\rho))$$

Therefore, in this case, the two dynamical equations (7.34) of the double track model become

$$m(\dot{\beta}u + \beta\dot{u} + u^2\rho) = Y(\beta, \rho; \delta_v, u^2\rho)$$
$$J_z(\dot{\rho}u + \rho\dot{u}) = N(\beta, \rho; \delta_v, u^2\rho) \tag{7.38}$$

where $|\dot{u}| \simeq 0$ and can be discarded. The dependence of Y and N on the lateral acceleration $u^2\rho$, and hence on the forward speed u, disappears if there is no roll steer. This is the main advantage in using β and ρ as state variables in the *double track model* for road cars.

Quite remarkably, we will see in (7.75) that in the *single track model*, when β and ρ are used as state variables, there is no dependence of Y and N on u even if roll steer is taken into account.

7.4 Vehicle in Steady-State Conditions

An essential step in understanding the behavior of a dynamical system, and therefore of a motor vehicle, is the determination of the steady-state (equilibrium) configurations (v_p, r_p). In physical terms, a vehicle is in steady-state conditions when, with fixed position δ_v of the steering wheel and at constant forward speed u, it goes around with *circular trajectories* of all of its points.

After having set $\dot{\delta}_v = 0$ and $\dot{u} = 0$, the mathematical conditions for the system being in steady state is to have $\dot{v} = 0$ and $\dot{r} = 0$ in (7.34). Accordingly, the *lateral acceleration* drops the $\dot{v}$ term and becomes at steady state

$$\tilde{a}_y = ur = u^2\rho = \frac{u^2}{R} \tag{7.39}$$

This equation was already introduced in (3.28).

Finding the equilibrium points (v_p, r_p), that is how the vehicle moves under given and constant δ_v and u, amounts to solving the system of two *algebraic* equations

$$\begin{aligned} mur &= Y(v, r; u, \delta_v) \\ 0 &= N(v, r; u, \delta_v) \end{aligned} \tag{7.40}$$

or, equivalently and more formally

$$\begin{aligned} 0 &= Y(v, r; u, \delta_v) - mur = f_v(v, r; u, \delta_v) \\ 0 &= N(v, r; u, \delta_v) = f_r(v, r; u, \delta_v) \end{aligned} \tag{7.41}$$

to get (v_p, r_p) such that

$$f_v(v_p, r_p; u, \delta_v) = 0 \quad \text{and} \quad f_r(v_p, r_p; u, \delta_v) = 0 \tag{7.42}$$

Because of the nonlinearity of the tire behavior, the number of possible solutions (v_p, r_p), for given (u, δ_v), is not known a priori. Typically, if more than one solution exists, at most only one is stable.

Equations (7.42) define implicitly the two *maps*

$$v_p = \hat{v}_p(u, \delta_v) \quad \text{and} \quad r_p = \hat{r}_p(u, \delta_v) \tag{7.43}$$

that is, the totality of steady-state (equilibrium) conditions as functions of the forward speed u and of the steering wheel angle δ_v. Given and kept constant the forward speed u and the steering wheel angle δ_v, after a while (a few seconds at most) the vehicle reaches the corresponding steady-state condition, characterized by a constant lateral speed v_p and a constant yaw rate r_p.

For a more "geometric", and hence more intuitive, analysis of the handling of vehicles, it is convenient to employ $\beta = v/u$ and $\rho = r/u$ instead of v and r, as done in Sect. 7.3.3. Therefore, (7.43) can be replaced by

$$\beta_p = \hat{\beta}_p(u, \delta_v) \quad \text{and} \quad \rho_p = \hat{\rho}_p(u, \delta_v) \tag{7.44}$$

The steady-state handling behavior is completely characterized by these *handling maps* of β and ρ, both as functions of two variables, namely, but not necessarily, u and δ_v

$$(u, \delta_v) \implies (\beta_p, \rho_p) \tag{7.45}$$

Indeed, it is common practice to employ $(\delta_v, \tilde{a}_y)$, instead of (u, δ_v), as parameters to characterize a steady-state condition. This is possible because

$$\tilde{a}_y = u\, r_p(u, \delta_v) \quad \text{which can be solved to get} \quad u = u(\delta_v, \tilde{a}_y) \tag{7.46}$$

Therefore, (7.43) becomes

$$\begin{aligned} v_p &= \hat{v}_p\big(\delta_v, u(\delta_v, \tilde{a}_y)\big) = v_p(\delta_v, \tilde{a}_y) \\ r_p &= \hat{r}_p\big(\delta_v, u(\delta_v, \tilde{a}_y)\big) = r_p(\delta_v, \tilde{a}_y) \end{aligned} \tag{7.47}$$

and, accordingly, (7.44) becomes

$$\begin{aligned} \beta_p &= \hat{\beta}_p\big(\delta_v, u(\delta_v, \tilde{a}_y)\big) = \beta_p(\delta_v, \tilde{a}_y) \\ \rho_p &= \hat{\rho}_p\big(\delta_v, u(\delta_v, \tilde{a}_y)\big) = \rho_p(\delta_v, \tilde{a}_y) \end{aligned} \tag{7.48}$$

At first it may look a bit odd to employ $(\delta_v, \tilde{a}_y)$ instead of (u, δ_v), but it is not, since it happens that in most road cars a few steady-state quantities are functions of $\tilde{a}_y$ only. This is quite a remarkable fact, but it should not be taken as a general rule.[5]

Similarly, we could use $(u, \tilde{a}_y)$ as parameters to characterize a steady-state condition.

Equations (7.44) or (7.48) provide a fairly general point of view that led to the new global approach that we called **Map of Achievable Performance (MAP)** in Chap. 6. Additional, relevant information are provided in Sects. 7.7 and 7.8. These MAPs can be obtained experimentally or through simulations. Therefore, they are not limited to mathematical models. Actually, as will be discussed in the next chapter,

---

[5] For instance, vehicles equipped with a locked differential and/or with relevant aerodynamic downforces always need (at least) two parameters.

they exist also for race cars, including Formula cars with very high aerodynamic downforces.

Now we address more classical topics, like the single track model and the associated handling diagram.

7.5 Single Track Model

The goal of this Section is to present a comprehensive analysis of the single track model [2, 5, 6, 8, 12, 17, 19], thus showing also its limitations. In many courses or books on vehicle dynamics (e.g., [8, p. 199]) the single track model, shown in Fig. 7.9, is proposed without explaining in detail why, despite its awful appearance, it can provide in some cases useful insights into vehicle handling, particularly for educational purposes. Vehicle engineers should be well aware of the steps taken to simplify the model, and hence realize that in some cases the single track model may miss some crucial phenomena, and the double track model should be used instead.

7.5.1 From Double Track to Single Track

The double track model is shown in Figs. 7.1 and 7.2. Of course, it has four wheels. In the corresponding schematic representations of the *single track model* shown in Fig. 7.9 it looks like there are only two wheels. It is not so. It is not a bicycle. In the

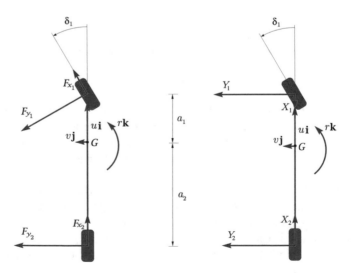

Fig. 7.9 Equivalent schematic representations of the single track model (with $\delta_2 = 0$)

single track model we look at each axle as a whole. Therefore, better names would be "four wheel" model and "two axle" model, respectively.

To go from the double track model to the single track model (Fig. 7.9) we need to further simplify (7.26): *the Ackermann corrections have to be set equal to zero*, that is

$$\varepsilon_1 = \varepsilon_2 = 0 \tag{7.49}$$

which is consistent with *small steering angles*. Indeed the Ackermann correction is a second order contribution in (7.26) and, important as it can be, it cannot be included in the single track model. You see that we are missing something.

This (not necessarily true) hypothesis (7.49) on the Ackermann coefficients, if combined with the simplified expressions (7.26), leads to the following (first-order) expressions for the lateral slips of the four wheels

$$
\begin{aligned}
\sigma_{y_{11}} &\simeq \left(\frac{v + ra_1}{u} - \tau_1 \delta_v \right) + \delta_1^0 - \Upsilon_1 \rho_1^s m \tilde{a}_y \\[2mm]
\sigma_{y_{12}} &\simeq \left(\frac{v + ra_1}{u} - \tau_1 \delta_v \right) - \delta_1^0 - \Upsilon_1 \rho_1^s m \tilde{a}_y \\[2mm]
\sigma_{y_{21}} &\simeq \left(\frac{v - ra_2}{u} - \tau_2 \delta_v \right) + \delta_2^0 - \Upsilon_2 \rho_2^s m \tilde{a}_y \\[2mm]
\sigma_{y_{22}} &\simeq \left(\frac{v - ra_2}{u} - \tau_2 \delta_v \right) - \delta_2^0 - \Upsilon_2 \rho_2^s m \tilde{a}_y
\end{aligned}
\tag{7.50}
$$

where we can still take into account the toe-in/toe-out terms δ_i^0, and also the roll steer contributions.

In (7.50) it is convenient to define what may be called the *apparent* slip angles α_1 and α_2 of the front and rear *axles*, respectively (Fig. 7.20)

$$
\begin{aligned}
\alpha_1 &= \tau_1 \delta_v - \frac{v + ra_1}{u} = \tau_1 \delta_v - \beta - \rho a_1 \\[2mm]
\alpha_2 &= \tau_2 \delta_v - \frac{v - ra_2}{u} = \tau_2 \delta_v - \beta + \rho a_2
\end{aligned}
\tag{7.51}
$$

Combining (7.50) and (7.51), we obtain that *both* front lateral slips $\sigma_{y_{11}}$ and $\sigma_{y_{12}}$ are *known* functions of the same *two* variables α_1 and $\tilde{a}_y$. Similarly, *both* rear lateral slips $\sigma_{y_{21}}$ and $\sigma_{y_{22}}$ are *known* functions of the same *two* variables α_2 and $\tilde{a}_y$

$$\sigma_{y_{11}} \simeq -\alpha_1 + \delta_1^0 - \Upsilon_1 \rho_1^s m \tilde{a}_y = \sigma_{y_{11}}(\alpha_1, \tilde{a}_y)$$

$$\sigma_{y_{12}} \simeq -\alpha_1 - \delta_1^0 - \Upsilon_1 \rho_1^s m \tilde{a}_y = \sigma_{y_{12}}(\alpha_1, \tilde{a}_y)$$

$$\sigma_{y_{21}} \simeq -\alpha_2 + \delta_2^0 - \Upsilon_2 \rho_2^s m \tilde{a}_y = \sigma_{y_{21}}(\alpha_2, \tilde{a}_y) \tag{7.52}$$

$$\sigma_{y_{22}} \simeq -\alpha_2 - \delta_2^0 - \Upsilon_2 \rho_2^s m \tilde{a}_y = \sigma_{y_{22}}(\alpha_2, \tilde{a}_y)$$

The two wheels of the same axle undergo the same apparent slip angle, but not necessarily the same lateral slip. The key point for the model to be single track is that the difference between left and right lateral slips must be a function only of $\tilde{a}_y = ur$. *This is the peculiar feature of the single track model* (cf. (7.26)). It is the fundamental brick for the next step.

But before doing that, it is worth noting the crucial difference between the *actual* slip angles α_{ij} of each *wheel*, defined in (3.58) (and also in (7.26)), and the *apparent* slip angles α_i of each *axle*, defined in (7.51).

$$\alpha_{11} = \alpha_1 - \delta_1^0 + \Upsilon_1 \rho_1^s mur + \varepsilon_1 \frac{t_1}{2l}(\tau_1 \delta_v)^2$$

$$\alpha_{12} = \alpha_1 + \delta_1^0 + \Upsilon_1 \rho_1^s mur - \varepsilon_1 \frac{t_1}{2l}(\tau_1 \delta_v)^2$$

$$\alpha_{21} = \alpha_2 - \delta_2^0 + \Upsilon_2 \rho_2^s mur + \varepsilon_2 \frac{t_2}{2l}(\tau_2 \delta_v)^2 \tag{7.53}$$

$$\alpha_{22} = \alpha_2 + \delta_2^0 + \Upsilon_2 \rho_2^s mur - \varepsilon_2 \frac{t_2}{2l}(\tau_2 \delta_v)^2$$

In general, the two apparent slip angles α_i can be defined only in the single track model (Fig. 7.20). In real vehicles there are four actual slip angles α_{ij}.

It is very common in traditional (oversimplified) vehicle dynamics not to take into account toe-in/toe-out and roll steering, thus having

$$\sigma_{y_{11}} \simeq \sigma_{y_{12}} \simeq -\alpha_1$$

$$\sigma_{y_{21}} \simeq \sigma_{y_{22}} \simeq -\alpha_2 \tag{7.54}$$

7.5.2 *"Forcing" the Lateral Forces*

Owing to (7.52), the first two equations in (7.30) become

$$Y_1 = F_{y_{11}}\left(Z_{11}(\tilde{a}_y), \gamma_{11}(\tilde{a}_y), \sigma_{y_{11}}(\alpha_1, \tilde{a}_y)\right) + F_{y_{12}}\left(Z_{12}(\tilde{a}_y), \gamma_{12}(\tilde{a}_y), \sigma_{y_{12}}(\alpha_1, \tilde{a}_y)\right)$$

$$= F_{y_{11}}(\alpha_1, \tilde{a}_y) + F_{y_{12}}(\alpha_1, \tilde{a}_y)$$

$$= F_{y_1}(\alpha_1, \tilde{a}_y);$$

$$Y_2 = F_{y_{21}}\left(Z_{21}(\tilde{a}_y), \gamma_{21}(\tilde{a}_y), \sigma_{y_{21}}(\alpha_2, \tilde{a}_y)\right) + F_{y_{22}}\left(Z_{22}(\tilde{a}_y), \gamma_{22}(\tilde{a}_y), \sigma_{y_{22}}(\alpha_2, \tilde{a}_y)\right)$$

$$= F_{y_{21}}(\alpha_2, \tilde{a}_y) + F_{y_{22}}(\alpha_2, \tilde{a}_y)$$

$$= F_{y_2}(\alpha_2, \tilde{a}_y),$$

$$(7.55)$$

while the third equation in (7.30) is set to zero because of the assumed very small steer angles

$$\Delta X_1 = 0 \qquad (7.56)$$

It is really crucial for a vehicle engineer to understand and keep in mind the differences between (7.30) and (7.55). In the final expressions of Y_i in (7.55) there appear only variables associated to the corresponding axle, not anymore to the single wheel.

As already obtained in (7.9) at the beginning of this chapter, we have that the lateral forces are basically linear functions of $\tilde{a}_y$ (open differential)

$$Y_1 \simeq \frac{ma_2}{l}\tilde{a}_y \quad \text{and} \quad Y_2 \simeq \frac{ma_1}{l}\tilde{a}_y \qquad (7.57)$$

Therefore, $F_{y_1}(\alpha_1, \tilde{a}_y)$ and $F_{y_2}(\alpha_2, \tilde{a}_y)$ must be such that

$$F_{y_1}(\alpha_1, \tilde{a}_y) = \frac{ma_2}{l}\tilde{a}_y \quad \text{and} \quad F_{y_2}(\alpha_2, \tilde{a}_y) = \frac{ma_1}{l}\tilde{a}_y \qquad (7.58)$$

which can be solved with respect to the lateral acceleration, to obtain[6]

$$\tilde{a}_y = g_1(\alpha_1) \quad \text{and} \quad \tilde{a}_y = g_2(\alpha_2) \qquad (7.59)$$

These relationships are affected by many setup parameters, like camber angles, roll steer, toe-in/toe-out, etc., as discussed in detail in Sect. 7.5.3.

The final, crucial, step is inserting (i.e., "forcing") these results back into (7.55), thus obtaining the *axle characteristics* of the single track model

$$Y_1(\alpha_1) = F_{y_1}(\alpha_1, g_1(\alpha_1)) \quad \text{and} \quad Y_2(\alpha_2) = F_{y_2}(\alpha_2, g_2(\alpha_2)) \qquad (7.60)$$

that is, two functions, one per axle, that give the axle lateral force as a function of *only* the corresponding *apparent* slip angle. In other words, each axle behaves formally as an equivalent single wheel with tire.

---

[6] This step would not be possible with F_{y_i} as in (7.30).

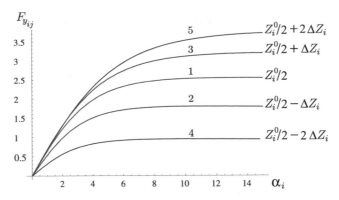

Fig. 7.10 Tire tested under symmetric vertical loads with respect to the static load $Z_0/2$

Forcing the lateral forces to be as in (7.60) is an approximation when the vehicle is in transient conditions. Moreover, to go from (7.30) to (7.55) we made assumptions (7.49) about the steer kinematics (parallel steering) and small steer angles.

The double track model provides more accurate results when running simulations. On the other hand, the single track model is a useful tool for educational purposes and for investigating steady-state conditions. It is less accurate, but more intuitive.

Equation (7.59) implies that in the single track model there is a link between α_1 and α_2

$$g_1(\alpha_1) = \tilde{a}_y = g_2(\alpha_2) \tag{7.61}$$

and that this link is not affected by u or δ_v. In a real vehicle this is not necessarily true. Vehicle engineers should be aware that the single track model is somehow an inconsistent model, albeit very appealing.

7.5.3   Axle Characteristics

As done in (7.60), by *axle characteristics* we mean two algebraic functions (one per axle) of the form

$$Y_i = F_{y_i} = Y_i(\alpha_i) \tag{7.62}$$

which provide the total lateral force as a function of the apparent slip angle only, with the effects, e.g., of the lateral load transfers already accounted for. They were obtained in (7.60), but the topic is so relevant to deserve an in-depth discussion.

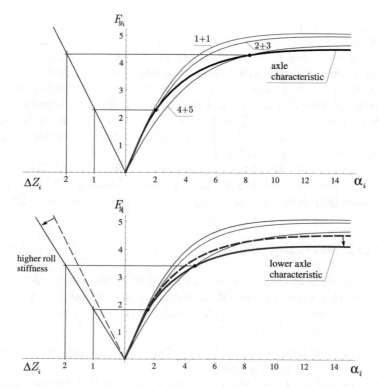

Fig. 7.11 Basic graphic construction of the axle characteristic and influence of changing the roll stiffness

7.5.3.1 The Basics

The basic procedure to obtain the axle characteristics is described here. The goal is to provide an intuitive and physical approach to the construction of the axle characteristics in the single track model. "Basic" means that only the effects of the lateral load transfers ΔZ_i are taken into account. Of course, lateral load transfers cannot be omitted. They must necessarily be included in the analysis (unless linear tire behavior is assumed, Fig. 7.62).

The first step is to test the tire under symmetric vertical loads with respect to the reference value $Z_i^0/2$, as shown in Fig. 7.10. "Symmetric" means that tests have to be carried out in pairs, that is with $F_z = Z_i^0/2 \pm \Delta Z_i$. In Fig. 7.10 two such pairs are shown.

The second step is to add the two tire curves obtained with symmetric vertical loads, as shown in Fig. 7.11(top), thus getting a sort of axle curve for each value of the lateral load transfer. To legitimate this second step it is mandatory that the inner wheel and the outer wheel of the same axle undergo the same apparent slip angle α_i. As expected, the higher the lateral load transfer ΔZ_i, the lower the corresponding axle curve.

The third step is to draw a straight line according to (7.13), to linearly relate the lateral load transfer ΔZ_i to the axle lateral force F_{y_i}.

The fourth and final step is to pick the *unique point* on each axle curve that corresponds to a real operating (steady-state) condition for the vehicle, as shown in Fig. 7.11(top). As a matter of fact, each axle curve was obtained testing the tire with given and constant $\pm \Delta Z_i$, but this amount of lateral load transfer requires a definite value of the lateral force F_{y_i} in the vehicle, and hence a definite value of α_i.

The sought *axle characteristic* $Y_i(\alpha_i)$ is just the curve connecting all these points, as schematically shown in Fig. 7.11.

Changing the value of η_i in (7.13) results in a different straight line and hence in different axle characteristics, as shown in Fig. 7.11(bottom). The axle curves are not affected by η_i, but the points corresponding to real operating conditions are.

7.5.3.2   The General Case

Now we are ready to address the construction of the axle characteristics with greater generality. It means that we will use the lateral acceleration $\tilde{a}_y$ as a parameter.

According to (7.30), (7.52), (7.55) and (7.57), the general framework, at steady state, for a given vehicle is that:

1. each axle lateral force Y_i is determined solely by the lateral acceleration $\tilde{a}_y$, see (7.9) (open differential);
2. there is a one-to-one correspondence between the lateral acceleration $\tilde{a}_y$ and the following quantities:

    - lateral load transfers ΔZ_i, see (7.10);
    - camber angles γ_{ij}, see (7.17) and (7.18);
    - roll steer angles $\Upsilon_i \phi_i^s \tilde{a}_y$, see (7.19);

3. both left and right tire lateral forces are known functions of the *lateral acceleration* $\tilde{a}_y$ and of the *same apparent slip angle* α_i, see (7.55).

Therefore, as discussed in Sect. 7.5.2, for any given value of $\tilde{a}_y$ we can obtain the corresponding load transfers, camber angles and roll steer angles. Consequently, we can plot (measure) the lateral forces $F_{y_{ij}}(\alpha_i)$ of each wheel as functions *only* of the apparent slip angle α_i

$$
\begin{aligned}
Y_{11}(\alpha_1) &= F_{y_{11}}\big(\alpha_1, g_1(\alpha_1)\big) \\[4pt]
Y_{12}(\alpha_1) &= F_{y_{12}}\big(\alpha_1, g_1(\alpha_1)\big) \\[4pt]
Y_{21}(\alpha_2) &= F_{y_{21}}\big(\alpha_2, g_2(\alpha_2)\big) \\[4pt]
Y_{22}(\alpha_2) &= F_{y_{22}}\big(\alpha_2, g_2(\alpha_2)\big)
\end{aligned}
\qquad (7.63)
$$

This way we can single out the contribution of each wheel. Typical curves for the inner and the outer tires of the same axle are shown in Fig. 7.12. As expected, the outer wheel (green plot) provides a lateral force larger than the inner wheel (red plot).

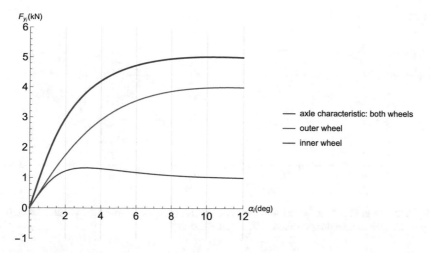

Fig. 7.12 Lateral forces exerted by the inner wheel (red), by the outer wheel (green) of the same axle, and resulting axle characteristic (blue). Maximum lateral force is 5 kN at 10.34 deg

Each *axle characteristic* is given by (Fig. 7.12, blue plot)

$$Y_i(\alpha_i) = Y_{i1}(\alpha_i) + Y_{i2}(\alpha_i) \tag{7.64}$$

Of course, any calculation of this type assumes the availability of tire data.

As shown in Fig. 7.12, the maximum lateral forces of the two wheels are not attained, in general, for the same apparent slip angle. This is not desirable, if we are interested in maximising the lateral acceleration. To mitigate this phenomenon we can resort on tuning some setup parameters. We have to understand the effects on the axle characteristics of changing these setup parameters. This extremely relevant practical topic is discussed hereafter, taking into account the effects of changing the:

1. lateral load transfer;
2. static camber angles;
3. roll camber;
4. toe-in/toe-out;
5. roll steer.

All plots in this section are for a car making a left turn ($\tilde{a}_y > 0$). In all plots in this section, the apparent slip angles are in degrees and the lateral forces are in kN.

7.5.3.3 Lateral Load Transfer ΔZ_i

Two additional basic examples are shown in Fig. 7.13. They are basic in the sense that it is assumed that the lateral acceleration $\tilde{a}_y$ affects only the lateral load transfer ΔZ_i. More precisely, it is assumed that $\gamma_{ij} = \delta_i^0 = \Upsilon_i = 0$.

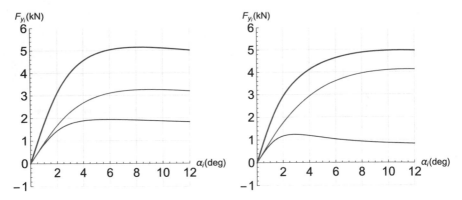

Fig. 7.13 As in Fig. 7.12, but with lower load transfer (left: max force 5.15 kN at 8.20 deg) or higher load transfer (right: max force 4.97 kN at 10.95 deg)

The two cases in Fig. 7.13 have different values of η_i, and hence different load transfers for the same lateral acceleration, with respect to the case in Fig. 7.12. One has lower η_i (left: max force 5.15 kN at 8.20 deg) and one has higher η_i (right: max force 4.97 kN at 10.95 deg).

A very relevant fact in vehicle dynamics, as stated in Sects 2.10.2 and 2.12, is that the lateral force exerted by a single tire grows *less than proportionally* with respect to the vertical load. This is clearly shown in Fig. 7.13, where the higher the lateral load transfer, the lower the resulting curve of Y_i.

7.5.3.4 Static Camber γ_i^0

The definition of static camber is given in Fig. 7.5 and in (7.17). The effects of negative and positive static camber angles, i.e. $\gamma_i^0 \neq 0$, are shown in Fig. 7.14, left and right, respectively. If the top of the wheel is farther out than the bottom (that is, away from the axle), it is called positive static camber. If the bottom of the wheel is farther out than the top, it is called negative camber. We see in Fig. 7.14 that there are lateral forces on each wheel (camber thrust) when the car is going straight. We also see that the peak (max) value of the *axle* lateral force is higher with negative static camber. The main reason is that the inner and outer wheel reach their peak values for apparent slip angles that are less far away from each other.

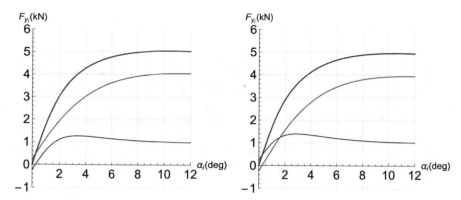

Fig. 7.14 As in Fig. 7.12, but with *negative static camber* (left: max 5.03 kN at 10.23 deg) or *positive static camber* (right: max 4.97 kN at 10.47 deg)

7.5.3.5 Roll Camber $\Delta\gamma_i$

As shown in Fig. 7.15 and in (7.18), roll camber $\Delta\gamma_i$ is an anti-symmetric setup modification. Therefore, it also affects the slope in the origin of the axle characteristic. Negative and positive camber variations due to roll motion are shown in Fig. 7.6. Also useful may be Fig. 7.7, which shows how the suspension architecture strongly affects roll camber.

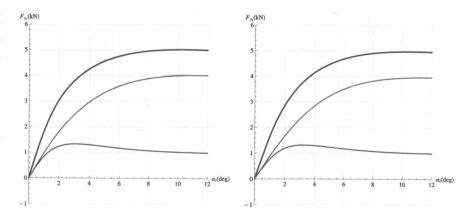

Fig. 7.15 As in Fig. 7.12, but with *negative roll camber* (left: max 5.02 kN at 10.29 deg) or *positive roll camber* (right: max 4.98 kN at 10.40 deg)

7.5.3.6   Toe-in/Toe-out δ_i^0

The definition of toe-in/toe-out is given in Fig. 3.15 and in (7.19). The effects of toe-in ($\delta_i^0 > 0$) and toe-out ($\delta_i^0 < 0$), are shown in Fig. 7.16, left and right, respectively. We see that also in this case there are lateral forces on the wheels when the car is going straight.

The beneficial cumulative effects of negative static camber, negative roll camber and toe-in are shown in Fig. 7.17. We see that the peak value from 5 kN reached 5.09 kN. This result was achieved because the inner and outer wheel reach their peak values for apparent slip angles that are now closer to each other.

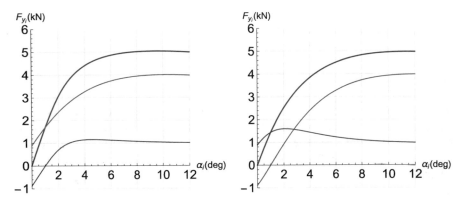

Fig. 7.16 As in Fig. 7.12, but with *toe-in* (left: max 5.03 kN at 9.40 deg) or *toe-out* (right: max 4.97 kN at 11.42 deg)

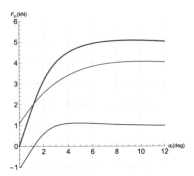

Fig. 7.17 As in Fig. 7.12, but with the cumulative effects of negative static camber, negative roll camber, and toe-in (max 5.09 kN at 9.24 deg)

7.5.3.7   Roll Steer $\Upsilon_i \rho_i^s m \tilde{a}_y$

Also interesting is the case of roll steer, i.e. $\Upsilon_i \neq 0$, shown in Fig. 7.18. While most effects are symmetric with respect to the vehicle axis, and hence the contributions of the two wheels cancel each other at low lateral acceleration, the roll steer is anti-symmetric, and hence it affects the axle characteristic also at low lateral accelerations. However, contrary to all other setup parameters here considered, it does not affect the peak value of the axle lateral force.

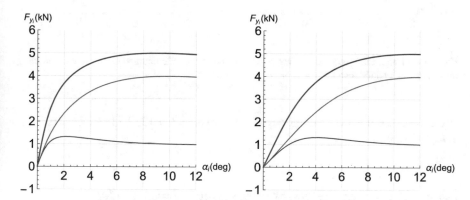

Fig. 7.18 As in Fig. 7.12, but with *positive roll steer* (left: max 5 kN at 9.12 deg) or *negative roll steer* (right: max 5 kN at 11.61 deg)

7.5.3.8   General (Mixed-up) Case

As already shown in Fig. 7.17, in general all these effects may very well coexist in a real car. The axle characteristics are what most characterize vehicle dynamics. They may differ in the initial slope (slip stiffness) and in the maximum value. Both aspects have a big influence on vehicle handling.

We remark that the axle characteristics, under an apparent simplicity, contain a lot of information about the vehicle features and setup (see also [12, Chap. 6]).

7.5.3.9   Look at the Peak Positions

For just a moment, let us assume the grip coefficient being not dependent on the vertical load. More precisely, set $a_1 = 0$ in (2.99). Even in this case, the amount of lateral load transfer does affect the peak value of the axle lateral force, as shown in Fig. 7.19. Therefore, we should do our best to keep the apparent slip angles of the inner and outer peaks as close together as possible. This issue is rarely discussed,

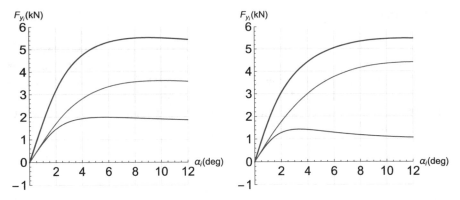

Fig. 7.19 Higher load transfers lower the peak value of the axle characteristic even if assuming (erroneously) load independent grip

maybe because it is unusual to draw Figures like 7.12, 7.13, 7.14, 7.15, 7.16, 7.17, 7.18 and 7.19.

7.5.4   Governing Equations of the Single Track Model

Summing up, the single track model is governed by the following three sets of fairly simple equations:

- two *equilibrium* equations (lateral and yaw), as in (7.5)

$$m(\dot{v} + ur) = Y = Y_1 + Y_2$$
$$J_z\dot{r} = N = Y_1a_1 - Y_2a_2 \tag{7.65}$$

- two *constitutive* equations (axle characteristics, which include the effects of several setup parameters), as in (7.60)

$$Y_1 = Y_1(\alpha_1)$$
$$Y_2 = Y_2(\alpha_2) \tag{7.66}$$

- two *congruence* equations (apparent slip angles), as in (7.51)

$$\alpha_1 = \tau_1\delta_v - \frac{v + ra_1}{u}$$
$$\alpha_2 = \tau_2\delta_v - \frac{v - ra_2}{u} \tag{7.67}$$

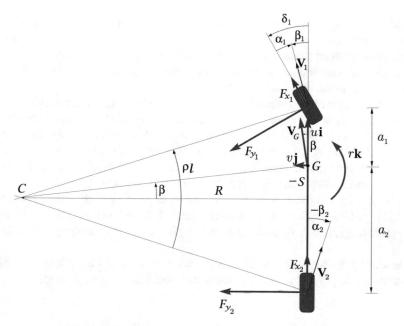

Fig. 7.20 Single track model

A comparison with the governing equations of the double track model (Sect. 7.3.1) shows that in the single track model:

- the term $\Delta X_1 t_1$ has disappeared from the equilibrium equations;
- there are two, instead of four, constitutive equations;
- there are two, instead of four, congruence equations.

A pictorial version of the single track model is shown in Fig. 7.20, where

$$\delta_1 = \tau_1 \delta_v = (1 + \kappa)\delta$$
$$\delta_2 = \tau_2 \delta_v = \chi \tau_1 \delta_v = \kappa \delta \tag{7.68}$$

with

$$\delta = \delta_1 - \delta_2 = (\tau_1 - \tau_2)\delta_v = (1 - \chi)\tau_1 \delta_v = \tau \delta_v \tag{7.69}$$

Reasonably, but also arbitrarily, we call δ_1 the steer angle of the front *axle*. A similar thing can be done for the rear axle.

The angle δ is called *net steer angle* of the (single track model of the) vehicle. Usually, $\kappa = 0$ and hence δ is just the steering angle δ_1 of the front axle. However, $\kappa \neq 0$ leaves room for rear steering δ_2 as well, without affecting δ.

In this single track model there is a one-to-one relationship between δ and δ_v, that is we have a *rigid* steering system. A vehicle model with *compliant* steering system is developed in Sect. 7.16.

Indeed, the equations governing the dynamical system of Fig. 7.20 are precisely (7.65), (7.66) and (7.67). Therefore, Fig. 7.20 can be used as a *shortcut* to quickly obtain the simplified equations of a vehicle. However, the vehicle model still has four wheels, lateral load transfers, camber and camber variations, roll steer, as discussed in Sect. 7.5.3 on the axle characteristics.

The main feature of this model is that the two wheels of the same axle undergo the same *apparent* slip angle α_i, and hence can be replaced by a sort of equivalent wheel, like in Fig. 7.20. However, that does not imply that the real slip angles of the two wheels of the same axle are the same. Neither are the camber angles, the roll steer angles, the vertical loads. Therefore, the single track model is not really single track! It retains many of the features of the double track model.

It is not necessary to assume that the center of mass G of the vehicle is at road level [11, p. 170], neither that the lateral forces of the left and right tires to be equal to each other [1, p. 53]. Actually, both assumptions would be strikingly false in any car.

Assuming the total mass to be concentrated at G, as if the vehicle were like a point mass [18, p. 223], is another unrealistic, and unnecessary, assumption.

7.5.5 Dynamical Equations of the Single Track Model

Among the governing equations, only the two equilibrium equations are differential equations, and both are first-order. The other four algebraic equations must be inserted into the equilibrium equations to ultimately obtain the two *dynamical equations* of the *single track model*

$$m(\dot{v} + ur) = Y_1\left(\delta_v \tau_1 - \frac{v + ra_1}{u}\right) + Y_2\left(\delta_v \tau_2 - \frac{v - ra_2}{u}\right)$$
$$J_z\dot{r} = a_1 Y_1\left(\delta_v \tau_1 - \frac{v + ra_1}{u}\right) - a_2 Y_2\left(\delta_v \tau_2 - \frac{v - ra_2}{u}\right) \tag{7.70}$$

or, more compactly

$$m(\dot{v} + ur) = Y(v, r; u, \delta_v)$$
$$J_z\dot{r} = N(v, r; u, \delta_v) \tag{7.71}$$

Therefore, the single track model is a dynamical system with two state variables, namely, but not necessarily, $v(t)$ and $r(t)$, as discussed in Sect. 7.5.6. The driver controls the steering wheel angle $\delta_v(t)$ and the forward speed u.

7.5.6 Alternative State Variables (β and ρ)

As already done in Sect. 7.3.3, instead of $v(t)$ and $r(t)$, we can use $\beta(t) = v/u$ and $\rho(t) = r/u$ to describe the handling of a vehicle.

The corresponding governing equations of the single track model become:

- equilibrium equations (cf. (7.65))

$$m(\dot{\beta}u + \beta\dot{u} + u^2\rho) = Y = Y_1 + Y_2$$
$$J_z(\dot{\rho}u + \rho\dot{u}) = N = Y_1a_1 - Y_2a_2$$

(7.72)

- constitutive equations (cf. (7.66))

$$Y_1 = Y_1(\alpha_1)$$
$$Y_2 = Y_2(\alpha_2)$$

(7.73)

- congruence equations (cf. (7.67))

$$\alpha_1 = \delta_v\tau_1 - \beta - \rho a_1$$
$$\alpha_2 = \delta_v\tau_2 - \beta + \rho a_2$$

(7.74)

Combining these three sets of equations, we obtain the dynamical equations, that is the counterpart of (7.70)

$$m(\dot{\beta}u + \beta\dot{u} + u^2\rho) = Y(\beta, \rho; \delta_v)$$
$$J_z(\dot{\rho}u + \rho\dot{u}) = N(\beta, \rho; \delta_v)$$

(7.75)

where $|\dot{u}| \simeq 0$.

It is worth noting that, differently from (7.38) of the double track model, the axle lateral forces Y_1 and Y_2, and hence also the total lateral force Y and the yaw moment N, do *not* depend explicitly on the forward speed u, even if roll steer is taken into account. All the effects of the lateral acceleration $\tilde{a}_y = ur = u^2\rho$ on Y and N are already included in the axle characteristics. Moreover, the expressions of Y and N in (7.75) are even simpler than those in (7.71).

7.5.7 Inverse Congruence Equations

The state variables v and r appear in both congruence equations (7.67). However, it is possible to invert these equations to obtain two other equivalent equations, with $\rho = r/u$ appearing only in the first equation and $\beta = v/u$ only in the second equation

$$\rho = \frac{r}{u} = \frac{\delta_1 - \delta_2}{l} - \frac{\alpha_1 - \alpha_2}{l}$$
$$\beta = \frac{v}{u} = \frac{\delta_1 a_2 + \delta_2 a_1}{l} - \frac{\alpha_1 a_2 + \alpha_2 a_1}{l}$$

(7.76)

where the more compact notation $\delta_1 = \delta_v\tau_1$ and $\delta_2 = \delta_v\tau_2$ has been used.

It is important to realize that all these inverse congruence equations are not limited to steady-state conditions, although they are mostly used for the evaluation of some steady-state features.

Another very common way to rewrite the first equation in (7.76) is as follows

$$\alpha_1 - \alpha_2 = (\delta_1 - \delta_2) - \frac{l}{R} = \delta - \frac{l}{R} \tag{7.77}$$

where $R = u/r$. Should $\alpha_1 = \alpha_2 = 0$ (very low speed), then $\delta = l/R$, which is often called Ackermann angle (not to be confused with Ackermann steering geometry, discussed in Sect. 3.4).

7.5.8 β_1 and β_2 as State Variables

Another useful set of state variables may be the vehicle slip angles at each axle midpoint (Fig. 7.20)

$$\beta_1 = \beta + \rho a_1 = \delta_1 - \alpha_1 = \tau_1 \delta_v - \alpha_1 = (1 + \kappa)\tau \delta_v - \alpha_1$$
$$\beta_2 = \beta - \rho a_2 = \delta_2 - \alpha_2 = \tau_2 \delta_v - \alpha_2 = \kappa \tau \delta_v - \alpha_2 \tag{7.78}$$

The inverse equations are

$$\rho = \frac{\beta_1 - \beta_2}{l} = \frac{1}{R}$$
$$\beta = \frac{\beta_1 a_2 + \beta_2 a_1}{l} \tag{7.79}$$

The corresponding governing equations of the single track model become:

- equilibrium equations

$$\dot{\beta}_1 u + \beta_1 \dot{u} + (\beta_1 - \beta_2)\frac{u^2}{l} = \frac{Y}{m} + \frac{N}{J_z} a_1$$
$$\dot{\beta}_2 u + \beta_2 \dot{u} + (\beta_1 - \beta_2)\frac{u^2}{l} = \frac{Y}{m} - \frac{N}{J_z} a_2 \tag{7.80}$$

- constitutive equations (from the axle characteristics)

$$Y_1 = Y_1(\alpha_1)$$
$$Y_2 = Y_2(\alpha_2) \tag{7.81}$$

- congruence equations

$$\alpha_1 = \delta_v \tau_1 - \beta_1 = \delta_1 - \beta_1$$
$$\alpha_2 = \delta_v \tau_2 - \beta_2 = \delta_2 - \beta_2$$

(7.82)

The two first-order differential equations (7.70) or (7.75), governing the dynamical system, become

$$\dot{\beta}_1 u + \beta_1 \dot{u} + (\beta_1 - \beta_2) \frac{u^2}{l} = \frac{J_z + ma_1^2}{m J_z} Y_1(\delta_v \tau_1 - \beta_1) + \frac{J_z - ma_1 a_2}{m J_z} Y_2(\delta_v \tau_2 - \beta_2)$$

$$\dot{\beta}_2 u + \beta_2 \dot{u} + (\beta_1 - \beta_2) \frac{u^2}{l} = \frac{J_z + ma_2^2}{m J_z} Y_2(\delta_v \tau_2 - \beta_2) + \frac{J_z - ma_1 a_2}{m J_z} Y_1(\delta_v \tau_1 - \beta_1)$$

(7.83)

where, again, the terms on the r.h.s. do not depend on u.

These equations highlight an interesting feature. The terms $(J_z - ma_1 a_2)$, which appear in both equations, are often very small in road cars, and could even be purposely set equal to zero. Therefore, the coupling between the two equations is fairly weak.

We observe that (7.77) becomes

$$\alpha_1 - \alpha_2 = (\delta_1 - \delta_2) - (\beta_1 - \beta_2)$$

(7.84)

and we also have

$$\alpha_1 a_2 + \alpha_2 a_1 = (\delta_1 a_2 + \delta_2 a_1) - (\beta_1 a_2 + \beta_2 a_1)$$

(7.85)

7.5.9 Driving Force

At the beginning of this chapter, and precisely in (7.2), we made the assumption of small longitudinal forces. But small does not mean zero. Indeed, a small amount of power is necessary even for keeping a vehicle in steady-state conditions. To make this statement quantitative, let us consider a rear-wheel-drive single track model (Fig. 7.20, with $F_{x_1} = 0$). The power balance

$$(F_{x_2} - F_{y_1} \delta_1) u + F_{y_1}(v_p + r_p a_1) + F_{y_2}(v_p - r_p a_2) - \left(\frac{1}{2} \rho S C_x u^2 \right) u = 0 \quad (7.86)$$

provides the following driving force F_{x_2}

$$F_{x_2} = F_{y_1}\left(\delta_1 - \frac{v_p + r_p a_1}{u}\right) + F_{y_2}\left(-\frac{v_p - r_p a_2}{u}\right) + \frac{1}{2}\rho S C_x u^2$$

$$= F_{y_1}\alpha_1 + F_{y_2}\alpha_2 + \frac{1}{2}\rho S C_x u^2 \tag{7.87}$$

and, assuming linear tire behavior:

$$= C_1\alpha_{1p}^2 + C_2\alpha_{2p}^2 + \frac{1}{2}\rho S C_x u^2$$

This force F_{x_2} has to counteract the aerodynamic drag (obvious) and also the drag due to tire slips (maybe not so obvious at first). Indeed, lateral axle forces F_{y_i} are not orthogonal to the corresponding velocity and hence absorb mechanical power.

That tire slips induce drag can be better appreciated from Fig. 7.21 (where, for simplicity, the aerodynamic drag is not considered). Points C and A do not coincide because of the apparent slip angles α_1 and α_2. Therefore, a longitudinal driving force F_{x_2} is required to achieve the dynamic equilibrium (cf. [12, p. 67]).

Of course, Fig. 7.21 is just a scheme. In real cases, slip angles are much smaller. The last line in (7.87), which is valid for about $|\alpha_i| < 0.05$ rad, clearly shows that F_{x_2} is much smaller than F_{y_i}, as assumed in (7.2).

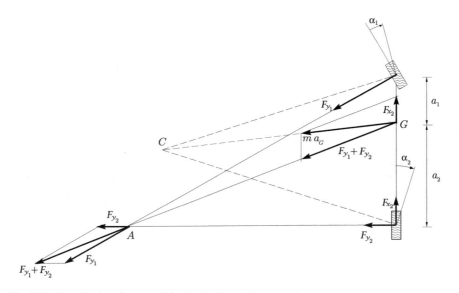

Fig. 7.21 Graphical evaluation of the driving force F_{x_2} at steady state

7.5.10 The Role of the Steady-State Lateral Acceleration

As already stated in Sect. 7.4, it is common practice to employ $(\delta_v, \tilde{a}_y)$, instead of (u, δ_v), as parameters to characterize a steady-state condition. In the single track model some steady-state quantities are functions of $\tilde{a}_y$ only.

The reason for such a fortunate coincidence in the case under examination is promptly explained. Just look at the equilibrium equations at steady state, with the inclusion of the constitutive equations (axle characteristics), that is for the single track model

$$m\tilde{a}_y = Y_1(\alpha_1) + Y_2(\alpha_2)$$
$$0 = Y_1(\alpha_1)a_1 - Y_2(\alpha_2)a_2 \tag{7.88}$$

They yield this result (already obtained in (7.9) and (7.57))

$$\frac{Y_1(\alpha_1)l}{ma_2} = \tilde{a}_y \quad \text{and} \quad \frac{Y_2(\alpha_2)l}{ma_1} = \tilde{a}_y \tag{7.89}$$

which can be more conveniently rewritten as

$$\frac{Y_1(\alpha_1)l}{mga_2} = \frac{Y_1(\alpha_1)}{Z_1^0} = \frac{\tilde{a}_y}{g} \quad \text{and} \quad \frac{Y_2(\alpha_2)l}{mga_1} = \frac{Y_2(\alpha_2)}{Z_2^0} = \frac{\tilde{a}_y}{g} \tag{7.90}$$

where Z_1^0 and Z_2^0 are the static vertical loads on each axle.

Therefore, if we take the monotone part of each axle characteristic, there is a one-to-one correspondence between $\tilde{a}_y$ and the apparent slip angles at steady state (Fig. 7.24)

$$\alpha_1 = \alpha_1(\tilde{a}_y) \quad \text{and} \quad \alpha_2 = \alpha_2(\tilde{a}_y) \tag{7.91}$$

This is the key fact for using $\tilde{a}_y$ as a parameter.

Both apparent slip angles α_1 and α_2 only "feel" the lateral acceleration, no matter if the vehicle has small u and large δ_v or, vice versa, large u and small δ_v. In other words, the radius of the circular trajectory of the vehicle does not matter at all (in this model). Only $\tilde{a}_y$ matters to the lateral forces and hence to the apparent slip angles. Actually, this very same property has been already used to build the axle characteristics. Equations (7.91) are just the inverse functions of (7.59).

We remark that (7.91) must not be taken as a general rule, but rather as a fortunate coincidence (it applies only to vehicles with two axles, open differential, no wings and parallel steering).

Another very important result comes directly from (7.90)

$$\frac{Y_1(\alpha_1)}{Z_1^0} = \frac{Y_2(\alpha_2)}{Z_2^0} \tag{7.92}$$

that is, at steady state, the lateral forces are always proportional to the corresponding static vertical loads. Therefore, the *normalized axle characteristics*

$$\hat{Y}_1(\alpha_1) = \frac{Y_1(\alpha_1)}{Z_1^0} \quad \text{and} \quad \hat{Y}_2(\alpha_2) = \frac{Y_2(\alpha_2)}{Z_2^0} \tag{7.93}$$

are what really matters in the vehicle dynamics of the single track model. The normalized axle characteristics are non-dimensional. Their maximum value is equal to the grip available in the lateral direction and is, therefore, a very relevant piece of information.

7.5.11 Slopes of the Axle Characteristics

It turns out that vehicle handling is pretty much affected by the slopes (derivatives) of the axle characteristics

$$\Phi_1 = \frac{dY_1}{d\alpha_1} \quad \text{and} \quad \Phi_2 = \frac{dY_2}{d\alpha_2} \tag{7.94}$$

Obviously, $\Phi_i > 0$ in the monotone increasing part of the axle characteristics.

According to (7.91), in the single track model we have that the slopes Φ_i of the axle characteristics are functions of the lateral acceleration only

$$\Phi_1 = \Phi_1(\alpha_1) = \Phi_1(\alpha_1(\tilde{a}_y))$$
$$\Phi_2 = \Phi_2(\alpha_2) = \Phi_2(\alpha_2(\tilde{a}_y)) \tag{7.95}$$

From (7.89)

$$\frac{d\tilde{a}_y}{d\alpha_1} = \frac{l\Phi_1}{ma_2} \quad \text{and} \quad \frac{d\tilde{a}_y}{d\alpha_2} = \frac{l\Phi_2}{ma_1} \tag{7.96}$$

and hence

$$\frac{d\alpha_1}{d\tilde{a}_y} = \frac{ma_2}{l\Phi_1} \quad \text{and} \quad \frac{d\alpha_2}{d\tilde{a}_y} = \frac{ma_1}{l\Phi_2} \tag{7.97}$$

7.6 Double Track, or Single Track?

Equations for the double track model and equations for the single track model are quite similar. Therefore, the effort for building a numerical model and running simulations is pretty much the same. Of course the double track model is more general. It does not require the assumptions of parallel steering and open differential. Moreover, it can deal with cases in which $u \gg |rt_i|$ does not hold.

The single track model is less realistic, but simpler, and hence more predictable for a human being. Almost all the complexity boils down to the axle characteristics. As already mentioned, the single track model can provide in many cases useful insights into vehicle handling, particularly for educational purposes. But "many cases" does not mean "all cases".

7.7 Steady-State Maps

We have already stated that the two functions (7.48) define *all* steady-state conditions of the double track model. However, the topic is so relevant to deserve additional attention and discussion.

From (7.46), (7.47), (7.68), (7.76) and (7.91) we have, at *steady state*, the following maps

$$
\begin{aligned}
\rho_p = \rho_p(\delta_v, \tilde{a}_y) = \frac{r_p}{u} = \left(\frac{\tau_1 - \tau_2}{l}\right)\delta_v - \frac{\alpha_1(\tilde{a}_y) - \alpha_2(\tilde{a}_y)}{l} \\
\beta_p = \beta_p(\delta_v, \tilde{a}_y) = \frac{v_p}{u} = \left(\frac{\tau_1 a_2 + \tau_2 a_1}{l}\right)\delta_v - \frac{\alpha_1(\tilde{a}_y)a_2 + \alpha_2(\tilde{a}_y)a_1}{l}
\end{aligned}
\tag{7.98}
$$

A vehicle-road system has unique functions $\rho_p(\delta_v, \tilde{a}_y)$ and $\beta_p(\delta_v, \tilde{a}_y)$. As will be shown, they tell us a lot about the global vehicle steady-state behavior. In other words, these two maps fully characterize *any* steady-state condition of the vehicle. Of course, the r.h.s. part of (7.98) is strictly related to the single track model, and it is useful to the vehicle engineer to understand how to modify the vehicle behavior.

The two functions $\rho_p(\delta_v, \tilde{a}_y)$ and $\beta_p(\delta_v, \tilde{a}_y)$ can also be obtained experimentally [4], once a prototype vehicle is available, by performing some rather standard tests on a flat proving ground. With the vehicle driven at almost constant speed u and a slowly increasing steering wheel angle δ_v (Slow Ramp Steer, often performed at 80 km/h), it suffices to measure the following quantities: $r_p, v_p, u, \tilde{a}_y$ and δ_v. It is worth noting that none of these quantities does require to know whether the vehicle has two axles or more, or how long the wheelbase is. In other words, they are all *well defined in any vehicle, including race cars*.

Fig. 7.22 Curves at constant ρ (1/m) in the plane $(\delta, \tilde{a}_y)$, for an understeer vehicle

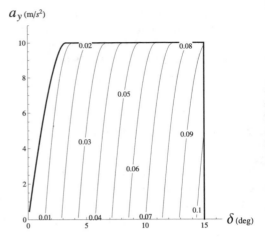

Fig. 7.23 Curves at constant β (deg) in the plane $(\delta, \tilde{a}_y)$, for an understeer vehicle

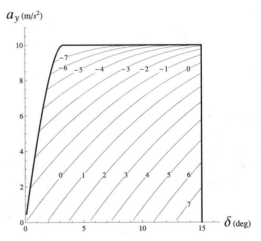

A key feature, confirmed by tests on real road cars (with open differential and no wings), is that the δ_v-dependence and the $\tilde{a}_y$-dependence are clearly *separated*.[7]

As shown in Figs. 7.22 and 7.23, both maps in (7.98) are (in the single track model) *linear* with respect to the steering wheel angle δ_v, whereas they are *nonlinear* with respect to the steady-state lateral acceleration $\tilde{a}_y$. The linear parts are totally under control, in the sense that both of them are simple functions of the steer gear ratios and of a_1 and a_2. The nonlinear parts are more challenging, coming directly from the interplay of the axle characteristics.

Figures 7.22 and 7.23, where $\delta = \tau \delta_v$, anticipate the **Map of Achievable Performance (MAP)** approach, discussed in Sect. 7.8.

---

[7] We remark that this is no longer true in vehicles with limited-slip differential and/or aerodynamic vertical loads.

7.7.1 Steady-State Gradients

It is informative, and hence quite useful, to define and compute/measure the *gradients* of the two maps $\beta_p(\delta_v, \tilde{a}_y)$ and $\rho_p(\delta_v, \tilde{a}_y)$, defined in (7.98)

$$
\begin{aligned}
\operatorname{grad} \rho_p &= \left(\frac{\partial \rho_p}{\partial \tilde{a}_y}, \frac{\partial \rho_p}{\partial \delta_v} \right) = (\rho_y, \rho_\delta) \\
\operatorname{grad} \beta_p &= \left(\frac{\partial \beta_p}{\partial \tilde{a}_y}, \frac{\partial \beta_p}{\partial \delta_v} \right) = (\beta_y, \beta_\delta)
\end{aligned}
\tag{7.99}
$$

As well known, gradients are vectors orthogonal to the level curves.

For the single track model, the explicit expressions of the components of the gradients $\operatorname{grad} \rho_p$ and $\operatorname{grad} \beta_p$ are as follows

$$
\begin{aligned}
\rho_y &= -\frac{m}{l^2} \left(\frac{\Phi_2 a_2 - \Phi_1 a_1}{\Phi_1 \Phi_2} \right) & \rho_\delta &= \frac{\tau_1 - \tau_2}{l} \\
\beta_y &= -\frac{m}{l^2} \left(\frac{\Phi_1 a_1^2 + \Phi_2 a_2^2}{\Phi_1 \Phi_2} \right) & \beta_\delta &= \tau \left(\frac{\tau_1 a_2 + \tau_2 a_1}{l} \right)
\end{aligned}
\tag{7.100}
$$

where, to compute β_y and ρ_y, we took into account (7.97).

It is worth noting that, for a given single track model of a vehicle, the two gradient components β_δ and ρ_δ are *constant*, whereas the other two gradient components β_y and ρ_y are functions of $\tilde{a}_y$ only.

As will be discussed shortly, only one out of four gradient components is usually employed in classical vehicle dynamics,[8] thus missing a lot of information. But this is not the only case in which classical vehicle dynamics turns out to be far from systematic and rigorous. This lack of generality of classical vehicle dynamics is the motivation for some of the next sections.

7.7.2 Alternative Steady-State Gradients

Although not commonly done, we evaluate the gradients of the front and rear slip angles $\beta_1(\delta_v, \tilde{a}_y)$ and $\beta_2(\delta_v, \tilde{a}_y)$, which were defined in (7.78)

---

[8] It is the well known understeer gradient K, defined in (7.117). Unfortunately, it is not a good parameter and should be replaced by the gradient components (7.99), as demonstrated in Sect. 7.14.1.

$$\beta_{1y} = -\frac{ma_2}{l\Phi_1} \qquad \beta_{1\delta} = (1 + \kappa)\tau$$

$$\beta_{2y} = -\frac{ma_1}{l\Phi_2} \qquad \beta_{2\delta} = \kappa\tau \tag{7.101}$$

A fairly obvious result, but that can turn out to be useful in some cases.

7.7.3 Understeer and Oversteer

For further developments, it is convenient to rewrite (7.98) in a more compact form

$$\rho_p = \rho_p(\delta_v, \tilde{a}_y) = \left(\frac{\tau_1 - \tau_2}{l}\right) \delta_v - f_\rho(\tilde{a}_y)$$

$$\beta_p = \beta_p(\delta_v, \tilde{a}_y) = \left(\frac{\tau_1 a_2 + \tau_2 a_1}{l}\right) \delta_v - f_\beta(\tilde{a}_y) \tag{7.102}$$

where, in the single track model

$$f_\rho(\tilde{a}_y) = \frac{\alpha_1(\tilde{a}_y) - \alpha_2(\tilde{a}_y)}{l}$$

$$f_\beta(\tilde{a}_y) = \frac{\alpha_1(\tilde{a}_y)a_2 + \alpha_2(\tilde{a}_y)a_1}{l} \tag{7.103}$$

The two known functions $f_\rho(\tilde{a}_y)$ and $f_\beta(\tilde{a}_y)$ are nonlinear functions, peculiar to a given road vehicle. They are called here *slip functions*.

Let us discuss this topic by means of a few examples.

First, let us consider the normalized axle characteristics (7.93) (multiplied by g) shown in Fig. 7.24(left). In this example, it has been assumed that both axles have the same lateral grip equal to 1. Moreover, to keep, for the moment, the analysis as simple as possible, we also assume that $\hat{Y}_1(x) = \hat{Y}_2(kx)$, with $k > 0$. When inverted, they provide the apparent slip angles $\alpha_1(\tilde{a}_y)$ and $\alpha_2(\tilde{a}_y)$ shown in Fig. 7.24(right). Assuming a wheelbase $l = 2.5$ m, $a_1 = 1.125$ m, and $a_2 = 1.375$ m, we get from (7.103) the two slip functions f_ρ and f_β of Fig. 7.25.

In all figures, angles are in degree, accelerations in m/s^2, and a_y should be read as $\tilde{a}_y$.

A vehicle with a monotone increasing slip function $f_\rho(\tilde{a}_y)$, as in Fig. 7.25, is said to be an *understeer* vehicle. A more precise definition is given in (7.108).

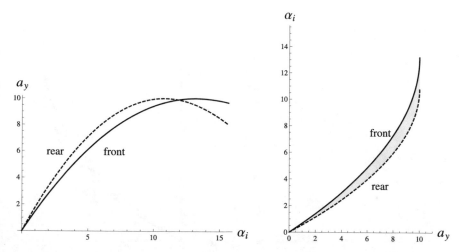

Fig. 7.24 Normalized axle characteristics (multiplied by g) of an *understeer* vehicle (left) and corresponding apparent slip angles (right)

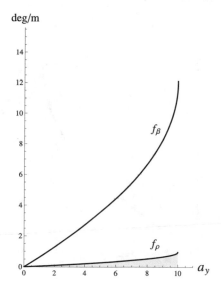

Fig. 7.25 Slip functions of an *understeer* vehicle

As a second example, let us consider the normalized axle characteristics (multiplied by g) shown in Fig. 7.26(left). They are like in Fig. 7.24, but interchanged. When inverted, they provide the two functions $\alpha_1(\tilde{a}_y)$ and $\alpha_2(\tilde{a}_y)$ shown in Fig. 7.26(right). In this case the two slip functions f_ρ and f_β are as in Fig. 7.27.

A vehicle with a monotone decreasing function $f_\rho(\tilde{a}_y)$, as in Fig. 7.27, is said to be an *oversteer* vehicle. A more precise definition is given in (7.110).

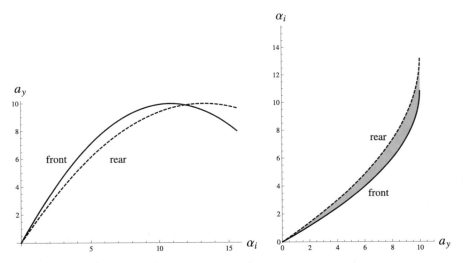

Fig. 7.26 Normalized axle characteristics (multiplied by g) of an *oversteer* vehicle (left) and corresponding apparent slip angles (right)

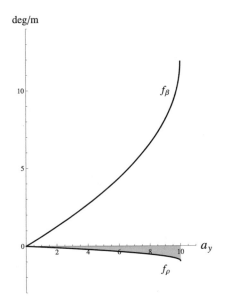

Fig. 7.27 Slip functions of an *oversteer* vehicle

7.7.4 Handling Diagram

Usually, only the function $f_\rho(\tilde{a}_y)$ is considered in classical vehicle dynamics, while $f_\beta(\tilde{a}_y)$ is neglected.

Since, at steady state, $\rho_p = \tilde{a}_y/u^2 = r/u = 1/R$, the first equation in (7.102) becomes

$$\frac{\tilde{a}_y}{u^2} = \left(\frac{\tau_1 - \tau_2}{l}\right)\delta_v - \frac{\alpha_1(\tilde{a}_y) - \alpha_2(\tilde{a}_y)}{l} \tag{7.104}$$

which, for given u and δ_v, is an equation for the unknown $\tilde{a}_y = \tilde{a}_y(u, \delta_v)$. See also (7.46).

Another, most classical, way to recast (7.104) is

$$\delta - \frac{l}{R} = \alpha_1(\tilde{a}_y) - \alpha_2(\tilde{a}_y) = f_\rho(\tilde{a}_y)l \tag{7.105}$$

where

$$\delta = (\tau_1 - \tau_2)\,\delta_v \tag{7.106}$$

is the *net steer angle*, already defined in (7.69).

It is customary [13–15] to rewrite (7.104) as a system of two equations

$$\begin{cases} y = \left(\dfrac{\tau_1 - \tau_2}{l}\right)\delta_v - \dfrac{\tilde{a}_y}{u^2} \\[2mm] y = f_\rho(\tilde{a}_y) = \dfrac{\alpha_1(\tilde{a}_y) - \alpha_2(\tilde{a}_y)}{l} \end{cases} \tag{7.107}$$

Solving this system amounts to obtaining the values of $(\tilde{a}_y, f_\rho)$ attained under the imposed operating conditions (u, δ_v). Geometrically, that can be seen as the intersection between a *straight line* (i.e., the first equation in (7.107)) and the so-called *handling curve* $y = f_\rho(\tilde{a}_y)$ (i.e., the second equation in (7.107)).

Together, the handling curve and the straight lines form the celebrated *handling diagram* [13–15]. Examples are shown in Figs. 7.28 and 7.29 (where a_y is indeed $\tilde{a}_y$ and y is in deg/m).

The handling curve $y = f_\rho(\tilde{a}_y)$ is peculiar to each vehicle-road system (in the single track model it depends on the normalized axle characteristics only). Therefore, for a given vehicle-road system it has to be drawn once and for all.

On the other hand, the straight line depends on the selected operating conditions (u, δ_v). For instance, in Fig. 7.28 the two intersecting lines correspond to two operating conditions with the same value of δ_v, while the two parallel lines share the same value of u.

Perhaps, the best way to understand the handling diagram (Figs. 7.28 and 7.29) is by assuming that the steering wheel angle δ_v is kept constant, while the forward speed u is (slowly) increased.

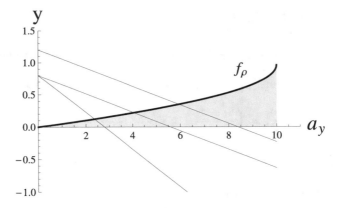

Fig. 7.28 Handling diagram of an understeer vehicle (y in deg/m)

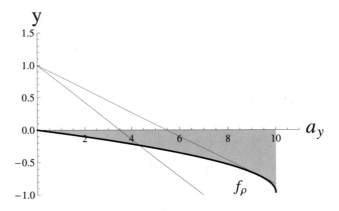

Fig. 7.29 Handling diagram of an oversteer vehicle (y in deg/m)

In Fig. 7.28, an increasing u, with constant δ_v, results also in an increasing y. Therefore, from (7.105) with constant δ_v (and hence constant δ), the higher the forward speed u, the larger the radius R of the trajectory of the vehicle. This is called *understeer behavior*. More precisely, we have understeer whenever

$$\frac{\mathrm{d}f_\rho}{\mathrm{d}\tilde{a}_y} > 0 \qquad (7.108)$$

where, in the single track model

$$\frac{\mathrm{d}f_\rho}{\mathrm{d}\tilde{a}_y} = \frac{m}{l^2}\left(\frac{\Phi_2 a_2 - \Phi_1 a_1}{\Phi_1 \Phi_2}\right) \qquad (7.109)$$

On the contrary, if the handling curve is, e.g., like in Fig. 7.29, the higher the forward speed u, with constant δ_v, the smaller the radius R. This is called *oversteer behavior*. More precisely, we have oversteer whenever

$$\frac{\mathrm{d} f_\rho}{\mathrm{d}\tilde{a}_y} < 0 \tag{7.110}$$

Actually, when the straight line becomes tangent to the handling curve, as shown in Fig. 7.29, the vehicle becomes unstable. It means that the vehicle has reached the *critical speed* associated to that value of δ_v. The concept of critical speed will be discussed in Sect. 7.13 in a more general framework.

A less classical, but maybe more interesting formula is (see (7.84))

$$\delta - (\alpha_1 - \alpha_2) = \beta_1 - \beta_2 \tag{7.111}$$

The relevance of this result is that it combines a very weak term $\delta - (\alpha_1 - \alpha_2)$ with a very robust term $\beta_1 - \beta_2$. The first one is not well defined in real vehicles, whereas the last one is. Understeer/oversteer should be defined and evaluated using the robust term.

Vehicles with aerodynamic devices and/or limited-slip differential do not exhibit a handling curve [13, p. 172], but a *handling surface* [7], instead. More precisely, (7.105) still holds true, but with $f_\rho(\tilde{a}_y, 1/R)$. Therefore, definitions (7.108) and (7.110) of understeer/ovesteer become meaningless. This topic is addressed in Sect. 8.5.2.

Classical vehicle dynamics stops about here. In the next section a fresh, more comprehensive, global approach is developed. It brings new insights into the global steady-state behavior of real vehicles, along with some new hints about the transient behavior.

7.8 Map of Achievable Performance (MAP)

The handling diagram [13–15], although noteworthy, does not provide a complete picture of the handling behavior. Just consider that the use of $\tilde{a}_y$ as input variable, that is one variable instead of two, hides some features of the vehicle handling behavior.

It would be better to have a more general approach, able to unveil *at a glance* the overall steady-state features of the vehicle under investigation, thus making it easier to distinguish between a "good" vehicle and a "not-so-good" one.

As already stated in (7.44), the steady-state handling behavior is completely described by the handling maps

$$\rho_p = \hat{\rho}_p(u, \delta_v) = \left(\frac{\tau_1 - \tau_2}{l}\right)\delta_v - \frac{\alpha_1(u, \delta_v) - \alpha_2(u, \delta_v)}{l}$$

$$\beta_p = \hat{\beta}_p(u, \delta_v) = \left(\frac{\tau_1 a_2 + \tau_2 a_1}{l}\right)\delta_v - \frac{\alpha_1(u, \delta_v)a_2 + \alpha_2(u, \delta_v)a_1}{l} \tag{7.112}$$

where the last terms are peculiar to the single track model.

In the single track model, it is convenient to define the *net steer angle* δ, as already done in (7.69) and in (7.106)

$$(1 + \kappa)\delta = \delta_1 = \tau_1 \delta_v$$
$$\kappa\delta = \delta_2 = \tau_2 \delta_v$$

(7.113)

Usually, $\kappa = 0$ and hence $\delta = \delta_1$ is just the steering angle of the front wheel. However, $\kappa \neq 0$ leaves room for direct rear steering as well. In general,

$$\delta = \delta_1 - \delta_2 = (\tau_1 - \tau_2)\delta_v$$

(7.114)

With this notation, the *handling maps* (7.112) become

$$\rho = \hat{\rho}(u, \delta) = \frac{\delta}{l} - \frac{\alpha_1(u, \delta) - \alpha_2(u, \delta)}{l}$$

$$\beta = \hat{\beta}(u, \delta) = \left(\frac{(1 + \kappa)a_2 + \kappa a_1}{l} \right) \delta - \frac{\alpha_1(u, \delta)a_2 + \alpha_2(u, \delta)a_1}{l}$$

(7.115)

where, for the sake of compactness, we dropped the subscript p.

These two maps fully characterize the steady-state behavior of the vehicle. This is a fairly general point of view that leads to the global approach presented in Chap. 6, that we called **Map of Achievable Performance (MAP)**.

Actually, under the acronym MAP we will present several types of possible graphical representations of the handling maps, each one on the corresponding *achievable region*. This is another key concept.

Figures in this section are for road cars with the following features: mass $m = 2000$ kg, wheelbase $l = 2.5$ m, $a_1 = 1.125$ m, $a_2 = 1.375$ m, grip coefficient $\mu = 1$, maximum speed $u_{max} = 40$ m/s, maximum steer angle of the front wheels $\delta_{max} = 15°$. The understeer version has normalized axle characteristics as in Fig. 7.24. The oversteer version has normalized axle characteristics as in Fig. 7.26. In all figures, angles are in degree, accelerations in m/s^2, and ρ in m^{-1}.

7.8.1 MAP Fundamentals

The main idea behind the MAP approach is simple: the driver controls (u, δ), the vehicle reacts with (ρ, β). That is

$$(u, \delta) \implies (\rho, \beta)$$

(7.116)

The input values (u, δ) that a given vehicle can really *achieve* are subject to three limitations:

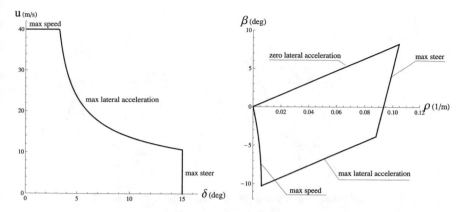

Fig. 7.30 Achievable *input* region (left) and achievable *output* region (right) for an understeer vehicle

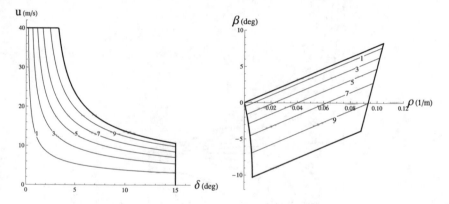

Fig. 7.31 Lines at constant lateral acceleration $\tilde{a}_y$ for an understeer vehicle

- maximum steer angle δ_{max};
- maximum speed u_{max}, or critical speed u_{cr}, if $u_{cr} < u_{max}$;
- maximum lateral acceleration (grip limited).

This is shown in Fig. 6.1 (left). Each achievable point (δ, u) results in the vehicle performing with precise values (ρ, β). Therefore, the *achievable input region* of Fig. 6.1 (left) is mapped onto the *achievable output region* shown in Fig. 6.1 (right).

Quite interesting are the MAPs (Maps of Achievable Performance) that can be drawn inside these achievable regions. For instance, curves at constant $\tilde{a}_y$ are drawn on both regions in Fig. 7.31. In an understeer vehicle without significant aerodynamic vertical loads, the grip-limited bound is just the curve at constant $\tilde{a}_y = \mu g$.

While the yaw rate r_p has typically the same sign as δ, the same does not apply to the lateral speed v_p. As shown in Fig. 7.32, in a left turn the vehicle slip angle $\beta = v_p/u$ can either be positive or negative. As a rule of thumb, at low forward speed

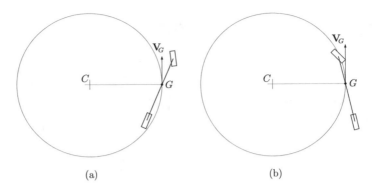

Fig. 7.32 Steady-state behavior: **a** nose-out (low speed), **b** nose-in (high speed)

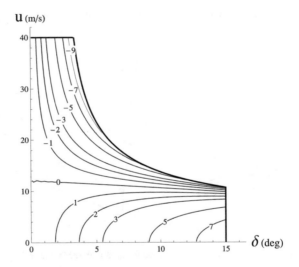

Fig. 7.33 u-δ MAP with curves at constant vehicle slip angle β for an understeer vehicle

the vehicle goes around "nose-out" ($\beta > 0$), whereas at high speed the vehicle goes around "nose-in" ($\beta < 0$). This statement can be made *quantitative* by drawing the curves at constant β on the achievable input region, as shown in Fig. 7.33. The almost horizontal line $\beta = 0$ clearly splits the region into a lower part with $\beta > 0$, and an upper part with $\beta < 0$.

Drawing curves at constant curvature ρ also highlights the overall understeer/ oversteer behavior of a vehicle. For instance, it is quite obvious that the pattern of Fig. 7.34 is typical of an understeer vehicle: the faster you go, the more you have to steer to keep ρ constant. Moreover, it is worth comparing Fig. 7.22, which is the contour plot of $\rho(\delta, \tilde{a}_y)$, and Fig. 7.34, which is the contour plot of $\rho(u, \delta)$. For instance, the first MAP is linear with respect to δ, whereas the second one is not. The reason is that the other independent variable is different: linear behavior with respect to δ requires constant lateral acceleration $\tilde{a}_y$, not constant forward speed u.

Fig. 7.34 u-δ MAP with curves at constant curvature ρ for an understeer vehicle

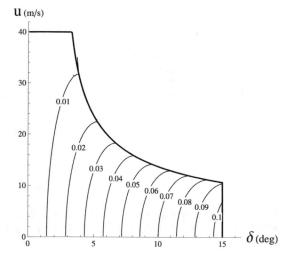

Fig. 7.35 ρ-β MAP with curves at constant u and lines at constant δ for an understeer vehicle

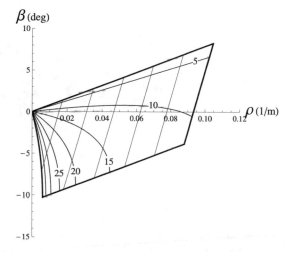

Curves at constant speed u, and also lines at constant steer angle δ, are shown in Fig. 7.35 for an understeer vehicle. As expected, moving top to bottom along each line at constant steer angle, that is with increasing speed, brings smaller values of the curvature ρ. Also interesting is to observe that at low speed the slip angle β grows with δ, whereas at high speed it is the other way around. The same phenomena can be observed more clearly in Fig. 7.33.

These MAPs can be obtained experimentally or through simulations. Therefore, they are not limited to the single track model. Actually, as will be discussed in the next chapter, they exist also for race cars, including cars with very high aerodynamic downforces.

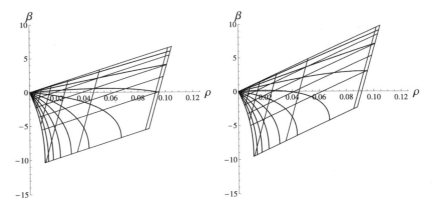

Fig. 7.36 Effects of rear steering on the achievable region: rear wheels turning opposite of the front wheels (left), rear wheels turning like the front wheels (right)

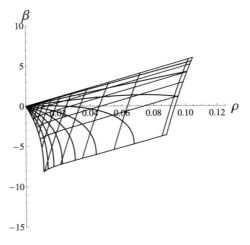

Fig. 7.37 ρ-β MAP for a vehicle with rear wheels turning opposite of the front wheels at low speed and like the front wheels at high speed

The effects of *rear steering* (in addition to front steering, of course) are shown in Fig. 7.36. The picture on the left is for the case of rear wheels turning opposite of the front wheels with $\delta_2 = -0.1\delta_1$, whereas the picture on the right is for rear wheels turning like the front wheels, with $\delta_2 = 0.1\delta_1$. The vehicle slip angle β is affected pretty much. Basically, a positive $\chi = \delta_2/\delta_1$ moves the achievable region upwards, and vice versa. On the other hand, rear steering does not impinge on the achievable region in the plane (δ, ρ), as will be discussed in Sect. 7.8.2.

Vehicles behave in a better way if the vehicle slip angle β spans a small range. To have a narrower achievable output region in the plane (ρ, β) we have to move down the upper part and move up the lower part. This is indeed the effect of a steering system with rear wheels turning opposite of the front wheels at low speed, and

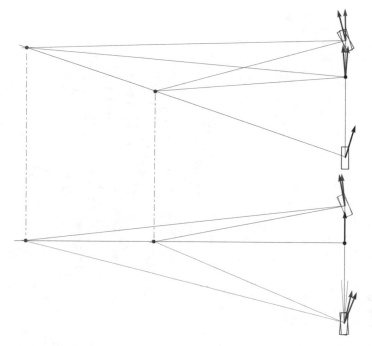

Fig. 7.38 Example of the effects of rear steering on β: front steering only (top); front and rear steering (bottom). All cases have the same $\tilde{a}_y$, and hence the same α_1 and α_2

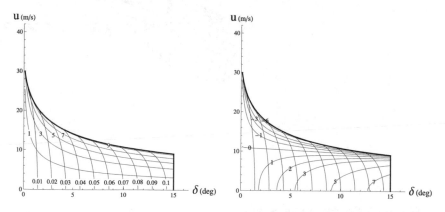

Fig. 7.39 Oversteer vehicle: u-δ MAPs with curves at constant ρ (both), constant $\tilde{a}_y$ (left) and constant β (right)

turning like the front wheels at high speed. That is a steering system (7.113) with, e.g., $\kappa(u) = -\kappa_0 \cos(\pi u/u_{\max})$. The net result can be appreciated by comparing Fig. 7.37 with Fig. 7.35. The MAP approach provides a better insight into rear steering effects than by looking at, e.g., Fig. 7.38.

Fig. 7.40 Oversteer vehicle:
ρ-β MAP with curves at
constant speed u and lines at
constant steer angle δ

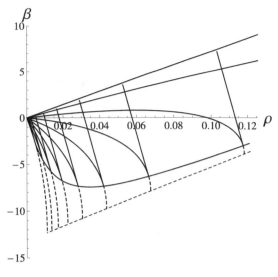

Fig. 7.41 Vehicle with too
much understeer: ρ-β MAP
with lines at constant u, $\tilde{a}_y$
and δ

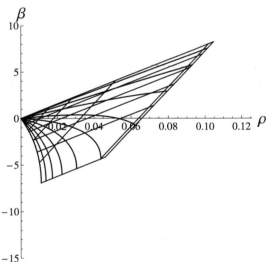

The achievable region in case of an *oversteer* vehicle is limited by the critical speeds, not by grip. A typical achievable input region, with noteworthy lines, is shown in Fig. 7.39.

The achievable region in the plane (ρ, β) for an *oversteer* vehicle is shown in Fig. 7.40, along with curves at constant speed u and lines at constant steer angle δ. As expected, moving top to bottom along the lines at constant steer angles, that is with increasing speed, entails larger values of the curvature ρ.

Very instructive is the comparison between Figs. 7.35 and 7.40, that is between an understeer and an oversteer vehicle. The two achievable regions have different

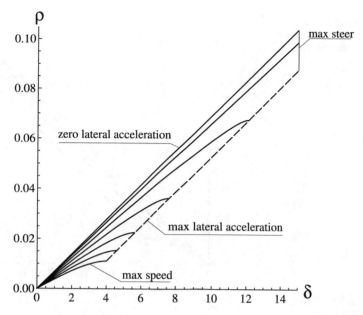

Fig. 7.42 Constant speed lines on the ρ-δ MAP for an understeer vehicle

shapes also because an oversteer vehicle becomes *unstable* for certain combinations of speed and steer angle. These critical combinations form a sort of stability boundary which collects all points where the u-curves and δ-lines are *tangent* to each other, as shown in Fig. 7.40.

On the opposite side, a vehicle with *too much understeer* has an achievable region like in Fig. 7.41 (see also Fig. 7.45 for a more intuitive MAP).

7.8.2 MAP Curvature ρ Versus Steer Angle δ

A central issue in vehicle dynamics is how a vehicle responds to the driver input commands (namely, the steering wheel angle δ_v and the forward speed u). Well, let us map it. The plane (δ, ρ) suits the purpose in a fairly intuitive and quantitative way.

Let us consider again a vehicle with the front and rear *normalized* axle characteristics (multiplied by g) shown in Fig. 7.24.[9] We recall that it is an understeer vehicle and that the corresponding slip functions and handling diagram are shown in Fig. 7.25 and Fig. 7.28, respectively.

If we draw the lines at constant speed u in the plane (δ, ρ), we get the plot shown in Fig. 7.42, if $\rho \geq 0$. In the same achievable region, we can draw the lines at constant

---

[9] To keep, for the moment, the analysis as simple as possible, we also assume that $\hat{Y}_1(x) = \hat{Y}_2(kx)$, with $k > 0$.

Fig. 7.43 Constant lateral acceleration lines on the ρ-δ MAP for an understeer vehicle

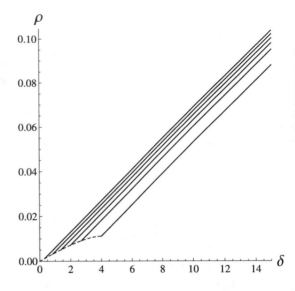

Fig. 7.44 ρ-δ MAP for an understeer vehicle

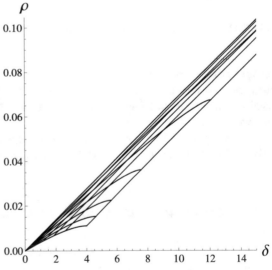

lateral acceleration $\tilde{a}_y$, as shown in Fig. 7.43. According to (7.98), they are parallel straight lines. In Fig. 7.44, both lines at constant u and constant $\tilde{a}_y$ are drawn on the whole achievable region.

The *achievable region* is bounded by:

1. maximum speed (dashed line in Fig. 7.43);
2. maximum lateral acceleration (dashed line in Fig. 7.42);
3. zero lateral acceleration;
4. maximum steer angle.

Fig. 7.45 ρ-δ MAP for a
vehicle with too much
understeer

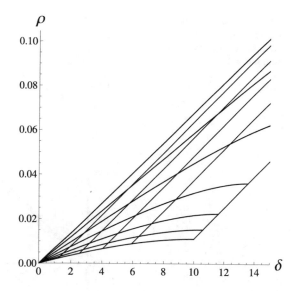

We see that the driver must act on both u and δ to control the vehicle, that is to drive it on a curve with curvature ρ and lateral acceleration $\tilde{a}_y$. But, the key feature is that it can be done fairly easily because the lines at constant speed are "well shaped", that is quite far apart from each other and neither too flat, nor too steep (Fig. 7.42).

In Fig. 7.44, all lines at constant speed intersect all lines at constant lateral acceleration. This is typical of all vehicles without significant aerodynamic vertical loads. This is another piece of information that is provided by this kind of maps on the achievable region.

An example of a not-so-nice achievable region is shown in Fig. 7.45. A vehicle with a map like in Fig. 7.45 shows too much understeer: the lines at high speed are too flat, showing that the driver can increase δ without getting a significant increase in ρ. Not a desirable behavior.

Another example of undesirable behavior, but for opposite reasons, is shown in Fig. 7.46. This is a vehicle with too little understeer. It has a very narrow achievable region, which means that the driver has a very heavy task in controlling the vehicle: the lines at zero and maximum lateral acceleration are very close together.

An oversteer vehicle (whose corresponding slip functions and handling diagram are shown in Fig. 7.27 and Fig. 7.29, respectively) has an achievable region as in Fig. 7.47. The lines at constant $\tilde{a}_y$, shown in Fig. 7.47, are quite far apart like in Fig. 7.43, but the lines at constant speed u are very badly shaped. At high speed they are too steep, meaning that a small variation of δ drastically changes ρ and $\tilde{a}_y$.

Moreover, the vehicle becomes unstable when the u-lines have vertical slope. Accordingly, the truly achievable region becomes smaller, as shown in Fig. 7.48, where the truly achievable region is bounded by the stability boundary (long-dashed line).

Fig. 7.46 Constant lateral acceleration lines on the ρ-δ MAP for a vehicle with too little understeer

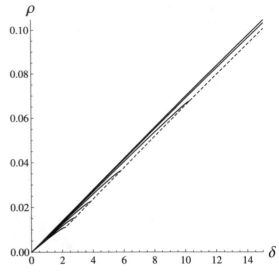

Fig. 7.47 Apparent achievable region on the ρ-δ MAP for an oversteer vehicle

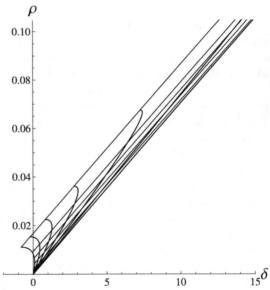

All these examples show how the map *curvature* versus *steer angle* provides a very clear and global picture of the vehicle handling behavior. It makes clear why a well tuned vehicle must be moderately understeer. Too much or too little understeer are not desirable because the vehicle becomes much more difficult to drive (for opposite reasons).

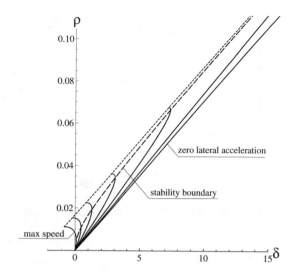

Fig. 7.48 Constant speed lines and truly achievable region on the ρ-δ MAP for an oversteer vehicle

The difference between understeer and oversteer is laid bare (Figs. 7.44 and 7.47). Both have far apart $\tilde{a}_y$-lines, but covering achievable regions on opposite sides. In fact, the u-lines are totally different.

The more one observes these handling MAPs on the corresponding achievable regions, the more the global handling behavior becomes clear.

7.8.3   Other Possible MAPs

So far we have discussed the fundamental MAPs (δ, u) and (ρ, β), and also the fairly intuitive, and very useful, MAP (ρ, δ).

Of course, several other MAPs are possible. For instance, in Fig. 7.49, curves at constant δ are drawn in the planes (ρ, u) and (β, u) for an understeer vehicle. The same kind of MAPs, but for an oversteer vehicle, are shown in Fig. 7.50. The onset of instability is clearly indicated, e.g., by the vertical tangent of the curves at constant δ in the plane (β, u).

Moreover, the MAP $(\delta, \tilde{a}_y)$ was introduced in Sect. 7.7 and is extensively employed in Sect. 7.10.

7.9   Weak Concepts in Classical Vehicle Dynamics

Some "fundamental" concepts in vehicle dynamics are indeed very weak if addressed with open mind. They are either not well defined, particularly when we look at real

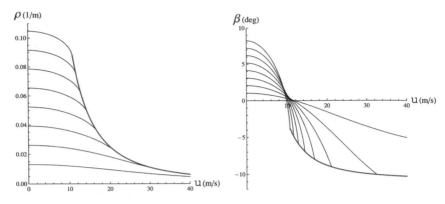

Fig. 7.49 Constant steer curves for an understeer vehicle

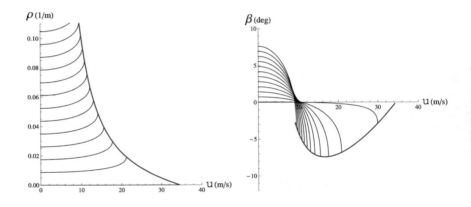

Fig. 7.50 Constant steer curves for an oversteer vehicle

vehicles, or they are commonly defined in an unsatisfactory way. This is a serious practical drawback that can lead to wrong results and conclusions.

7.9.1 The Understeer Gradient

According to the SAE J266 Standard, *Steady-State Directional Control Test Procedures For Passenger Cars and Light Trucks*

> understeer/oversteer gradient K is defined as the difference between steer angle gradient and Ackermann steer angle gradient.

This definition of K is equivalent to the following formula

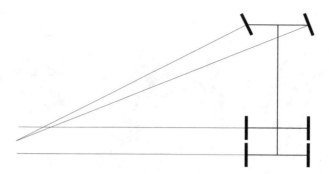

Fig. 7.51 Case not covered by the classical theory

$$K = \frac{\mathrm{d}}{\mathrm{d}\tilde{a}_y}\left(\delta - \frac{l}{R}\right) = l\frac{\mathrm{d}f_\rho(\tilde{a}_y)}{\mathrm{d}\tilde{a}_y} \qquad (7.117)$$

which comes directly from (7.105). See also (7.109).

Therefore, to compute/measure K we need both the *net steer angle* δ and the *Ackermann steer angle* l/R. Unfortunately, *none of them is clearly defined in a real vehicle*. In fact, they are well defined only in the single track model, as it is done, e.g., in Figure A1 in the SAE J266 Standard.

In a real vehicle, the two front wheels have typically different steer angles (Fig. 7.51). Therefore, the net steer angle δ is not precisely defined.

The Ackermann steer angle l/R also gets in trouble whenever a vehicle has three or more axles, as the wheelbase l is no longer a clear concept (Fig. 7.51). One may object that almost all cars have two axles. Nonetheless, we cannot ground a theory on such a weak concept.

The understeer gradient K has been an important performance metric in analyzing the handling behavior of vehicles. Unfortunately, it should not have been. It will be demonstrated in Sect. 7.14.1 that *it is not a good parameter to measure the handling behavior of a vehicle*. Nor even of a single track model. A much better parameter is ρ_y, discussed in Sect. 7.13.

7.9.2 Popular Definitions of Understeer/Oversteer

Perhaps, the most astonishing case of use of unclear concepts is the popular way to "define" understeer and oversteer:

> Oversteer is what occurs when a car steers by more than the amount commanded by the driver. Conversely, understeer is what occurs when a car steers less than the amount commanded by the driver.[10]

---

[10] From Wikipedia.

Fig. 7.52 What did the driver intend to do?

Understeer: a tendency of an automobile to turn less sharply than the driver intends (or would expect).

The term understeer means that you have to give your car more steering input than the corner should require to get it to go around.

What is the "amount commanded by the driver"? What is the scientific, quantitative meaning of what "the driver intends"? What does "than the corner should require" mean?

Figure 7.52 exemplifies this paradoxical situation. Three different curves, three identical trajectories, only one is fine in each case. What about the understeer/oversteer behavior of the car? What did the driver intend to do?

7.10   Double Track Model in Transient Conditions

Steady-state analysis cannot be the whole story. Indeed, a vehicle is quite often in transient conditions, that is with time-varying quantities (forces, speeds, yaw rate, etc.). Addressing the transient behavior is, of course, more difficult than "simply" analyzing the steady state. More precisely, the steady-state conditions (also called trim conditions) are just the equilibrium points from which a transient behavior can start or can end.

The general way to study the transient behavior of any dynamical system is through in-time simulations. However, this approach has some drawbacks. Even after a large number of simulations it is quite hard to predict beforehand what the outcome of the next simulation will be.

One way to simplify the analysis of a non-linear dynamical system is to consider only small perturbations (oscillations) about steady-state (trim) conditions. This idea leads to the approach based on *stability derivatives* and *control derivatives* (as they are called in aerospace engineering [12, p. 151]).

The nonlinear equations of motion of the *double track* model of the vehicle are (cf. (7.38))

$$m(u\dot{\beta} + \dot{u}\beta + u^2\rho) = Y(\beta, \rho; u, \delta_v)$$
$$J_z(u\dot{\rho} + \dot{u}\rho) = N(\beta, \rho; u, \delta_v)$$

(7.118)

We prefer to use (ρ, β) as state variables, instead of (v, r), because they provide a more "geometric" description of the vehicle motion. Since $\beta = v/u$ and $\rho = r/u$, it is pretty much like having normalized with respect to the forward speed u.

7.10.1 Equilibrium Points

At steady state we have, by definition, $\dot{v} = \dot{r} = 0$, that is $\dot{\beta} = \dot{\rho} = 0$. The driver has direct control on u and δ_v, which are kept constant and whose trim values are named u_a and δ_{va}. The subscript a is introduced here to distinguish clearly between the generic and the trim values (i.e., *assigned* values).

The equations of motion (7.118) become

$$mu_a^2\rho = Y(\beta, \rho; u_a, \delta_{va})$$
$$0 = N(\beta, \rho; u_a, \delta_{va})$$

(7.119)

which can be solved to get the steady-state maps (exactly like in (7.44) or (7.112))

$$\beta_p = \hat{\beta}_p(u_a, \delta_{va}) = \frac{v_p(u_a, \delta_{va})}{u_a}$$
$$\rho_p = \hat{\rho}_p(u_a, \delta_{va}) = \frac{r_p(u_a, \delta_{va})}{u_a}$$

(7.120)

These maps have been thoroughly discussed in Sect. 7.8, where the new concept of MAP (Map of Achievable Performance) was introduced.

Actually, when applying the MAP approach to the vehicle transient behavior it is more convenient to do like in (7.98), that is to use $\tilde{a}_y = u_a r_p(u_a, \delta_{va})$, which provides (exactly like in (7.46))

$$u_a = u_a(\delta_{va}, \tilde{a}_y)$$

(7.121)

and hence

$$\beta_p = \beta_p(\delta_{va}, \tilde{a}_y) = \hat{\beta}_p(u_a(\delta_{va}, \tilde{a}_y), \delta_{va})$$
$$\rho_p = \rho_p(\delta_{va}, \tilde{a}_y) = \hat{\rho}_p(u_a(\delta_{va}, \tilde{a}_y), \delta_{va})$$

(7.122)

An example of achievable region in $(\delta, \tilde{a}_y)$ is shown in Fig. 7.53 for an understeer vehicle, along with lines at constant β (left) and constant ρ (right). In a real vehicle, these maps can be obtained by means of classical steady-state tests. Therefore, they do not require departing from the traditional way of vehicle testing.

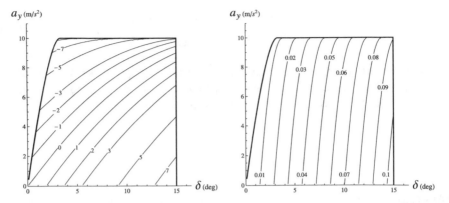

Fig. 7.53 MAPs in the plane $(\delta, \tilde{a}_y)$ with curves at constant β (left) and constant ρ (right) for an understeer vehicle

7.10.2   Free Oscillations (No Driver Action)

The basic idea is to linearize around an equilibrium point to obtain information in its neighborhood about the dynamical behavior. It is a standard approach for almost any kind of nonlinear dynamical systems.

Assuming that the driver takes no action (i.e., both $u = u_a$ and $\delta_v = \delta_{va}$ are constant in time), the first-order Taylor series expansion of the equations of motion (7.118) around the equilibrium point (7.120) are as follows

$$m(u_a\dot{\beta} + u_a^2\rho) = Y_0 + Y_\beta(\beta - \beta_p) + Y_\rho(\rho - \rho_p)$$
$$J_z u_a\dot{\rho} = N_0 + N_\beta(\beta - \beta_p) + N_\rho(\rho - \rho_p)$$
$$(7.123)$$

where

$$Y_0 = Y(\beta_p, \rho_p; u_a, \delta_{va}) = mu_a^2\rho_p, \qquad N_0 = N(\beta_p, \rho_p; u_a, \delta_{va}) = 0 \quad (7.124)$$

The *stability derivatives* Y_β, Y_ρ, N_β and N_ρ are simply the partial derivatives

$$Y_\beta = \frac{\partial Y}{\partial \beta}, \qquad Y_\rho = \frac{\partial Y}{\partial \rho}, \qquad N_\beta = \frac{\partial N}{\partial \beta}, \qquad N_\rho = \frac{\partial N}{\partial \rho}, \qquad (7.125)$$

all evaluated at the equilibrium (trim) conditions $(\beta_p, \rho_p; u_a, \delta_{va})$. Like Y and N, each stability derivative depends on the whole set of chosen coordinates. When evaluated at an equilibrium point, they depend ultimately on the two input coordinates.

According to Maxwell's Reciprocal Theorem

$$Y_\rho = N_\beta \qquad (7.126)$$

and hence there are only three independent stability derivatives. See Sect. 7.14 for an example.

It is convenient to introduce the *shifted coordinates*

$$\beta_t(t) = \beta(t) - \beta_p \quad \text{and} \quad \rho_t(t) = \rho(t) - \rho_p \quad (7.127)$$

into the linearized system of Eqs. (7.123), thus getting

$$
\begin{aligned}
mu_a\dot{\beta}_t &= Y_\beta\beta_t + (Y_\rho - mu_a^2)\rho_t \\
J_z u_a\dot{\rho}_t &= N_\beta\beta_t + N_\rho\rho_t
\end{aligned}
\quad (7.128)
$$

where $\dot{\beta} = \dot{\beta}_t$ and $\dot{\rho} = \dot{\rho}_t$. The shifted coordinates are just the distance of the current values from the selected trim values.

The same system of two first-order linear differential equations with constant coefficients can be rewritten in matrix notation as

$$
\begin{bmatrix} \dot{\beta}_t \\ \dot{\rho}_t \end{bmatrix} =
\begin{bmatrix} \dfrac{Y_\beta}{mu_a} & \dfrac{Y_\rho - mu_a^2}{mu_a} \\ \dfrac{N_\beta}{J_z u_a} & \dfrac{N_\rho}{J_z u_a} \end{bmatrix}
\begin{bmatrix} \beta_t \\ \rho_t \end{bmatrix} = \mathbf{A} \begin{bmatrix} \beta_t \\ \rho_t \end{bmatrix}
\quad (7.129)
$$

where the matrix $\mathbf{A}$ is not time dependent.

As a possible further analytical step, we can reformulate the problem as *two identical* second order linear differential equations, with constant coefficients, one in $\rho_t(t)$ and the other in $\beta_t(t)$ (see Sect. 7.18.6 for details)

$$
\begin{aligned}
\ddot{\rho}_t + \dot{\rho}_t &\left(\frac{-mN_\rho - J_z Y_\beta}{J_z m u_a} \right) + \rho_t \left(\frac{Y_\beta N_\rho - Y_\rho N_\beta + mu_a^2 N_\beta}{J_z m u_a^2} \right) \\
&= \ddot{\rho}_t - \text{tr}(\mathbf{A})\dot{\rho}_t + \det(\mathbf{A})\rho_t \\
&= \ddot{\rho}_t + 2\zeta\omega_n\dot{\rho}_t + \omega_n^2\rho_t = 0 \\
&= \ddot{\beta}_t + 2\zeta\omega_n\dot{\beta}_t + \omega_n^2\beta_t = 0
\end{aligned}
\quad (7.130)
$$

where

$$
\begin{aligned}
2\zeta\omega_n &= -\text{tr}(\mathbf{A}) = -\frac{mN_\rho + J_z Y_\beta}{J_z m u_a} = -(\lambda_1 + \lambda_2) \\
\omega_n^2 &= \det(\mathbf{A}) = \frac{(Y_\beta N_\rho - Y_\rho N_\beta) + mu_a^2 N_\beta}{J_z m u_a^2} = \lambda_1\lambda_2
\end{aligned}
\quad (7.131)
$$

The solutions of (7.129) depend on two initial conditions, i.e. $\beta_t(0)$ and $\rho_t(0)$. From the system of equations (7.128) we get $\dot{\beta}(0)$ and $\dot{\rho}(0)$, which are the two additional initial conditions needed in (7.130). Therefore, the two state variables have identical dynamic behavior (i.e., same ζ and ω_n) and are not independent from each other.

From (7.130) and (7.131) we see that the vehicle behaves as a mechanical vibrating system with

$$
\begin{array}{lll}
\text{equivalent mass} & = & J_z m u_a^2 \\[4pt]
\text{equivalent damping} & = & -u_a(m N_\rho + J_z Y_\beta) \\[4pt]
\text{equivalent stiffness} & = & (Y_\beta N_\rho - Y_\rho N_\beta) + m u_a^2 N_\beta
\end{array}
\tag{7.132}
$$

The derivatives Y_β and N_ρ are always negative and act as viscous dampers. Apparently, the quantity $Y_\beta N_\rho - Y_\rho N_\beta$ is always positive (see (7.194)). The derivatives $Y_\rho = N_\beta$ can be positive or negative. Understeer vehicles have $N_\beta > 0$, oversteer vehicles have $N_\beta < 0$. It is important to understand the physical significance of each stability derivative [12, p. 151].

The matrix $\mathbf{A}$ in (7.129) has two eigenvalues

$$
\lambda_j = -\zeta \omega_n \pm \omega_n \sqrt{\zeta^2 - 1}, \quad j = 1, 2
\tag{7.133}
$$

From (7.131) we can obtain the damping ratio ζ

$$
\zeta = -\frac{m N_\rho + J_z Y_\beta}{2\sqrt{J_z m}\sqrt{Y_\beta N_\rho - (Y_\rho - m u_a^2) N_\beta}}
\tag{7.134}
$$

If $\zeta < 1$, the two eigenvalues are complex conjugate

$$
\lambda_j = -\zeta \omega_n \pm i\omega_n \sqrt{1 - \zeta^2} = -\zeta \omega_n \pm i\omega_s
\tag{7.135}
$$

and the system has a damped oscillation with natural angular frequency ω_s

$$
\omega_s = \omega_n \sqrt{1 - \zeta^2}
\tag{7.136}
$$

It is kind of interesting to observe that all these relevant dynamic parameters ζ, ω_n and ω_s depend on the following four quantities

$$
Y_\beta N_\rho - Y_\rho N_\beta + m u_a^2 N_\beta \qquad m N_\rho + J_z Y_\beta \qquad J_z m \qquad u_a
\tag{7.137}
$$

Of course, the eigenvalues depend on (u_a, δ_{va}), as shown in Figs. 7.54 and 7.55 for (the single track model of) an understeer vehicle. In these figures, the real part (gray lines) and the imaginary parts (black lines) are plotted as functions of the forward speed u_a. In Fig. 7.54 the car is going straight, that is with $\delta = 0$. In Fig. 7.55 the car has a net steer angle $\delta = 5°$ (defined in (7.68)). In both cases, the eigenvalues are complex conjugate for speeds higher than about $4\,\mathrm{m/s}$.

Fig. 7.54 Real and imaginary parts of the two eigenvalues (7.135), for $\delta = 0$

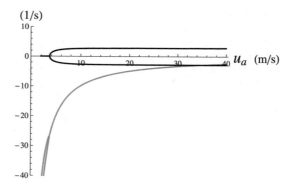

Fig. 7.55 Real and imaginary parts of the two eigenvalues (7.135), for $\delta = 5°$

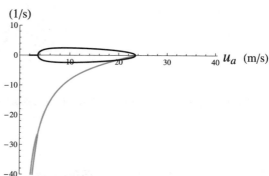

Interestingly enough, when the car goes straight (Fig. 7.54), the real part $-\zeta \omega_n$ and the imaginary part ω_s are almost constant for $u_a > 25\,\text{m/s}$, that is for about $u_a > 90\,\text{km/h}$. Indeed, it is at $u_a \simeq 100\,\text{km/h}$ that car makers typically perform the steering harmonic sweep test, in which the steer input is a harmonic function but with a slowly increasing frequency.

As expected, Fig. 7.55 is almost like Fig. 7.54 for low speeds, say $u_a < 10\,\text{m/s}$. For higher speeds, the two figures are very different. The maximum speed is limited by grip when a vehicle is making a turn.

A clearer picture of the global dynamical features of the vehicle is provided by the MAP approach (7.122) when applied to the damping ratio $\zeta(\delta, \tilde{a}_y)$ and to the damped natural frequency $\omega_s(\delta, \tilde{a}_y)$, as in Fig. 7.56. It immediately arises that the closer the vehicle is to the grip limit (maximum lateral acceleration), the lower both ζ and ω_s. Therefore, the dynamical behavior of the vehicle changes significantly. Perhaps, an expert driver may take advantage of these phenomena to "feel" how close the vehicle is to the grip limit.

Summing up, we have seen that the dynamical features of the vehicle in the neighborhood of an equilibrium point depend on the *four stability derivatives* (7.125), besides m, J_z and u_a. Actually, we know that $Y_\rho = N_\beta$, and hence there are only three independent stability derivatives. See also Sect. 7.14.

Fig. 7.56 MAP in the plane $(\delta, \tilde{a}_y)$ with curves at constant damping ratio ζ (dashed lines) and constant damped natural angular frequency ω_s (solid lines)

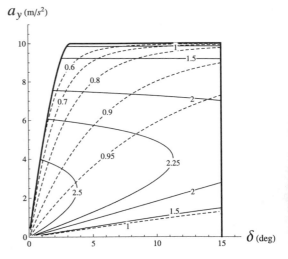

The characterization of the vehicle requires knowledge of these stability derivatives.

7.10.3 Stability of the Equilibrium

An equilibrium point can be either stable or unstable. The typical way to assess whether there is stability or not is by looking at the eigenvalues (7.133). As well known

$$\text{stability} \quad \Longleftrightarrow \quad \text{Re}(\lambda_1) < 0 \quad \text{and} \quad \text{Re}(\lambda_2) < 0 \qquad (7.138)$$

that is, both eigenvalues must have a negative real part. A convenient way to check this condition without computing the two eigenvalues is

$$\text{stability} \quad \Longleftrightarrow \quad \left(\lambda_1 + \lambda_2 = \text{tr}(\mathbf{A})\right) < 0 \quad \text{and} \quad \left(\lambda_1 \lambda_2 = \det(\mathbf{A})\right) > 0 \quad (7.139)$$

Typically, vehicles may become *unstable* because one of the two real eigenvalues becomes positive. From (7.131), the mathematical condition is

$$(Y_\beta N_\rho - Y_\rho N_\beta) + mu_a^2 N_\beta < 0 \qquad (7.140)$$

As already mentioned, instability may occur only if $N_\beta < 0$.

7.10.4 Forced Oscillations (Driver Action)

Linearized systems can also be used to study the effect of small driver actions on the forward speed and/or on the steering wheel angle to control the vehicle. More precisely, we have $u = u_a + u_t$ and $\delta_v = \delta_{va} + \delta_{vt}$.

The linearized inertial terms in (7.118) are

$$m(u\dot{\beta} + \dot{u}\beta + u^2\rho) \simeq m(u_a\dot{\beta} + \dot{u}\beta_p + u_a^2\rho_p + u_a^2\rho_t + 2u_au_t\rho_p)$$
$$J_z(u\dot{\rho} + \dot{u}\rho) \simeq J_z(u_a\dot{\rho} + \dot{u}\rho_p) \qquad (7.141)$$

where $mu_a^2\rho_p = Y_0$, according to (7.119).

The linearized system becomes

$$m(u_a\dot{\beta}_t + \dot{u}\beta_p + u_a^2\rho_t + 2u_a\rho_pu_t) = Y_\beta\beta_t + Y_\rho\rho_t + Y_uu_t + Y_\delta\delta_{vt}$$
$$J_z(u_a\dot{\rho}_t + \dot{u}\rho_p) = N_\beta\beta_t + N_\rho\rho_t + N_uu_t + N_\delta\delta_{vt} \qquad (7.142)$$

where, in addition to the four stability derivatives (7.125), there are also four *control derivatives*

$$Y_\delta = \frac{\partial Y}{\partial \delta_v} > 0, \quad N_\delta = \frac{\partial N}{\partial \delta_v} > 0, \quad Y_u = \frac{\partial Y}{\partial u} \simeq 0, \quad N_u = \frac{\partial N}{\partial u} \simeq 0 \quad (7.143)$$

evaluated, like the others, at the equilibrium point $(\beta_p, \rho_p; u_a, \delta_{va})$. A better way to write (7.142) is

$$mu_a\dot{\beta}_t = Y_\beta\beta_t + (Y_\rho - mu_a^2)\rho_t + (Y_u - 2mu_a\rho_p)u_t + Y_\delta\delta_{vt} - m\beta_p\dot{u}_t$$
$$J_zu_a\dot{\rho}_t = N_\beta\beta_t + N_\rho\rho_t + N_uu_t + N_\delta\delta_{vt} - J_z\rho_p\dot{u}_t \qquad (7.144)$$

which generalizes (7.128).

The most intuitive case is the driver acting only on the steering wheel, which is described by the simplified set of equations

$$mu_a\dot{\beta}_t = Y_\beta\beta_t + (Y_\rho - mu_a^2)\rho_t + Y_\delta\delta_{vt}$$
$$J_zu_a\dot{\rho}_t = N_\beta\beta_t + N_\rho\rho_t + N_\delta\delta_{vt} \qquad (7.145)$$

since $u_t = \dot{u} = 0$. Moreover, $\dot{u} = 0$ is consistent with the assumptions made at the beginning of this chapter.

In matrix notation, (7.144) become

$$\begin{bmatrix} \dot{\beta}_t \\ \dot{\rho}_t \end{bmatrix} = \mathbf{A}\begin{bmatrix} \beta_t \\ \rho_t \end{bmatrix} + \mathbf{B}\begin{bmatrix} u_t \\ \delta_{vt} \\ \dot{u}_t \end{bmatrix} = \mathbf{A}\begin{bmatrix} \beta_t \\ \rho_t \end{bmatrix} + \mathbf{b} \qquad (7.146)$$

or, in an even more compact notation

$$\dot{\mathbf{w}} = \mathbf{A}\mathbf{w} + \mathbf{b} \tag{7.147}$$

where the entries of matrix $\mathbf{A}$ are, exactly as in (7.129)

$$a_{11} = Y_\beta/(mu_a) \qquad\qquad a_{12} = (Y_\rho - mu_a^2)/(mu_a) \tag{7.148}$$

$$a_{21} = N_\beta/(J_z u_a) \qquad\qquad a_{22} = N_\rho/(J_z u_a)$$

and the components of vector $\mathbf{b}$ are

$$b_1 = \frac{1}{mu_a}[(Y_u - 2mu_a\rho_p)u_t + Y_\delta\delta_{vt} - m\beta_p\dot{u}_t]$$

$$b_2 = \frac{1}{J_z u_a}[N_u u_t + N_\delta\delta_{vt} - J_z\rho_p\dot{u}_t] \tag{7.149}$$

Like in (7.130), we can recast the problem (7.144) as two second-order linear differential equations, only apparently independent from each other

$$\ddot{\beta}_t + 2\zeta\omega_n\dot{\beta}_t + \omega_n^2\beta_t = -a_{22}b_1 + a_{12}b_2 + \dot{b}_1 = F_\beta$$

$$\ddot{\rho}_t + 2\zeta\omega_n\dot{\rho}_t + \omega_n^2\rho_t = a_{21}b_1 - a_{11}b_2 + \dot{b}_2 = F_\rho \tag{7.150}$$

where

$$\dot{b}_1 = \frac{1}{mu_a}[(Y_u - 2mu_a\rho_p)\dot{u}_t + Y_\delta\dot{\delta}_v - m\beta_p\ddot{u}_t]$$

$$\dot{b}_2 = \frac{1}{J_z u_a}[N_u\dot{u}_t + N_\delta\dot{\delta}_v - J_z\rho_p\ddot{u}_t] \tag{7.151}$$

Again, if the driver acts only on the steering wheel, like in (7.145), all these expressions become much simpler. More precisely

$$b_1 = \frac{Y_\delta}{mu_a}\delta_{vt}, \quad b_2 = \frac{N_\delta}{J_z u_a}\delta_{vt}, \quad \dot{b}_1 = \frac{Y_\delta}{mu_a}\dot{\delta}_v, \quad \dot{b}_2 = \frac{N_\delta}{J_z u_a}\dot{\delta}_v \tag{7.152}$$

and hence

$$F_\beta = \left(\frac{-N_\rho Y_\delta + (Y_\rho - mu_a^2)N_\delta}{mJ_z u_a^2}\right)\delta_{vt} + \frac{Y_\delta}{mu_a}\dot{\delta}_v$$

$$F_\rho = \left(\frac{N_\beta Y_\delta - Y_\beta N_\delta}{mJ_z u_a^2}\right)\delta_{vt} + \frac{N_\delta}{J_z u_a}\dot{\delta}_v \tag{7.153}$$

The two differential equations (7.150) have identical values of ζ and ω_n, but different forcing terms F_β and F_ρ. However, in (7.153) we still find the four quantities listed in (7.137).

The fundamental result of this analysis is that the transient dynamics of a vehicle in the neighborhood of an equilibrium point is fully characterized by a *finite number* of normalized stability derivatives and control derivatives:

- normalized stability derivatives Y_β/m, Y_ρ/m, N_β/J_z, and N_ρ/J_z;
- normalized control derivatives Y_u/m, Y_δ/m, N_u/J_z, and N_δ/J_z.

It will be discussed shortly that in most cases $Y_u = N_u = 0$, thus leaving six derivatives. It is worth noting that the equality $Y_\rho = N_\beta$ does not reduce the number of relevant derivatives to five. Indeed, we still have $Y_\rho/m \neq N_\beta/J_z$.

The key point is how to measure (identify) all the stability derivatives and all the control derivatives. Their knowledge would be very relevant practical information. The next section presents indeed a novel method to extract these data from the results of steady-state tests. This approach appears to be simpler and more reliable than direct measurements.

7.11 Relationship Between Steady-State Data and Transient Behavior

Most classical vehicle dynamics deals with steady-state data. Understeer and oversteer are steady-state concepts. Or they are not? This is a crucial question. What does a professional driver mean when he/she complains about his/her car being understeer or oversteer? Does it have anything to do with the classical definition of understeer/oversteer as discussed in Sect. 7.7?

Two aspects should be carefully taken into account. While the concepts of velocity, acceleration, mass, stability etc. arise in any branch of mechanics, why do the concepts of understeer and oversteer only belong to vehicle dynamics? This is rather surprising. Why are vehicles so special dynamical systems that they need concepts conceived uniquely for them?

The other aspect is somehow more practical. Why should steady-state tests tell us anything about the transient behavior of a vehicle? In more technical terms, why should steady-state data be related to stability derivatives? Are they or not? If they are related, what is the relationship? Indeed, in [3] it is admitted "Transient responses are related to understeer to some extent, but there is no one-to-one relationship between steady-state and transient response".

This section is devoted to the investigation of the link between the universe of steady-state data and the universe of the dynamical, hence transient, behavior of a vehicle. It will be shown that a link does indeed exist, but it is not direct, not to mention obvious.

It is worth noting that this section is not strictly related to the single track model. The theory developed here is applicable to real road vehicles.

7.11.1 Stability Derivatives from Steady-State Gradients

The starting point is a sort of mathematical trick. At steady state, the lateral force Y and the yawing moment N have very simple values

$$Y_0 = m\tilde{a}_y \quad \text{and} \quad N_0 = 0 \tag{7.154}$$

Nevertheless, by combining (7.119), (7.121), and (7.122), they can be given, as functions, the following expressions

$$Y_0(\delta_{va}, \tilde{a}_y) = Y\big(\beta_p(\delta_{va}, \tilde{a}_y), \rho_p(\delta_{va}, \tilde{a}_y); u_a(\delta_{va}, \tilde{a}_y), \delta_{va}\big) = m\tilde{a}_y$$
$$N_0(\delta_{va}, \tilde{a}_y) = N\big(\beta_p(\delta_{va}, \tilde{a}_y), \rho_p(\delta_{va}, \tilde{a}_y); u_a(\delta_{va}, \tilde{a}_y), \delta_{va}\big) = 0 \tag{7.155}$$

Now, the key idea is to take the partial derivatives of the just defined function $Y_0(\delta_{va}, \tilde{a}_y)$, thus obtaining

$$\frac{\partial Y_0}{\partial \tilde{a}_y} = Y_\beta \frac{\partial \beta_p}{\partial \tilde{a}_y} + Y_\rho \frac{\partial \rho_p}{\partial \tilde{a}_y} + Y_u \frac{\partial u_a}{\partial \tilde{a}_y} \qquad = m\frac{\partial \tilde{a}_y}{\partial \tilde{a}_y} = m$$

$$\frac{\partial Y_0}{\partial \delta_{va}} = Y_\beta \frac{\partial \beta_p}{\partial \delta_{va}} + Y_\rho \frac{\partial \rho_p}{\partial \delta_{va}} + Y_u \frac{\partial u_a}{\partial \delta_{va}} + Y_\delta = m\frac{\partial \tilde{a}_y}{\partial \delta_{va}} = 0 \tag{7.156}$$

The same steps can be taken for the yawing moment $N_0(\delta_{va}, \tilde{a}_y)$, getting

$$\frac{\partial N_0}{\partial \tilde{a}_y} = N_\beta \frac{\partial \beta_p}{\partial \tilde{a}_y} + N_\rho \frac{\partial \rho_p}{\partial \tilde{a}_y} + N_u \frac{\partial u_a}{\partial \tilde{a}_y} \qquad = 0$$

$$\frac{\partial N_0}{\partial \delta_{va}} = N_\beta \frac{\partial \beta_p}{\partial \delta_{va}} + N_\rho \frac{\partial \rho_p}{\partial \delta_{va}} + N_u \frac{\partial u_a}{\partial \delta_{va}} + N_\delta = 0 \tag{7.157}$$

In a road vehicle, that is without significant aerodynamic vertical loads, it is reasonable to assume

$$Y_u = N_u = 0 \tag{7.158}$$

if we take β and ρ as state variables to describe the vehicle motion.[11] In other words, Y and N do not change if we modify only u, keeping constant β, ρ and δ_v, that is keeping constant α_1 and α_2 (cf. (7.51)). It would not be so in Formula cars, that is in cars with aerodynamic devices.

The two equations in (7.156), with $Y_u = N_u = 0$, yield the system of linear equations

---

[11] Actually, as discussed right after (7.38), these partial derivatives are not zero if there is roll steer in a double track model. However, they should be very small. See also (7.75).

$$\begin{cases} Y_\beta \dfrac{\partial \beta_p}{\partial \tilde{a}_y} + Y_\rho \dfrac{\partial \rho_p}{\partial \tilde{a}_y} = m \\[4mm] Y_\beta \dfrac{\partial \beta_p}{\partial \delta_{va}} + Y_\rho \dfrac{\partial \rho_p}{\partial \delta_{va}} = -Y_\delta \end{cases} \qquad (7.159)$$

and, similarly, from (7.157)

$$\begin{cases} N_\beta \dfrac{\partial \beta_p}{\partial \tilde{a}_y} + N_\rho \dfrac{\partial \rho_p}{\partial \tilde{a}_y} = 0 \\[4mm] N_\beta \dfrac{\partial \beta_p}{\partial \delta_{va}} + N_\rho \dfrac{\partial \rho_p}{\partial \delta_{va}} = -N_\delta \end{cases} \qquad (7.160)$$

These two systems of equations

$$\begin{bmatrix} \beta_y & \rho_y \\ \beta_\delta & \rho_\delta \end{bmatrix} \begin{bmatrix} Y_\beta \\ Y_\rho \end{bmatrix} = \begin{bmatrix} m \\ -Y_\delta \end{bmatrix} \quad \text{and} \quad \begin{bmatrix} \beta_y & \rho_y \\ \beta_\delta & \rho_\delta \end{bmatrix} \begin{bmatrix} N_\beta \\ N_\rho \end{bmatrix} = \begin{bmatrix} 0 \\ -N_\delta \end{bmatrix} \qquad (7.161)$$

have the same matrix, whose coefficients are the four components of the *gradients* defined in (7.99)

$$\operatorname{grad} \rho_p = \left(\dfrac{\partial \rho_p}{\partial \tilde{a}_y}, \dfrac{\partial \rho_p}{\partial \delta_{va}} \right) = (\rho_y, \rho_\delta)$$

$$\operatorname{grad} \beta_p = \left(\dfrac{\partial \beta_p}{\partial \tilde{a}_y}, \dfrac{\partial \beta_p}{\partial \delta_{va}} \right) = (\beta_y, \beta_\delta) \qquad (7.99')$$

of the two *steady-state maps* (7.122). After having performed the standard steady-state tests, all these gradient components (already introduced in Sect. 7.7.1) are known functions.

The four *stability derivatives* are the solution of the two systems of equations (7.161)

$$Y_\beta = \dfrac{Y_\delta \rho_y + m \rho_\delta}{\beta_y \rho_\delta - \beta_\delta \rho_y} \qquad Y_\rho = -\dfrac{Y_\delta \beta_y + m \beta_\delta}{\beta_y \rho_\delta - \beta_\delta \rho_y}$$

$$N_\beta = \dfrac{N_\delta \rho_y}{\beta_y \rho_\delta - \beta_\delta \rho_y} \qquad N_\rho = -\dfrac{N_\delta \beta_y}{\beta_y \rho_\delta - \beta_\delta \rho_y} \qquad (7.162)$$

Therefore, they are known functions of the *gradient components* and of the *control derivatives* Y_δ and N_δ. This is a fundamental original result, as it shows why steady-state data can indeed provide information about the transient behavior, although not in an obvious way.

Moreover, from (7.126) (i.e., $Y_\rho = N_\beta$) and (7.162) we have that

$$\beta_y Y_\delta + \rho_y N_\delta = -m\beta_\delta$$

which means

$$(7.163)$$

$$Y_\delta = -\frac{N_\delta \rho_y + m\beta_\delta}{\beta_y} \quad \text{or} \quad N_\delta = -\frac{Y_\delta \beta_y + m\beta_\delta}{\rho_y}$$

The transient behavior of the vehicle is characterized by the stability derivatives. This is well known. What is new is that the stability derivatives are strictly related to the gradients of steady-state maps. This result opens up new perspectives in the objective evaluation of the handling of vehicles (cf. [10]).

7.11.2 Equations of Motion

Now, we can go back to the linearized equations of motion (7.145). The stability derivatives can be replaced by the expressions in (7.162), thus obtaining

$$mu_a \dot{\beta}_t = \left(\frac{Y_\delta \rho_y + m\rho_\delta}{\beta_y \rho_\delta - \beta_\delta \rho_y}\right) \beta_t + \left(-\frac{Y_\delta \beta_y + m\beta_\delta}{\beta_y \rho_\delta - \beta_\delta \rho_y} - mu_a^2\right) \rho_t + Y_\delta \delta_{vt}$$

$$J_z u_a \dot{\rho}_t = \left(\frac{N_\delta \rho_y}{\beta_y \rho_\delta - \beta_\delta \rho_y}\right) \beta_t + \left(-\frac{N_\delta \beta_y}{\beta_y \rho_\delta - \beta_\delta \rho_y}\right) \rho_t + N_\delta \delta_{vt}$$

$$(7.164)$$

where β_t and ρ_t are the shifted coordinates defined in (7.127).

In some cases it is convenient to define and use the generalized control derivatives

$$\hat{Y}_\delta = \frac{Y_\delta}{m} \quad \text{and} \quad \hat{N}_\delta = \frac{N_\delta}{J_z}$$

$$(7.165)$$

thus obtaining

$$u_a \dot{\beta}_t = \left(\frac{\hat{Y}_\delta \rho_y + \rho_\delta}{\beta_y \rho_\delta - \beta_\delta \rho_y}\right) \beta_t + \left(-\frac{\hat{Y}_\delta \beta_y + \beta_\delta}{\beta_y \rho_\delta - \beta_\delta \rho_y} - u_a^2\right) \rho_t + \hat{Y}_\delta \delta_{vt}$$

$$u_a \dot{\rho}_t = \left(\frac{\hat{N}_\delta \rho_y}{\beta_y \rho_\delta - \beta_\delta \rho_y}\right) \beta_t + \left(-\frac{\hat{N}_\delta \beta_y}{\beta_y \rho_\delta - \beta_\delta \rho_y}\right) \rho_t + \hat{N}_\delta \delta_{vt}$$

$$(7.166)$$

This is quite a remarkable (and original) result. It shows how the equations of motion can be given in terms of data collected in steady-state tests. It is the link between the realm of steady-state gradients and the realm of transient behavior.

7.11.3 Estimation of the Control Derivatives

The control derivatives $\hat{Y}_\delta$ and $\hat{N}_\delta$ can be estimated by means of standard dynamic tests. For instance, let us consider a generalized *step steering input*, that is a sudden increase δ_{vt} of the steering wheel angle δ_v applied to a vehicle in a steady-state (equilibrium) configuration. We say "generalized" since it should and can be done from any steady-state configuration, not necessarily from a straight-line trajectory. Since, by definition $\beta_t(0) = 0$ and $\rho_t(0) = 0$, from (7.166) we obtain

$$\hat{Y}_\delta = \frac{u_a \dot{\beta}_t(0)}{\delta_{vt}} \quad \text{and} \quad \hat{N}_\delta = \frac{u_a \dot{\rho}_t(0)}{\delta_{vt}} \tag{7.167}$$

Combining this result with (7.163), we also get that in a step steering input

$$\beta_y \dot{\beta}_t(0) + \frac{J_z}{m}\rho_y \dot{\rho}_t(0) = -\frac{\delta_{vt}}{u_a}\beta_\delta \tag{7.168}$$

7.11.4 Objective Evaluation of Car Handling

The two coefficients $2\zeta\omega_n = -(\lambda_1 + \lambda_2)$ and $\omega_n^2 = \lambda_1\lambda_2$ of the differential equations (7.130), can now be expressed as combinations of steady-state gradient components and control derivatives

$$2\zeta\omega_n = \frac{1}{u_a(\beta_y\rho_\delta - \beta_\delta\rho_y)}\left[\left(\hat{N}_\delta\beta_y - \hat{Y}_\delta\rho_y\right) - \rho_\delta\right] = -\operatorname{tr}(\mathbf{A}) = n_1(\delta_{va}, \tilde{a}_y)$$

$$\omega_n^2 = \frac{\hat{N}_\delta}{(\beta_y\rho_\delta - \beta_\delta\rho_y)}\left(\rho_y - \frac{1}{u_a^2}\right) = \det(\mathbf{A}) = n_2(\delta_{va}, \tilde{a}_y)$$

$$\tag{7.169}$$

or, equivalently

$$2\zeta\omega_n = \frac{1}{J_z m u_a}\frac{(mN_\delta\beta_y - J_zY_\delta\rho_y) - J_zm\rho_\delta}{(\beta_y\rho_\delta - \beta_\delta\rho_y)} = -\operatorname{tr}(\mathbf{A}) = n_1(\delta_{va}, \tilde{a}_y)$$

$$\tag{7.170}$$

$$\omega_n^2 = \frac{1}{J_z m u_a}\frac{mN_\delta(u_a^2\rho_y - 1)}{u_a(\beta_y\rho_\delta - \beta_\delta\rho_y)} = \det(\mathbf{A}) = n_2(\delta_{va}, \tilde{a}_y)$$

Exactly like in (7.132), we have the physical interpretation as a vibrating system

$$\text{equivalent mass} \quad = \quad J_z m u_a^2$$

$$\text{equivalent damping} \quad = \quad u_a \frac{(m N_\delta \beta_y - J_z Y_\delta \rho_y) - J_z m \rho_\delta}{\beta_y \rho_\delta - \beta_\delta \rho_y} \tag{7.171}$$

$$\text{equivalent stiffness} \quad = \quad \frac{m N_\delta (u_a^2 \rho_y - 1)}{\beta_y \rho_\delta - \beta_\delta \rho_y}$$

Once again, the dynamic features of the vehicle are strictly related to the gradients of data obtained in steady-state tests.

From (7.169), we see that the vehicle becomes unstable ($\det(\mathbf{A}) < 0$) if

$$\rho_y - \frac{1}{u_a^2} > 0 \tag{7.172}$$

which requires $\rho_y > 0$ (oversteer). This condition is completely equivalent to (7.140) (see also Sect. 7.13). From (7.162), we obtain that N_β and ρ_y have opposite signs

$$N_\beta = \frac{N_\delta \rho_y}{\beta_y \rho_\delta - \beta_\delta \rho_y} \tag{7.173}$$

since $N_\delta > 0$ and $(\beta_y \rho_\delta - \beta_\delta \rho_y) < 0$.

Following the same path of reasoning, the two forcing terms F_β and F_ρ in (7.153) can be rewritten as

$$F_\beta = -\frac{\hat{N}_\delta}{u_a^2} \left(\frac{\beta_\delta}{\beta_y \rho_\delta - \beta_\delta \rho_y} + u_a^2 \right) \delta_{vt} + \frac{\hat{Y}_\delta}{u_a} \dot{\delta}_v$$

$$= n_3(\delta_{va}, \tilde{a}_y) \delta_{vt} + n_4(\delta_{va}, \tilde{a}_y) \dot{\delta}_v \tag{7.174}$$

and

$$F_\rho = -\frac{\hat{N}_\delta}{u_a^2} \left(\frac{\rho_\delta}{\beta_y \rho_\delta - \beta_\delta \rho_y} \right) \delta_{vt} + \frac{\hat{N}_\delta}{u_a} \dot{\delta}_v$$

$$= n_5(\delta_{va}, \tilde{a}_y) \delta_{vt} + n_6(\delta_{va}, \tilde{a}_y) \dot{\delta}_v \tag{7.175}$$

Typical patterns are shown in the MAP in Fig. 7.57 for F_β, and in the MAP of Fig. 7.58 for F_ρ.

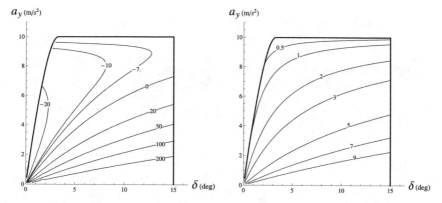

Fig. 7.57 MAP in the plane $(\delta, \tilde{a}_y)$ for F_β, with curves at constant n_3 (left) and constant n_4 (right)

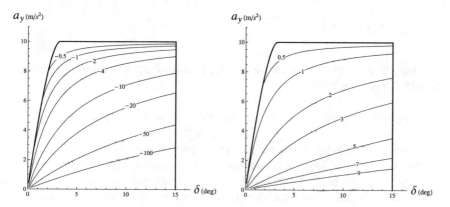

Fig. 7.58 MAP in the plane $(\delta, \tilde{a}_y)$ for F_ρ, with curves at constant n_5 (left) and constant n_6 (right)

7.11.4.1   Vehicle "DNA"

Equations (7.169), (7.174) and (7.175) show that the dynamical behavior of a road vehicle in the neighborhood of any equilibrium point is fully described by six maps $n_i(\delta_{va}, \tilde{a}_y)$. These maps (functions) can be seen as a sort of "DNA" of the vehicle, in the sense that they determine the vehicle transient behavior. To help the reader, these six maps are listed below:

$$n_1(\delta_{va}, \tilde{a}_y) = \frac{1}{u_a(\beta_y \rho_\delta - \beta_\delta \rho_y)} \left[\left(\hat{N}_\delta \beta_y - \hat{Y}_\delta \rho_y\right) - \rho_\delta\right] = 2\zeta \omega_n$$

$$n_2(\delta_{va}, \tilde{a}_y) = \frac{\hat{N}_\delta}{(\beta_y \rho_\delta - \beta_\delta \rho_y)} \left(\rho_y - \frac{1}{u_a^2}\right) = \omega_n^2$$

$$n_3(\delta_{va}, \tilde{a}_y) = -\frac{\hat{N}_\delta}{u_a^2} \left(\frac{\beta_\delta}{\beta_y \rho_\delta - \beta_\delta \rho_y} + u_a^2\right)$$

$$n_4(\delta_{va}, \tilde{a}_y) = \frac{\hat{Y}_\delta}{u_a} \tag{7.176}$$

$$n_5(\delta_{va}, \tilde{a}_y) = -\frac{\hat{N}_\delta}{u_a^2} \left(\frac{\rho_\delta}{\beta_y \rho_\delta - \beta_\delta \rho_y}\right)$$

$$n_6(\delta_{va}, \tilde{a}_y) = \frac{\hat{N}_\delta}{u_a}$$

However, all these quantities are, ultimately, combinations of the following six fundamental "handling bricks":

$$s_1 = \beta_y, \quad s_2 = \rho_y, \quad s_3 = \beta_\delta, \quad s_4 = \rho_\delta, \quad s_5 = \hat{N}_\delta, \quad s_6 = \hat{Y}_\delta \tag{7.177}$$

all of them, in general, functions of two variables like, e.g., $\tilde{a}_y$ and δ_v.

Two vehicles with the same s_i, and hence with the same n_i, have *identical transient handling behavior*, notwithstanding their size, weight, etc. In other words, the two vehicles react in exactly the same way to given driver input. Therefore, there is indeed a strong relationship between data collected in steady-state tests and the transient dynamical behavior of a vehicle.

Objective measures of car handling should be based on the quantities defined in (7.176).

On the practical side, we see that the components of the gradients (7.99) of the steady-state maps $\beta_p(\delta_v, \tilde{a}_y)$ and $\rho_p(\delta_v, \tilde{a}_y)$ provide four out of six "handling bricks", the other two being the generalized control derivatives. Basically, we have found a more feasible way, based on the gradient components of the steady-state MAPs, to measure the six stability and control derivatives listed on Sect. 7.11.

7.12 Stability (Again)

According to (7.139), an equilibrium point is stable if and only if $\text{tr}(\mathbf{A}) < 0$ and $\det(\mathbf{A}) > 0$. These two conditions, after (7.169), can be expressed in terms of the six fundamental handling bricks (7.177) and the forward speed. However, provided $\Phi_1 > 0$ and $\Phi_2 > 0$, the onset of instability is given by (7.172).

7.13   New Understeer Gradient

Let us discuss in detail the component ρ_y of the new understeer gradient introduced in (7.99). In general it is a function of two variables

$$\rho_y = \rho_y(\delta_{va}, \tilde{a}_y) \tag{7.178}$$

except in some special cases, like the single track model with open differential, where, according to (7.102), $\rho_y = \rho_y(\tilde{a}_y) = -\mathrm{d}f_\rho/\mathrm{d}\tilde{a}_y$.
   More explicitly,

$$\rho_y = \frac{\partial \rho_p}{\partial \tilde{a}_y} = \frac{\partial}{\partial \tilde{a}_y}\left(\frac{1}{R}\right) = -\frac{K}{l} \tag{7.179}$$

This is similar to the definition (7.117) of the classical understeer gradient K, but with a few *fundamental differences*.
   The definition of ρ_y does not involve any weak concept, like the wheelbase l or the Ackermann steer angle, as discussed in Sect. 7.9. Therefore, it is much more general. This new understeer gradient is defined for *any* vehicle.
   Moreover, it is the *correct* measure of understeer/oversteer, while K is not. This may look surprising, but that is the way it is, as will be shown in Sect. 7.14.1 (see in particular Table 7.1).
   Of course, the partial derivative in (7.179) requires the steer angle to be kept constant, according to (7.178).[12]
   But there are other reasons that support ρ_y as a good handling parameter. Let us consider a *constant steering wheel test* and monitor the yaw rate $r_p = r_p(u_a; \delta_{va})$ as a function of the forward speed u_a, keeping constant the steering wheel angle δ_{va}. For brevity, let $r_p' = \mathrm{d}r_p/\mathrm{d}u_a$. Equation (7.179) can be rewritten as

$$\frac{\mathrm{d}\rho_p}{\mathrm{d}\tilde{a}_y} = \frac{\mathrm{d}(r_p/u_a)}{\mathrm{d}(r_p u_a)} = \frac{\mathrm{d}(r_p/u_a)}{\mathrm{d}u_a}\left(\frac{\mathrm{d}(r_p u_a)}{\mathrm{d}u_a}\right)^{-1} = \frac{1}{u_a^2}\left(\frac{r_p' u_a - r_p}{r_p' u_a + r_p}\right) = \rho_y(u_a; \delta_{va}) \tag{7.180}$$

This general equation provides a way to obtain the critical speed and the characteristic speed.
   If $\rho_y < 0$ (understeer), the *characteristic speed* u_{ch} is, by definition [12, pp. 181–185], the speed at which $r_p' = 0$, that is the *yaw velocity gain* $r_p(u_a; \delta_{va})$ is maximum. By letting $r_p' \to 0$ in (7.180), we obtain that the characteristic speed must satisfy the following equation

$$\frac{1}{u_a^2} = -\rho_y(u_a; \delta_{va}) \quad \text{that is} \quad u_{\mathrm{ch}}(\delta_{va}) = \sqrt{-\frac{1}{\rho_y(u_{\mathrm{ch}}; \delta_{va})}} \tag{7.181}$$

---

[12] Tests with constant steer angle are the most general: they can be performed on any kind of vehicle.

Similarly, if $\rho_y > 0$ (oversteer), the *critical speed* u_{cr} is, by definition [12, p. 177], the speed at which $r'_p \to \infty$, which means

$$\frac{1}{u_a^2} = \rho_y(u_a; \delta_{va}) \qquad \text{that is} \qquad u_{cr}(\delta_{va}) = \sqrt{\frac{1}{\rho_y(u_{cr}; \delta_{va})}} \qquad (7.182)$$

Summing up:

- ρ_y has been defined without any recourse to weak concepts, like a reference vehicle having Ackermann steering [16];
- ρ_y can be easily measured in constant steering wheel tests;
- the critical speed and the characteristic speed come out naturally as special cases.[13]

A similar treatment applies to the other gradient component β_y. In this case $v_p = v_p(u_a; \delta_{va})$, thus obtaining

$$\beta_y = \frac{\mathrm{d}\beta_p}{\mathrm{d}\tilde{a}_y} = \frac{\mathrm{d}(v_p/u_a)}{\mathrm{d}(u_a r_p)} = \frac{1}{u_a^2}\left(\frac{v'_p u_a - v_p}{r'_p u_a + r_p}\right) \qquad (7.183)$$

In general

$$\beta_y = \beta_y(\delta_{va}, \tilde{a}_y) \qquad (7.184)$$

except in cases like the single track model with open differential, where, according to (7.102), $\beta_y = \beta_y(\tilde{a}_y) = -\mathrm{d}f_\beta/\mathrm{d}\tilde{a}_y$.

7.14 The Nonlinear Single Track Model Revisited

The general approach presented in Sect. 7.11, which explains why steady-state data are also relevant for the transient behavior, is applied here to the single track model. The goal is to clarify the matter by a significant worked-out example.

For simplicity, we assume $u = u_a$ and $\dot{u} = 0$ and hence start with the linearized equations of motion (7.145).

In the single track model (with open differential), the *stability derivatives* (7.125) can be obtained directly (cf. (7.88)), taking into account the congruence equations (7.51) and the axle characteristics (7.73)

---

[13] Actually, the real critical speed can be lower than the value predicted by (7.182), as shown in Sect. 7.16.3 [9, pp. 216–219]. Basically, (7.182) may not predict the right value because in real vehicles we control the longitudinal force, not directly the forward speed. Therefore, a real vehicle is a system with three state variables, not just two. This additional degree of freedom does affect the critical speed, unless the vehicle is going straight.

$$Y_\beta = \frac{dY_1}{d\alpha_1}\frac{\partial\alpha_1}{\partial\beta} + \frac{dY_2}{d\alpha_2}\frac{\partial\alpha_2}{\partial\beta} = -\frac{dY_1}{d\alpha_1} - \frac{dY_2}{d\alpha_2} = -\Phi_1 - \Phi_2$$

$$Y_\rho = \frac{dY_1}{d\alpha_1}\frac{\partial\alpha_1}{\partial\rho} + \frac{dY_2}{d\alpha_2}\frac{\partial\alpha_2}{\partial\rho} = -a_1\frac{dY_1}{d\alpha_1} + a_2\frac{dY_2}{d\alpha_2} = -a_1\Phi_1 + a_2\Phi_2$$

(7.185)

and

$$N_\beta = a_1\frac{dY_1}{d\alpha_1}\frac{\partial\alpha_1}{\partial\beta} - a_2\frac{dY_2}{d\alpha_2}\frac{\partial\alpha_2}{\partial\beta} = -a_1\frac{dY_1}{d\alpha_1} + a_2\frac{dY_2}{d\alpha_2} = -a_1\Phi_1 + a_2\Phi_2$$

$$N_\rho = a_1\frac{dY_1}{d\alpha_1}\frac{\partial\alpha_1}{\partial\rho} - a_2\frac{dY_2}{d\alpha_2}\frac{\partial\alpha_2}{\partial\rho} = -a_1^2\frac{dY_1}{d\alpha_1} - a_2^2\frac{dY_2}{d\alpha_2} = -a_1^2\Phi_1 - a_2^2\Phi_2$$

(7.186)

where

$$\Phi_1 = \frac{dY_1}{d\alpha_1} \quad \text{and} \quad \Phi_2 = \frac{dY_2}{d\alpha_2}$$

(7.187)

are the *slopes* of the axle characteristics *at the equilibrium point*, defined in (7.94). Obviously, $\Phi_i > 0$ in the monotone increasing part of the axle characteristics. These slopes are simple to be defined, but not so simple to be measured directly.

It is also worth recalling that

$$Y_\rho = N_\beta$$

(7.188)

To proceed further, as already done in (7.68), let

$$\delta_1 = (1 + \kappa)\tau\delta_v \quad \text{and} \quad \delta_2 = \kappa\tau\delta_v$$

(7.189)

thus linking the rear steer angle δ_2 to the front steer angle δ_1 in such a way to keep constant the net steer angle $\tau\delta_v = \delta_1 - \delta_2 = \delta$. To have front steering only it suffices to set $\kappa = 0$.

We can now obtain also the explicit expressions of the *control derivatives*

$$Y_\delta = [(1 + \kappa)\Phi_1 + \kappa\Phi_2]\tau, \qquad N_\delta = [(1 + \kappa)\Phi_1 a_1 - \kappa\Phi_2 a_2]\tau$$

(7.190)

In this vehicle model, all stability derivatives and all control derivatives are functions of $\tilde{a}_y$ only, that is $Y_\beta = Y_\beta(\tilde{a}_y)$, and so on.

The linearized equations of motions (7.145) become

$$m(u_a\dot{\beta}_t + u_a^2\rho_t) = -(\Phi_1 + \Phi_2)\beta_t - (\Phi_1 a_1 - \Phi_2 a_2)\rho_t + ((1+\kappa)\Phi_1 + \kappa\Phi_2)\tau\delta_{vt}$$

$$J_z u_a\dot{\rho}_t = -(\Phi_1 a_1 - \Phi_2 a_2)\beta_t - (\Phi_1 a_1^2 + \Phi_2 a_2^2)\rho_t + ((1+\kappa)\Phi_1 a_1 - \kappa\Phi_2 a_2)\tau\delta_{vt}$$

(7.191)

Similarly, (7.131) becomes, in this case

$$2\zeta\omega_n = -\operatorname{tr}(\mathbf{A}) = \frac{1}{u_a}\left(\frac{\Phi_1 + \Phi_2}{m} + \frac{\Phi_1 a_1^2 + \Phi_2 a_2^2}{J_z}\right)$$
$$= \frac{\Phi_1(J_z + ma_1^2) + \Phi_2(J_z + ma_2^2)}{J_z m u_a} \tag{7.192}$$

and

$$\omega_n^2 = \det(\mathbf{A}) = \frac{1}{J_z m u_a^2}\left[\Phi_1\Phi_2(a_1 + a_2)^2 - mu_a^2(\Phi_1 a_1 - \Phi_2 a_2)\right] \tag{7.193}$$

Quite remarkable is that in (7.131)

$$Y_\beta N_\rho - Y_\rho N_\beta = \Phi_1\Phi_2(a_1 + a_2)^2 \tag{7.194}$$

Therefore this quantity is always positive, provided the vehicle operates with positive Φ_1 and Φ_2.

The damping ratio (7.134) has the following expression

$$\zeta = \frac{(\Phi_1 + \Phi_2)J_z + (\Phi_1 a_1^2 + \Phi_2 a_2^2)m}{2\sqrt{J_z m}\sqrt{\Phi_1\Phi_2(a_1 + a_2)^2 - mu_a^2(\Phi_1 a_1 - \Phi_2 a_2)}} \tag{7.195}$$

and the natural angular frequency (7.136) becomes

$$\omega_s^2 = \frac{\Phi_2 a_2 - \Phi_1 a_1}{J_z}$$
$$- \frac{1}{(2J_z m u_a)^2}\left[(\Phi_1 + \Phi_2)^2 J_z^2 + 2(\Phi_2 a_2 - \Phi_1 a_1)^2 J_z m\right.$$
$$\left. - 2(a_1 + a_2)^2 \Phi_1\Phi_2 J_z m + (\Phi_1 a_1^2 + \Phi_2 a_2^2)^2 m^2\right] \tag{7.196}$$

or, equivalently

$$\omega_s^2 = -\frac{\Phi_1 a_1}{J_z} + \frac{\Phi_2 a_2}{J_z}$$
$$- \Phi_1\Phi_2\left[\frac{J_z^2 - (a_1^2 + 4a_1 a_2 + a_2^2)J_z m + a_1^2 a_2^2 m^2}{2(J_z m u_a)^2}\right]$$
$$- \Phi_1^2\left(\frac{J_z + ma_1^2}{2J_z m u_a}\right)^2 - \Phi_2^2\left(\frac{J_z + ma_2^2}{2J_z m u_a}\right)^2 \tag{7.197}$$

These parameters characterize the handling behavior in the neighborhood of an equilibrium point. More explicitly, like in (7.132), we have that the vehicle behaves as a mechanical vibrating system with

$$\text{equivalent mass} \quad = \quad J_z m u_a^2$$

$$\text{equivalent damping} \quad = \quad u_a[\Phi_1(J_z + m a_1^2) + \Phi_2(J_z + m a_2^2)] \qquad (7.198)$$

$$\text{equivalent stiffness} \quad = \quad \Phi_1 \Phi_2 (a_1 + a_2)^2 - m u_a^2 (\Phi_1 a_1 - \Phi_2 a_2)$$

It is a kind of interesting comparing (7.132), (7.171), and (7.198). They provide the same information with very different tools.

In the single track model, the explicit expressions of the two forcing functions (7.153) can also be obtained

$$
\begin{aligned}
F_\beta &= \left[\frac{(a_1 + a_2)\Phi_1 \Phi_2 ((1 + \kappa)a_2 + \kappa a_1)}{J_z m u_a^2} - \frac{(1 + \kappa)a_1 \Phi_1 - \kappa a_2 \Phi_2}{J_z} \right] \tau \delta_{vt} \\
&\quad + \frac{(1 + \kappa)\Phi_1 + \kappa \Phi_2}{m u_a} \tau \dot{\delta}_v
\end{aligned}
$$

$$
F_\rho = \frac{(a_1 + a_2)\Phi_1 \Phi_2}{J_z m u_a^2} \tau \delta_{vt} + \frac{(1 + \kappa)a_1 \Phi_1 - \kappa a_2 \Phi_2}{J_z u_a} \tau \dot{\delta}_v
$$

$$(7.199)$$

with obvious simplifications if $\kappa = 0$ (front steering only).

All the equations obtained in this section show that for a single track model there are *seven* design parameters

$$
\frac{\Phi_1}{m}, \quad \frac{\Phi_2}{m}, \quad a_1, \quad a_2, \quad \frac{J_z}{m}, \quad \kappa, \quad \tau \qquad (7.200)
$$

in addition to the control parameters u and $\delta_v(t)$, with constant $u = u_a$.

Now, we can relate these design parameters to the *six* fundamental "handling bricks" of (7.177).

The components of the gradients grad β_p and grad ρ_p, defined in (7.99), have been obtained for the single track model in (7.100)

$$
\beta_y = -\frac{m}{l^2} \left(\frac{\Phi_1 a_1^2 + \Phi_2 a_2^2}{\Phi_1 \Phi_2} \right) \qquad \beta_\delta = \tau \left(\frac{(1 + \kappa)a_2 + \kappa a_1}{a_1 + a_2} \right)
$$

$$(7.100')$$

$$
\rho_y = -\frac{m}{l^2} \left(\frac{\Phi_2 a_2 - \Phi_1 a_1}{\Phi_1 \Phi_2} \right) \qquad \rho_\delta = \tau \frac{(1 + \kappa) - \kappa}{a_1 + a_2}
$$

As already stated, all these components can be measured experimentally from standard steady-state tests, and without having to bother about Ackermann steer angle and the like.

Also interesting is that

$$
\beta_y \rho_\delta - \beta_\delta \rho_y = -\tau m \frac{(1 + \kappa)\Phi_1 a_1 - \kappa \Phi_2 a_2}{\Phi_1 \Phi_2 (a_1 + a_2)^2} = -\frac{m \, N_\delta}{Y_\beta N_\rho - Y_\rho N_\beta} < 0 \qquad (7.201)
$$

In case of no rear steering (i.e., $\kappa = 0$) it becomes

$$\beta_y \rho_\delta - \beta_\delta \rho_y = -\frac{\tau m a_1}{\Phi_2(a_1 + a_2)^2} \tag{7.202}$$

The generalized control derivatives $\hat{Y}_\delta = Y_\delta/m$ and $\hat{N}_\delta = N_\delta/J_z$ are immediately obtained from (7.190).

Summing up, for the single track model the six "handling bricks" $s_i(\delta_{va}, \tilde{a}_y)$ in (7.177) are

$$s_1 = \beta_y = -\frac{m}{(a_1 + a_2)^2}\left(\frac{\Phi_2 a_2^2 + \Phi_1 a_1^2}{\Phi_1 \Phi_2}\right)$$

$$s_2 = \rho_y = -\frac{m}{(a_1 + a_2)^2}\left(\frac{\Phi_2 a_2 - \Phi_1 a_1}{\Phi_1 \Phi_2}\right)$$

$$s_3 = \beta_\delta = \tau \frac{(1+\kappa)a_2 + \kappa a_1}{a_1 + a_2}$$

$$s_4 = \rho_\delta = \tau \frac{(1+\kappa) - \kappa}{a_1 + a_2} = \frac{\tau}{a_1 + a_2} \tag{7.203}$$

$$s_5 = \hat{N}_\delta = \tau \frac{(1+\kappa)\Phi_1 a_1 - \kappa \Phi_2 a_2}{J_z}$$

$$s_6 = \hat{Y}_\delta = \tau \frac{(1+\kappa)\Phi_1 + \kappa \Phi_2}{m}$$

Therefore, we have six "handling bricks" depending on seven design parameters. This means that there exist infinitely many different vehicles sharing the same handling transient behavior. This observation opens up many new paths of reasoning.

One of these paths of reasoning is worked out in the next section. The results are quite surprising.

7.14.1   Very Different Vehicles with Identical Handling

As a test of the new theory presented in Sect. 7.14, we are going to compare the transient handling behavior of, say, three linear single track models. These vehicles will be very different, and identical at the same time. How is it possible?

These three vehicles will share exactly the same values of all the six handling bricks listed in (7.203). Therefore, they will have the same handling behavior. However, they need not to be exactly alike, since we can play with seven design parameter to fulfill the six handling requirements.

A good test is to define a first vehicle with front steer only, a second vehicle with also negative rear steer, and a third one with also positive rear steer. This can be easily done by means of parameter κ, introduced in (7.189)

$$\delta_1 = (1 + \kappa)\tau\delta_v \quad \text{and} \quad \delta_2 = \kappa\tau\delta_v \tag{7.189'}$$

Parameter κ controls the amount of rear steer with respect to front steer, while keeping constant the net steer angle $\delta = \tau\delta_v = \delta_1 - \delta_2$. The rear wheels turn opposite to the front wheels if $\kappa < 0$, while both front and rear wheels turn alike if $\kappa > 0$. Typically, $|\kappa| < 0.1$, that is the rear wheels cannot turn as far as the front wheels.

But, let us do some numerical examples. Let us consider a vehicle with front steering only ($\kappa = 0$), with the following features:

- $\tau = 1/20$;
- $m = 1300$ kg;
- $J_z = 2000$ kgm^2;
- $a_1 = 1$ m;
- $a_2 = 1.60$ m;
- $\Phi_1 = \Phi_1(0) = 70000$ N/rad;
- $\Phi_2 = \Phi_2(0) = 90000$ N/rad.

From (7.203) we can compute all six handling bricks s_i for this vehicle, and then use them for the other two vehicles. This way, it is possible to create vehicles that look very different, but which ultimately have exactly the same handling behavior.

The vehicle features for $\kappa \pm 0.1$, that is two very high amounts of rear steer, are shown in Table 7.1. The three vehicles there reported are strikingly different (Fig. 7.71), yet they have the same handling behavior, and not limited to steady state. For the driver, they behave exactly the same way under any transient conditions.

For instance, starting from a straight trajectory, let us impose a step steering input $\delta_v = 60°$, the forward velocity being $u = 30$ m/s. Figures 7.59 and 7.60 show the lateral velocity $v(t)$ and the yaw rate $r(t)$, respectively. They are identical for the three vehicles, thus confirming the theoretical claims.

Of course, the slip angles are not identical, as shown in Fig. 7.61. The three vehicles are indeed different. It is left to the reader to figure out which curve is for $\kappa = 0.1$, etc.

Just out of curiosity, the most extreme vehicles that can be obtained with this algorithm are shown in Table 7.2. Of course, we are not suggesting that they are feasible vehicles. They are reported here because they provide some rigorous

Table 7.1 Design parameters of vehicles with different amounts of rear steering κ, but with identical transient handling behavior. Note that the classical understeer gradient K conveys misleading information

κ	Φ_1	Φ_2	a_1	a_2	J_z	m	τ	K	$-\rho_y$
[−]	[N/rad]	[N/rad]	[m]	[m]	[kg m^2]	[kg]	[−]	[deg/g]	[deg/(mg)]
−0.10	86332	73668	0.73	1.86	2000	1300	0.99/20	3.28	1.27
0.00	70000	90000	1.00	1.60	2000	1300	1.00/20	3.30	1.27
+0.10	49065	110935	1.48	1.32	2000	1300	1.08/20	3.55	1.27

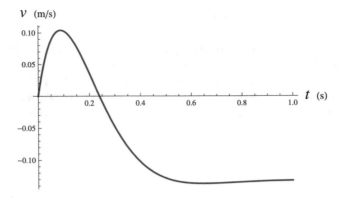

Fig. 7.59 Lateral velocity $v(t)$ of any of the three vehicles after a step steering input

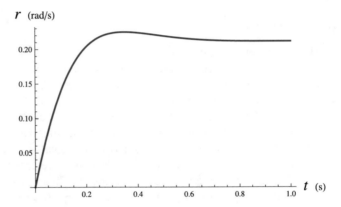

Fig. 7.60 Yaw rate $r(t)$ of any of the three vehicles after a step steering input

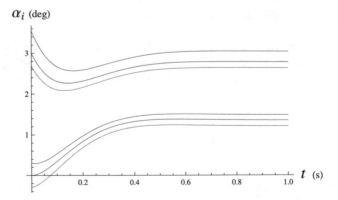

Fig. 7.61 Front and rear slip angles of the three vehicles after a step steering input

Table 7.2 Design parameters of vehicles with extreme amounts of rear steering κ, but with identical transient handling behavior

κ	Φ_1	Φ_2	a_1	a_2	J_z	m	τ	K	$-\rho_y$
[−]	[N/rad]	[N/rad]	[m]	[m]	[kg m^2]	[kg]	[−]	[deg/g]	[deg/(mg)]
−0.60	141316	18684	0.01	4.01	2000	1300	1.55/20	5.10	1.27
0.00	70000	90000	1.00	1.60	2000	1300	1.00/20	3.30	1.27
+0.166	22426	137574	2.73	0.98	2000	1300	1.42/20	4.71	1.27

evidence that rear steer must be kept small to have good handling behavior, as intuitively everybody knows.

But perhaps the most astonishing result obtained in this section is that all these vehicles of Tables 7.1 and 7.2, although with identical handling behavior, do not have the same classical understeer gradient K. Just have a look at the next to last column in Table 7.1. In other words, they would have been classified as very different if evaluated in terms of their classical understeer gradient K [16].

The conclusion is that the classical understeer gradient K is *not* a good parameter and should be abandoned. It should be replaced by the gradient components proposed in (7.99) and discussed in Sect. 7.13, which have proven to really provide a measure of the dynamic features of a vehicle. In particular, the gradient component ρ_y, shown in the last column in Tables 7.1 and 7.2, is the real measure of understeer/oversteer.

7.15   Linear Single Track Model

The simplest dynamical systems are those governed by *linear* ordinary differential equations with *constant* coefficients. The single track model of Fig. 7.20 is governed by the *nonlinear* ordinary differential equations (7.162), unless the axle characteristics are replaced by linear functions

$$Y_1 = C_1\alpha_1 \quad \text{and} \quad Y_2 = C_2\alpha_2 \tag{7.204}$$

where

$$C_1 = \left.\frac{\mathrm{d}Y_1}{\mathrm{d}\alpha_1}\right|_{\alpha_1=0} = \Phi_1(0) \quad \text{and} \quad C_2 = \left.\frac{\mathrm{d}Y_2}{\mathrm{d}\alpha_2}\right|_{\alpha_2=0} = \Phi_2(0) \tag{7.205}$$

The *axle lateral slip stiffness* C_i is usually equal to twice the tire lateral slip stiffness, firstly introduced in (2.88). It is affected by the static vertical load (Fig. 2.23), but not by the load transfer, neither by the amount of grip. The influence of roll steer is quite peculiar (Fig. 7.18).

However, as shown in Fig. 7.62, this linear approximation is acceptable only if $|\alpha_i| < 2°$, that is for very low values of $\tilde{a}_y$.

Fig. 7.62 Linear
approximation of the axle
characteristics

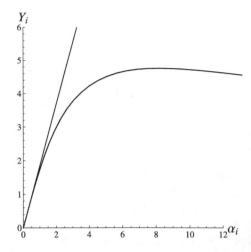

The main advantage of the *linear* single track model lies in its simplicity, the main disadvantage is that it does not model the vehicle behavior at all, unless the lateral acceleration is really small (typically, $\tilde{a}_y < 0.2g$ on dry asphalt). In some sense, it is a "dangerous" model because one may be tempted to use it outside its range of validity. Indeed, too often it is the only handling model that is presented and discussed in detail.

However, in some cases it is useful to have a model where everything can be obtained analytically. For this reason, the linear single track model is included in this book as well, albeit not in a prominent position.

7.15.1 Governing Equations

The *linear* single track model differs from the more general nonlinear model only in its constitutive equations. However, we list here all relevant equations, that is equilibrium equations (7.5)

$$m(\dot{v} + ur) = Y = Y_1 + Y_2$$
$$J_z\dot{r} = N = Y_1a_1 - Y_2a_2$$

(7.206)

congruence equations (7.67) (with $|\chi| \ll 1$, and often equal to zero)

$$\alpha_1 = \tau_1\delta_v - \frac{v + ra_1}{u}$$
$$\alpha_2 = \chi\tau_1\delta_v - \frac{v - ra_2}{u}$$

(7.207)

and the just defined *linear* constitutive equations (7.204) [12, Chap. 5]

$$
\begin{aligned}
Y_1 &= C_1 \alpha_1 \\
Y_2 &= C_2 \alpha_2
\end{aligned}
\tag{7.208}
$$

Combining congruence and constitutive equations we get

$$
\begin{aligned}
Y_1 &= C_1 \alpha_1 = C_1 \left(\tau_1 \delta_v - \frac{v + r a_1}{u} \right) \\
Y_2 &= C_2 \alpha_2 = C_2 \left(\tau_1 \chi \delta_v - \frac{v - r a_2}{u} \right)
\end{aligned}
\tag{7.209}
$$

which are linear in v and r, but not in u.

Inserting these equations into the equilibrium equations, we obtain the governing equations, that is two *linear* differential equations

$$
\begin{aligned}
\dot{v} &= -\left(\frac{C_1 + C_2}{mu} \right) v - \left(\frac{C_1 a_1 - C_2 a_2}{mu} + u \right) r + \frac{C_1 + \chi C_2}{m} \tau_1 \delta_v \\
\dot{r} &= -\left(\frac{C_1 a_1 - C_2 a_2}{J_z u} \right) v - \left(\frac{C_1 a_1^2 + C_2 a_2^2}{J_z u} \right) r + \frac{C_1 a_1 - \chi C_2 a_2}{J_z} \tau_1 \delta_v
\end{aligned}
\tag{7.210}
$$

In matrix notation, (7.210) becomes

$$
\dot{\mathbf{w}} = \mathbf{A} \mathbf{w} + \mathbf{b} \delta_v
\tag{7.211}
$$

where $\mathbf{w}(t) = \big(v(t), r(t) \big)$ is the vector of state variables, the r.h.s. known vector is

$$
\mathbf{b}(t) = \tau_1
\begin{bmatrix}
\dfrac{C_1 + \chi C_2}{m} \\[2ex]
\dfrac{C_1 a_1 - \chi C_2 a_2}{J_z}
\end{bmatrix}
\tag{7.212}
$$

and

$$
\mathbf{A} = \mathbf{A}(u(t)) = -
\begin{bmatrix}
\dfrac{C_1 + C_2}{mu} & \dfrac{C_1 a_1 - C_2 a_2}{mu} + u \\[2ex]
\dfrac{C_1 a_1 - C_2 a_2}{J_z u} & \dfrac{C_1 a_1^2 + C_2 a_2^2}{J_z u}
\end{bmatrix}
\tag{7.213}
$$

is the coefficient matrix. It is important to note that $\mathbf{A}$ depends on the forward speed u, but not on the steer angle δ_v, which multiplies the known vector $\mathbf{b}$.

7.15.2   Solution for Constant Forward Speed

As well known, the general solution $\mathbf{w}(t)$ of (7.211) is given by the solution $\mathbf{w}_o$ of the homogeneous equation plus a particular solution $\mathbf{w}_p$

$$\mathbf{w}(t) = \mathbf{w}_o(t) + \mathbf{w}_p(t) \tag{7.214}$$

Unfortunately, analytical solutions are not available if $u(t) \neq$ const.

If u is *constant* ($\dot{u} = 0$), the system (7.211) has constant coefficients and the homogeneous solution must fulfill

$$\dot{\mathbf{w}}_o = \mathbf{A}\mathbf{w}_o \tag{7.215}$$

with a *constant* matrix $\mathbf{A}$. Assuming constant u is therefore a very relevant assumption. We look for a solution among the exponential functions

$$\mathbf{w}_o(t) = \big(v_o(t), r_o(t)\big) = \mathbf{x}e^{\lambda t} \tag{7.216}$$

which implies $\dot{\mathbf{w}}_o(t) = \lambda \mathbf{x}e^{\lambda t}$, and consequently yields an eigenvalue problem for the matrix $\mathbf{A}$

$$\mathbf{A}\mathbf{x} = \lambda \mathbf{x} \tag{7.217}$$

The eigenvalues are the solutions of the characteristic equation

$$\det(\mathbf{A} - \lambda \mathbf{I}) = 0 \tag{7.218}$$

which, for a (2×2) matrix, becomes

$$\lambda^2 - \mathrm{tr}(\mathbf{A})\lambda + \det(\mathbf{A}) = 0 \tag{7.219}$$

The two eigenvalues λ_1 and λ_2 are

$$\lambda_{1,2} = \frac{\mathrm{tr}(\mathbf{A}) \pm \sqrt{\mathrm{tr}(\mathbf{A})^2 - 4\det(\mathbf{A})}}{2} = -\zeta \omega_n \pm \omega_n \sqrt{\zeta^2 - 1} \tag{7.220}$$

If the discriminant is negative, that is if $\zeta < 1$, the dynamical system is underdamped and the eigenvalues are complex conjugates.

From (7.213) we get the trace

$$\mathrm{tr}(\mathbf{A}) = -\frac{1}{u} \left(\frac{C_1 + C_2}{m} + \frac{C_1 a_1^2 + C_2 a_2^2}{J_z} \right) < 0 \tag{7.221}$$

and the determinant

$$\det(\mathbf{A}) = \frac{1}{u^2 m J_z} \left[C_1 C_2 (a_1 + a_2)^2 - m u^2 (C_1 a_1 - C_2 a_2) \right] \qquad (7.222)$$

These two quantities are very important because they provide handy information about the two eigenvalues λ_1 and λ_2 of $\mathbf{A}$, since

$$\mathrm{tr}(\mathbf{A}) = \lambda_1 + \lambda_2 \qquad (7.223)$$
$$\det(\mathbf{A}) = \lambda_1 \lambda_2 \qquad (7.224)$$

These two relationships can be obtained easily writing the characteristic equation as $(\lambda - \lambda_1)(\lambda - \lambda_2) = 0$.

Once the two eigenvalues have been obtained, we can compute the two eigenvectors $\mathbf{x}_1$ and $\mathbf{x}_2$.

Therefore, the solution of the homogeneous system is

$$\mathbf{w}_o(t) = \gamma_1 \mathbf{x}_1 e^{\lambda_1 t} + \gamma_2 \mathbf{x}_2 e^{\lambda_2 t} \qquad (7.225)$$

where γ_1 and γ_2 are constants still to be determined. In components we have

$$v_o(t) = \gamma_1 x_{11} e^{\lambda_1 t} + \gamma_2 x_{12} e^{\lambda_2 t}$$
$$r_o(t) = \gamma_1 x_{21} e^{\lambda_1 t} + \gamma_2 x_{22} e^{\lambda_2 t} \qquad (7.226)$$

where $\mathbf{x}_1 = (x_{11}, x_{21})$ and $\mathbf{x}_2 = (x_{12}, x_{22})$.

The particular integral $\mathbf{w}_p(t) = (v_p(t), r_p(t))$ depends on the known vector $\mathbf{b}$ and on the steering wheel angle $\delta_v(t)$. The simplest case is for constant δ_v, but analytical solutions are available also when $\delta_v(t)$ is a polynomial or trigonometric function.

Summing up, the general solution of the system (7.211) is

$$\mathbf{w}(t) = \mathbf{w}_o(t) + \mathbf{w}_p(t) = \gamma_1 \mathbf{x}_1 e^{\lambda_1 t} + \gamma_2 \mathbf{x}_2 e^{\lambda_2 t} + \mathbf{w}_p(t) \qquad (7.227)$$

in which the two constants γ_1 and γ_2 are to be determined from the initial conditions $\mathbf{w}(0) = (v(0), r(0))$, that is solving the system

$$\mathbf{S} \mathbf{y} = \mathbf{w}(0) - \mathbf{w}_p(0) \qquad (7.228)$$

where $\mathbf{y} = (\gamma_1, \gamma_2)$ and $\mathbf{S}$ is the matrix whose columns are the two eigenvectors of $\mathbf{A}$.

7.15.3 Critical Speed

The two parts $\mathbf{w}_o$ and $\mathbf{w}_p$ of the general solution have distinct physical meanings. The particular integral is what the vehicle does asymptotically, that is basically at steady-state. The solution of the homogeneous system shows how the vehicle behaves before reaching the steady-state condition, if the vehicle is stable.

As already discussed in Sect. 7.10.3, the stability of the vehicle is completely determined by the two eigenvalues λ_1 and λ_2, or better, by the sign of their real parts $\text{Re}(\lambda_1)$ and $\text{Re}(\lambda_2)$. The rule is very simple: the system is asymptotically stable if and only if both eigenvalues have negative real parts

$$\text{stability} \quad \Longleftrightarrow \quad \text{Re}(\lambda_1) < 0 \quad \text{and} \quad \text{Re}(\lambda_2) < 0 \tag{7.229}$$

If just one eigenvalue has a positive real part, the corresponding exponential solution grows without bound in time, and the system is unstable.

Fortunately, we can check the stability without computing the two eigenvalues explicitly, but simply looking at (7.223) and (7.224). To have an asymptotically stable vehicle it suffices to check that

$$\text{stability} \quad \Longleftrightarrow \quad \text{tr}(\mathbf{A}) < 0 \quad \text{and} \quad \det(\mathbf{A}) > 0 \tag{7.230}$$

From (7.221) we see immediately that $\text{tr}(\mathbf{A}) < 0$ is always fulfilled. Stability is therefore completely due to the second condition in (7.230). Setting $\det(\mathbf{A}) = 0$ in (7.193) yields an equation in the unknown forward speed u, whose solution, if it exists, is the *critical speed* u_{cr}

$$u_{\text{cr}} = \sqrt{\frac{C_1 C_2 l^2}{m(C_1 a_1 - C_2 a_2)}}. \tag{7.231}$$

Beyond the critical speed the vehicle becomes unstable. It is worth noting that u_{cr} does not depend on J_z.

In the linear single track model, the critical speed exists if and only if

$$C_1 a_1 - C_2 a_2 > 0 \tag{7.232}$$

that is, if the vehicle is oversteer. In this vehicle model (which, we recall, has a very limited range of applicability), the critical speed is not affected by the steer angle.

7.15.4 Transient Vehicle Behavior

It may be of some interest to know how the eigenvalues evolve as speed changes. To this end, it is useful to plot $\text{tr}(\mathbf{A})$ vs $\det(\mathbf{A})$, which, according to (7.221) and (7.222), can be compactly expressed as[14]

---

[14] Here α, β and γ are just constants. They have no connection with slip and camber angles.

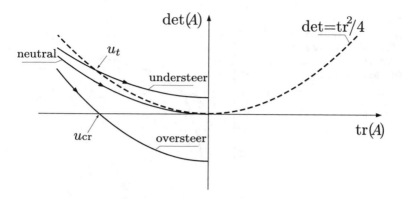

Fig. 7.63 Evolution of $\det(A)$ and $\text{tr}(A)$ when u grows

$$\det(\mathbf{A}) = \frac{\alpha}{u^2} + \beta, \qquad \text{tr}(\mathbf{A}) = -\frac{\gamma}{u} \qquad (7.233)$$

where α and γ are always positive, while $\beta = (C_2 a_2 - C_1 a_1)/J_z$ can be either positive or negative, depending on the vehicle being understeer or oversteer, respectively.

Both functions are monotone in u (if $u > 0$). They can be combined to get

$$\det(\mathbf{A}) = \frac{\alpha}{\gamma^2} \, \text{tr}(\mathbf{A})^2 + \beta. \qquad (7.234)$$

Moreover, it is easy to show that

$$\lim_{u \to +\infty} \text{tr}(\mathbf{A}) = 0^-, \qquad \lim_{u \to +\infty} \det(\mathbf{A}) = \beta \qquad (7.235)$$

Therefore, as u grows, we draw parabolas, as shown in Fig. 7.63, up to their vertex in $(0, \beta)$.

Also plotted in Fig. 7.63 is the parabola $\det = \text{tr}^2/4$. According to (7.220), it corresponds to the points where $\lambda_1 = \lambda_2$. Below this parabola, i.e. $u < u_t$, the two eigenvalues are real, whereas above it they are complex conjugates.

It can be shown that

$$\left(\frac{\alpha}{\gamma^2} = \frac{C_1 C_2 k^2 l^2}{[k^2(C_1 + C_2) + C_1 a_1^2 + C_2 a_2^2]^2} \right) \leq \frac{1}{4} \qquad (7.236)$$

where $J_z = mk^2$. Since it attains its maximum value $1/4$ when $C_1 a_1 = C_2 a_2$ (neutral vehicle) and $J_z = m a_1 a_2$, we see that all vehicles at sufficiently low speed have real negative eigenvalues.

As the speed increases, the following evolutions are possible. An oversteer vehicle (actually, an oversteer linear single track model) has always two real eigenvalues.

Fig. 7.64 Evolution of the real part and of the imaginary part of λ_1 and λ_2 as functions of the forward speed u, for an understeer vehicle

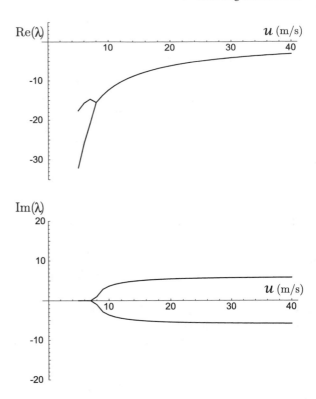

When the parabola in Fig. 7.63 crosses the horizontal axis (det $= 0$), one eigenvalue becomes positive and the vehicle becomes unstable. That happens for $u = u_{cr}$.

An understeer vehicle has two negative real eigenvalues at low speed. For speeds higher than $u = u_t$, they become complex conjugate with negative real parts (Fig. 7.63): $\lambda_1 = -\zeta\omega_n + i\omega_n\sqrt{1-\zeta^2}$, $\lambda_2 = -\zeta\omega_n - i\omega_n\sqrt{1-\zeta^2}$. Therefore, at sufficiently high speed, the transient motion is a damped oscillation (very damped, indeed). The speed u_t is given by

$$u_t = \sqrt{\frac{\gamma^2 - 4\alpha}{4\beta}} = \sqrt{\frac{[J_z(C_1 + C_2) + m(C_1a_1^2 + C_2a_2^2)]^2 - 4J_zmC_1C_2l^2}{4m^2J_z(C_2a_2 - C_1a_1)}}$$

(7.237)

From Fig. 7.64, we see that the imaginary part of the eigenvalues, that is the angular frequency $\omega_s = \omega_n\sqrt{1-\zeta^2}$, is almost constant up to relatively high speeds. This is typical and makes the classical sine sweep test quite insensitive to the selected speed.

The general solution is given by (7.227). However, when the eigenvalues are complex conjugates, also the eigenvectors $\mathbf{x}_1$ and $\mathbf{x}_2$ and the constants γ_1 and γ_2 are complex conjugates. Having to deal with so many complex numbers to eventually get a real function $\mathbf{w}(t)$ is not very convenient. Fortunately, we can rearrange it in a way that it involves only real numbers. As well known, $e^{(\zeta + i\omega)t} = e^{\zeta t}[\cos(\omega t) + i\sin(\omega t)]$, and the general solution can be written as

$$
\begin{aligned}
\mathbf{w}(t) &= \mathbf{w}_o(t) + \mathbf{w}_p(t) \\
&= \gamma_1 \mathbf{x}_1 e^{\lambda_1 t} + \gamma_2 \mathbf{x}_2 e^{\lambda_2 t} + \mathbf{w}_p(t) \\
&= e^{-\zeta \omega_n t}[(\gamma_1 \mathbf{x}_1 + \gamma_2 \mathbf{x}_2)\cos(\omega_s t) + i(\gamma_1 \mathbf{x}_1 - \gamma_2 \mathbf{x}_2)\sin(\omega_s t)] + \mathbf{w}_p(t) \\
&= e^{-\zeta \omega_n t}[\mathbf{z}_1 \cos(\omega_s t) + \mathbf{z}_2 \sin(\omega_s t)] + \mathbf{w}_p(t)
\end{aligned}
$$

$$(7.238)$$

where $\omega_s = \omega_n\sqrt{1 - \zeta^2}$.

To obtain $\mathbf{z}_1$ and $\mathbf{z}_2$ we can proceed as follows. Vector $\mathbf{z}_1$ is simply obtained setting $t = 0$ in the last expression in (7.238)

$$
\mathbf{z}_1 = \mathbf{w}(0) - \mathbf{w}_p(0) \tag{7.239}
$$

where $\mathbf{w}(0)$ is the vector of the initial conditions. To obtain the other vector, just consider that

$$
\dot{\mathbf{w}}_o(0) = \mathbf{A}\mathbf{w}_o(0) = -\zeta \omega_n \mathbf{z}_1 + \omega_s \mathbf{z}_2 = \mathbf{z}_1 \tag{7.240}
$$

and hence

$$
\mathbf{z}_2 = \frac{1}{\omega_s}(\mathbf{A} + \zeta \omega_n \mathbf{I})\mathbf{z}_1 \tag{7.241}
$$

7.15.5 Steady-State Behavior: Steering Pad

As already stated, the particular integral $\mathbf{w}_p(t) = (v_p(t), r_p(t))$ is determined, in this linear model, by the known vector $\mathbf{b}$, and hence by the function $\delta_v(t)$. The simplest case is when $\delta_v = \text{const.}$

Keeping the steering wheel in a fixed position and driving at constant speed makes the vehicle go round in a circle. This is called steering pad. To obtain the steady-state solution, we have to solve the system

$$
- \mathbf{A}\mathbf{w}_p = \mathbf{b}\delta_v \tag{7.242}
$$

thus getting

$$
\begin{aligned}
v_p &= \frac{[C_1 C_2 l(a_2 + a_1\chi) - mu^2(C_1 a_1 - C_2 a_2\chi)]u}{m J_z u^2 \det(\mathbf{A})}\tau_1\delta_v, \\
r_p &= \frac{C_1 C_2 l(1 - \chi)u}{m J_z u^2 \det(\mathbf{A})}\tau_1\delta_v = \frac{C_1 C_2 l(1 - \chi)u}{C_1 C_2 l^2 - mu^2(C_1 a_1 - C_2 a_2)}\tau_1\delta_v.
\end{aligned}
$$

$$(7.243)$$

Once we have obtained v_p and r_p, we can easily compute all other relevant quantities, like the vehicle slip angle β_p and the curvature ρ_p

$$\beta_p = \frac{v_p}{u} = \left(\frac{a_2 + a_1\chi}{l}\right)\tau_1\delta_v - \frac{m}{l^2}\left(\frac{C_1a_1^2 + C_2a_2^2}{C_1C_2}\right)\tilde{a}_y = \frac{S_p}{R_p}$$

$$\rho_p = \frac{r_p}{u} = \left(\frac{1-\chi}{l}\right)\tau_1\delta_v - \frac{m}{l^2}\left(\frac{C_2a_2 - C_1a_1}{C_1C_2}\right)\tilde{a}_y = \frac{1}{R_p}$$

(7.244)

According to (7.207), we can compute the steady-state front and rear slip angles

$$\alpha_{1p} = \tau_1\delta_v - \frac{v_p + r_pa_1}{u} = \frac{ma_2}{lC_1}\tilde{a}_y$$

$$\alpha_{2p} = \chi\tau_1\delta_v - \frac{v_p - r_pa_2}{u} = \frac{ma_1}{lC_2}\tilde{a}_y$$

(7.245)

A non-zero lateral speed v_p at steady state may look a bit strange, at first sight. It simply means that the trajectory of G is not tangent to the vehicle longitudinal axis, as shown in Fig. 7.32.

The speed u_β that makes $\beta_p = v_p = 0$ is given by (7.243) and is equal to (if $\chi = 0$)

$$u_\beta = \sqrt{\frac{C_2a_2l}{a_1m}}$$

(7.246)

It is called *tangent speed* [12, p. 174].

7.15.6 Lateral Wind Gust

It is of some practical interest to study the behavior of a vehicle (albeit a very linear one) when suddenly subjected to a lateral force, like the force due to a lateral wind gust hitting the car when, e.g., exiting a tunnel. As shown in Sect. 7.15.7, the same mathematical problem also covers the case of a car going straight along a banked road.

We have only to modify the equilibrium equations (7.206) by adding a lateral force $\mathbf{F}_l = -F_l\mathbf{j}$, applied at a distance x from G

$$m(\dot{v} + ur) = F_{y_1} + F_{y_2} - F_l$$
$$J_z\dot{r} = F_{y_1}a_1 - F_{y_2}a_2 - F_lx.$$

(7.247)

where $x > 0$ if $\mathbf{F}_l$ is applied between G and the front axle. The other equations are not affected directly by $\mathbf{F}_l$.

The equations of motion are like in (7.211), with the only difference that the term

$$\mathbf{b}_F = -\begin{bmatrix} 1/m \\ x/J_z \end{bmatrix}F_l$$

(7.248)

must be added to the known vector.

If we assume $\delta_v = 0$, the steady-state conditions $\mathbf{w}_p$ are obtained, as usual, by solving the system of equations $-\mathbf{A}\mathbf{w}_p = \mathbf{b}_F$, with $\mathbf{A}$ as given in (7.213). Accordingly, we have the following quantities at steady-state

$$v_p = \frac{[x(C_1 a_1 - C_2 a_2 + mu^2) - (C_1 a_1^2 + C_2 a_2^2)]u}{C_1 C_2 l^2 - mu^2(C_1 a_1 - C_2 a_2)} F_l,$$

$$r_p = \frac{[C_1 a_1 - C_2 a_2 - x(C_1 + C_2)]u}{C_1 C_2 l^2 - mu^2(C_1 a_1 - C_2 a_2)} F_l = -(x - e)\frac{(C_1 + C_2)u}{C_1 C_2 l^2 - mu^2(C_1 a_1 - C_2 a_2)} F_l,$$

$$(7.249)$$

where

$$e = \frac{C_1 a_1 - C_2 a_2}{C_1 + C_2} \tag{7.250}$$

Should the steer angle be non-zero, it suffices to superimpose the effects. This is legitimate because of the linearity of the equations.

This quantity e in (7.250) is often called *static margin*. The yaw rate is zero, that is $r_p = 0$, if and only if the lateral force is applied at a distance e from G. This is the distance that makes the vehicle translate diagonally under the action of a lateral force, as shown in Fig. 7.65. The point N_p on the axis of the vehicle at a distance e from G is called *neutral steer point*.

Fig. 7.65 Lateral force applied at the neutral point N_p (i.e., $x = e$)

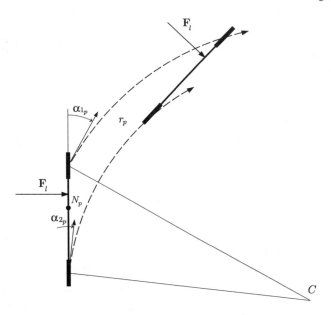

Fig. 7.66 Lateral force applied at a point ahead of the neutral point ($x > e$)

Obviously, the condition $r_p = 0$ with $\delta_v = 0$ is equivalent to $\alpha_{1p} = \alpha_{2p} = \alpha_p$. Inserting this condition into (7.247) we get

$$
\begin{aligned}
0 &= (C_1 + C_2)\alpha_p - F_l \\
0 &= (C_1 a_1 - C_2 a_2)\alpha_p - F_l e,
\end{aligned}
\tag{7.251}
$$

which provide another way to obtain e.

An oversteer vehicle has $e > 0$, whereas $e < 0$ in an understeer vehicle.

If $\delta_v = 0$, the steady-state distance R_p is

$$
R_p = \frac{u}{r_p} = \frac{C_1 C_2 l^2 - mu^2(C_1 a_1 - C_2 a_2)}{-(x - e)(C_1 + C_2)F_l}.
\tag{7.252}
$$

The numerator is always positive if $u < u_{\mathrm{cr}}$. Therefore, $R_p > 0$ if $x < e$, and vice versa.

If the point of application of the lateral force is located ahead of the neutral point N_p, the vehicle behaves like in Fig. 7.66, turning in the same direction as the lateral force. This is commonly considered good behavior.

If the point of application of the lateral force is behind the neutral point N_p, the vehicle behaves like in Fig. 7.67. This is commonly considered bad behavior.

Of course, since an oversteer vehicle has the neutral point N_p ahead of G, the likelihood that a wind gust applies a force behind the neutral point is higher, much higher, than in an understeer vehicle.

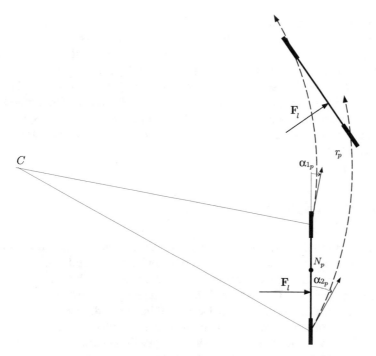

Fig. 7.67 Lateral force applied at a point behind the neutral point ($x < e$)

Fig. 7.68 Lateral force applied by means of a rocket (General Motors Corporation, circa 1960)

To understand why the first case is considered good, while the second is considered bad, we have to look at the lateral forces that the tires have to exert. In the first case, the inertial effects counteract the wind gust, thus alleviating the tire job. In the second case, the inertial effects add to the lateral force, making the tire job harder.

Figures 7.66 and 7.67 show a lateral force constantly perpendicular to the vehicle axis, pretty much like if a rocket were strapped on the side of the car. Indeed, in some cases a rocket has been really employed as shown in Fig. 7.68, taken from a presentation by Tom Bundorf at the SAE Automotive Dynamics and Stability Conference (2000).

7.15.7   Banked Road

A car going straight on a banked road is subject to a lateral force due to its own weight. Therefore, it is a situation somehow similar to a lateral wind gust, but not equal. The main difference is that the lateral force is now applied at G.

Understeer and oversteer vehicles behave differently, as shown in Fig. 7.69. Both axles must exert lateral forces directed uphill to counteract the weight force $mg \sin \varepsilon$. Therefore, both must work with positive slip angles α_1 and α_2, if the banking is like in Fig. 7.69. However, due to the different locations of the neutral point N_p with respect to G, the two front axles cannot have the same slip angle. To go straight, we must steer the front wheels uphill in an understeer vehicle and (apparently) downhill in an oversteer vehicle, as shown in Fig. 7.69. More precisely, in both cases $\alpha_1 - \delta_1 = \alpha_2$, where $\delta_1 > 0$ if the vehicle is understeer, while $\delta_1 < 0$ if the vehicle is oversteer.

7.16   Compliant Steering System

Many modern cars use rack and pinion steering mechanisms. The steering wheel turns the pinion gear, which moves the rack, thus converting rotational motion into linear motion. This motion applies steering torque to the front wheels via tie rods and a short lever arm called the steering arm.

So far we have assumed the steering system to be perfectly rigid, as stated on Sects. 3.1 and 7.5.4. More precisely, equations (3.210) have been used to relate the steer angles δ_{ij} of each wheel to the angle δ_v of the steering wheel.

In the single track model (Fig. 7.20) we have taken a further step, assuming that the left and right gear ratios of the steering system are almost equal, that is

$$(\tau_{11} = \tau_{12}) = \tau_1 \quad \text{and} \quad (\tau_{21} = \tau_{22}) = \tau_2 \tag{7.49'}$$

thus getting (7.68)

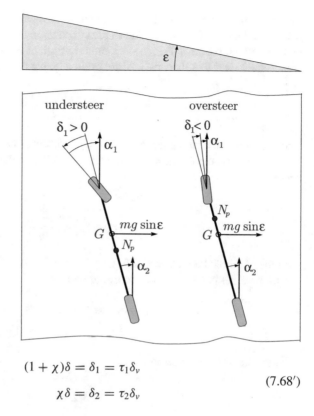

Fig. 7.69 Understeer and oversteer vehicles going straight on a banked road

$$(1 + \chi)\delta = \delta_1 = \tau_1 \delta_v$$

$$\chi\delta = \delta_2 = \tau_2 \delta_v \qquad (7.68')$$

Now, in the framework of the *linear* single track model, we relax the assumption of rigid steering system. This means to make a few changes in the congruence equations (7.207), since δ_1 and $\tau_1 \delta_v$ are no longer equal to each other.

7.16.1 Governing Equations

As shown in Fig. 7.70, the steering system now has a finite angular stiffness k_{s_1} with respect to the axis about which the front wheel steers. In a turn, the lateral force Y_1

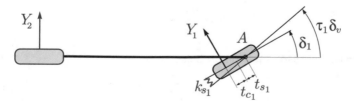

Fig. 7.70 Single track model with compliant steering system

exerts a vertical moment with respect to the steering axis A because of the pneumatic trail t_{c_1} and also of the trail t_{s_1} due to the suspension layout (see Fig. 3.1). The effect of this vertical moment $Y_1(t_{c_1} + t_{s_1})$ on a compliant steering system is to make the front wheel steer less than $\tau_1 \delta_v$. More precisely, we have that (Fig. 7.70)

$$\delta_1 = \tau_1 \delta_v - \frac{Y_1(t_{c_1} + t_{s_1})}{k_{s_1}} \tag{7.253}$$

The computation of the pneumatic trail t_{c_1} is discussed on page 512.

Accordingly, the congruence equations (7.207) of the linear single track model become

$$\alpha_1 = \delta_1 - \frac{v + ra_1}{u}$$
$$\alpha_2 = \chi \tau_1 \delta_v - \frac{v - ra_2}{u} \tag{7.254}$$

with the additional equation (7.253).

On the other hand, the equilibrium equations

$$m(\dot{v} + ur) = Y = Y_1 + Y_2$$
$$J_z \dot{r} = N = Y_1 a_1 - Y_2 a_2 \tag{7.206'}$$

and the constitutive equations

$$Y_1 = C_1 \alpha_1$$
$$Y_2 = C_2 \alpha_2 \tag{7.208'}$$

do not change at all.

7.16.2 Effects of Steer Compliance

Equation (7.253) can be rewritten taking the first equation in (7.208) into account

$$\delta_1 = \tau_1 \delta_v - \frac{C_1(t_{c_1} + t_{s_1})}{k_{s_1}} \alpha_1 = \tau_1 \delta_v - \varepsilon \alpha_1 \tag{7.255}$$

where

$$\varepsilon = \frac{C_1(t_{c_1} + t_{s_1})}{k_{s_1}} \tag{7.256}$$

The first congruence equation becomes

$$(1 + \varepsilon)\alpha_1 = \tau_1 \delta_v - \frac{v + a_1 r}{u} \tag{7.257}$$

which leads naturally to define a fictitious slip angle

$$\tilde{\alpha}_1 = (1 + \varepsilon)\alpha_1 \qquad (7.258)$$

and, consequently, a fictitious slip stiffness

$$\tilde{C}_1 = \frac{C_1}{1 + \varepsilon} \qquad (7.259)$$

Summing up, the linear single track model with *compliant* steering system is governed by the set of equations

$$
\begin{aligned}
m(\dot{v} + ur) &= Y = Y_1 + Y_2 \\
J_z \dot{r} &= N = Y_1 a_1 - Y_2 a_2 \\
\tilde{\alpha}_1 &= \tau_1 \delta_v - \frac{v + r a_1}{u} \\
\alpha_2 &= \chi \tau_1 \delta_v - \frac{v - r a_2}{u} \\
Y_1 &= \tilde{C}_1 \tilde{\alpha}_1 \\
Y_2 &= C_2 \alpha_2
\end{aligned}
\qquad (7.260)
$$

which is formally identical to the set governing the single track model with rigid steering system. Therefore, the analysis developed in Sect. 7.15 applies entirely, provided we take into account that $C_1 \to \tilde{C}_1$ and $\alpha_1 \to \tilde{\alpha}_1$.

Since $\tilde{C}_1 < C_1$, a compliant steering system makes the vehicle behavior more understeer.

7.16.3 There Is Something Unsafe

Apparently, the critical speed of the linear single track model is not affected by the steer angle, as shown in Sect. 7.15.3. However, according to (7.243), if $\det(\mathbf{A}) = 0$ the model predicts unlimited values of lateral velocity v_p and yaw rate r_p, unless $\delta_v = 0$. These unrealistic results suggest that in the analysis something relevant is missing.

Indeed, instead of imposing the forward speed u, it would more realistic to impose the longitudinal force F_{x_2}, thus having also u as a state variable. Of course, the steady-state results (7.243) do not change. But what about stability?

This more general single track model, with three state variables (u, v, r), is governed by the following differential equations

$$\dot{u} = \frac{1}{m}\left[mvr + C_1\left(\frac{v+a_1 r}{u}\right)\delta_1 - \frac{1}{2}\rho S_a C_x u^2 + F_{x_2} - C_1\delta_1^2\right]$$

$$\dot{v} = \frac{1}{m}\left[-mur - (C_1 + C_2)\frac{v}{u} + (C_2 a_2 - C_1 a_1)\frac{r}{u} + C_1\delta_1\right] \quad (7.261)$$

$$\dot{r} = \frac{1}{J}\left[(C_2 a_2 - C_1 a_1)\frac{v}{u} - (C_1 a_1^2 + C_2 a_2^2)\frac{r}{u} + C_1 a_1\delta_1\right]$$

Even with linear tires, the governing equations are no longer linear.
The counterpart of matrix $\mathbf{A}$ in Sect. 7.10.2 is this matrix $\mathbf{B}$

$$\mathbf{B} = \begin{bmatrix} -\frac{C_1\delta_1(v_p+a_1 r_p)+\rho S_a C_x u_p^3}{mu_p^2} & \frac{C_1\delta_1+mu_p r_p}{mu_p} & \frac{C_1 a_1\delta_1+mu_p v_p}{mu_p} \\ \frac{(C_1+C_2)v_p-(C_2 a_2-C_1 a_1+mu_p^2)r_p}{mu_p^2} & -\frac{C_1+C_2}{mu_p} & \frac{C_2 a_2-C_1 a_1-mu_p^2}{mu_p} \\ \frac{-(C_2 a_2-C_1 a_1)v_p+(C_1 a_1^2+C_2 a_2^2)r_p}{J_z u_p^2} & \frac{C_2 a_2-C_1 a_1}{J_z u_p} & -\frac{C_1 a_1^2+C_2 a_2^2}{J_z u_p} \end{bmatrix} \quad (7.262)$$

which contains $\mathbf{A}$ as a submatrix. It is worth noting that the new entries (first raw
and column) depend on v_p, r_p, and δ_1.

Stability requires the real part of all eigenvalues of $\mathbf{B}$ to be negative.

In case of straight running $\delta_1 = 0$ and, obviously, $v_p = r_p = 0$. Two eigenvalues
are exactly the same of $\mathbf{A}$. The third eigenvalue $\lambda_3 = -\rho SC_x u_p$ is always negative.
Therefore, the classical critical speed u_{cr} in (7.231) is confirmed.

What happens when $\delta_1 \neq 0$? The characteristic equation is something like

$$b_0\lambda^3 + b_1\lambda^2 + b_2\lambda + b_3 = 0, \quad \text{with} \quad b_0 > 0 \quad (7.263)$$

According to Routh criterion, all eigenvalues have negative real part if and only if
$b_1 > 0$, $b_3 > 0$, and $b_1 b_2 - b_0 b_3 > 0$.

Since analytic expressions are very complex, we prefer to perform just a numerical
test. For instance, let $m = 1000$ kg, $a_1 = 1.4$ m, $a_2 = 1.2$ m, $J_z = 1680$ kg m^2,
$C_1 = C_2 = 100000$ N/rad. Moreover, $\rho = 1.3$ kg/m^3, $S_a = 1.8$ m^2, $C_x = 0.35$.

This vehicle has, in straight running, a critical speed $u_{cr} = 58.14$ m/s $= 209$ km/h.
However, if, e.g., $\delta_1 = 6$ deg, the quantity $b_1 b_2 - b_0 b_3$ becomes negative at the speed
$\tilde{u}_{cr} = 53.21$ m/s $= 191$ km/h, which is quite lower than u_{cr}. Therefore, imposing u
may not be a safe assumption!

7.17 Road Vehicles with Locked or Limited Slip Differential

The handling of cars equipped with either a locked or a limited-slip differential is
addressed in Chap. 8, that is in the chapter devoted to the handling behavior of race
cars. This has been done because the limited-slip differential is a peculiarity of almost
all race cars, whereas very few road cars have it.

7.18 Exercises

7.18.1 Camber Variations

As shown in (7.18) and in Fig. 7.7, camber variations due to vehicle roll motion are
determined by some suspension parameters. Given the track length t_i, find the values
of c_i to have:

1. $\Delta\gamma_i/\phi_i^s = -1$;
2. $\Delta\gamma_i/\phi_i^s = 0$;
3. $\Delta\gamma_i/\phi_i^s = 1$.

Solution

It is a simple calculation to obtain

1. $c_i = t_i/4$;
2. $c_i = t_i/2$;
3. $c_i = +\infty$.

Quite a big difference.

7.18.2 Ackermann Coefficient

According to (7.19), and assuming $\delta_1^0 = 0$, $l = 2.6$ m, $t_1 = 1.6$ m, and $\varepsilon_1 = 1$ (Ackermann steering), compute δ_{11} and δ_{12} when $\tau_1\delta_v$ is equal to 5 deg, 10 deg, and 15 deg.

Solution

It is a simple calculation to obtain

1. $\delta_{11} = 5.13$ deg, $\delta_{12} = 4.87$ deg;
2. $\delta_{11} = 10.54$ deg, $\delta_{12} = 9.46$ deg;
3. $\delta_{11} = 16.21$ deg, $\delta_{12} = 13.79$ deg.

We see that the Ackermann correction is relevant, with respect to parallel steering,
only for not so small steer angles.

7.18.3 Toe-In

Repeat the previous calculations now with 1 deg of toe-in.

Solution

1. $\delta_{11} = 4.13$ deg, $\delta_{12} = 5.87$ deg;
2. $\delta_{11} = 9.54$ deg, $\delta_{12} = 10.46$ deg;
3. $\delta_{11} = 15.21$ deg, $\delta_{12} = 14.79$ deg.

Quite influential.

7.18.4 Steering Angles

With reference to (7.68), obtain the relationship between χ and κ for any steering system.

Solution

From the following system of equations

$$(1 + \kappa)\tau = \tau_1$$
$$\kappa\tau = \chi\tau_1$$

(7.264)

we obtain

$$\chi = \frac{\kappa}{1 + \kappa}$$

(7.265)

7.18.5 Axle Characteristics

Axle characteristics are very important in vehicle dynamics. In Sect. 7.5.3, the effects of the following setup parameters were discussed (not in this order):

1. *roll stiffness;*
2. *static camber angles;*
3. *roll camber;*
4. *roll steer;*
5. *toe-in/toe-out.*

Some of these parameters have similar effects on the axle characteristics. Before going back to Sect. 7.5.3, think about the physics of each parameter and try to figure out the similarities.

Solution

Have a look at Sect. 7.5.3.

7.18.6 *Playing with Linear Differential Equations*

Find out how to go from (7.128) to (7.130), that is, from a system of two first-order linear differential equations with constant coefficients to two second-order equations.

Solution

Like in (7.128), the starting point is

$$
\begin{aligned}
\dot{\beta}_t &= a_{11}\beta_t + a_{12}\rho_t \\
\dot{\rho}_t &= a_{21}\beta_t + a_{22}\rho_t
\end{aligned}
\tag{7.266}
$$

where a_{ij} are the entries of matrix **A**, as in (7.148).

We can see (7.266) as a system of two algebraic equations and solve it with respect to $\dot{\beta}_t$ and β_t, thus getting

$$
\begin{aligned}
\beta_t &= \frac{-a_{22}\rho_t + \dot{\rho}_t}{a_{21}} \\
\dot{\beta}_t &= \frac{(a_{12}a_{21} - a_{11}a_{22})\rho_t + a_{11}\dot{\rho}_t}{a_{21}}
\end{aligned}
\tag{7.267}
$$

Differentiating the first equation in (7.267) and setting it equal to the second equation in (7.267) provides the sought second-order linear differential equation

$$
\ddot{\rho}_t - (a_{11} + a_{22})\dot{\rho}_t + (a_{11}a_{22} - a_{12}a_{21})\rho_t = 0
\tag{7.268}
$$

exactly like in (7.130).

7.18.7 *Static Margin*

Compute the static margin *for the single track model defined on Sect. 7.14.1.*

Solution

To compute the static margin we have to use (7.250). The result is $e = -0.46\,\mathrm{m}$. A negative value is typical of understeer vehicles.

7.18.8   Banked Road

The same vehicle is travelling on a straight road with 6 deg *of banking. Compute the steering wheel angle required to have a trajectory parallel to the road (that is to go straight ahead).*

Solution

With the aid of Fig. 7.69, we see that the rear axle has to counteract a lateral force $Y_2 = mg \sin(6\deg)a_1/l = 452.8\,\text{N}$. That means that the rear axle operates with a slip angle $\alpha_2 = Y_2/\Phi_2(0) = 0.29\,\text{deg}$.

Similarly, the front axle has to balance a force $Y_1 = mg \sin(6\deg)a_2/l = 724.4\,\text{N}$, which needs a slip angle $\alpha_1 = Y_1/\Phi_1(0) = 0.59\,\text{deg}$.

Therefore, the front steer angle has to be $\delta_1 = 0.59 - 0.29 = 0.3\,\text{deg}$. The steering wheel angle is $\delta_v = 20 \times 0.3 = 6\,\text{deg}$.

Of course, the vehicle slip angle is $\beta = -\alpha_2 = -0.29\,\text{deg}$.

7.18.9   Rear Steer

Repeat the calculations of the banked road for the two vehicles with rear steer whose features are listed in Table 7.1.

Solution

First we consider the vehicle with $\kappa = -0.1$. We have $Y_2 = 331.8\,\text{N}$ and hence $\alpha_2 = 0.26\,\text{deg}$. Similarly, $Y_1 = 845.4\,\text{N}$, and $\alpha_1 = 0.56\,\text{deg}$.

To obtain the net steer angle δ we have to solve the equation

$$\alpha_1 - (1+\kappa)\delta = \alpha_2 - \kappa\delta \tag{7.269}$$

with $\kappa = -0.1$, which provides $\delta = 0.3\,\text{deg}$, and hence a steering wheel angle $\delta_v = 0.3 \times 20/0.99 = 6.0\,\text{deg}$.

The vehicle slip angle is $\beta = -(0.26 + 0.1 \times 0.3) = -0.29\,\text{deg}$.

Then we consider the vehicle with $\kappa = 0.1$. We have $Y_2 = 622.2\,\text{N}$ and hence $\alpha_2 = 0.32\,\text{deg}$. Similarly $Y_1 = 555.0\,\text{N}$, and $\alpha_1 = 0.65\,\text{deg}$.

To obtain the net steer angle δ we have to solve (7.269), with $\kappa = 0.1$, which provides $\delta = 0.33\,\text{deg}$, and hence a steering wheel angle $\delta_v = 0.3 \times 20/0.99 = 6.0\,\text{deg}$.

The vehicle slip angle is $\beta = -(0.32 - 0.1 \times 0.33) = -0.29\,\text{deg}$.

As expected, for the driver the three vehicles behave exactly the same way: same steer wheel angle δ_v, same vehicle slip angle β. The three vehicles also have the same static margin $e = -0.46\,\text{m}$.

Fig. 7.71 Comparison of the
three vehicles of Table 7.1
under a lateral wind gust

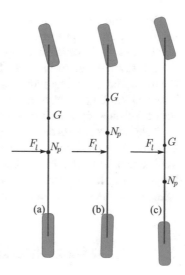

7.18.10   Wind Gust

Are the three vehicles of Table 7.1 fully equivalent with respect to a lateral wind gust?

Solution

These three vehicles are compared in Fig. 7.71. The point of application of a lateral
force F_l due to a wind gust depends on the shape of the vehicle. However, we can
reasonably assume F_l be applied like in Fig. 7.71. Should this be the case, the three
vehicles would behave very differently.

Vehicle (a), which has $\kappa = 0$, would do like in Fig. 7.65. Vehicle (b), which has
$\kappa = -1$, would do like in Fig. 7.67. Vehicle (c), which has $\kappa = 1$, would do like in
Fig. 7.66.

Therefore, the three vehicles are not equivalent with respect to a lateral wind gust.
Actually, their behaviors can be completely different.

7.19   Summary

Road cars are characterized by having an open differential and no significant aerodynamic downforces. These two aspects allow for some substantial simplifications of
the vehicle model. With the additional assumption of equal gear ratios of the steering
system for both front wheels, we have been able to formulate the single track model.

Quite contrary to common belief, we have shown that the axle characteristics can take into account many vehicle features, like toe in/out, roll steering, camber angles and camber angle variations.

The steady-state analysis has been carried out first using the classical handling diagram. Then, the new global approach MAP (Map of Achievable Performance), based on handling maps on achievable regions has been introduced and discussed in detail. This new approach shows the overall vehicle behavior at a glance.

Stability and control derivatives have been introduced to study the vehicle transient behavior. Moreover, the relationship between data collected in steady-state tests and vehicle transient behavior has been thoroughly analyzed in a systematic framework. To prove the effectiveness of these results, a number of apparently different vehicles with almost the same handling characteristics have been generated.

7.20   List of Some Relevant Concepts

Section 7.1.1—road cars are normally equipped with an open differential;
Section 7.4—some steady-state quantities are functions of the lateral acceleration only because of the open differential and no significant downforces;
Section 7.5—to go from the double track to the single track model we need the following additional assumption: the left and right gear ratios of the steering system are almost equal;
Section 7.5.4—the main feature of the single track model is that the two wheels of the same axle undergo the same apparent slip angle;
Section 7.9—some "fundamental" concepts in classical vehicle dynamics are indeed very weak if addressed with open mind;
Section 7.12—the classical understeer gradient is not a good parameter and should be dismissed.

7.21   Key Symbols

a_1	distance of G from the front axle
a_2	distance of G from the rear axle
a_n	centripetal acceleration
a_t	tangential acceleration
a_x	longitudinal acceleration
a_y	lateral acceleration
$\tilde{a}_y$	steady-state lateral acceleration
C	velocity center
C_i	lateral slip stiffness of ith axle
C_x, C_y, C_z	aerodynamic coefficients
d	diameter of the inflection circle

F_l	lateral force (wind gust)
$F_{x_{ij}}$	tire longitudinal force
$F_{y_{ij}}$	tire lateral force
$F_{z_{ij}}$	tire vertical force
g	gravitational acceleration
G	center of mass
h	height of G
J_x, J_y, J_z	moments of inertia
K	acceleration center
K	classical understeer gradient
k_ϕ	total roll stiffness
k_{ϕ_i}	global roll stiffness of ith axle
$k_{\phi_i}^p$	tire roll stiffness
$k_{\phi_i}^s$	suspension roll stiffness
l	wheelbase
m	mass
N	yaw moment
N_β, N_ρ	stability derivatives
N_δ	control derivative
q_1	height of the front no-roll center
Q_1	front no-roll center
q_2	height of the rear no-roll center
Q_2	rear no-roll center
r	yaw rate
R	lateral coordinate of C
r_i	rolling radii
S	longitudinal coordinate of C
S_a	frontal area
t_1	front track
t_2	rear track
u	longitudinal velocity
v	lateral velocity
X	longitudinal force
X_a	aerodynamic drag
Y	lateral force
Y_i	lateral force on the ith axle
Y_β, Y_ρ	stability derivatives
Y_δ	control derivative
Z	vertical force
Z_i	vertical load on ith axle
Z_i^0	static vertical load on ith axle
Z_i^a	aerodynamic vertical load on ith axle
ΔZ	longitudinal load transfer
ΔZ_i	lateral load transfer on ith axle

α_i	apparent slip angles
α_{ij}	tire slip angles
β	ratio v/u
$\hat{\beta}$	vehicle slip angle
β_t	shifted coordinate
(β_y, β_δ)	gradient components
γ_{ij}	camber angles
δ_{ij}	steer angle of the wheels
δ_v	steering wheel angle of rotation
ε_1	Ackermann coefficient
ζ	damping ratio
η_h	internal efficiency of the differential housing
ρ	ratio r/u
ρ_a	air density
ρ_t	shifted coordinate
(ρ_y, ρ_δ)	gradient components
$\sigma_{x_{ij}}$	tire longitudinal slips
$\sigma_{y_{ij}}$	tire lateral slips
τ	steer gear ratio
ϕ	roll angle
Φ_i	slope of the axle characteristics
φ_{ij}	spin slips
ψ	yaw angle
ω_{ij}	angular velocity of the rims
ω_n	natural angular frequency
ω_s	damped natural angular frequency

References

1. Abe M (2009) Vehicle handling dynamics. Butterworth-Heinemann, Oxford
2. Bastow D, Howard G, Whitehead JP (2004) Car suspension and handling, 4th edn. SAE International, Warrendale
3. Bundorf RT (1968) The influence of vehicle design parameters on characteristic speed and understeer. SAE Trans 76:548–560
4. Chindamo D, Lenzo B, Gadola M (2018) On the vehicle sideslip angle estimation: a literature review of methods, models, and innovations. Appl Sci 8(3)
5. Dixon JC (1991) Tyres, suspension and handling. Cambridge University Press, Cambridge
6. Font Mezquita J, Dols Ruiz JF (2006) La Dinámica del Automóvil. Editorial de la UPV, Valencia
7. Frendo F, Greco G, Guiggiani M, Sponziello A (2007) The handling surface: a new perspective in vehicle dynamics. Veh Syst Dyn 45:1001–1016
8. Gillespie TD (1992) Fundamentals of vehicle dynamics. SAE International, Warrendale
9. Guiggiani M (2007) Dinamica del Veicolo. CittaStudiEdizioni, Novara
10. Mastinu G, Ploechl M (eds) (2014) Road and off-road vehicle system dynamics handbook. CRC Press

11. Meywerk M (2015) Vehicle dynamics. Wiley
12. Milliken WF, Milliken DL (1995) Race car vehicle dynamics. SAE International, Warrendale
13. Pacejka HB (1973) Simplified analysis of steady-state turning behaviour of motor vehicles, part 1. handling diagrams of simple systems. Veh Syst Dyn 2:161–172
14. Pacejka HB (1973) Simplified analysis of steady-state turning behaviour of motor vehicles, part 2: stability of the steady-state turn. Veh Syst Dyn 2:173–183
15. Pacejka HB (1973) Simplified analysis of steady-state turning behaviour of motor vehicles, part 3: more elaborate systems. Veh Syst Dyn 2:185–204
16. Pascarella R, Tandy DF, Colborn J, Bae JC, Coleman C (2015) The true definition and measurement of oversteer and understeer. SAE Int J Commer Veh 8:160–181
17. Popp K, Schiehlen W (2010) Ground vehicle dynamics. Springer, Berlin
18. Schramm D, Hiller M, Bardini R (2014) Vehicle dynamics. Springer, Berlin
19. Wong JY (2001) Theory of ground vehicles. Wiley, New York

Chapter 8
Handling of Race Cars

Race cars come in a number of shapes, sizes, engines, types of wings, etc. However, most of them share the following features relevant to handling:

1. four wheels (two axles);
2. two-wheel drive;
3. *aerodynamic devices* (and hence, significant aerodynamic downforces, along with significant aerodynamic drag);
4. *limited-slip differential*;
5. often no intervention by electronic active safety systems like ABS or ESP.

The handling analysis of race cars is more involved than that of road cars (Chap. 7). The non-open differential makes the vehicle behavior very sensitive to the *turning radius*, while the aerodynamic effects make the vehicle handling behavior very sensitive to the *forward speed*.

8.1 Assumptions for Race Car Handling

The analysis developed here is based on the vehicle model introduced in Chap. 3 and summarized in Sect. 3.14. However, it is recommended to read also Chaps. 5–7.

For definiteness, let us suppose to deal with a *rear-wheel-drive* vehicle. Owing to the presence of a *limited-slip differential* and of relevant *aerodynamic loads* (high downforce and hence high drag), the tires of the driven axle undergo significant longitudinal slips under almost all operating conditions

$$\sigma_{x_{21}} \neq 0 \quad \text{and} \quad \sigma_{x_{22}} \neq 0 \tag{8.1}$$

The original version of this chapter was revised. Equation 8.64 was corrected. The correction to this chapter is available at https://doi.org/10.1007/978-3-031-06461-6_13

© The Author(s), under exclusive license to Springer Nature Switzerland AG 2023, corrected publication 2023
M. Guiggiani, *The Science of Vehicle Dynamics*,
https://doi.org/10.1007/978-3-031-06461-6_8

Therefore, it does not make much sense to restrict the analysis to steady state since the very beginning.

To highlight the role of the limited-slip differential, we do not consider the vehicle while braking,[1] but only during power-on/power-off conditions. Therefore, we have negligible longitudinal tire forces at the front axle

$$F_{x_{11}} = F_{x_{12}} = 0 \tag{8.2}$$

and hence, still at the front axle, negligible longitudinal tire slips

$$\sigma_{x_{11}} = \sigma_{x_{12}} = 0 \tag{8.3}$$

8.1.1 Aerodynamic Downforces and Drag

Many race cars have wings and an underbody diffuser to create downforces that press the race car against the surface of the track. Therefore, the vertical loads Z_1^a and Z_2^a acting on the tires may be very speed dependent.

Aerodynamic forces have been discussed in Sects. 3.7.2 and 4.12. The overall aerodynamic load can be correctly and conveniently represented as in Fig. 8.1 (see also Fig. 8.62). At high speeds, Z_1^a and Z_2^a, and also the aerodynamic drag X_a, have fairly high positive values.

8.1.2 Limited-Slip Differential

Race cars are usually equipped with a *limited-slip differential*, that is a differential with a torque bias, which can become totally locked[2] or totally open in some cases (Sect. 3.13).

Torque bias means that the torques applied to the left and right shafts may not be equal to each other. Therefore, as shown in Fig. 8.2, we may have

$$F_{x_{21}} \neq F_{x_{22}} \tag{8.4}$$

In a curve, counterintuitive as it may appear, the inside wheel has not necessarily an angular speed lower than the outside wheel. Just consider a race car accelerating while exiting a curve: in some cases, due to the still high lateral acceleration, its inside

---

[1] Braking of formula cars is discussed in Sect. 4.12.

[2] A *locked differential* is actually not a differential. Indeed, a differential mechanism must convey power from a single shaft to two shafts while permitting different rotation speeds. A locked differential no longer has this degree of freedom and the *two wheels must rotate at the same angular speed*.

Fig. 8.1 Aerodynamic drag and downforces (all positive)

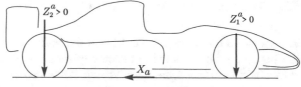

Fig. 8.2 Road-tire grip forces for a car with limited-slip differential (cf. Fig. 7.1)

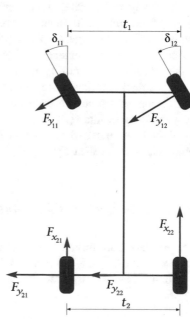

wheel is barely touching the ground, and hence it is probably spinning faster than the outer wheel (Fig. 3.56c). This phenomenon is one of the main reasons that makes a limited-slip differential almost mandatory in a race car. Otherwise, that is with an open differential, the car would not accelerate much, as the maximum longitudinal force would be limited by the inner wheel (the one barely touching the ground). On the other hand, if a vehicle is turning at low lateral acceleration, the inside wheel will be turning slower than the outside wheel, and hence it will receive more torque (Fig. 3.56a).

To make the torques applied to the left and right shafts not equal to each other, limited-slip differentials are built to have some sort of *friction* inside the housing. Indeed, a limited-slip differential is characterized by its internal efficiency $\eta_h \ll 1$, and hence by a Torque Bias Ratio (TBR $= 1/\eta_h) \gg 1$. The mechanics of differential mechanisms has been discussed in Sect. 3.13, where the relevant equations have been obtained.

8.2 Vehicle Model for Race Car Handling

The equations collected in Sect. 3.14 for the fairly general vehicle model described in Chap. 3 are now tailored to the case of race cars (limited-slip differential, non-constant forward speed, and aerodynamic downforces).

As in any dynamical system, there are input (known) functions and output (to be found) functions. Perhaps, the most natural way to set up the problem is to assign as input functions the angular speed $\omega_h(t)$ of the housing of the differential mechanism (Sect. 3.13) and the angular position $\delta_v(t)$ of the steering wheel. Imposing $\omega_h(t)$ is more realistic than imposing directly the forward velocity $u(t)$ (cf. Chap. 7).

The vehicle motion is the sought output. According to Chap. 3, to monitor the vehicle motion we can use, for instance, the forward velocity $u(t)$, the lateral velocity $v(t)$ and the yaw rate $r(t)$.

To link the input to the output we have to build and solve a system of differential-algebraic equations (DAE).

8.2.1 Equilibrium Equations

The in-plane equilibrium equations are the most intuitive, and we start with them, with the aid of Figs. 8.1, 8.2, and 8.3.

For a rear-wheel-drive race car, the in-plane equilibrium equations (3.94) become

Fig. 8.3 Vehicle model

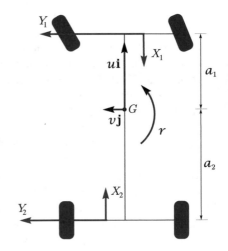

$$ma_x = X$$

$$ma_y = Y \tag{8.5}$$

$$J_z \dot{r} = N$$

or, more explicitly

$$m(\dot{u} - vr) = X_1 + X_2 - X_a$$

$$m(\dot{v} + ur) = Y_1 + Y_2 \tag{8.6}$$

$$J_z \dot{r} = Y_1 a_1 - Y_2 a_2 + N_X = N_Y + N_X$$

where

$$X_1 = -F_{y_{11}} \sin(\delta_{11}) - F_{y_{12}} \sin(\delta_{12})$$

$$X_2 = F_{x_{21}} + F_{x_{22}}$$

$$X_a = \frac{1}{2} \rho S C_x u^2$$

$$Y_1 = F_{y_{11}} \cos(\delta_{11}) + F_{y_{12}} \cos(\delta_{12})$$

$$Y_2 = F_{y_{21}} + F_{y_{22}} \tag{8.7}$$

$$N_X = \Delta X_1 t_1 + \Delta X_2 t_2 = N_f + N_d$$

with

$$\Delta X_1 = \frac{1}{2} \left[F_{y_{11}} \sin(\delta_{11}) - F_{y_{12}} \sin(\delta_{12}) \right]$$

$$\Delta X_2 = \frac{1}{2} \left(F_{x_{22}} - F_{x_{21}} \right)$$

Of course X_a is always positive (drag). It is kind of interesting to compare these equations with (7.4).

In any two-axle car, the yawing moment

$$N_Y = Y_1 a_1 - Y_2 a_2 \tag{8.8}$$

is always present in the yaw equation, that is in the third equation in (8.6).

The other yawing moment N_X collects two very different contributions.

The contribution $N_f = \Delta X_1 t_1$ comes from the difference between the longitudinal components of the front lateral forces (Fig. 8.4). Therefore, it becomes significant only when the front steer angles are not small, like in FSAE competitions. In other competitions, the front steer angles are usually below 0.2 rad (11°) and hence $\Delta X_1 t_1$ is probably very small.

The other contribution $N_d = \Delta X_2 t_2$ comes from the limited-slip differential (Fig. 8.2). It involves the difference between the rear tire longitudinal forces. It

Fig. 8.4 Origin of ΔX_1
(steer angles of about 30°)

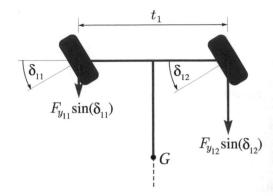

can be quite relevant, depending on the type of differential and on the operating conditions (lateral acceleration, power-on/power-off, steer angle).

8.2.2 Lateral Forces for Dynamic Equilibrium

Regardless of (8.7), the last two equations in (8.6) can be already solved with respect to the axle lateral forces Y_1 and Y_2, thus obtaining (a result already given in (3.102) and (3.156))

$$Y_1 = \frac{1}{l}\left[ma_2\,a_y + (J_z\dot{r} - N_X)\right] = Y_1(a_y, \dot{r}, N_X)$$

$$Y_2 = \frac{1}{l}\left[ma_1\,a_y - (J_z\dot{r} - N_X)\right] = Y_2(a_y, \dot{r}, N_X) \tag{8.9}$$

where $a_y = \dot{v} + ur$. The key point is that we can have $(N_Y = J_z\dot{r} - N_X) \neq 0$, even if $\dot{r} = 0$. Therefore, differently from (7.9), Y_1 and Y_2 depend, in general, also on the yawing moment N_X.

8.2.3 Gyroscopic Torques

As already obtained in (3.77), each wheel of the ith axle requires a gyroscopic torque

$$- L_{w_i}\,\mathbf{i} = -\frac{J_{w_i}}{r_i}ur\,\mathbf{i} \simeq -\frac{J_{w_i}}{r_i}a_y\,\mathbf{i} \tag{8.10}$$

Since the suspensions employed in race cars have nearly parallel arms (Fig. 8.5), the four gyroscopic torques $L_{w_i}\,\mathbf{i}$ of the rotating wheels go almost entirely into the

Fig. 8.5 Gyroscopic torques
as inertial moments (front
view, left turn)

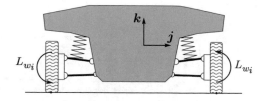

chassis, to be redistributed, according to the roll balance, to each wheel as an increase in lateral load transfer.

8.2.4   Roll Angles

In most race cars the suspension roll stiffnesses $k^s_{\phi_i}$ and the tire roll stiffnesses $k^p_{\phi_i}$ are not very different. Therefore, assuming rigid tires is not quite correct.

We have the following relationships between the axle lateral forces and the roll angles. From (3.146), with the addition of the gyroscopic torque L_w of the wheels, defined in (3.77), the roll angles (front and rear) due to the tires are

$$\phi^p_1 = \frac{1}{k^p_{\phi_1}} \frac{k_{\phi_1}k_{\phi_2}}{k_\phi} \left[\frac{(Y_1 + Y_2)(h - q^b) + L_w}{k_{\phi_2}} + \frac{Y_1 q_1}{k^s_{\phi_1}} + \frac{Y_1 q_1}{k^s_{\phi_2}} + \frac{Y_1 q_1 + Y_2 q_2}{k^p_{\phi_2}} \right]$$

$$\phi^p_2 = \frac{1}{k^p_{\phi_2}} \frac{k_{\phi_1}k_{\phi_2}}{k_\phi} \left[\frac{(Y_1 + Y_2)(h - q^b) + L_w}{k_{\phi_1}} + \frac{Y_2 q_2}{k^s_{\phi_1}} + \frac{Y_2 q_2}{k^s_{\phi_2}} + \frac{Y_1 q_1 + Y_2 q_2}{k^p_{\phi_1}} \right]$$

$$(8.11)$$

Similarly, from (3.147) plus L_w, the roll angles due to the suspension springs are

$$\phi^s_1 = \frac{1}{k^s_{\phi_1}} \frac{k_{\phi_1}k_{\phi_2}}{k_\phi} \left[\frac{(Y_1 + Y_2)(h - q^b) + L_w}{k_{\phi_2}} - \frac{Y_1 q_1}{k^p_{\phi_1}} + \frac{Y_2 q_2}{k^p_{\phi_2}} \right]$$

$$\phi^s_2 = \frac{1}{k^s_{\phi_2}} \frac{k_{\phi_1}k_{\phi_2}}{k_\phi} \left[\frac{(Y_1 + Y_2)(h - q^b) + L_w}{k_{\phi_1}} - \frac{Y_2 q_2}{k^p_{\phi_2}} + \frac{Y_1 q_1}{k^p_{\phi_1}} \right]$$

$$(8.12)$$

Of course

$$\phi^p_1 + \phi^s_1 = \phi^p_2 + \phi^s_2 = \phi \qquad (8.13)$$

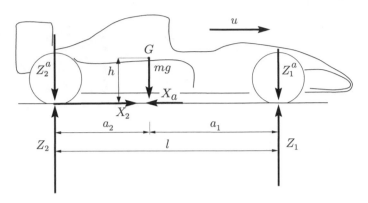

Fig. 8.6 Forces acting on a Formula car

8.2.5 Vertical Loads on Each Wheel

As shown in (3.108), the vertical load acting on each wheel is the algebraic sum of four contributions (Fig. 8.6):

1. the static load (weight);
2. the aerodynamic load;
3. the longitudinal load transfer;
4. the lateral load transfer.

More explicitly, the expressions (3.109) for the vertical loads on each tire must be taken in full, except for the $J_{zx}r^2$ term, which is almost certainly negligible. In compact form, (3.109) can be recast as (cf. (7.14))

$$Z_{11} = \frac{1}{2}\left(\frac{mga_2}{l} + \frac{1}{2}\rho_a S_a C_{z1} u^2 - \frac{ma_x h}{l}\right) - \Delta Z_1 = Z_{11}(u, a_x, \Delta Z_1)$$

$$Z_{12} = \frac{1}{2}\left(\frac{mga_2}{l} + \frac{1}{2}\rho_a S_a C_{z1} u^2 - \frac{ma_x h}{l}\right) + \Delta Z_1 = Z_{12}(u, a_x, \Delta Z_1)$$

$$\tag{8.14}$$

$$Z_{21} = \frac{1}{2}\left(\frac{mga_1}{l} + \frac{1}{2}\rho_a S_a C_{z2} u^2 + \frac{ma_x h}{l}\right) - \Delta Z_2 = Z_{21}(u, a_x, \Delta Z_2)$$

$$Z_{22} = \frac{1}{2}\left(\frac{mga_1}{l} + \frac{1}{2}\rho_a S_a C_{z2} u^2 + \frac{ma_x h}{l}\right) + \Delta Z_2 = Z_{22}(u, a_x, \Delta Z_2)$$

where $a_x = \dot{u} - vr$. A race car with wings has $C_{zi} > 0$ (Fig. 8.1).

It is interesting to compare (8.14) with (7.14). There are two important differences:

1. the speed-dependent aerodynamic vertical loads;
2. the longitudinal load transfer due to the longitudinal acceleration a_x.

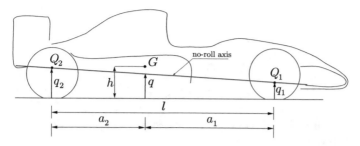

Fig. 8.7 No-roll axis

Moreover, there is the effect of the yaw moment N_d on the lateral forces and hence on the lateral load transfers, as discussed hereafter.

8.2.6 Lateral Load Transfers

The lateral load transfers ΔZ_i were obtained in (3.151) as linear functions of the axle lateral forces Y_1 and Y_2 (Fig. 8.7). Here we may add the contribution of the gyroscopic torque L_w of the wheels, defined in (3.77)

$$\Delta Z_1 = \frac{1}{t_1} \left\{ \frac{k_{\phi_1}}{k_\phi} \left[(Y_1 + Y_2)\left(h - q^b\right) + L_w \right] + Y_1 q_1 + \frac{k_{\phi_1} k_{\phi_2}}{k_\phi} \left(\frac{Y_2 q_2}{k_{\phi_2}^p} - \frac{Y_1 q_1}{k_{\phi_1}^p} \right) \right\}$$

$$\Delta Z_2 = \frac{1}{t_2} \left\{ \frac{k_{\phi_2}}{k_\phi} \left[(Y_1 + Y_2)\left(h - q^b\right) + L_w \right] + Y_2 q_2 + \frac{k_{\phi_1} k_{\phi_2}}{k_\phi} \left(\frac{Y_1 q_1}{k_{\phi_1}^p} - \frac{Y_2 q_2}{k_{\phi_2}^p} \right) \right\}$$

$$(8.15)$$

where $q^b \simeq q$ (Fig. 8.7).

In Formula cars, the contribution of L_w to $\Delta Z_1 t_1 + \Delta Z_2 t_2$ can even be of about 10%. Definitely not small.

Owing to (8.9), that is to the moment N_X, mainly due to the limited-slip differential, the lateral load transfers no longer depend only on a_y. This is quite a big difference for the complexity of the vehicle dynamic behavior, if compared to (7.10).

8.2.7 In-Plane Tire Forces

According to the tire constitutive equations (2.82), and taking (8.2) into account, the front tire forces can be expressed as

$$F_{x_{11}} = 0$$

$$F_{y_{11}} = F_{y_{11}}(Z_{11}, \gamma_{11}, \sigma_{y_{11}})$$

$$F_{x_{12}} = 0$$ \hfill (8.16)

$$F_{y_{12}} = F_{y_{12}}(Z_{12}, \gamma_{12}, \sigma_{y_{12}})$$

where Z_{1j} are the vertical loads, γ_{1j} are the camber angles and $\sigma_{y_{1j}}$ are the lateral theoretical slips.

The rear tires are under combined slip conditions and, therefore, also the longitudinal slips $\sigma_{x_{2j}}$, that is the angular speed of rotation ω_{2j} of each wheel, have to be taken into account

$$F_{x_{21}} = F_{x_{21}}(Z_{21}, \gamma_{21}, \sigma_{x_{21}}, \sigma_{y_{21}})$$

$$F_{y_{21}} = F_{y_{21}}(Z_{21}, \gamma_{21}, \sigma_{x_{21}}, \sigma_{y_{21}})$$

$$F_{x_{22}} = F_{x_{22}}(Z_{22}, \gamma_{22}, \sigma_{x_{22}}, \sigma_{y_{22}})$$ \hfill (8.17)

$$F_{y_{22}} = F_{y_{22}}(Z_{22}, \gamma_{22}, \sigma_{x_{22}}, \sigma_{y_{22}})$$

Here we are assuming that we know the grip available in the contact patch. Of course, this is a rather unrealistic assumption, but in this analysis we cannot afford to model also the phenomenon of grip generation [9].

The constitutive (tire) Eqs. (8.16) and (8.17) need other algebraic equations for the longitudinal and lateral slips $\sigma_{x_{ij}}$ and $\sigma_{y_{ij}}$, for the camber angles γ_{ij}, and for the vertical loads Z_{ij}, as shown below.

8.2.8 Tire Slips

In general, the *rear* (driven) tires apply both longitudinal and lateral forces to the vehicle. Therefore, we need all slip components. According to (3.61)

$$\sigma_{x_{21}} = \frac{(u - rt_2/2) - \omega_{21} r_2}{\omega_{21} r_2} \qquad \sigma_{y_{21}} = \frac{v - ra_2}{\omega_{21} r_2}$$

$$\sigma_{x_{22}} = \frac{(u + rt_2/2) - \omega_{22} r_2}{\omega_{22} r_2} \qquad \sigma_{y_{22}} = \frac{v - ra_2}{\omega_{22} r_2}$$ \hfill (8.18)

where r_2 is the rolling radius.

In compact form, as in (3.211), we have

$$\sigma_{x_{21}} = \sigma_{x_{21}}(u, r, \omega_{21}) \qquad \sigma_{y_{21}} = \sigma_{y_{21}}(v, r, \omega_{21})$$

$$\sigma_{x_{22}} = \sigma_{x_{22}}(u, r, \omega_{22}) \qquad \sigma_{y_{22}} = \sigma_{y_{22}}(v, r, \omega_{22})$$ \hfill (8.19)

where

$$\omega_{21} = \omega_h - \Delta\widetilde{\omega} \quad \text{and} \quad \omega_{22} = \omega_h + \Delta\widetilde{\omega} \tag{8.20}$$

Of course, $\Delta\widetilde{\omega}(t)$ is unknown (in the sense that it is not an input quantity).

At the *front* axle we have longitudinal pure rolling and, accordingly, we can rely on the expressions (7.21)

$$\sigma_{y_{11}} = \frac{(v + ra_1)\cos(\delta_{11}) - (u - rt_1/2)\sin(\delta_{11})}{(u - rt_1/2)\cos(\delta_{11}) + (v + ra_1)\sin(\delta_{11})}$$
$$\sigma_{y_{12}} = \frac{(v + ra_1)\cos(\delta_{12}) - (u + rt_1/2)\sin(\delta_{12})}{(u + rt_1/2)\cos(\delta_{12}) + (v + ra_1)\sin(\delta_{12})} \tag{8.21}$$

In compact form

$$\sigma_{x_{11}} = 0 \quad \sigma_{y_{11}} = \sigma_{y_{11}}(u, v, r, \delta_{11})$$
$$\sigma_{x_{12}} = 0 \quad \sigma_{y_{12}} = \sigma_{y_{12}}(u, v, r, \delta_{12}) \tag{8.22}$$

The steer angles δ_{ij} need additional algebraic equations, as discussed in Sect. 8.2.10.

8.2.9 Camber Angles

As discussed in Sect. 3.10.3, the camber angles of the two wheels of the same axle are given by the sum of three contributions

$$\gamma_{i1} = -\gamma_i^0 + \Delta\gamma_i^r + \Delta\gamma_i^z$$
$$\gamma_{i2} = \quad \gamma_i^0 + \Delta\gamma_i^r - \Delta\gamma_i^z \tag{7.17'}$$

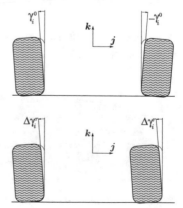

Fig. 8.8 Positive static camber γ_i^0 (front view)

Fig. 8.9 Positive camber variations $\Delta\gamma_i^r$ due to roll motion (front view)

Fig. 8.10 Suspension
first-order parameters that
affect camber variations

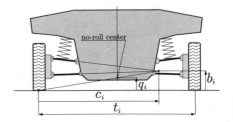

$\mp \gamma_i^0$ are the camber angles under static conditions (Fig. 8.8). Of course, they are constant by definition.

$\Delta \gamma_i^r$ is the camber variation due to the roll motion *only* (Fig. 8.9). The roll camber variation $\Delta \gamma_i^r$ depends on the roll angles ϕ_i^s and ϕ_i^p and on the suspension kinematics (Fig. 8.10)

$$\Delta \gamma_i^r \simeq \left(\frac{c_i - t_i/2}{c_i} \right) \phi_i^s + \phi_i^p = \Delta \gamma_i^r (\phi_i^s, \phi_i^p) \tag{8.23}$$

$\pm \Delta \gamma_i^z$ depend on the height variation z_i^s, and hence on the variations of the vertical loads $Z_i - Z_i^0$, and on the suspension kinematics (Sect. 3.9.1 and Fig. 8.10). More precisely, according to Sect. 3.10.6,

$$\Delta \gamma_i^z = -\frac{z_i^s}{c_i} = \frac{Z_i - Z_i^0}{k_{z_i}^s c_i} = \Delta \gamma_i^z (Z_i - Z_i^0) \tag{8.24}$$

As can be appreciated in Fig. 8.11, one contribution to the camber variations is antisymmetric (same sign) and the other is symmetric (opposite sign).

Different suspensions with the same no-roll center share only the same value of q_i (Fig. 8.10). Therefore they behave differently with respect to camber.

8.2.10 Steer Angles

According to (3.210) we have the following (simplified) expressions for the steering angles of the front wheels

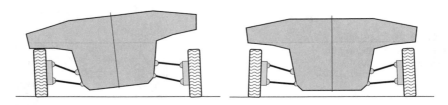

Fig. 8.11 Antisymmetric and symmetric camber variations (front view, left turn)

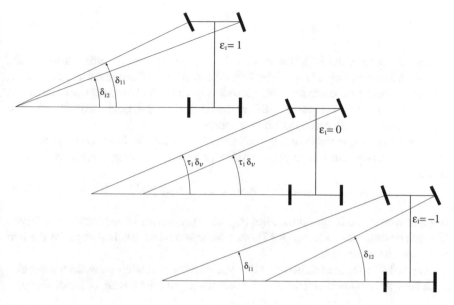

Fig. 8.12 Ackermann steering (top), parallel steering (middle), anti-Ackermann steering (bottom). Any other value of ε_1 is possible

$$\delta_{11} = -\delta_1^0 + \tau_1\delta_v + \varepsilon_1\frac{t_1}{2l}(\tau_1\delta_v)^2 + \Upsilon_1\phi_1^s = \delta_{11}(\delta_v, \phi_1^s)$$

$$\delta_{12} = \quad \delta_1^0 + \tau_1\delta_v - \varepsilon_1\frac{t_1}{2l}(\tau_1\delta_v)^2 + \Upsilon_1\phi_1^s = \delta_{12}(\delta_v, \phi_1^s)$$

(8.25)

which are functions of the steering wheel rotation δ_v and of the front suspension roll angle ϕ_1^s.

In (8.25), as discussed in Sect. 3.4, δ_1^0 is the static toe angle, τ_1 is the gear ratio of the whole steering system, ε_1 is the Ackermann coefficient for dynamic toe (Fig. 8.12), and Υ_1 is the roll steer coefficient. Most cars have $\tau_2 = \varepsilon_2 = 0$, that is no direct steering of the rear wheels.

Actually, (8.25) is a Taylor series expansion. We believe it is a good way to classify and compare steering geometries. It shows in a quantitative, yet simple, way how much the steering system differs from parallel steering (Fig. 8.12).

8.2.11   Simple Model of a Limited-Slip Differential

In a car equipped with a limited-slip differential, the two longitudinal forces $F_{x_{21}}$ and $F_{x_{22}}$, exerted by the rear tires on the vehicle, are not necessarily equal to each other (Fig. 8.2). Therefore we have a *yawing moment* N_d coming from the longitudinal forces acting on the vehicle

$$N_d = \Delta X_2 t_2 = \frac{1}{2}(F_{x_{22}} - F_{x_{21}})t_2 \tag{8.26}$$

When compared with (7.1), that is with the case of open differential, it looks like a small difference, but it is not. The limited-slip differential does affect quite a bit the vehicle handling behavior, and, accordingly, the vehicle model becomes much more involved when compared with the model of a vehicle equipped with an open differential, even with aerodynamic downforces.

Another consequence is that we have significant longitudinal forces at the rear wheels and thus significant longitudinal slips, even when turning at constant forward speed

$$F_{x_{2j}} \neq 0 \quad \text{and hence} \quad \sigma_{x_{2j}} \neq 0 \tag{8.27}$$

In other words, the longitudinal slips $\sigma_{x_{21}}$ and $\sigma_{x_{22}}$ cannot be neglected, and hence the tire constitutive equations (8.17) must include the longitudinal slips for the two wheels of the driven axle.

Any differential mechanism provides the same relationship between the angular velocities of the wheels and the angular velocity ω_h of the housing of the differential

$$\omega_{21} = \omega_l = \omega_h - \Delta\widetilde{\omega} \quad \text{and} \quad \omega_{22} = \omega_r = \omega_h + \Delta\widetilde{\omega} \tag{8.28}$$

As discussed in Sect. 3.13.13, a limited-slip differential, with internal efficiency η_h, provides a link between the tire longitudinal forces due to the engine power. For a rear driven vehicle we can model it as

$$F_{x_{22}} = \eta_h^{\zeta(t)} F_{x_{21}} \tag{8.29}$$

where

$$\zeta(t) = \frac{\arctan(\chi\,\Delta\widetilde{\omega}(t)\,\mathrm{sign}(F_{x_{21}}(t) + F_{x_{22}}(t)))}{\pi/2} \tag{8.30}$$

with χ a positive *big number*, something around 1000 s. This way, the limited-slip differential action is activated whenever $\Delta\widetilde{\omega}(t)$ has significant values, with a smooth transition through the locked state of the differential ($\Delta\widetilde{\omega}(t) \simeq 0$). Figures 3.67–3.73 are applications of this model. By setting $\eta_h = 1$ in (3.196) or (8.29) we obtain the behavior of the open differential.

Summing up, a mathematical model of a vehicle with a limited-slip differential is definitely more involved than one of a vehicle with an open differential.

8.2.12 Reducing the Number of Equations

To define this vehicle model for race car handling we have introduced three differential equations and several algebraic equations. It is possible, and convenient, to
reduce the number of algebraic equations by combining them.

From (8.9) we have that the axle lateral forces Y_i are known (linear) functions of
the lateral acceleration a_y, of the yaw acceleration $\dot{r}$, and of the moment N_X.

These functions $Y_i(a_y, \dot{r}, N_X)$ can be plugged directly into the expressions of the
lateral load transfers ΔZ_i, of the suspension roll angles ϕ_i^s, and of the tire roll angles
ϕ_i^p.

Then, $\Delta Z_i\big(Y_1(a_y, \dot{r}, N_X), Y_2(a_y, \dot{r}, N_X), a_y\big)$ go into the expressions of
$Z_{ij}(u, a_x, \Delta Z_i)$, while the roll angles and the suspensions vertical displacements
have to be inserted into the camber angle equations $\gamma_{ij}(\phi_i^s, \phi_i^p, z_i^s)$ and, possibly,
into the steer angles $\delta_{ij}(\delta_v, \phi_i^s)$.

The steer angles just obtained have to be plugged into the expressions of the front
lateral slips $\sigma_{1j}(u, v, r, \delta_{1j})$.

The rear slips involve the angular velocities of the wheels. It maybe be better to
set $\omega_{2j} = \omega_h \pm \Delta\tilde{\omega}$, like in (8.28). This way, since ω_h is given, we have that only $\Delta\tilde{\omega}$
is unknown.[3]

The just obtained expressions of the vertical loads, of the camber angles, and of
the slip components go into the tire constitutive equations (8.16) and (8.17).

Ultimately, all tire force components can be *explicitly*, and easily, set as functions
of:

1. the state variables $(u(t), v(t), r(t))$;
2. the derivatives $(\dot{u}(t), \dot{v}(t), \dot{r}(t))$;
3. the angular velocity $\Delta\tilde{\omega}(t)$;
4. the moment $N_X(t)$;
5. the given angular velocity $\omega_h(t)$ of the differential housing;
6. the given angle $\delta_v(t)$ of the steering wheel.

That is to say that we know all the following algebraic functions

$$F_{x_{ij}} = F_{x_{ij}}\big(u, v, r, \dot{u}, \dot{v}, \dot{r}, \Delta\tilde{\omega}, N_X; \omega_h, \delta_v\big)$$
$$F_{y_{ij}} = F_{y_{ij}}\big(u, v, r, \dot{u}, \dot{v}, \dot{r}, \Delta\tilde{\omega}, N_X; \omega_h, \delta_v\big) \tag{8.31}$$

It is very important to note that among the arguments of these functions there is the
moment N_X, which is defined in (8.7) in terms of some of the tire force components

$$\begin{aligned} N_X &= \Delta X_1 t_1 + \Delta X_2 t_2 \\ &= \frac{t_1}{2}\big[F_{y_{11}} \sin(\delta_{11}) - F_{y_{12}} \sin(\delta_{12})\big] + \frac{t_2}{2}\big(F_{x_{22}} - F_{x_{21}}\big) \end{aligned} \tag{8.32}$$

---

[3] The small difference between the angular velocities of the wheels has negligible relevance for the
gyroscopic torque.

Moreover, Eq. (8.29) governing the behavior of the limited-slip differential has to be included in the mathematical model

$$F_{x_{22}} = \eta_h^{\zeta(t)} F_{x_{21}} \tag{8.33}$$

where $\zeta(t)$ can be defined, e.g., as in (8.32).

The number of algebraic equations has been drastically reduced, but two of them are still there and will be there. Indeed, we have three state variables and three differential equations, but also two other unknown functions ($\Delta\widetilde{\omega}(t)$ and $N_X(t)$), which require two additional algebraic equations. The two unknown functions "survived" the equation reduction process because, in general, there is no way to "extract" them analytically. Therefore, it is convenient to solve numerically a system of $3 + 2$ differential-algebraic equations (DAE).

8.3   Double Track Race Car Model

After a bit of work, we are now ready to set up the fundamental governing equations for the *transient* handling of a race car equipped with a *limited-slip differential* and with *aerodynamic devices*

$$
\begin{aligned}
m(\dot{u} - vr) &= X(u, v, r, \dot{u}, \dot{v}, \dot{r}, \Delta\widetilde{\omega}, N_X; \omega_h, \delta_v) \\
m(\dot{v} + ur) &= Y(u, v, r, \dot{u}, \dot{v}, \dot{r}, \Delta\widetilde{\omega}, N_X; \omega_h, \delta_v) \\
J_z\dot{r} &= N_X + N_Y(u, v, r, \dot{u}, \dot{v}, \dot{r}, \Delta\widetilde{\omega}, N_X; \omega_h, \delta_v) \\
N_X &= \frac{t_1}{2}\left[F_{y_{11}}\sin(\delta_{11}) - F_{y_{12}}\sin(\delta_{12})\right] + \frac{t_2}{2}\left(F_{x_{22}} - F_{x_{21}}\right) \\
F_{x_{22}} &= \eta_h^{\zeta(t)} F_{x_{21}}
\end{aligned}
\tag{8.34}
$$

In case of open differential, just set $\eta_h = 1$.

As already stated, a fairly practical way to set up the problem is to assign the angular speed $\omega_h(t)$ of the housing of the differential (Sect. 3.13) and the angular position $\delta_v(t)$ of the steering wheel, along with the initial conditions, and then to numerically solve this system of *five* differential-algebraic equations in the five unknown functions ($u(t), v(t), r(t), \Delta\widetilde{\omega}(t), N_X(t)$). Imposing $\omega_h(t)$ is more realistic than imposing the forward speed $u(t)$ directly.

It is a system of differential and algebraic equations because there are no derivatives of $\Delta\widetilde{\omega}(t)$ and $N_X(t)$. .

The model (8.34) for race cars is a generalization of the model for road cars presented in Chap. 7. Wings make the vertical loads strongly dependent on the vehicle speed. The limited-slip differential provides a yawing moment very sensitive to the lateral acceleration and to the steer angle. None of these phenomena can be found

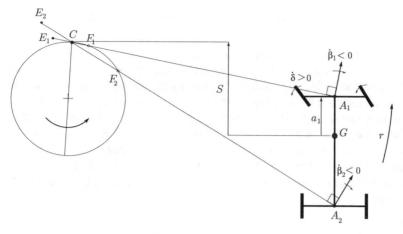

Fig. 8.13 Possible (usual?) kinematics of a race car *entering* a bend properly

in (most) road cars. On the other hand, torque vectoring can be activated in modern road cars, whereas driver-assistance systems are usually prohibited in competitions.

The comparison of (8.34) with (7.34), that is with the governing equation of a double track model for road vehicles, clearly shows the increased complexity of the model. That is no surprise: a race car exhibits indeed a much richer handling behavior. However, the effort for building and running a race car model is pretty much the same as for road car model.

Running a complex vehicle model can be very interesting, but often we would like to get a more global understanding of the vehicle handling behavior.

8.4 Kinematics of Race Cars when Cornering

The analysis developed in Chap. 5 applies to race cars as well. However, most often in competitive driving we have that

$$S > a_1 \tag{8.35}$$

as in Fig. 8.13. Only in tight hairpin bends we have $S < a_1$.

Therefore, according to (8.35), race cars may exhibit a peculiar kinematics when negotiating a bend. It is left to the reader to catch the differences between Fig. 5.18 (road) and Fig. 8.13 (race).

8.5   Handling of Race Cars with Open Differential

It is in the handling of Formula cars that aerodynamics really comes into play
(Fig. 8.6). Thanks to well-designed aerodynamic devices, very high downforces
are generated at high speeds, although at the expense of high drag as well. A mathematical model that takes aerodynamics into account has been developed in Sect. 8.1.

Although Formula cars have a limited-slip differential, at the apex of a bend, that
is when the vehicle is more or less close to steady state, the (electronically controlled)
differential is basically set open. Therefore, the steady-state analysis is more realistic
if done with *open differential*, leaving the locked one for power-off and power-on.

Here we discuss some of the main phenomena that make the handling of this kind
of cars so peculiar.

8.5.1   Single Track Model for Race Cars

Although we should refrain from using the single track model as much as possible, a
question that naturally arises at this point is whether we can go "single track" or not,
as it was done for road cars in Sect. 7.5. To answer this question we should recall that
by single track [1–3, 7, 15] we meant a vehicle model having two axle characteristics
(7.66), that is two constitutive equations $Y_i(\alpha_i)$, one per axle, involving, in that case,
a single kinematic variable each (namely, the axle apparent slip angle α_i).

A *generalized single track* model can still be formulated, as it was for road cars,
provided that:

1. the race car has an *open differential* ($\eta_h = 1$), and hence $\Delta X_2 = 0$;
2. the race car has *parallel steering* ($\varepsilon_1 = 0$);
3. it is assumed $\Delta X_1 \simeq 0$;
4. it is assumed $\sigma_{x_{ij}} \simeq 0$;
5. it is $a_x \simeq 0$;
6. the driver controls directly the forward velocity u, in addition to the steering
   wheel angle δ_v.

The steps to formulate this generalized single track model are like in Sect. 7.5. The
main difference is that now the vertical load Z_{ij} acting on each tire is strongly *speed
dependent*, as shown in (8.14). Therefore, also the lateral forces are speed dependent
(cf. (7.55)):

$$Y_1 = F_{y_{11}}\big(Z_{11}(u, 0, \tilde{a}_y), \gamma_{11}(\tilde{a}_y), \sigma_{y_{11}}(\alpha_1, \tilde{a}_y)\big) + F_{y_{12}}\big(Z_{12}(u, 0, \tilde{a}_y), \gamma_{12}(\tilde{a}_y), \sigma_{y_{12}}(\alpha_1, \tilde{a}_y)\big)$$

$$= F_{y_{11}}(\alpha_1, u, \tilde{a}_y) + F_{y_{12}}(\alpha_1, u, \tilde{a}_y)$$

$$= F_{y_1}(\alpha_1, u, \tilde{a}_y);$$

$$Y_2 = F_{y_{21}}\big(Z_{21}(u, 0, \tilde{a}_y), \gamma_{21}(\tilde{a}_y), \sigma_{y_{21}}(\alpha_2, \tilde{a}_y)\big) + F_{y_{22}}\big(Z_{22}(u, 0, \tilde{a}_y), \gamma_{22}(\tilde{a}_y), \sigma_{y_{22}}(\alpha_2, \tilde{a}_y)\big)$$

$$= F_{y_{21}}(\alpha_2, u, \tilde{a}_y) + F_{y_{22}}(\alpha_2, u, \tilde{a}_y)$$

$$= F_{y_2}(\alpha_2, u, \tilde{a}_y),$$

$$(8.36)$$

As already obtained in (7.9), also in this case the lateral forces are basically linear functions of the lateral acceleration $\tilde{a}_y = ur$ (open differential)

$$Y_1 \simeq \frac{ma_2}{l}\tilde{a}_y \quad \text{and} \quad Y_2 \simeq \frac{ma_1}{l}\tilde{a}_y \tag{8.37}$$

Therefore, $F_{y_1}(\alpha_1, u, \tilde{a}_y)$ and $F_{y_2}(\alpha_2, u, \tilde{a}_y)$ in (8.36) must be such that

$$F_{y_1}(\alpha_1, u, \tilde{a}_y) = \frac{ma_2}{l}\tilde{a}_y \quad \text{and} \quad F_{y_2}(\alpha_2, u, \tilde{a}_y) = \frac{ma_1}{l}\tilde{a}_y \tag{8.38}$$

which can be solved with respect to the (steady-state) lateral acceleration $\tilde{a}_y$, to obtain

$$\tilde{a}_y = g_1(\alpha_1, u) \quad \text{and} \quad \tilde{a}_y = g_2(\alpha_2, u) \tag{8.39}$$

From (8.38) we can also obtain

$$\begin{aligned}
\alpha_1 &= \alpha_1(\tilde{a}_y, u) = \tau_1 \delta_v - \beta - \rho a_1 \\
\alpha_2 &= \alpha_2(\tilde{a}_y, u) = \tau_2 \delta_v - \beta + \rho a_2
\end{aligned} \tag{8.40}$$

The final step is inserting (8.39) back into (8.36), thus obtaining the *axle characteristics* of the single track model for race cars with aerodynamic downforces and open differential

$$\begin{aligned}
Y_1 &= Y_1(\alpha_1, u) = F_{y_1}\big(\alpha_1, u, g_1(\alpha_1, u)\big) \\
Y_2 &= Y_2(\alpha_2, u) = F_{y_2}\big(\alpha_2, u, g_2(\alpha_2, u)\big)
\end{aligned} \tag{8.41}$$

That said, we remark, once again, that with a limited-slip differential it is not possible to obtain the axle characteristics, nor even for the front axle, since there is a strong interaction between lateral and longitudinal tire forces. More precisely, the analysis developed in Sect. 7.5.10 about the role of lateral acceleration is no longer applicable. Therefore, we cannot end up with a single track model for vehicles equipped with a limited-slip differential. We have to stick to a more general double track model. However, this is somehow good news, as the double track model is much more realistic, and only a little more complex, than the single track model.

8.5.1.1   Partial Derivatives of the Axle Characteristics

It turns out that vehicle handling is pretty much affected by the slopes (derivatives) of the axle characteristics. Here we have to generalize (7.94), which becomes

$$
\Phi_1 = \frac{\partial Y_1}{\partial \alpha_1}, \qquad \Psi_1 = \frac{\partial Y_1}{\partial u}
$$

$$
\Phi_2 = \frac{\partial Y_2}{\partial \alpha_2}, \qquad \Psi_2 = \frac{\partial Y_2}{\partial u}
$$

$$(8.42)$$

Obviously, $\Phi_i > 0$ in the monotone increasing part of the axle characteristics.

In the single track model for race cars we have that the slopes Φ_i of the axle characteristics are not anymore functions of the lateral acceleration only

$$
\Phi_1 = \Phi_1(\alpha_1, u), \qquad \Psi_1 = \Psi_1(\alpha_1, u)
$$

$$
\Phi_2 = \Phi_2(\alpha_2, u), \qquad \Psi_2 = \Psi_2(\alpha_2, u)
$$

$$(8.43)$$

From (7.89)

$$
\frac{d\tilde{a}_y}{d\alpha_1} = \frac{l\Phi_1}{ma_2} \quad \text{and} \quad \frac{d\tilde{a}_y}{d\alpha_2} = \frac{l\Phi_2}{ma_1}
$$

$$(8.44)$$

and hence

$$
\frac{d\alpha_1}{d\tilde{a}_y} = \frac{ma_2}{l\Phi_1} \quad \text{and} \quad \frac{d\alpha_2}{d\tilde{a}_y} = \frac{ma_1}{l\Phi_2}
$$

$$(8.45)$$

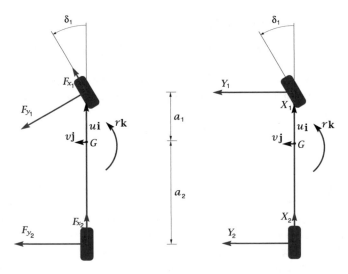

Fig. 8.14 Equivalent schematic representations of the single track model (with $\delta_2 = 0$)

8.5.1.2 Governing Equations

As expected, the single track model is still governed by three sets of equations (Fig. 8.14):

- two *equilibrium* equations (lateral and yaw), as in (7.65)

$$m(\dot{v} + ur) = Y = Y_1 + Y_2$$
$$J_z \dot{r} = N = Y_1 a_1 - Y_2 a_2$$

(8.46)

- two *congruence* equations (apparent slip angles), as in (7.67)

$$\alpha_1 = \tau_1 \delta_v - \frac{v + r a_1}{u} = \tau_1 \delta_v - \beta - \rho a_1$$
$$\alpha_2 = \tau_2 \delta_v - \frac{v - r a_2}{u} = \tau_2 \delta_v - \beta + \rho a_2$$

(8.47)

- two *constitutive* equations (axle characteristics), similar but not equal to (7.66)

$$Y_1 = Y_1(\alpha_1, u)$$
$$Y_2 = Y_2(\alpha_2, u)$$

(8.48)

Differently from the classical single track model defined in Sect. 7.5.4, the constitutive equations are now functions also of the forward velocity u. The axle characteristics are no longer curves. They are surfaces.

Inserting (8.47) into (8.48), and then into (8.46), we obtain the *dynamical* equations of this generalized single track model

$$m(\dot{v} + ur) = Y(v, r; u, \delta_v)$$
$$J_z \dot{r} = N(v, r; u, \delta_v)$$

(8.49)

formally identical to (7.71). However, the counterpart of (7.75) is

$$m(\dot{\beta} u + \beta \dot{u} + u^2 \rho) = Y(\beta, \rho; u, \delta_v)$$
$$J_z(\dot{\rho} u + \rho \dot{u}) = N(\beta, \rho; u, \delta_v)$$

(8.50)

We see that, notwithstanding the use of $\beta = v/u$ and $\rho = r/u = 1/R$, and differently from (7.75), there is still an *explicit* dependence of Y and N on the forward velocity u (aerodynamic downforces and drag).

8.5.2 What About Understeer/Oversteer?

Maybe the most striking effect of aerodynamic downloads is that the definition
(7.117) of the classical understeer gradient K becomes meaningless. As a matter of
fact, in addition to the minor problems discussed in Sect. 7.9, here we have a major
drawback for the application of (7.117).

More precisely, (7.105) still holds true, but with $\alpha_i(\tilde{a}_y, u)$, and hence $f_\rho(\tilde{a}_y, u)$,
being functions of two variables (e.g., $\tilde{a}_y$, plus another one)

$$\delta - \frac{l}{R} = \alpha_1(\tilde{a}_y, u) - \alpha_2(\tilde{a}_y, u) = f_\rho(\tilde{a}_y, u)l \qquad (8.51)$$

where

$$\delta = (\tau_1 - \tau_2)\, \delta_v \qquad (8.52)$$

is the *net steer angle*, already defined in (7.69).

Owing to aerodynamic loads, the speed matters a lot when a car is making a turn.
The faster the car, the higher the lateral acceleration that can be achieved, assuming
the same physical grip between the tires and the road. Therefore, if we try to get
the classical handling curve we will end up with a different curve for each testing
condition.

Each test at constant forward speed and variable steer angle will yield a different
curve for each speed, as shown in Fig. 8.15.

Tests at constant steer angle and variable forward speed will produce a different
set of curves, as shown in Fig. 8.16.

Fig. 8.15 Formula car with
open differential: different
handling curves obtained
from constant-speed,
variable-steer tests

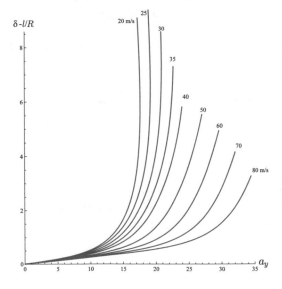

Fig. 8.16 Formula car with *open* differential: handling curves obtained in constant speed, variable-steer tests (green) and constant-steer, variable-speed tests (black)

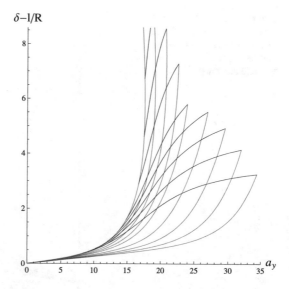

The definition (7.117) of the understeer gradient K as the slope of the handling curve cannot be applied: there is not a single handling curve anymore. An apparently fundamental concept, like K, has proven to be very weak indeed. As a result, we do not have an official definition of understeer/oversteer for a car with aerodynamic downforces.

Incidentally, beware that the counterpart of (7.92) is

$$\frac{Y_1(\alpha_1, u)}{Z_1^0} = \frac{Y_2(\alpha_2, u)}{Z_2^0} = \frac{\tilde{a}_y}{g} \tag{8.53}$$

where the denominators still include the *static* loads acting on each axle.

Locking the differential completely affects these curves, but not much, as shown in Fig. 8.17 (the aerodynamics is more influential). The main difference between Figs. 8.15 and 8.17 is, perhaps, that all curves in case of open differential share the same slope near the origin of the reference system, whereas in case of locked differential each one has a different slope.

8.6 Steady-State Handling Analysis

It is customary in vehicle dynamics to start with the steady-state analysis, that is with all time-derivatives in the governing equations (8.34) set equal to zero. That means having the vehicle going round along a circle of constant radius at constant forward speed, lateral speed and yaw rate. In practice, it is much more convenient to do a *slowly increasing steer* maneuver, also called *constant-speed, variable-steer*

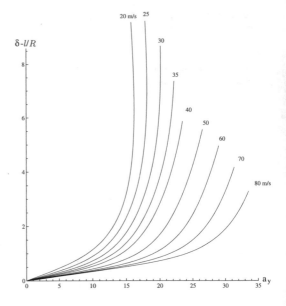

Fig. 8.17 Formula car with *locked* differential: different handling curves obtained in constant-speed, variable-steer tests

test. The vehicle is almost in steady-state conditions, but the test procedure is much faster.

Exactly like in Sect. 7.7, the whole steady-state behavior is described by the maps

$$\rho = \frac{r}{u} = \rho_p(\delta_v, \tilde{a}_y) \quad \text{and} \quad \beta = \frac{v}{u} = \beta_p(\delta_v, \tilde{a}_y) \tag{8.54}$$

which, besides being important by themselves, make it also possible to unambiguously define the *gradients*

$$\text{grad } \rho_p(\delta_v, \tilde{a}_y) = \left(\frac{\partial \rho_p}{\partial \tilde{a}_y}, \frac{\partial \rho_p}{\partial \delta_v} \right) = (\rho_y, \rho_\delta)$$

$$\text{grad } \beta_p(\delta_v, \tilde{a}_y) = \left(\frac{\partial \beta_p}{\partial \tilde{a}_y}, \frac{\partial \beta_p}{\partial \delta_v} \right) = (\beta_y, \beta_\delta) \tag{7.99'}$$

All these quantities are well defined in any real vehicle, including race cars.

Equivalently, we can use the maps

$$\rho = \hat{\rho}_p(u, \delta_v) \quad \text{and} \quad \beta = \hat{\beta}_p(u, \delta_v) \tag{8.55}$$

with *gradients*

$$\mathrm{grad}\,\hat{\rho}_p(u, \delta_v) = \left(\frac{\partial \hat{\rho}_p}{\partial u}, \frac{\partial \hat{\rho}_p}{\partial \delta_v}\right) = (\hat{\rho}_u, \hat{\rho}_\delta)$$

$$\mathrm{grad}\,\hat{\beta}_p(u, \delta_v) = \left(\frac{\partial \hat{\beta}_p}{\partial u}, \frac{\partial \hat{\beta}_p}{\partial \delta_v}\right) = (\hat{\beta}_u, \hat{\beta}_\delta)$$

(8.56)

After performing the standard steady-state tests, all these gradient components are known functions.

Let us consider a specific steady-state (trim) condition, and compute the gradient components there. It is worth noting that, in general, $\hat{\beta}_\delta(u, \delta_v) \neq \beta_\delta(\delta_v, \tilde{a}_y)$ and $\hat{\rho}_\delta(u, \delta_v) \neq \rho_\delta(\delta_v, \tilde{a}_y)$. The derivatives with respect to δ_v with u kept constant are not the same as those with $\tilde{a}_y$ kept constant. This is quite an intuitive result.

In (8.54) we have omitted, with respect to (7.98) and (7.112), the r.h.s. terms, that is those terms involving the apparent slip angles α_1 and α_2 (7.51) and the steering angles. This has been done for greater generality, because α_1 and α_2 are not well defined, unless we assume $\varepsilon_1 = \varepsilon_2 = 0,$[4] as in (7.49). But the key point is that the apparent slip angles α_1 and α_2, even if well defined, are no longer functions of the lateral acceleration $\tilde{a}_y$ only. This aspect has a lot of important consequences. For instance, as already mentioned in Sect. 8.5.2, the classical *handling diagram* [10–12] does not exist any more. At the very least, it has to be replaced by the *handling surface*, first defined and discussed in [4–6].

However, the new global approach to handling evaluation, called **MAP**, introduced in Chap. 6 and also discussed in Sect. 7.8, turns out to be more general, and very informative for race cars as well, as will be shown shortly. The analysis is particularly interesting when aerodynamics is taken into account.

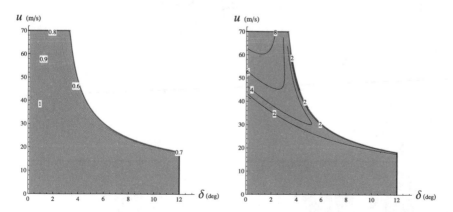

Fig. 8.18 Level curves in the plane (δ, u) for constant damping ratio ζ (left) and constant damped natural frequency ω_s (right)

---

[4] However, many race cars do have $\varepsilon_1 = \varepsilon_2 = 0$, that is parallel steering.

We remark that *level curves* of any physical quantity can be drawn on input achievable regions. For instance, in Figs. 8.18 and 8.19, we see level curves for constant damping ratio ζ, defined in (7.134), and constant damped natural frequency ω_s (rad/s), defined in (7.136). We have at a glance a clear and complete picture of how the dynamic features of the vehicle evolve when changing the steer angle δ, the forward speed u, and the lateral acceleration a_y. In the lower part of the achievable regions, the vehicle behavior is overdamped and hence not oscillatory. On the other hand, when approaching the a_y-limited boundary the damping ratio tends to zero.

To better appreciate the interplay between ω_s and ζ, we can draw the achievable region in the plane (ω_s, ζ) (Fig. 8.20), with red lines at constant speed and black lines at constant steer angle. Of course, this achievable region only covers the oscillatory behavior of the vehicle.

8.6.1 Map of Achievable Performance (MAP)

The global approach MAP was introduced in Chap. 6, and applied in Sect. 7.8 to road cars, that is cars without any significant aerodynamic downforces and with open differential. However, this approach is completely general, and its application to race cars is straightforward. More precisely, MAPs for road cars and race cars are qualitatively the same, differing only quantitatively.

The basic idea, as discussed on Sect. 7.8, is to employ the maps $\mathbb{R}^2 \to \mathbb{R}^2$ to monitor the vehicle at steady state. This is a more general point of view than the handling surface (not to mention the handling diagram).

The maps in this section are typical for a Formula 1 car, year 2013. As usual, all quantities are in SI units, except angles that are in degrees.

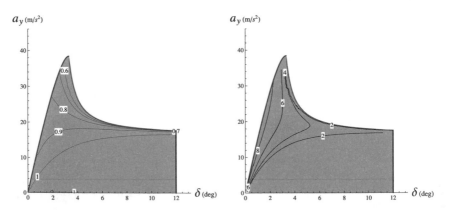

Fig. 8.19 Level curves in the plane $(\delta, \tilde{a}_y)$ for constant damping ratio ζ (left) and constant damped natural frequency ω_s (right)

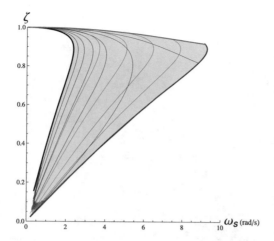

Fig. 8.20 Level curves for constant δ (black lines) and constant u (red lines) in the plane (ω_s, ζ)

8.6.1.1 ρ–δ MAP (Curvature–Steer Angle)

A fairly practical, intuitive MAP is, perhaps, the curvature $\rho = r/u$ vs wheel steer angle δ (although we could employ the steering wheel angle δ_v as well). In Fig. 8.21 we can see the lines at constant speed u, ranging from 20 to 80 m/s, and also the lines at constant lateral acceleration $\tilde{a}_y$, in case of open differential. In Fig. 8.22, we have the same picture, but for locked differential.

Lines at constant speed for open and locked differential are compared in Fig. 8.23. As expected, the locked differential makes the car turn on bigger radii (hence smaller values of ρ).

The strong influence of aerodynamics on the handling of the vehicle is highlighted by the pattern of the lines at constant lateral acceleration. Going back to Fig. 7.44, that is to the map for an ordinary road vehicle, we see that each line at constant $\tilde{a}_y$ intersects all lines at constant u. That means that the level of lateral acceleration that can be achieved is not affected by the forward speed (no wings). On the other hand, in Figs. 8.21 and 8.22, only lines up to about 16 m/s^2 intersect all constant speed lines. The lines for $\tilde{a}_y > 16$ m/s^2 only intersect lines for sufficiently high speed. Indeed, 1.6 is about the grip coefficient between the tire and the road, that is the "physical grip". The grip that does not need any aerodynamic contribution. Higher values of apparent grip do indeed need aerodynamic downforce and hence they can be achieved only for sufficiently high values of the forward speed u. The map shows this fact, and does so in a clear and global way. A close-up is shown in Fig. 8.24 for better clarity.

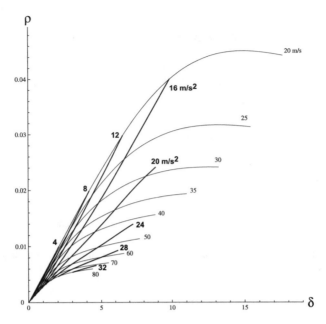

Fig. 8.21 ρ–δ MAP of a *Formula 1* car with *open* differential. Curves at constant speed u and curves at constant lateral acceleration $\tilde{a}_y$

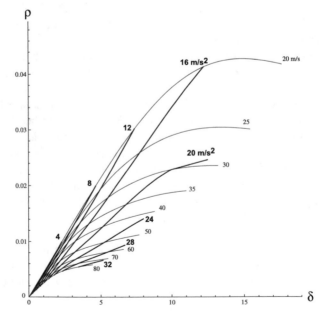

Fig. 8.22 ρ–δ MAP of a *Formula 1* car with *locked* differential. Curves at constant speed u and curves at constant lateral acceleration $\tilde{a}_y$

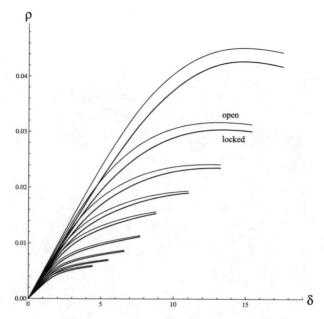

Fig. 8.23 Comparison between Figs. 8.21 and 8.22 for lines at constant speed

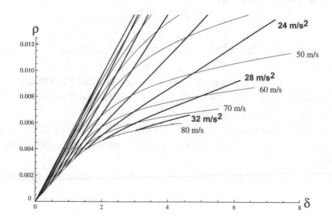

Fig. 8.24 Close-up of Fig. 8.21

8.6.1.2 β–ρ MAP (Vehicle Slip Angle–Curvature)

Also interesting is the handling β–ρ MAP, that is vehicle slip angle vs curvature. The lines at constant speed u and the lines at constant lateral acceleration $\tilde{a}_y$ are shown in Fig. 8.25. Again, only lines for $\tilde{a}_y < 16\,\text{m/s}^2$ intersect all lines at constant speed, thus indicating that 1.6 is indeed the physical grip (of course we could be more precise

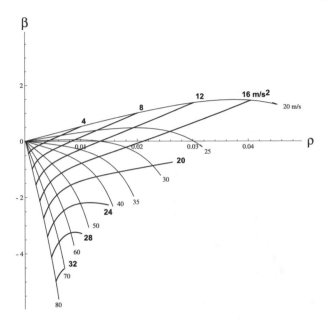

Fig. 8.25 β–ρ MAP for a Formula 1 car with open differential. Curves at constant speed u and curves at constant lateral acceleration $\tilde{a}_y$

by drawing more lines). Therefore, we have a tool to obtain a good approximation of the physical grip.

Looking at the overall picture, we see how the control parameters u and δ are related to curvature and vehicle slip angle. For instance, if $u > 30\,\text{m/s}$, we have basically $\beta \leq 0$ (in a left turn) at any speed.

Lines at constant steer angle are shown in Fig. 8.26. Looking at the slope of these curves, it immediately arises that the vehicle is more understeer at low speeds than at high speeds.

To help the reader catch other features in this MAP, all lines are shown in Fig. 8.27.

8.6.1.3 Comparison of Setups

Another interesting application of the MAPs is to compare setups. This is done in Figs. 8.28 and 8.29 for two setups which have different aero balances. The second setup (dashed lines) has higher aerodynamic load on the front axle and less aerodynamic load on the rear axle.

Very interesting is to observe that the lines at constant $\tilde{a}_y$ that are more affected are precisely those that need higher aerodynamic downforces to be achieved (Fig. 8.28).

From Fig. 8.29 we see that the new aero balance does not affect the lines at constant δ in a uniform way. This may help understand which setup is faster for a given circuit.

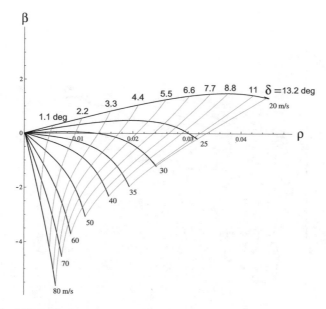

Fig. 8.26 β–ρ MAP or a Formula 1 car with open differential. Curves at constant speed u and curves at constant steer angle δ

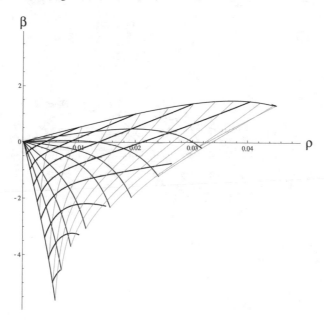

Fig. 8.27 β–ρ MAP for a Formula 1 car with open differential. Superimposition of Figs. 8.25 and 8.26

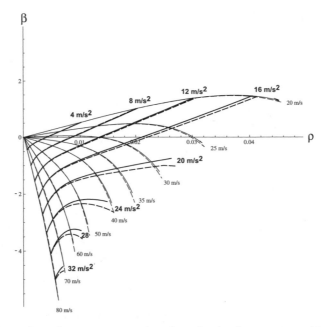

Fig. 8.28 Comparison of curves at constant lateral acceleration for two setups with different aero balance

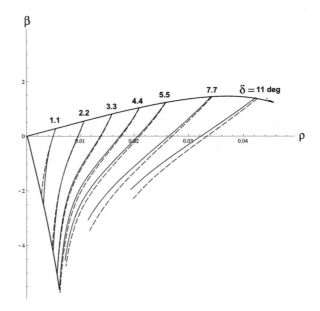

Fig. 8.29 Comparison of curves at constant steer angle for two setups with different aerodynamic balance (higher for dashed lines)

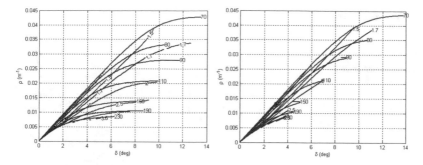

Fig. 8.30 Effects of different roll balance [13]

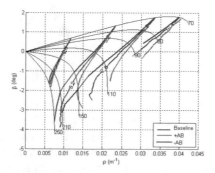

Fig. 8.31 Effects of different aero balance [13]

8.6.2 MAPs from Real Cases

Here we present a few MAPs, obtained with very sophisticated vehicle models [13], just to show the effects of different roll balance or different aero balance (Figs. 8.30 and 8.31).

8.6.3 Power-Off and Power-On

So far we have considered steady-state conditions. However, a Formula car is almost always under transient conditions, with the driver acting on the gas and/or brake pedals. The MAPs can be useful to monitor what is going on also during these more general working conditions. The trick is to do, e.g., constant-speed, variable-steer simulations as if the car were constantly going uphill or downhill. This way, we have, strictly speaking, steady-state conditions, but the loads on the tires are pretty much like if the car were accelerating or slowing down with the engine (no braking), that is during power-on and power-off conditions.

During power-off and power-on, the differential of a Formula 1 car is locked. Therefore, all figures in this section are for locked differential.

Fig. 8.32 Curves at constant
speed in the ρ–δ MAP for a
Formula 1 car during
power-off (dashed lines) and
power-on (solid lines)

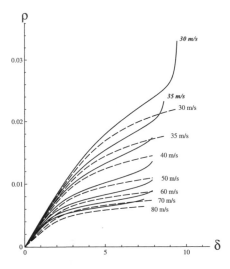

A few figures are provided to show how the MAPs can be used to have a global
view of the vehicle behavior even under pseudo-transient conditions. Figure 8.32
shows the ρ–δ map with lines at constant speed during power-off (dashed lines) and
power-on (solid lines). Speeds below 30 m/s have been omitted. The two cases are for
a longitudinal acceleration of ± 0.5 m/s^2. Figure 8.33 shows the comparison between
power-off (dashed lines) and power-on (solid lines) in the plane β–ρ. At high steer
angles and relatively low speeds there are, as expected, very big differences.

During power-on, the locked differential generates a yawing moment that can
have either the same sign as the yaw rate (Fig. 8.34) or opposite sign (Fig. 8.35),
depending on the operating conditions of the vehicle.

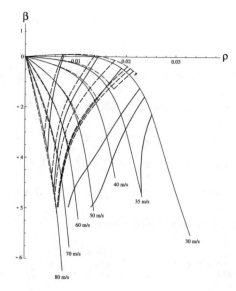

Fig. 8.33 Lines at constant u and constant δ in the β–ρ MAP for a Formula 1 car during power-off (dashed lines) and power-on (solid lines)

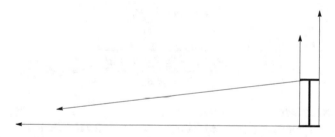

Fig. 8.34 Power-on with locked differential: forces received from the road at $u = 40\,\text{m/s}$ and $\delta = 7°$

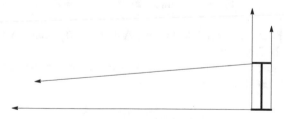

Fig. 8.35 Power-on with locked differential: forces received from the road at $u = 40\,\text{m/s}$ and $\delta = 5°$

8.7 Stability Derivatives and Control Derivatives

The analysis about the transient behavior developed in Sect. 7.10 applies entirely, even for cars with significant aerodynamic loads and, possibly, limited-slip differential.

We recall that the vehicle behaves as a mechanical vibrating system with

$$\text{equivalent mass} \qquad m_e = J_z m u_a^2$$

$$\text{equivalent damping} \quad c_e = -u_a(m N_\rho + J_z Y_\beta) \tag{8.57}$$

$$\text{equivalent stiffness} \quad k_e = (Y_\beta N_\rho - Y_\rho N_\beta) + m u_a^2 N_\beta$$

Therefore, the vehicle has a damping ratio ζ

$$\zeta = \frac{c_e}{2\sqrt{m_e k_e}} = -\frac{m N_\rho + J_z Y_\beta}{2\sqrt{J_z m}\sqrt{(Y_\beta N_\rho - Y_\rho N_\beta) + m u_a^2 N_\beta}} \tag{8.58}$$

If $\zeta < 1$, the system has a free oscillation with damped natural angular frequency ω_s

$$\omega_s = \omega_n \sqrt{1 - \zeta^2} \tag{8.59}$$

where

$$\omega_n = \sqrt{\frac{k_e}{m_e}} = \sqrt{\frac{(Y_\beta N_\rho - Y_\rho N_\beta) + m u_a^2 N_\beta}{J_z m u_a^2}} = \sqrt{\det(\mathbf{A})} \tag{8.60}$$

is the undamped natural angular frequency.

There is an important difference with respect to road cars. Owing to (8.48), here we have (cf. (7.143))

$$Y_u \neq 0 \quad \text{and, maybe} \quad N_u \neq 0 \tag{8.61}$$

That is, the tire forces are speed dependent. A possible (typical?) pattern of Y_u for a F1 car is shown in Fig. 8.36.

We recall an important result concerning the stability of the car. From (7.131), the mathematical condition for instability is

$$\omega_n^2 J_z m u_a^2 = \det(\mathbf{A}) J_z m u_a^2$$
$$= (Y_\beta N_\rho - Y_\rho N_\beta) + m u_a^2 N_\beta < 0 \tag{8.62}$$

where $(Y_\beta N_\rho - Y_\rho N_\beta) > 0$. Therefore, instability may occur only if $N_\beta < 0$. The sign of N_β can be taken as a reliable way to define understeer ($N_\beta > 0$, as in Fig. 8.37) and oversteer ($N_\beta < 0$, as in Fig. 8.38).

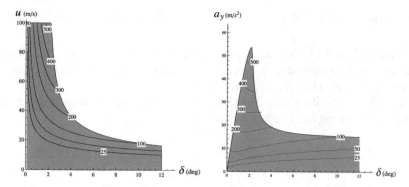

Fig. 8.36 Understeer F1 car: MAPs of Y_u

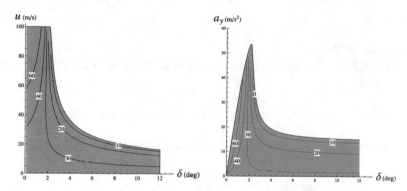

Fig. 8.37 Understeer F1 car: MAPs of $N_\beta/1000$

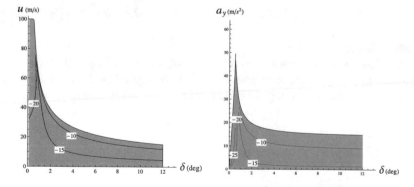

Fig. 8.38 Oversteer F1 car: MAPs of $N_\beta/1000$

It is also worth noting the different shapes of the achievable regions in Figs. 8.37 and 8.38, particularly in the $(\delta, \tilde{a}_y)$ plane.

8.8 Stability Derivatives from Steady-State Gradients

We refer to Sect. 7.11 for a general introduction to this crucial topic. Very briefly, here we look for a rigorous link between the dynamic behavior (time histories) and the steady-state results (no time involved).

According to (8.61), we have to generalize (7.162). The solutions of the systems of Eqs. (7.156) and (7.157) are as follows [8]

$$Y_\beta = \frac{Y_\delta \rho_y + m\rho_\delta + Y_u(u_\delta \rho_y - u_y \rho_\delta)}{\beta_y \rho_\delta - \beta_\delta \rho_y} \qquad Y_\rho = -\frac{Y_\delta \beta_y + m\beta_\delta + Y_u(u_\delta \beta_y - u_y \beta_\delta)}{\beta_y \rho_\delta - \beta_\delta \rho_y}$$

$$N_\beta = \frac{N_\delta \rho_y + N_u(u_\delta \rho_y - u_y \rho_\delta)}{\beta_y \rho_\delta - \beta_\delta \rho_y} \qquad N_\rho = -\frac{N_\delta \beta_y + N_u(u_\delta \beta_y - u_y \beta_\delta)}{\beta_y \rho_\delta - \beta_\delta \rho_y}$$

$$(8.63)$$

where $u_y = \partial u / \partial \tilde{a}_y$ etc.

We know that $Y_\rho = N_\beta$ (Maxwell's Reciprocal Theorem). Therefore, strange as it may appear,

$$-(Y_\delta \beta_y + m\beta_\delta + Y_u(u_\delta \beta_y - u_y \beta_\delta)) = N_\delta \rho_y + N_u(u_\delta \rho_y - u_y \rho_\delta) \qquad (8.64)$$

The mathematical condition (8.62) for instability becomes

$$\frac{1}{\beta_y \rho_\delta - \beta_\delta \rho_y} \left\{ N_\delta \left[m(u_a^2 \rho_y - 1) + Y_u u_y \right] \right.$$

$$\left. + N_u \left[u_\delta m(u_a^2 \rho_y - 1) - u_y(Y_\delta + mu_a^2 \rho_\delta) \right] \right\} < 0 \quad (8.65)$$

where, typically, $N_u \simeq 0$.

Since $Y_u u_y > 0$, we see that we may get in trouble even with $\rho_y < 0$, that is with an apparently understeer vehicle (see Sect. 7.13). This is quite an unexpected outcome, particularly if compared with (7.172) for road cars (i.e., when setting $Y_u = N_u = 0$).

8.8.1 Alternative Independent Variables

Instead of $(\delta_{va}, \tilde{a}_y)$, someone may prefer to use (u_a, δ_{va}) as independent variables, like in (8.55).

At steady state we have

$$Y_0(u_a, \delta_{va}) = Y(\hat{\beta}_p(u_a, \delta_{va}), \hat{\rho}_p(u_a, \delta_{va}); u_a, \delta_{va}) = mu_a^2 \hat{\rho}_p(u_a, \delta_{va})$$

$$(8.66)$$

$$N_0(u_a, \delta_{va}) = N(\hat{\beta}_p(u_a, \delta_{va}), \hat{\rho}_p(u_a, \delta_{va}); u_a, \delta_{va}) = 0$$

Taking the partial derivatives we obtain

$$\frac{\partial Y_0}{\partial u_a} = Y_\beta \frac{\partial \hat{\beta}_p}{\partial u_a} + Y_\rho \frac{\partial \hat{\rho}_p}{\partial u_a} + Y_u = m \left(2 u_a \rho + u_a^2 \frac{\partial \hat{\rho}_p}{\partial u_a} \right)$$

$$\frac{\partial Y_0}{\partial \delta_{va}} = Y_\beta \frac{\partial \hat{\beta}_p}{\partial \delta_{va}} + Y_\rho \frac{\partial \hat{\rho}_p}{\partial \delta_{va}} + Y_\delta = m u_a^2 \frac{\partial \hat{\rho}_p}{\partial \delta_{va}}$$

(8.67)

and

$$\frac{\partial N_0}{\partial u_a} = N_\beta \frac{\partial \hat{\beta}_p}{\partial u_a} + N_\rho \frac{\partial \hat{\rho}_p}{\partial u_a} + N_u = 0$$

(8.68)

$$\frac{\partial N_0}{\partial \delta_{va}} = N_\beta \frac{\partial \hat{\beta}_p}{\partial \delta_{va}} + N_\rho \frac{\partial \hat{\rho}_p}{\partial \delta_{va}} + N_\delta = 0$$

These two systems of equations

$$\begin{bmatrix} \hat{\beta}_u & \hat{\rho}_u \\ \hat{\beta}_\delta & \hat{\rho}_\delta \end{bmatrix} \begin{bmatrix} Y_\beta \\ Y_\rho \end{bmatrix} = \begin{bmatrix} m(2u_a\rho + u_a^2\hat{\rho}_u) - Y_u \\ mu_a^2\hat{\rho}_\delta - Y_\delta \end{bmatrix}$$

and

$$\begin{bmatrix} \hat{\beta}_u & \hat{\rho}_u \\ \hat{\beta}_\delta & \hat{\rho}_\delta \end{bmatrix} \begin{bmatrix} N_\beta \\ N_\rho \end{bmatrix} = \begin{bmatrix} -N_u \\ -N_\delta \end{bmatrix}$$

(8.69)

have the same matrix, whose coefficients are the four components of the *gradients* (8.56). After performing the standard steady-state tests on a vehicle, all these gradient components are known functions.

The four *stability derivatives* at steady state are the solutions of the two systems of Eq. (8.69) [8]

$$Y_\beta = \frac{Y_\delta \hat{\rho}_u - Y_u \hat{\rho}_\delta + 2m u_a \rho \hat{\rho}_\delta}{\hat{\beta}_u \hat{\rho}_\delta - \hat{\beta}_\delta \hat{\rho}_u}$$

$$Y_\rho = -\frac{Y_\delta \hat{\beta}_u - Y_u \hat{\beta}_\delta + m u_a [2\rho \hat{\beta}_\delta - u_a(\hat{\beta}_u \hat{\rho}_\delta - \hat{\beta}_\delta \hat{\rho}_u)]}{\hat{\beta}_u \hat{\rho}_\delta - \hat{\beta}_\delta \hat{\rho}_u}$$

and

(8.70)

$$N_\beta = \frac{N_\delta \hat{\rho}_u - N_u \hat{\rho}_\delta}{\hat{\beta}_u \hat{\rho}_\delta - \hat{\beta}_\delta \hat{\rho}_u}$$

$$N_\rho = -\frac{N_\delta \hat{\beta}_u - N_u \hat{\beta}_\delta}{\hat{\beta}_u \hat{\rho}_\delta - \hat{\beta}_\delta \hat{\rho}_u}$$

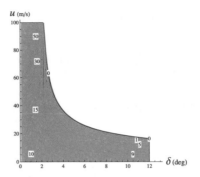

Fig. 8.39 Understeer F1 car: MAP of $\det(\mathbf{A})\, J_z m u_a^2 \times 10^{-11}$, that is of (8.62), (8.65), and (8.72)

where, again, $Y_\rho = N_\beta$. Therefore, it has to be

$$Y_\delta \hat{\beta}_u - Y_u \hat{\beta}_\delta + m u_a [2\rho \hat{\beta}_\delta - u_a (\hat{\beta}_u \hat{\rho}_\delta - \hat{\beta}_\delta \hat{\rho}_u)] = N_\delta \hat{\rho}_u - N_u \hat{\rho}_\delta \qquad (8.71)$$

In the present formulation, the mathematical condition (8.62) for instability becomes

$$\frac{Y_u N_\delta - N_u Y_\delta - 2 m u_a \rho N_\delta}{\hat{\beta}_u \hat{\rho}_\delta - \hat{\beta}_\delta \hat{\rho}_u} < 0 \qquad (8.72)$$

where the denominator is < 0.

Strangely enough, (8.72) looks completely different from (8.65). It involves other functions. Obviously, all three Eqs. (8.62), (8.65), and (8.72) provide the same results. An example of a MAP of these results is shown in Fig. 8.39 for an understeer Formula car.

From a practical point of view, we observe that (8.62), (8.65), and (8.72) involve very different quantities, which may not be equally measurable. Therefore, these equations may not be equivalent on track.

Maybe a more intuitive and physical way to understand the dynamic behavior of a Formula car, including its stability, is to look at the damped natural frequency $\omega_s/(2\pi)$. An example is shown in Fig. 8.40.

As well known, $\omega_s = \omega_n \sqrt{1 - \zeta^2}$. Therefore, we also provide the corresponding MAPs of $\omega_n/(2\pi)$ in Fig. 8.41, and of ζ in Fig. 8.42.

It is worth noting the very different pattern of ω_s and ω_n. This is due to the very high values of ζ.

Stability is an important issue, particularly for oversteer cars. Indeed, we see in Fig. 8.43 that increasing the level of oversteer makes the achievable region in the plane $(u, \tilde{a}_y)$ smaller and smaller, with strong limitations on the achievable performances. On the contrary, achievable regions for understeer vehicles are much less sensitive, as shown in Fig. 6.9.

Once again, we stress that three different, fairly general, approaches have been provided to obtain and to monitor important parameters concerning the dynamic

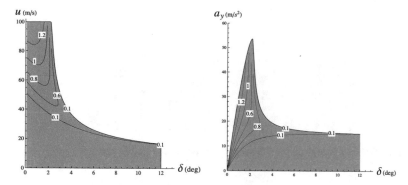

Fig. 8.40 Understeer F1 car: MAPs of the damped natural frequency $\omega_s/(2\pi)$

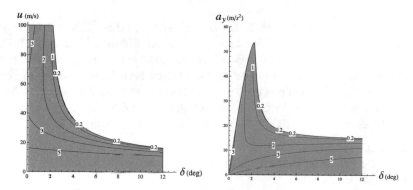

Fig. 8.41 Understeer F1 car: MAPs of the undamped natural frequency $\omega_n/(2\pi)$

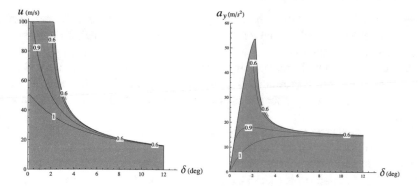

Fig. 8.42 Understeer F1 car: MAPs of the damping ratio ζ

behavior of race cars. Moreover, the MAP approach provides at a glance the whole range of achievable performance/behavior.

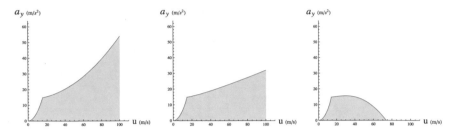

Fig. 8.43   Achievable regions for increasingly oversteer cars (left to right)

8.9   Comparison of Limited-Slip Differentials

Most race cars are equipped with limited-slip differentials (LSD). The relevant equations for limited-slip differentials were obtained and discussed in Sect. 3.13. Here we compare different types of differentials in constant-speed, slowly-increasing steer maneuvers. The goal is to provide some examples that can help understand intuitively the behavior of differential mechanisms. Of course, we ultimately have to resort on the equations presented in Sect. 3.13, but trying to guess the result is good exercise.

Four types of differentials are considered (notation as in Sect. 3.13):

1. open differential: $M_l = M_r$;
2. geared LSD with *constant* internal efficiency $\eta_h = 0.5$, that is TBR $= 2$ (Fig. 3.59);
3. clutch-pack LSD with *constant* difference of torques $\Delta M = 400\,\text{Nm}$ (Fig. 3.60);
4. locked differential: $\omega_l = \omega_r$.

They were selected because they span a wide range of differentials, and because of their relatively simple behavior. Usually, more sophisticated differentials are employed in competitions.

A double track vehicle model, with parallel steering, is used to mimic a Formula car. However, the results presented here have only a qualitative relevance. We do not claim to be quantitative. There are far too many parameters to accurately model a Formula car.

We perform constant-speed, slowly-increasing steer maneuvers at 30, 40 and 50 m/s. In each test we monitor the vehicle behavior as a function of the lateral acceleration a_y. In particular, we plot the longitudinal forces $F_{x_{21}}$ and $F_{x_{22}}$, the difference of torques $\Delta \widetilde{M}$ as defined in (3.183), the difference of angular speeds $\Delta \widetilde{\omega}$ as defined in (3.164), and also the power loss W_d as in (3.174) and (3.180).

It is kind of interesting to compare the behavior of the four differentials listed above.

8.9.1 Yaw Moment

According to (3.184), the yaw moment $\Delta X_2 t_2$ is directly related to the difference of torques $\Delta \widetilde{M}$. The yaw moment $\Delta X_2 t_2$ is the most relevant effect of non-open differentials.

In Fig. 8.44 we compare $\Delta \widetilde{M}$ as obtained in the test at 30 m/s. At low values of a_y all non-open differentials are locked. Pretty soon, the geared differential unlocks (green line). After a while, also the clutch-pack differential unlocks (red line). We clearly see that, in a wide range, the values of $\Delta \widetilde{M}$ are much lower with respect to the locked differential (blue line). At high values of a_y, all differentials lock again.

Similar results are obtained in the test at 40 m/s, as shown in Fig. 8.45, but with the important difference that in the final part, where all differentials are locked, $\Delta \widetilde{M}$ becomes positive: the outer wheel pushes more than the inner wheel.

The test at 50 m/s (Fig. 8.46) shows that the geared differential is almost always locked.

Obviously, with an open differential we have $\Delta \widetilde{M} = 0$.

Fig. 8.44 Test at 30 m/s. Difference of torques: (2—green) geared differential, (3—red) clutch-pack differential, (4—blue) locked differential

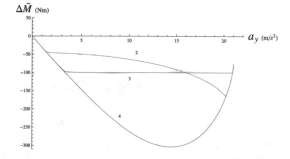

Fig. 8.45 Test at 40 m/s. Difference of torques: (2—green) geared differential, (3—red) clutch-pack differential, (4—blue) locked differential

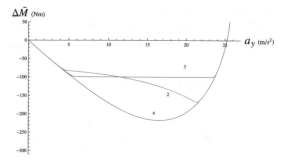

Fig. 8.46 Test at 50 m/s.
Difference of torques:
(2—green) geared
differential, (3—red)
clutch-pack differential,
(4—blue) locked differential

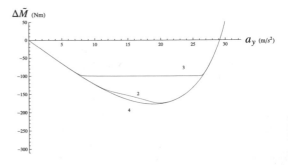

Fig. 8.47 Test at 40 m/s.
Difference of angular
speeds: (1—black) open
differential, (2—green)
geared differential, (3—red)
clutch-pack differential

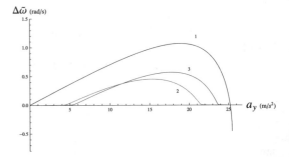

8.9.2 Difference of Angular Speeds

The difference $\Delta\tilde{\omega}$ of angular speeds in the test at 40 m/s is shown in Fig. 8.47. As
expected, the open differential (black line) exhibits the higher values of $\Delta\tilde{\omega}$. It is
worth noting that, in the final part, the open differential lets the inner wheel rotate
faster than the outer wheel.

Obviously, with a locked differential we have $\Delta\tilde{\omega} = 0$.

8.9.3 Internal Power Loss

Whenever the differential action is activated ($\Delta\tilde{\omega} \neq 0$), LSD are characterized by an
internal power loss $W_d = \Delta\tilde{M}\,\Delta\tilde{\omega}$. In the cases under investigation, similar values of
W_d are obtained at 30 m/s (Fig. 8.48) and at 40 m/s (Fig. 8.49). At 50 m/s the geared
LSD is almost always locked (Fig. 8.46) and hence $W_d \simeq 0$. However, $W_d < 100$ W
in all cases considered here.

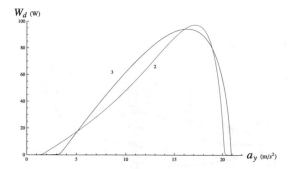

Fig. 8.48 Test at 30 m/s. Power loss: (2—green) geared differential, (3—red) clutch-pack differential

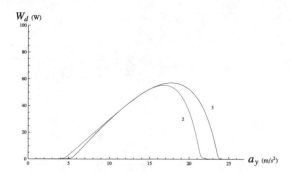

Fig. 8.49 Test at 40 m/s. Power loss: (2—green) geared differential, (3—red) clutch-pack differential

8.9.4 Longitudinal Forces

The longitudinal forces in the test at 40 m/s are shown in Fig. 8.50. It is interesting to compare the behavior of different types of differential. Noteworthy is the common point at $a_y \simeq 25 \, \text{m/s}^2$.

Examples of longitudinal and lateral forces are shown in Figs. 8.51, 8.52, 8.53 and 8.54. In all cases, the forward speed $u = 40 \, \text{m/s}$ and the front steer angle $\delta = 3$ deg. Of course, the lateral acceleration is not the same because the trajectories are a little different.

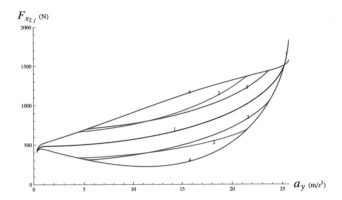

Fig. 8.50 Test at 40 m/s. Longitudinal forces: (1—black) open differential, (2—green) geared differential, (3—red) clutch-pack differential, (4—blue) locked differential

Fig. 8.51 Open differential. Test at 40 m/s with $\delta = 3$ deg

Fig. 8.52 Geared LSD. Test at 40 m/s with $\delta = 3$ deg

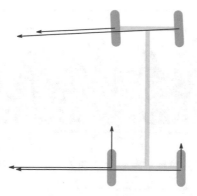

Fig. 8.53 Clutch-pack LSD. Test at 40 m/s with $\delta = 3$ deg

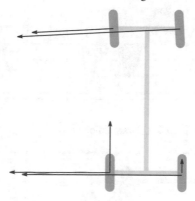

Fig. 8.54 Locked differential. Test at 40 m/s with $\delta = 3$ deg

8.10 Exercises

8.10.1 Vehicle Kinematic Equations

The vehicle kinematic equations, introduced in Sect. 3.2, are relationships between kinematic quantities. Some of these kinematic quantities (telemetry data) are measured directly on the vehicle. They can then be combined to compute other quantities (mathematical channels). We remind that kinematics is the branch of mechanics that describes the motion of objects, but not the forces involved.

Playing with real world quantities helps develop quantitative reasoning. Therefore, let us consider the telemetry data measured directly in a race car (Dallara GP2) during one lap of the Barcelona circuit (sample rate is 100 Hz):

- forward velocity u of G (Fig. 8.55);
- vehicle slip angle $\hat{\beta}$ at G (Fig. 8.56);
- yaw rate r (Fig. 8.57);

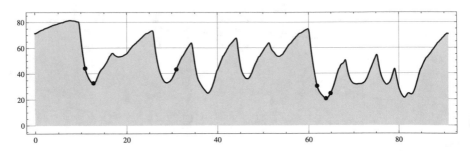

Fig. 8.55 Forward velocity u, in m/s, versus time, in s

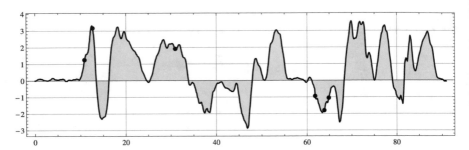

Fig. 8.56 Vehicle slip angle $\hat{\beta}$, in deg, versus time

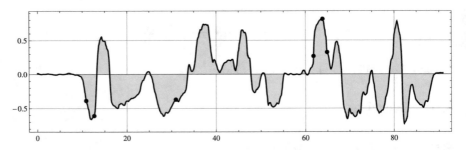

Fig. 8.57 Vehicle yaw rate r, in rad/s, versus time

- longitudinal acceleration a_x of G (Fig. 8.59);
- lateral acceleration a_y of G (Fig. 8.60);
- front wheel steer angle δ_1 (Fig. 8.61).

All data in Figs. 8.55, 8.56, 8.57, 8.58, 8.59, 8.60 and 8.61 were filtered to reduce noise.

According to (3.16) and (3.18), the lateral velocity $v(t)$ of G is promptly obtained as

$$v(t) = u(t) \tan \hat{\beta}(t) \tag{8.73}$$

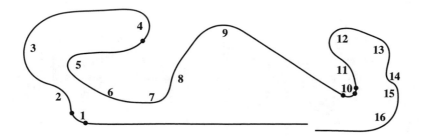

Fig. 8.58 Computed trajectory of the center of mass of a GP2 car. Also shown the six positions of G corresponding to the instants of time of Tables 8.1 and 8.2

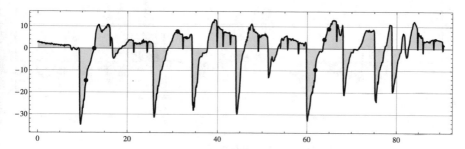

Fig. 8.59 Longitudinal acceleration a_x, in m/s^2, versus time

As shown in (3.8), the integral of $r(t)$ provides the vehicle yaw angle $\psi(t)$. Then, as discussed in Sect. 3.2.2, the trajectory of G can be obtained. The final result is shown in Fig. 8.58.

Typically, the computed trajectory is not a closed curve. This is due to the accumulation of small errors and noise which inevitably affect the measured data. However, in this case the result is pretty good. In Fig. 8.58 all turns of the Barcelona circuit are numbered sequentially, as customary.

The accelerations a_x and a_y are also measured directly, thus avoiding the very unreliable computation of $\dot{u}$ and $\dot{v}$, if we had to employ (3.26) and (3.27).

The plot of the longitudinal acceleration $a_x(t)$ is shown in Fig. 8.59 (negative values mean braking). It is worth noting how sharp the transitions are whenever braking begins to be applied.

The plot of the lateral acceleration $a_y(t)$ is shown in Fig. 8.60. As expected after (3.27), this plot is similar to that of $r(t)$.

Although not strictly necessary in this framework, also the plot of the front wheel steer angle δ_1 is shown (Fig. 8.61). Quite interestingly, $\hat{\beta}$ and δ_1 always have opposite signs.

As discussed in Sect. 3.2, the measured telemetry data shown in Figs. 8.55, 8.56, 8.57, 8.58, 8.59, 8.60 and 8.61 allow the computation, among others, of the following kinematic quantities as mathematical channels:

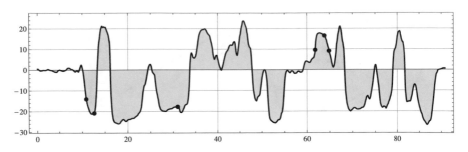

Fig. 8.60 Lateral acceleration a_y, in m/s^2, versus time

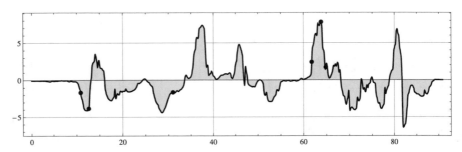

Fig. 8.61 Front wheel steer angle δ_1, in deg, versus time

Table 8.1 Samples of measured telemetry data of a GP2 car

	t (s)	turn no	u (m/s)	$\hat{\beta}$ (deg)	r (rad/s)	a_x (m/s^2)	a_y (m/s^2)	δ_1 (deg)
1	10.87	1	44.21	1.26	−0.39	−14.63	−14.05	−1.68
2	12.66	1	32.67	3.16	−0.61	−0.05	−20.86	−3.84
3	30.98	4	43.16	1.92	−0.37	7.72	−17.77	−1.65
4	61.84	10	30.26	−0.94	0.27	−9.52	9.56	2.41
5	63.87	10	20.64	−1.80	0.81	4.22	16.53	7.81
6	64.87	10	24.60	−1.05	0.33	9.05	9.12	1.71

- lateral velocity v of G (Eq. (8.73));
- coordinates S and R of the velocity center C in the vehicle frame (Eqs. (3.14) and (3.13));
- radius of curvature R_G of the trajectory of the center of mass G (Eq. (3.36));
- ratios β and ρ (Eqs. (3.16) and (3.17));
- tangential acceleration a_t of G (Eq. (3.34));
- centripetal acceleration a_n of G (Eq. (3.35)).

Values of telemetry data measured at six different instants of time during the same lap are listed in Table 8.1. The corresponding positions of the vehicle on the track are marked by black points in Figs. 8.55, 8.56, 8.57, 8.58, 8.59, 8.60 and 8.61. More

Table 8.2 Computed values (mathematical channels) from the telemetry data of Table 8.1

	t (s)	turn no	v (m/s)	S (m)	R (m)	R_G (m)	β (deg)	a_t (m/s^2)	a_n (m/s^2)	$\dot{r}$ (rad/s^2)
1	10.87	1	0.96	2.49	−114.30	−142.50	1.25	−14.93	−13.73	−0.32
2	12.66	1	1.80	2.97	−53.70	−51.40	3.16	−1.20	−20.83	0.31
3	30.98	4	1.45	3.90	−116.40	−103.50	1.92	7.12	−18.02	0.02
4	61.84	10	−0.49	1.83	113.10	97.37	−0.93	−9.67	9.41	1.53
5	63.87	10	−0.65	0.80	25.41	25.60	−1.81	3.69	16.65	−0.08
6	64.87	10	−0.45	1.38	75.67	65.21	−1.04	8.88	9.28	−0.52

precisely, there are two points on turn 1, one point on turn 4 and three points on turn 10. The corresponding values of the mathematical channels listed above are given in Table 8.2. It is strongly recommended to try to figure out what is going on at each of these instants of time.

Values of the angular acceleration $\dot{r}(t)$ are reported in the last column of Table 8.2. As already stated in Sect. 3.2.8, it would be desirable to have sensors to measure $\dot{r}$ directly. For the moment we have to compute it as the numerical derivative of the filtered signal $r(t)$. Quite an arbitrary process, as shown, e.g., in Fig. 5.23.

After this long introduction, here is the exercise.

Discuss, from a kinematic point of view, the data given in Tables 8.1 and 8.2.

Solution

Points 1 and 4 have significant negative values of a_x. Moreover, at both points $|a_x| \simeq |a_y|$. Therefore the driver is still braking while entering the turn.

Point 2 is at the so-called apex of the corner, characterized by $|a_x| \simeq 0$ and maximum $|a_y|$.

Points 3, 5 and 6 have positive values of a_x. Therefore, the vehicle is accelerating. It is also exiting the turn, as confirmed by the values of a_y.

In all cases $|\hat{\beta}|$ is very small, and hence $\hat{\beta} \simeq \beta$. Indeed, the lateral velocity v is much lower than u. Nonetheless, (a_x, a_y) are similar, but not almost equal to (a_t, a_n).

The coordinate S of the velocity center C is always positive. Therefore, the vehicle slip angle $\hat{\beta}$ and the front wheel steer angle δ_1 always have opposite signs. The coordinate R of C is usually quite different from the radius of curvature R_G of the trajectory of G. They get closer to each other when the vehicle is near the apex of the corner.

8.10.2 Spin Slip Contributions

According to (3.62), there are three contributions to the spin slip φ_{ij}. Discuss their relevance in a GP2 car.

Table 8.3 Coordinates of K and components of $\mathbf{a}_C$, both in the vehicle reference frame, according to the telemetry data of Table 8.1 and to the last column in Table 8.2

	t (s)	turn no	GK_x (m)	GK_y (m)	a_{C_x} (m/s^2)	a_{C_y} (m/s^2)
1	10.87	1	-53.74	20.65	-51.45	2.26
2	12.66	1	27.62	-33.50	15.31	-0.08
3	30.98	4	70.08	-121.20	9.02	-1.70
4	61.84	10	-6.52	-5.91	-182.70	4.26
5	63.87	10	9.23	23.96	5.68	-0.30
6	64.87	10	20.39	-13.35	47.97	0.41

Solution

From Fig. 8.57, we see that the yaw rate $|r|$ is always lower than 1 rad/s. From Fig. 8.61, we can estimate that the steer angle rate $|\dot{\delta}_{ij}|$ does not exceed 0.5 rad/s, and it is usually much lower. With a camber reduction factor ε_i of about 0.5 and a camber angle γ_{ij} of, say, 4°, the third term ranges between 1 rad/s and 5 rad/s, depending on the value of the wheel angular speed ω_{ij}.

8.10.3 Acceleration Center and Acceleration of the Velocity Center

Employing the measured values shown in Table 8.1, along with the values of $\dot{r}$ reported in the last column in Table 8.2, compute the coordinates of the acceleration center K and the components of $\mathbf{a}_C$ in the vehicle reference frame.

Solution

We can use (3.46) and (3.48). Results are given in Table 8.3. It is worth noting that, in most cases, K and C are quite far apart.

8.10.4 Aerodynamic Downforces

A GP2 race car has the following features (notation as in Sect. 3.7.2):

- $m = 680\,\mathrm{kg}$;
- $a_1/a_2 = 1.27$, *that is weight distribution front/rear of 0.44/0.56;*
- $S_a\,C_{z1} = 1.5\,\mathrm{m}^2$;
- $S_a\,C_{z2} = 2.1\,\mathrm{m}^2$;
- $S_a\,C_x = 1.1\,\mathrm{m}^2$.

Compute the vertical loads acting on the front axle and on the rear axle when the car is stationary, and when it is running straight at 150 km/h *and at 300* km/h.

Solution

The car total weight is $mg = 6670.8$ N. Therefore, according to (3.105) or (8.14), the vertical static loads are $Z_1^0 = 0.44 \times 6670.8 = 2935.15$ N for the front axle, and $Z_2^0 = 0.56 \times 6670.8 = 3735.65$ N for the rear axle.

We can now employ (3.82) to evaluate the aerodynamic downforces Z_1^a and Z_2^a when the car has a speed of 150 km/h $= 41.67$ m/s. The air density is assumed to be $\rho_a = 1.25$ kg/m^3. After a simple calculation we get $Z_1^a = 1627.47$ N and $Z_2^a = 2278.46$ N.

Therefore, at 150 km/h, according to (3.103) or (8.14), the total vertical load Z_1 acting on the front axle amounts to $Z_1 = Z_1^0 + Z_1^a = 2935.15 + 1627.47 = 4562.62$ N, while for the rear axle we obtain $Z_2 = Z_2^0 + Z_2^a = 3735.65 + 2278.46 = 6014.11$ N.

The drag force X_a at 150 km/h is equal to 1193.48 N.

Now, let us repeat the computation for a speed of 300 km/h $= 83.33$ m/s. For the aerodynamic downforces on each axle we get $Z_1^a = 6509.90$ N and $Z_2^a = 9113.85$ N. Doubling the speed makes the aerodynamic loads four times as much.

At 300 km/h, the total vertical load Z_1 acting on the front axle amounts to $Z_1 = Z_1^0 + Z_1^a = 2935.15 + 6509.90 = 9445.05$ N, while for the rear axle we obtain $Z_2 = Z_2^0 + Z_2^a = 3735.65 + 9113.85 = 12849.50$ N.

It is interesting to check the front/rear balance of the total loads. It was 0.44/0.56 at zero speed, to become 0.43/0.57 at 150 km/h, and 0.42/0.58 at 300 km/h. Indeed, the front/rear balance of the aerodynamic loads alone is 0.42/0.58.

8.10.5 Roll Stiffnesses in Formula Cars

Formula cars, including FSAE cars, have rather flexible tires in the radial direction. The goal of this exercise is to appreciate how much the radial stiffness of the tires can affect the vehicle roll stiffness, and hence the roll motion of the car. This topic is addressed in Sect. 3.10.12. The data are as follows:

- *mass $m = 305$ kg;*
- *front track $t_1 = 1.21$ m;*
- *rear track $t_2 = 1.11$ m;*
- *$a_1 = 0.816$ m;*
- *$a_2 = 0.724$ m;*
- *center of mass height $h = 0.32$ m;*
- *front no-roll center height $q_1 = 0.025$ m;*
- *rear no-roll center height $q_2 = 0.045$ m;*
- *front suspension roll stiffness $k_{\phi_1}^s = 21740.6$ Nm/rad $= 379.4$ Nm/deg;*

- *rear suspension roll stiffness* $k^s_{\phi_2} = 22322.2\,\text{Nm/rad} = 389.6\,\text{Nm/deg}$;
- *tire radial stiffness* $p_1 = p_2 = 85000\,\text{N/m}$;
- *grip* $\mu = 1.4$.

Solution

According to (3.124), we compute the front and rear tire roll stiffnesses $k^p_{\phi_1} = 62224.3\,\text{Nm/rad} = 1086.0\,\text{Nm/deg}$ and $k^p_{\phi_2} = 52364.3\,\text{Nm/rad} = 913.9\,\text{Nm/deg}$, respectively.

We see that $k^p_{\phi_1} > k^p_{\phi_2}$ because $t_1 > t_2$. Moreover, as expected, tire roll stiffnesses are bigger than suspension roll stiffnesses, but not that much. More precisely $k^p_{\phi_1}/k^s_{\phi_1} = 2.86$ and $k^p_{\phi_2}/k^s_{\phi_2} = 2.35$.

The roll stiffnesses k_{ϕ_1} and k_{ϕ_2} of the front and rear axles can be computed using (3.122). We obtain with a simple calculation $k_{\phi_1} = 16111.4\,\text{Nm/rad} = 281.2\,\text{Nm/deg}$ and $k_{\phi_2} = 15650.6\,\text{Nm/rad} = 273.2\,\text{Nm/deg}$. Adding these two quantities, as in (3.148), we obtain the vehicle global roll stiffness $k_\phi = 31762.0\,\text{Nm/rad} = 554.4\,\text{Nm/deg}$.

For simplicity, we assume to apply a lateral force, say, $Y = \mu m g = 4188.9\,\text{N}$ at the center of mass G. The height $q = 0.036\,\text{m}$ of the no-roll axis under the center of mass is given by (3.140). How large is the vehicle roll angle ϕ?

We are now ready to make a mistake. As a matter of fact, we are tempted to employ the very simple equation (3.157) to estimate the vehicle roll angle ϕ under a lateral force Y. The (wrong) result would be $\phi = 2.15\,\text{deg}$.

The correct equation is (3.149), which needs $Y_1 = Y a_2/(a_1 + a_2) = 1968.8\,\text{N}$ and $Y_2 = Y a_1/(a_1 + a_2) = 2220.1\,\text{N}$, and provides the vehicle roll angle $\phi = 2.23\,\text{deg}$.

Moreover, by means of (3.146) or (8.11) we can compute the tire roll angles $\phi^p_1 = 0.61\,\text{deg}$ and $\phi^p_2 = 0.74\,\text{deg}$, and, by means of (3.147) or (8.12), the suspension roll angles $\phi^s_1 = 1.61\,\text{deg}$ and $\phi^s_2 = 1.48\,\text{deg}$. Of course, they must fulfill (3.133).

With the (wrong) assumption of rigid tires, as in (3.157) with $k_\phi = k^s_{\phi_1} + k^s_{\phi_2} = 44062.8\,\text{Nm/rad} = 769.0\,\text{Nm/deg}$, the (wrong) vehicle roll angle would have been $\phi = 1.55\,\text{deg}$.

8.10.6   Lateral Load Transfers in Formula Cars

With the data and results of the previous exercise, compute the lateral load transfers ΔZ_1 *and* ΔZ_2.

Solution

Since, as already stated, the tires cannot be assumed as rigid, we have to use (8.15) for computing the lateral load transfers. The results are $\Delta Z_1 = 547.4\,\text{N}$ and $\Delta Z_2 = 610.9\,\text{N}$, which makes $\Delta Z_1/(\Delta Z_1 + \Delta Z_2) = 0.47$.

We know that part of these load transfers comes from the suspension links and part from the suspension and tire stiffnesses. Employing (3.152) and (3.153) we obtain $\Delta Z_1^Y = 40.7\,\text{N}$, $\Delta Z_1^L = 506.7\,\text{N}$, $\Delta Z_2^Y = 90.0\,\text{N}$ and $\Delta Z_2^L = 520.9\,\text{N}$.

Of course, $Yh = \Delta Z_1 t_1 + \Delta Z_2 t_2$.

According to (3.105), the static load on the front axle is $Z_1^0 = 1406.3\,\text{N}$, while it is $Z_2^0 = 1585.8\,\text{N}$ on the rear axle. We remind that the static load on each wheel is not necessarily 50% of the axle load (see Sect. 3.9.2).

For comparison, we repeat the computation assuming (erroneously) rigid tires and get $\Delta Z_1 = 526.5\,\text{N}$ and $\Delta Z_2 = 633.7\,\text{N}$, which makes $\Delta Z_1/(\Delta Z_1 + \Delta Z_2) = 0.45$. This result confirms that tire stiffness has to be taken into account.

We invite the reader to figure out what can be done on the car to end up with the ratio $\Delta Z_1/\Delta Z_2 > 1$, as it should be to have an understeer vehicle.

8.10.7 Centrifugal Force Not Applied at the Center of Mass

Going from turn 1 to turn 2 of the Barcelona circuit requires a sharp change in direction, which means fairly high values of $\dot{r}$. For instance, at the end of turn 1 a Formula car had, at a given instant, $\dot{r} = 1.56\,\text{rad/s}^2$ and $a_y = -8.34\,\text{m/s}^2$. Assuming $m = 680\,\text{kg}$ and $J_z = 700\,\text{kgm}^2$, compute how far was the lateral force Y from the center of mass G. We also know that the steer angle δ_1 of the front wheels was only 0.21 deg.

Solution

To answer this question we can rely on (3.101). Indeed, it is precisely x_N the sought distance. Therefore, we need the lateral force $Y = ma_y = 5671.2\,\text{N}$ and the vertical moment $N = J_z\dot{r} = 1092\,\text{Nm}$. We can assume $N_X \simeq 0$ because the car was going almost straight and hence the limited-slip differential had no effect.

The distance of the lateral (centrifugal) force Y from G is $x_N = N/Y = 0.19\,\text{m}$. As expected, the centrifugal force does *not* act through the center of mass (cf. [14, p. 133]).

8.10.8 Global Aerodynamic Force

Combine the three aerodynamic forces shown in Fig. 8.1 to obtain the line of action and the magnitude of the global aerodynamic force $\mathbf{F}_a$.

Solution

We prefer to use a graphic approach. Since forces are applied vectors, we can redraw them only along their line of action. As shown in Fig. 8.62, we first combine X_a and

Fig. 8.62 Vectorial sum of the aerodynamic drag and axle downforces to obtain the global aerodynamic force with its line of action

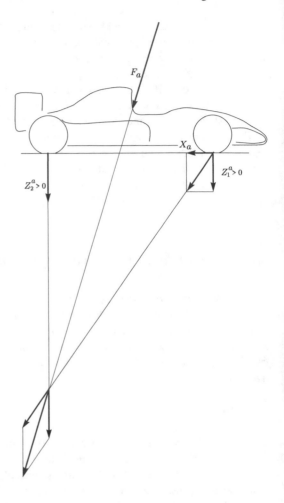

Z_1^a. The resulting vector is then added to Z_2^a, again keeping each force on its line of action, thus obtaining the global aerodynamic force F_a.

It is interesting to compare this result with the analogous result for a road car, shown in Fig. 3.22.

8.11   Summary

Limited-slip differentials and aerodynamic devices are typical of race cars. Both greatly impact on vehicle handling (otherwise they would not be used). Therefore, the first part of this Chapter has been devoted to the formulation of a suitable double track vehicle model.

In case of open differential, it has been shown that often a single track model can still be meaningful. However, in all cases the definition of the classical understeer gradient becomes meaningless.

The new global approach to handling evaluation, called MAP, turns out to be more general, and very informative for race cars.

The relationship between steady-state gradients and stability derivatives is investigated in detail. This analysis leads to three different, albeit equivalent, criteria for the vehicle stability.

The behaviors of different types of limited-slip differentials are compared.

8.12   List of Some Relevant Concepts

A non-open differential makes the vehicle behavior very sensitive also to the turning radius. Aerodynamic effects make the vehicle handling behavior very sensitive to the forward speed;

Section 8.6.1—by means of the Map of Achievable Performance (MAP) it is possible to single out the physical grip;

Section 8.6.3—the yawing moment due to the limited-slip differential can be either positive or negative.

8.13   Key Symbols

a_1	distance of G from the front axle
a_2	distance of G from the rear axle
a_n	centripetal acceleration
a_t	tangential acceleration
a_x	longitudinal acceleration
a_y	lateral acceleration
$\tilde{a}_y$	steady-state lateral acceleration
C	velocity center
C_i	lateral slip stiffness of ith axle
C_x, C_y, C_z	aerodynamic coefficients
d	diameter of the inflection circle
$F_{x_{ij}}$	tire longitudinal force
$F_{y_{ij}}$	tire lateral force
$F_{z_{ij}}$	tire vertical force
g	gravitational acceleration
G	center of mass
h	height of G
J_x, J_y, J_z	moments of inertia
K	acceleration center
K	classical understeer gradient

k_ϕ	total roll stiffness
k_{ϕ_i}	global roll stiffness of ith axle
$k_{\phi_i}^p$	tire roll stiffness
$k_{\phi_i}^s$	suspension roll stiffness
l	wheelbase
L_{w_i}	gyroscopic torque
m	mass
N	yaw moment
N_β, N_ρ	stability derivatives
N_δ	control derivative
q_1	height of the front no-roll center
Q_1	front no-roll center
q_2	height of the rear no-roll center
Q_2	rear no-roll center
r	yaw rate
R	lateral coordinate of C
r_i	rolling radii
S	longitudinal coordinate of C
S_a	frontal area
t_1	front track
t_2	rear track
u	longitudinal velocity
v	lateral velocity
X	longitudinal force
X_a	aerodynamic drag
Y	lateral force
Y_i	lateral force on the ith axle
Y_β, Y_ρ	stability derivatives
Y_δ	control derivative
Z	vertical force
Z_i	vertical load on ith axle
Z_i^0	static vertical load on ith axle
Z_i^a	aerodynamic vertical load on ith axle
ΔZ	longitudinal load transfer
ΔZ_i	lateral load transfer on ith axle
α_{ij}	tire slip angles
β	ratio v/u
$\hat{\beta}$	vehicle slip angle
β_t	shifted coordinate
$(\hat{\beta}_u, \hat{\beta}_\delta)$	gradient components
(β_y, β_δ)	gradient components
γ_{ij}	camber angles
δ_{ij}	steer angle of the wheels

δ_v	steering wheel angle of rotation
ε_1	Ackermann coefficient
ζ	exponent
ζ	damping ratio
η_h	internal efficiency of the differential housing
ρ	ratio r/u
ρ_a	air density
ρ_t	shifted coordinate
$(\hat{\rho}_u, \hat{\rho}_\delta)$	gradient components
(ρ_y, ρ_δ)	gradient components
$\sigma_{x_{ij}}$	tire longitudinal slips
$\sigma_{y_{ij}}$	tire lateral slips
τ	steer gear ratio
ϕ	roll angle
ϕ_i	slope of the axle characteristics
φ_{ij}	spin slips
ψ	yaw angle
ω_h	angular velocity of the differential housing
ω_{ij}	angular velocity of the rims
ω_n	natural angular frequency
ω_s	damped natural angular frequency

References

1. Bastow D, Howard G, Whitehead JP (2004) Car suspension and handling, 4th edn. SAE International, Warrendale
2. Dixon JC (1991) Tyres, suspension and handling. Cambridge University Press, Cambridge
3. Font Mezquita J, Dols Ruiz JF (2006) La Dinámica del Automóvil. Editorial de la UPV, Valencia
4. Frendo F, Greco G, Guiggiani M (2006) Critical review of handling diagram and understeer gradient for vehicles with locked differential. Veh Syst Dyn 44:431–447
5. Frendo F, Greco G, Guiggiani M, Sponziello A (2007) The handling surface: a new perspective in vehicle dynamics. Veh Syst Dyn 45:1001–1016
6. Frendo F, Greco G, Guiggiani M, Sponziello A (2008) Evaluation of the vehicle handling performances by a new approach. Veh Syst Dyn 46:857–868
7. Gillespie TD (1992) Fundamentals of vehicle dynamics. SAE International, Warrendale
8. Loppini D (2017) Tools for handling analysis of race cars. Master's thesis, Università di Pisa
9. Michelin (2001) The tyre encyclopedia. Part 1: grip. Société de Technologie Michelin, Clermont–Ferrand, [CD-ROM]
10. Pacejka HB (1973) Simplified analysis of steady-state turning behaviour of motor vehicles, part 1. Handling diagrams of simple systems. Veh Syst Dyn 2:161–172
11. Pacejka HB (1973) Simplified analysis of steady-state turning behaviour of motor vehicles, part 2: stability of the steady-state turn. Veh Syst Dyn 2:173–183
12. Pacejka HB (1973) Simplified analysis of steady-state turning behaviour of motor vehicles, part 3: more elaborate systems. Veh Syst Dyn 2:185–204

13. Senni F (2013) Identificazione del "DNA" di una Formula 1 per valutarne l'assetto e le prestazioni. Master's thesis, Università di Pisa
14. Seward D (2014) Race car design. Palgrave, London
15. Wong JY (2001) Theory of ground vehicles. Wiley, New York

Chapter 9
Handling with Roll Motion

So far we have investigated the handling behavior of a vehicle under the assumption of negligible roll. Actually, we have not completely discarded roll angles, as they are absolutely necessary for evaluating, e.g., lateral load transfers. But we have not considered, for instance, the inertial effects of roll motion.

In this chapter, the roll motion is taken into account (Fig. 9.1). It is hard work, as the analysis becomes more involved [10]. However, it also sheds light onto one of the most controversial concepts in vehicle dynamics: the roll axis [1, 3, 5, 6, 11], in this book renamed no-roll axis. This concept has been already discussed in Sect. 3.10.9, but it will be considered again here.

We state from the very beginning what the outcome of our analysis will be: the roll axis, as that axis about which the vehicle rolls, does not exist. Or, in other words, the concept of an axis about which the vehicle rolls is meaningless (cf. [9, p. 115]). We understand it sounds harsh, but that is the way it is. There is no such thing as an axis about which the vehicle rolls, albeit the vehicle rolls indeed. A similar conclusion was obtained also in [7] and in [2, p. 400]. However, the no-roll axis, as defined here in Sect. 3.10.9, maintains its validity.

9.1 Vehicle Position and Orientation

Defining the position and orientation of a vehicle when roll is assumed to be zero is a simple matter. As shown in Fig. 3.3, the motion is two-dimensional and hence it suffices to know, with respect to a ground-fixed reference system, the two coordinates of the center of mass G and the yaw angle ψ.

Including roll (and, perhaps, also pitch) means having to deal with a full three-dimensional problem. Therefore, we must employ more sophisticated tools. Quite paradoxically, it turns out that it is easier to define unambiguously the *orientation* of the vehicle body, rather than the *position* of the vehicle. The reason is that the

© The Author(s), under exclusive license to Springer Nature Switzerland AG 2023
M. Guiggiani, *The Science of Vehicle Dynamics*,
https://doi.org/10.1007/978-3-031-06461-6_9

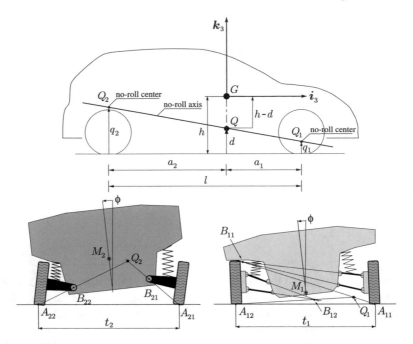

Fig. 9.1 Vehicle basic scheme including roll angle ϕ (lateral and front views)

concept of "position of the vehicle" is not so clear anymore. As a matter of fact, roll causes point G to move sideways with respect to the wheels, but this movement does not change the "position of the vehicle" directly. In other words, we pretend that the lateral velocity v of the vehicle does not contain any roll contribution. We will address this important aspect shortly. First, some other concepts need to be introduced.

9.2 Yaw, Pitch and Roll

Although everybody has an intuitive notion of roll, pitch and yaw of a vehicle, we need a more precise definition at this stage. The goal is to know the *orientation* of the vehicle body (assumed to be a rigid body) with respect to a ground-fixed reference system S_0. A typical approach is to give a *sequence* of three *elemental* rotations, that is rotations about the axes of a chain of coordinate systems.

The three elemental rotations must follow a definite order. In other words, the same rotations in a different order provide a different orientation. This aspect can be appreciated by a simple example. In Fig. 9.2a, a parallelepiped is rotated by 90° about the axis **i** and then by −90° about the axis **j**. In Fig. 9.2b, the same parallelepiped is

Fig. 9.2 Finite rotations are not commutative (i.e., their order is important)

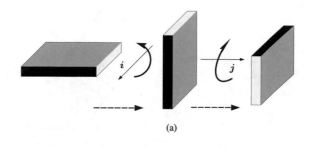

(a)

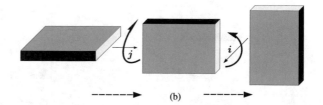

(b)

subject to the same two rotations, but in reverse order. The final orientation is totally different, thus confirming that finite rotations are not commutative.[1]

Human beings are comfortable with two-dimensional rotations, and Euler was, perhaps, no exception when he invented the technique of three *elemental* rotations, often referred to as Euler angles. The basic idea is to generate a sequence of four Cartesian reference systems S_i, each one sharing one axis with the preceding system and another axis with the next one. Therefore, we can go from one system to the next by means of a two-dimensional rotation about their common axis.[2]

In vehicle dynamics it is convenient to use the following sequence of reference systems (Fig. 9.3)

$$(\mathbf{i}_0, \mathbf{j}_0, \mathbf{k}_0) \xrightarrow[\mathbf{k}_0=\mathbf{k}_1]{\psi} (\mathbf{i}_1, \mathbf{j}_1, \mathbf{k}_1) \xrightarrow[\mathbf{j}_1=\mathbf{j}_2]{\theta} (\mathbf{i}_2, \mathbf{j}_2, \mathbf{k}_2) \xrightarrow[\mathbf{i}_2=\mathbf{i}_3]{\phi} (\mathbf{i}_3, \mathbf{j}_3, \mathbf{k}_3) \qquad (9.1)$$

to go from the ground-fixed reference system S_0, with directions $(\mathbf{i}_0, \mathbf{j}_0, \mathbf{k}_0)$, to the vehicle-fixed reference system S_3, with directions $(\mathbf{i}_3, \mathbf{j}_3, \mathbf{k}_3)$. This vehicle-fixed reference system has been already introduced in Fig. 1.4, although with a slightly different notation (no subscripts). When the vehicle is at rest, direction $\mathbf{k}_3 = \mathbf{k}$ is orthogonal to the road (hence directed like $\mathbf{k}_0$) and direction $\mathbf{i}_3 = \mathbf{i}$ is parallel to the road and pointing forward (hence like $\mathbf{i}_1$, Fig. 9.1).

During the vehicle motion, S_3 moves accordingly. At any instant of time, the key step is the definition of the auxiliary direction $\mathbf{j}_1 = \mathbf{j}_2$

---

[1] Rotation matrices are a tool to represent finite rotation. As well known, the product of matrices is not commutative, in general.

[2] More precisely, the axis must share the same direction. The origin can be different.

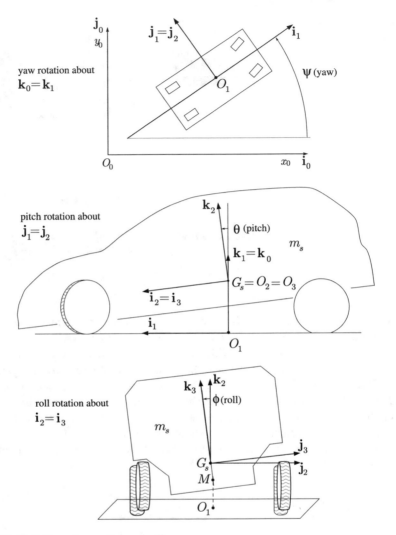

Fig. 9.3 Definition of yaw, pitch and roll

$$\mathbf{j}_1 = \mathbf{j}_2 = \frac{\mathbf{k}_0 \times \mathbf{i}_3}{|\mathbf{k}_0 \times \mathbf{i}_3|} = \frac{\mathbf{k}_1 \times \mathbf{i}_2}{|\mathbf{k}_1 \times \mathbf{i}_2|} \qquad (9.2)$$

often called the *line of nodes*, which is *orthogonal* to both $\mathbf{k}_0 = \mathbf{k}_1$ and $\mathbf{i}_2 = \mathbf{i}_3$. This direction $\mathbf{j}_1 = \mathbf{j}_2$ is the link between the ground-fixed and the vehicle-fixed reference systems. This way, we have that we can go from $\mathbf{S}_0$ to $\mathbf{S}_1$ with an elemental rotation ψ about $\mathbf{k}_0 = \mathbf{k}_1$, and so on. Any two consecutive reference systems differ by a two-dimensional rotation, as shown in (9.1).

More precisely, as shown in Fig. 9.3, the first rotation ψ (yaw) is about the third axis $\mathbf{k}_0 = \mathbf{k}_1$, which $\mathbf{S}_0$ and $\mathbf{S}_1$ have in common, the second rotation θ (pitch) is about the second axis $\mathbf{j}_1 = \mathbf{j}_2$, shared by S_1 and S_2, and the third rotation ϕ (roll) is about the first common axis $\mathbf{i}_2 = \mathbf{i}_3$ of S_2 and S_3. This is why this sequence of elemental rotations is marked $(3, 2, 1)$, or *yaw, pitch and roll*.[3] In vehicle dynamics, the pitch and roll angles are very small.

9.3 Angular Velocity

With this sequence of reference systems, the angular velocity of the vehicle body $\boldsymbol{\Omega}$ is given by

$$\boldsymbol{\Omega} = \dot{\phi}\ \mathbf{i}_2(\psi, \theta) + \dot{\theta}\ \mathbf{j}_1(\psi) + \dot{\psi}\ \mathbf{k}_0 \tag{9.3}$$

This is a simple and intuitive equation, but it has the drawback that the three unit vectors are not mutually orthogonal (Fig. 9.3). Therefore, our goal is to obtain the following equation[4]

$$\boldsymbol{\Omega} = p\ \mathbf{i}_3 + q\ \mathbf{j}_3 + r\ \mathbf{k}_3 \tag{9.4}$$

where the vector $\boldsymbol{\Omega}$ is expressed in terms of its components in the vehicle-fixed reference system S_3.[5]

The expressions of p, q and r can be easily obtained by means of the rotation matrices

$$
\begin{aligned}
\begin{bmatrix} p \\ q \\ r \end{bmatrix} &= \mathbf{R}_1(\phi) \begin{bmatrix} \dot{\phi} \\ 0 \\ 0 \end{bmatrix} + \mathbf{R}_1(\phi)\mathbf{R}_2(\theta) \begin{bmatrix} 0 \\ \dot{\theta} \\ 0 \end{bmatrix} + \mathbf{R}_1(\phi)\mathbf{R}_2(\theta)\mathbf{R}_3(\psi) \begin{bmatrix} 0 \\ 0 \\ \dot{\psi} \end{bmatrix} \\
&= \begin{bmatrix} \dot{\phi} \\ 0 \\ 0 \end{bmatrix} + \mathbf{R}_1(\phi) \begin{bmatrix} 0 \\ \dot{\theta} \\ 0 \end{bmatrix} + \mathbf{R}_1(\phi)\mathbf{R}_2(\theta) \begin{bmatrix} 0 \\ 0 \\ \dot{\psi} \end{bmatrix}
\end{aligned} \tag{9.5}
$$

where, as well known, the rotation matrices for elemental rotations are as follows, for a generic angle α
- rotation around the first axis

$$\mathbf{R}_1(\alpha) = \begin{bmatrix} 1 & 0 & 0 \\ 0 & \cos\alpha & \sin\alpha \\ 0 & -\sin\alpha & \cos\alpha \end{bmatrix} \tag{9.6}$$

- rotation around the second axis

---

[3] Classical Euler angles use the sequence $(3, 1, 3)$.

[4] In this chapter the symbol q is a component of $\boldsymbol{\Omega}$. Therefore, we use the symbol d for the height of the no-roll center Q (Fig. 9.1).

[5] The components p, q and r of $\boldsymbol{\Omega}$ cannot be given, in general, as time derivatives of an angle.

$$\mathbf{R}_2(\alpha) = \begin{bmatrix} \cos\alpha & 0 & -\sin\alpha \\ 0 & 1 & 0 \\ \sin\alpha & 0 & \cos\alpha \end{bmatrix} \tag{9.7}$$

- rotation around the third axis

$$\mathbf{R}_3(\alpha) = \begin{bmatrix} \cos\alpha & \sin\alpha & 0 \\ -\sin\alpha & \cos\alpha & 0 \\ 0 & 0 & 1 \end{bmatrix} \tag{9.8}$$

The final result is

$$p = \dot{\phi} - \dot{\psi}\,\sin\theta$$

$$q = \dot{\theta}\,\cos\phi + \dot{\psi}\,\sin\phi\cos\theta \tag{9.9}$$

$$r = \dot{\psi}\,\cos\phi\cos\theta - \dot{\theta}\,\sin\phi$$

which can be simplified in

$$p \simeq \dot{\phi} - \dot{\psi}\,\theta$$

$$q \simeq \dot{\theta} + \dot{\psi}\,\phi \tag{9.10}$$

$$r \simeq \dot{\psi}$$

because of the small values of pitch and roll. Therefore, the angular velocity of the vehicle body can be expressed as

$$\mathbf{\Omega} \simeq (\dot{\phi} - \dot{\psi}\,\theta)\,\mathbf{i}_3 + (\dot{\theta} + \dot{\psi}\,\phi)\,\mathbf{j}_3 + \dot{\psi}\,\mathbf{k}_3 \tag{9.11}$$

in the vehicle-fixed reference system.

Moreover, if there is no pitch, that is $\theta = \dot{\theta} = 0$, we have a further simplification

$$p \simeq \dot{\phi}$$

$$q \simeq \dot{\psi}\,\phi \tag{9.12}$$

$$r \simeq \dot{\psi}$$

A lot of work for getting such a simple result.

This definition of roll, pitch and yaw is quite general. It only needs the reasonable assumption that the vehicle body be considered as perfectly rigid. It is worth remarking that what matters in the definition of roll, pitch and yaw are only the *directions* of the axes of the four reference systems S_i. Their positions, that is the positions of their origins O_i, have no relevance at all.

It is useful to obtain the expressions of the unit vectors ($\mathbf{i}_3$, $\mathbf{j}_3$, $\mathbf{k}_3$) in terms of ($\mathbf{i}_1$, $\mathbf{j}_1$, $\mathbf{k}_1$)

$$\mathbf{i}_3 = \cos(\theta)\,\mathbf{i}_1 - \sin(\theta)\,\mathbf{k}_1$$

$$\mathbf{j}_3 = \sin(\phi)[\sin(\theta)\,\mathbf{i}_1 + \cos(\theta)\,\mathbf{k}_1] + \cos(\phi)\,\mathbf{j}_1 \tag{9.13}$$

$$\mathbf{k}_3 = \cos(\phi)[\sin(\theta)\,\mathbf{i}_1 + \cos(\theta)\,\mathbf{k}_1] - \sin(\phi)\,\mathbf{j}_1$$

which can be simplified into

$$\mathbf{i}_3 \simeq \mathbf{i}_1 - \theta\,\mathbf{k}_1$$

$$\mathbf{j}_3 \simeq \mathbf{j}_1 + \phi\,\mathbf{k}_1 \tag{9.14}$$

$$\mathbf{k}_3 \simeq \theta\,\mathbf{i}_1 - \phi\,\mathbf{j}_1 + \mathbf{k}_1$$

9.4 Angular Acceleration

The angular acceleration $\dot{\boldsymbol{\Omega}}$ is promptly obtained by differentiating (9.4) with respect to time

$$\dot{\boldsymbol{\Omega}} = \dot{p}\,\mathbf{i}_3 + \dot{q}\,\mathbf{j}_3 + \dot{r}\,\mathbf{k}_3 + \boldsymbol{\Omega} \times \boldsymbol{\Omega}$$
$$= \dot{p}\,\mathbf{i}_3 + \dot{q}\,\mathbf{j}_3 + \dot{r}\,\mathbf{k}_3 \tag{9.15}$$

where, according to (9.10)

$$\dot{p} \simeq \ddot{\phi} - \ddot{\psi}\,\theta - \dot{\psi}\,\dot{\theta}$$

$$\dot{q} \simeq \ddot{\theta} + \ddot{\psi}\,\phi + \dot{\psi}\,\dot{\phi} \tag{9.16}$$

$$\dot{r} \simeq \ddot{\psi}$$

9.5 Vehicle Lateral Velocity

The vehicle lateral velocity v was introduced in (3.1) in the case of negligible roll motion. Now we need to extend that definition when the roll motion is taken into account. This task is not as simple as it may seem. Intuitively, we would like to obtain an expression of v independent of ϕ. Therefore, we are looking for a point which, broadly speaking, follows the vehicle motion, without being subject to roll. A point that is like G, except that it does not roll. More precisely, we are looking for the origin O_1 of the reference system S_1 in Fig. 9.3, that is a reference system which yaws, but does not pitch and roll.

For simplicity, we assume the tires are perfectly rigid in this chapter.

Fig. 9.4 Roll rotations about different points and comparison of the relative contact patch positions

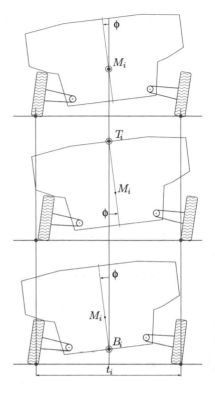

9.5.1 Track Invariant Points

Roll motion is part of vehicle dynamics. However, it is useful to start with a purely kinematic analysis to get an idea of the several effects of roll motion. This kinematic analysis should be seen as a primer for better investigating roll dynamics.

Figure 3.35 shows how to determine the no-roll centers Q_i for a swing arm suspension and a double wishbone suspension. The same method is applied in Fig. 3.37 to a MacPherson strut. In all these cases, the vehicle is in its reference configuration (no roll). When the vehicle rolls, the no-roll centers Q_i migrate with respect to the vehicle body. They can be obtained, as shown in Fig. 9.1, using the same procedure of Fig. 3.35, i.e., as the intersection of the two lines passing through points A_{ij} and B_{ij}.

However, determining the current position of Q_i has little relevance in this context. Much more important are the following definitions.

We define point M_1 as the point *of the vehicle body* that coincides with Q_1 in the vehicle reference configuration (Fig. 9.1). The same idea, applied to the rear axle, leads to the definition of M_2. These points are called here *track invariant points*. Let us investigate their properties.

In Fig. 9.4, the vehicle body is rotated, in turn, by the same roll angle ϕ about three different points, namely M_i, T_i, and B_i. We see that in all cases the track length

Fig. 9.5 Roll rotations about the track invariant point M_i for three different suspension layouts (top to bottom): swing arm, MacPherson strut, double wishbone

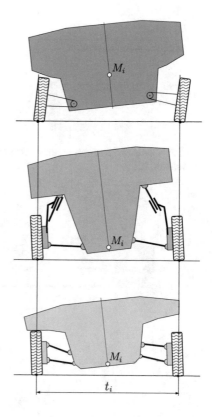

t_i is almost constant. However, in general, the two contact patches move sideways with respect to the point (see also [2, p. 400]). The only exception is with point M_i, which remains midway between the two contact patches (see also [4, p. 97]). This is the reason why it has been called *track invariant point*.

The property that a roll rotation about the track invariant point M_i does not affect the positions of the tire contact patches with respect to M_i itself holds true for any suspension type, as shown in Fig. 9.5.[6]

However, the vehicle does not care much about which point we applied the roll rotation. This is demonstrated in Fig. 9.6, where we superimposed the three vehicle rotations shown in Fig. 9.4. They are almost indistinguishable, suggesting that the notion of a roll axis about which the vehicle rolls is meaningless. For the vehicle, all points between, say, T_i and B_i are pretty much equivalent.

In general, in addition to roll, there may be some suspension jacking, which results in a vehicle vertical displacement z_i, as discussed in Sect. 3.10.10. Figure 9.7 shows the same axle with and without suspension jacking. The roll angle is the same. It is

---

[6] In Fig. 9.5 it is also quite interesting to note the camber variations due to pure roll in each type of suspension. This topic has been addressed in Sect. 3.10.3.

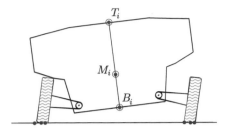

Fig. 9.6 Comparison of roll rotations about different points: they have almost the same effect on the vehicle

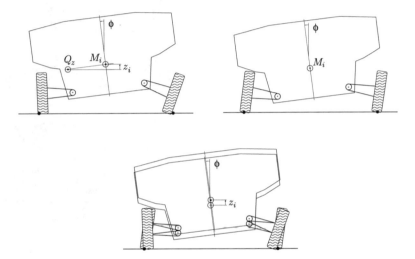

Fig. 9.7 Roll rotations with and without suspension jacking

evident, particularly when comparing the two cases, as it is done in Fig. 9.7-bottom, that the combination of roll and suspension jacking is like a rotation about a point Q_z.

We recall that suspension jacking occurs whenever the lateral forces exerted by the two tires of the same axle are not equal, which is always the case, indeed. However, it is a small effect that can be safely neglected, particularly in more sophisticated suspensions, like the double wishbone or the MacPherson strut.

9.5.2 Vehicle Invariant Point (VIP)

Now let us look at both axles together, that is at the vehicle as a whole, as done in Fig. 9.8. For simplicity, let us assume the front and rear tracks to be equal to each other, that is $t_1 = t_2$, and that they are not affected by roll (no suspension jacking).

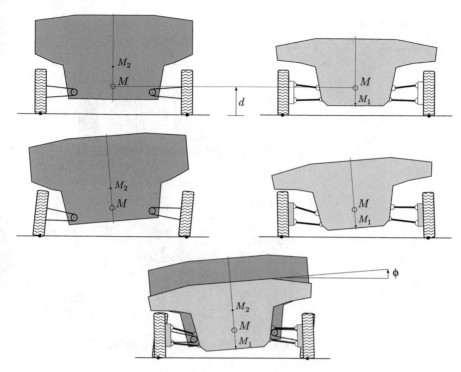

Fig. 9.8 Roll motion explained without the recourse to any roll axis, and definition of the vehicle invariant point M

Points M_1 and M_2 have, in general, different heights. Therefore, roll motion makes the front and rear tracks "slide" a little bit with respect to each other (Fig. 9.8). We remark that we know the *direction* $\mathbf{i}_3$ about which the vehicle (by definition) rolls, but we cannot say anything about an elusive *axis* about which the vehicle rolls.

We are looking for a point of the vehicle body that, regardless of the roll angle ϕ, remains most centered with respect to the four contact patches. Figure 9.8 suggests that the point that is less sensitive to roll is indeed a point M between M_1 and M_2.

Therefore, we define point M as the point *of the vehicle body* that, in the reference configuration, coincides with the no-roll center Q. We call M *vehicle invariant point* (VIP). Point O_1 is the point on the ground always below M, as shown in Fig. 9.9.

The selection of point M as the best suited to represent the vehicle position purged by the roll motion, is reasonable (we believe), but nonetheless arbitrary.[7] However, this is what is commonly done in vehicle dynamics, although often without providing an explicit explanation. We repeat that point M, and hence also O_1, are basically in the middle of the vehicle, even when it rolls. This is the reason that makes them the best option to monitor the vehicle position.

---

[7] The use of the center of mass G to represent the vehicle position in Chaps. 3–8 was arbitrary as well.

Fig. 9.9 Definition of the
lateral velocity $v\,\mathbf{j}_1$ of the
vehicle (front view)

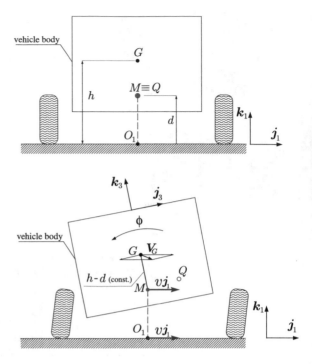

9.5.3 Lateral Velocity and Acceleration

The vehicle velocity is, by definition, that of the *vehicle invariant point M*. Therefore, pretty much like in (3.1)

$$\mathbf{V}_M = u\,\mathbf{i}_1 + v\,\mathbf{j}_1 \tag{9.17}$$

where u is the forward velocity and v is the lateral velocity. We recall that we have assumed the tires to be rigid, and hence there is no roll motion due to tire deformation.

The vehicle acceleration is given by a formula identical to (3.24)

$$\begin{aligned}
\mathbf{a}_M &= (\dot{u} - v\dot{\psi})\,\mathbf{i}_1 + (\dot{v} + u\dot{\psi})\,\mathbf{j}_1 \\
&\simeq (\dot{u} - vr)\,\mathbf{i}_1 + (\dot{v} + ur)\,\mathbf{j}_1
\end{aligned} \tag{9.18}$$

Actually, point M may also have a vertical velocity, due to uneven road or suspension jacking. Here we assume the road to be perfectly flat and suspension jacking to be negligible.

Point M inherits almost everything that was obtained for G in Chaps. 3–8, in the sense that now we have to use M (or O_1) to define the vehicle slip angle $\hat{\beta}$, trajectory, etc.

9.6   Three-Dimensional Vehicle Dynamics

We have assumed the vehicle sprung mass m_s to be a rigid body. If roll motion is taken into account, it has a three-dimensional dynamics. For simplicity, at least at the beginning, it is useful to suppose the unsprung mass m_n to be negligible (i.e., $m = m_s$).

Like in (3.72), the classical force and torque equations for the dynamics of a single rigid body are [8]

$$m\, \mathbf{a}_G = \mathbf{F}$$

$$\dot{\mathbf{K}}^r_G = \mathbf{M}_G \tag{9.19}$$

where $m = m_s$ is the total mass of the vehicle, $\mathbf{a}_G$ is the acceleration of its center of mass, $\mathbf{F}$ is the resultant of all forces applied to the vehicle body, $\dot{\mathbf{K}}^r_G$ is the rate of change of the angular momentum of the vehicle body with respect to $G = G_s$, and $\mathbf{M}_G$ is the global moment (torque) of all forces, again with respect to G.

If the second equation is written with respect to any other point, like, e.g., the freshly defined vehicle invariant point M, it generalizes into

$$\dot{\mathbf{K}}^r_G + MG \times m\, \mathbf{a}_G = \mathbf{M}_M \tag{9.20}$$

9.6.1   Velocity and Acceleration of G

Dynamics cannot get rid of G. We have to compute its velocity and acceleration.

Both points M and G belong to the same rigid body. Therefore, we can use again the fundamental formula (5.1) to relate the velocity of G to the velocity of M, plus the roll contribution

$$\mathbf{V}_G = \mathbf{V}_M + \boldsymbol{\Omega} \times MG \tag{9.21}$$

where, by definition

$$MG = (h - d)\, \mathbf{k}_3$$

$$\simeq (h - d)(\theta\, \mathbf{i}_1 - \phi\, \mathbf{j}_1 + \mathbf{k}_1) \tag{9.22}$$

Therefore

$$\mathbf{V}_G = u\, \mathbf{i}_1 + v\, \mathbf{j}_1 - p(h - d)\, \mathbf{j}_3 + q(h - d)\, \mathbf{i}_3$$

$$= u\, \mathbf{i}_1 + v\, \mathbf{j}_1 - (\dot{\phi} - \dot{\psi}\theta)(h - d)\, \mathbf{j}_3 + (\dot{\theta} + \dot{\psi}\phi)(h - d)\, \mathbf{i}_3 \tag{9.23}$$

where in the last equation we employed the approximate expression (9.11).

In this chapter the symbol q is a component of $\boldsymbol{\Omega}$. Therefore, we use the symbol d for the height of the no-roll center Q (Fig. 9.1).

We can proceed in a similar way for accelerations, that is using the fundamental formula (5.4)

$$\mathbf{a}_G = \mathbf{a}_M + \dot{\mathbf{\Omega}} \times MG + \mathbf{\Omega} \times (\mathbf{\Omega} \times MG) \tag{9.24}$$

that is

$$
\begin{aligned}
\mathbf{a}_G = \; & (\dot{u} - v\dot{\psi}) \, \mathbf{i}_1 + (\dot{v} + u\dot{\psi}) \, \mathbf{j}_1 \\
& - \dot{p}(h - d) \mathbf{j}_3 + \dot{q}(h - d) \, \mathbf{i}_3 \\
& + (h - d)[-p(p \, \mathbf{k}_3 - r \, \mathbf{i}_3) + q(r \, \mathbf{j}_3 - q \, \mathbf{k}_3)]
\end{aligned}
\tag{9.25}
$$

which can be rewritten as

$$
\begin{aligned}
\mathbf{a}_G = \; & (\dot{u} - v\dot{\psi}) \, \mathbf{i}_1 + (\dot{v} + u\dot{\psi}) \, \mathbf{j}_1 \\
& + (h - d)[-\dot{p} \, \mathbf{j}_3 + \dot{q} \, \mathbf{i}_3] \\
& + (h - d)[r(p \, \mathbf{i}_3 + q \, \mathbf{j}_3) - (p^2 + q^2) \, \mathbf{k}_3]
\end{aligned}
\tag{9.26}
$$

Each term has a clear physical meaning. The acceleration $\mathbf{a}_G$ is one of the fundamental bricks in the force equation in (9.19).

The acceleration $\mathbf{a}_G$ can be expressed in S_1

$$
\begin{aligned}
\mathbf{a}_G = \; & (\dot{u} - v\dot{\psi}) \, \mathbf{i}_1 + (\dot{v} + u\dot{\psi}) \, \mathbf{j}_1 \\
& + (h - d)[-\dot{p}(\mathbf{j}_1 + \phi \, \mathbf{k}_1) + \dot{q}(\mathbf{i}_1 - \theta \, \mathbf{k}_1)] \\
& + (h - d)\{r[p(\mathbf{i}_1 - \theta \, \mathbf{k}_1) + q(\mathbf{j}_1 + \phi \, \mathbf{k}_1)] - (p^2 + q^2)(\theta \, \mathbf{i}_1 - \phi \mathbf{j}_1 + \mathbf{k}_1)\}
\end{aligned}
\tag{9.27}
$$

which can be rearranged as

$$
\begin{aligned}
\mathbf{a}_G = \; & (\dot{u} - v\dot{\psi}) \, \mathbf{i}_1 + (\dot{v} + u\dot{\psi}) \, \mathbf{j}_1 \\
& + (h - d)[\dot{q} + rp - (p^2 + q^2)\theta] \, \mathbf{i}_1 \\
& + (h - d)[-\dot{p} + rq + (p^2 + q^2)\phi] \, \mathbf{j}_1 \\
& + (h - d)[-\dot{p}\phi - \dot{q}\theta - rp\theta + rq\phi - (p^2 + q^2)] \, \mathbf{k}_1
\end{aligned}
\tag{9.28}
$$

Taking (9.16) into account, and discarding the small terms, we get

$$
\begin{aligned}
\mathbf{a}_G \simeq \; & (\dot{u} - v\dot{\psi}) \, \mathbf{i}_1 + (\dot{v} + u\dot{\psi}) \, \mathbf{j}_1 \\
& + (h - d)[(\ddot{\theta} + \ddot{\psi}\,\phi + \dot{\psi}\,\dot{\phi}) + \dot{\psi}(\dot{\phi} - \dot{\psi}\theta)] \, \mathbf{i}_1 \\
& + (h - d)[-(\ddot{\phi} - \ddot{\psi}\,\theta - \dot{\psi}\,\dot{\theta}) + \dot{\psi}(\dot{\theta} + \dot{\psi}\phi)] \, \mathbf{j}_1
\end{aligned}
\tag{9.29}
$$

If also $\dot{\psi}$ and $\ddot{\psi}$ are small

$$\mathbf{a}_G \simeq (\dot{u} - v\dot{\psi})\,\mathbf{i}_1 + (\dot{v} + u\dot{\psi})\,\mathbf{j}_1 + (h - d)[\ddot{\theta}\,\mathbf{i}_1 - \ddot{\phi}\,\mathbf{j}_1] \tag{9.30}$$

9.6.2 Rate of Change of the Angular Momentum

It is very convenient to use, as already done in Sect. 9.2, a reference system S_3 attached to the vehicle body and with its origin in the center of gravity G_s of the sprung mass.

As already stated, when the vehicle is at rest, direction $\mathbf{k}_3$ of S_3 is orthogonal to the road and direction $\mathbf{i}_3$ is parallel to the road pointing forward (like in Fig. 1.4, where the body-fixed axes do not have the subscript 3, or in Fig. 9.1). Therefore, in general, S_3 is not directed as the principal axes of inertia, and the inertia tensor

$$\mathbf{J} = \begin{bmatrix} J_x & -J_{xy} & -J_{xz} \\ -J_{yx} & J_y & -J_{yz} \\ -J_{zx} & -J_{zy} & J_z \end{bmatrix} \tag{9.31}$$

is not diagonal.

Consequently, the expression of $\dot{\mathbf{K}}_G^r$ is a little involved (see also (3.75))

$$\begin{aligned}
\dot{\mathbf{K}}_G^r = \; & [J_x\dot{p} - (J_y - J_z)qr - J_{xy}(\dot{q} - rp) - J_{yz}(q^2 - r^2) - J_{zx}(\dot{r} + pq)]\,\mathbf{i}_3 \\
& + [J_y\dot{q} - (J_z - J_x)rp - J_{yz}(\dot{r} - pq) - J_{zx}(r^2 - p^2) - J_{xy}(\dot{p} + qr)]\,\mathbf{j}_3 \\
& + [J_z\dot{r} - (J_x - J_y)pq - J_{zx}(\dot{p} - qr) - J_{xy}(p^2 - q^2) - J_{yz}(\dot{q} + rp)]\,\mathbf{k}_3
\end{aligned} \tag{9.32}$$

Actually, most vehicles have $(J_{xy} = J_{yz}) \simeq 0$, and hence we can use the simplified expression

$$\begin{aligned}
\dot{\mathbf{K}}_G^r = \; & [J_x\dot{p} - (J_y - J_z)qr - J_{zx}(\dot{r} + pq)]\,\mathbf{i}_3 \\
& + [J_y\dot{q} - (J_z - J_x)rp - J_{zx}(r^2 - p^2)]\,\mathbf{j}_3 \\
& + [J_z\dot{r} - (J_x - J_y)pq - J_{zx}(\dot{p} - qr)]\,\mathbf{k}_3
\end{aligned} \tag{9.33}$$

This very same quantity can be expressed in S_1, if (9.14) is taken into account

$$\begin{aligned}
\dot{\mathbf{K}}_G^r = \; & [J_x\dot{p} - (J_y - J_z)qr - J_{zx}(\dot{r} - pq)](\mathbf{i}_1 - \theta\,\mathbf{k}_1) \\
& + [J_y\dot{q} - (J_z - J_x)rp - J_{zx}(r^2 - p^2)](\mathbf{j}_1 + \phi\,\mathbf{k}_1) \\
& + [J_z\dot{r} - (J_x - J_y)pq - J_{zx}(\dot{p} - qr)](\theta\,\mathbf{i}_1 - \phi\,\mathbf{j}_1 + \mathbf{k}_1)
\end{aligned} \tag{9.34}$$

That is, with some further simplifications because θ, ϕ, p and q are small

$$
\begin{aligned}
\dot{\mathbf{K}}_G^r = \ & [J_x \dot{p} - (J_y - J_z)qr - J_{zx}\dot{r} + J_z\dot{r}\theta]\,\mathbf{i}_1 \\
& + [J_y \dot{q} - (J_z - J_x)rp - J_{zx}r^2 - J_z\dot{r}\phi]\,\mathbf{j}_1 \\
& + [J_z \dot{r} + J_{zx}(\dot{r}\theta - r^2\phi - \dot{p} + qr)]\,\mathbf{k}_1
\end{aligned}
\tag{9.35}
$$

And finally, taking (9.16) into account (cf. (3.75))

$$
\begin{aligned}
\dot{\mathbf{K}}_G^r = \ & [J_x(\ddot{\phi} - \ddot{\psi}\theta - \dot{\psi}\dot{\theta}) - (J_y - J_z)(\dot{\theta} + \dot{\psi}\phi)\dot{\psi} - J_{zx}\ddot{\psi} + J_z\ddot{\psi}\theta]\,\mathbf{i}_1 \\
& + [J_y(\ddot{\theta} + \ddot{\psi}\phi + \dot{\psi}\dot{\phi}) - (J_z - J_x)\dot{\psi}(\dot{\phi} - \dot{\psi}\theta) - J_{zx}\dot{\psi}^2 - J_z\ddot{\psi}\phi]\,\mathbf{j}_1 \\
& + [J_z\ddot{\psi} + J_{zx}(2\ddot{\psi}\theta - \ddot{\phi} + 2\dot{\psi}\dot{\theta})]\,\mathbf{k}_1
\end{aligned}
\tag{9.36}
$$

Of course, all inertia terms J_x, J_{xz}, etc. are constant because the reference system S_3 is fixed to the vehicle body. We see that the definition of roll, pitch and yaw is crucial in these equations.

9.6.3   Completing the Torque Equation

Once that $\mathbf{a}_G$ has been obtained, we can also compute the term $MG \times m\,\mathbf{a}_G$ in the torque equation (9.20). To keep the analysis fairly simple, we employ the simplified expressions (9.22) and (9.30)

$$
\begin{aligned}
MG \times m\,\mathbf{a}_G \simeq \ & \\
m\{[(h - d)(\theta\,\mathbf{i}_1 & - \phi\,\mathbf{j}_1 + \mathbf{k}_1)] \times [(\dot{u} - v\dot{\psi})\,\mathbf{i}_1 + (\dot{v} + u\dot{\psi})\,\mathbf{j}_1 + (h - d)(\ddot{\theta}\,\mathbf{i}_1 - \ddot{\phi}\,\mathbf{j}_1)]\}
\end{aligned}
\tag{9.37}
$$

which provides

$$
\begin{aligned}
MG \times m\,\mathbf{a}_G \simeq m\{ & [(h - d)^2\ddot{\phi} - (h - d)(\dot{v} + u\dot{\psi})]\,\mathbf{i}_1 \\
& + [(h - d)^2\ddot{\theta} + (h - d)(\dot{u} - v\dot{\psi})]\,\mathbf{j}_1 \\
& + (h - d)\dot{u}\phi\,\mathbf{k}_1 \}
\end{aligned}
\tag{9.38}
$$

9.6.4   Equilibrium Equations

We have obtained all inertia terms of the force and torque equations (left hand side terms). Considering (9.30), (9.37), and (9.38), we get the following explicit (linearized) form of the equilibrium equations (9.19) and (9.20) for a vehicle that can roll and pitch

$$m[(\dot{u} - vr) + (h - d)\ddot{\theta}] = ma_x = X$$

$$m[(\dot{v} + ur) - (h - d)\ddot{\phi}] = ma_y = Y$$

$$0 = Z$$

$$[J_x + m(h - d)^2]\ddot{\phi} - J_{zx}\dot{r} - m(h - d)(\dot{v} + ur) = L_M$$

$$[J_y + m(h - d)^2]\ddot{\theta} + m(h - d)(\dot{u} - vr) = M_M$$

$$J_z\dot{r} - J_{zx}\ddot{\phi} + m(h - d)\dot{u}\phi = N_M = N$$

(9.39)

where, according to (9.10), we set $r = \dot{\psi}$, and (L_M, M_M, N_M) are the three components of the torque acting with respect to point M. It is useful to compare these equations with (3.94) and (3.95), that is with the equilibrium equations obtained when the inertial effects of pitch and roll are neglected.

Interestingly enough, the last three equations in (9.39) can be rewritten as

$$J_x\ddot{\phi} - J_{zx}\dot{r} - ma_y(h - d) = L_M$$

$$J_y\ddot{\theta} + ma_x(h - d) = M_M$$

(9.40)

$$J_z\dot{r} - J_{zx}\ddot{\phi} + ma_x(h - d)\phi = N_M = N$$

Formally, L_M, M_M and N_M are like L, M and N in (3.92), of course with d replacing h.

Of course, everything looks like the car rolls about point M, but it is not so. Actually, the car rolls about the point M as it does with respect to *any other of its points* (Fig. 9.6). It is just the fundamental law (9.21) of the kinematics of rigid bodies. Therefore, we should avoid sentences like "the car rolls about the roll axis", simply because they have no physical meaning at all.

9.6.5 Including the Unsprung Mass

If the unsprung mass m_n cannot be neglected, Eqs. (9.39) become

$$m(\dot{u} - vr) + m_s(h - d)\ddot{\theta} = X$$

$$m(\dot{v} + ur) - m_s(h - d)\ddot{\phi} = Y$$

$$0 = Z$$

$$[J_x + m_s(h - d)^2]\ddot{\phi} - \tilde{J}_{zx}\dot{r} - m_s(h - d)(\dot{v} + ur) = L_M$$

$$[J_y + m_s(h - d)^2]\ddot{\theta} + m_s(h - d)(\dot{u} - vr) = M_M$$

$$\tilde{J}_z\dot{r} - \tilde{J}_{zx}\ddot{\phi} + m_s(h - d)\dot{u}\phi = N$$

(9.41)

where $\tilde{J}_z$ and $\tilde{J}_{zx}$ take into account both m_s and m_n.

9.7   Handling with Roll Motion

The analysis carried out in Chap. 3 can now be extended taking roll and pitch into account. However, as already stated, we assume here that the tires are rigid, as in Sect. 3.10.14. Otherwise, the theory would become too involved, and some physical aspects would not be clear enough.

9.7.1   Equilibrium Equations

The inertia terms of the equilibrium equations have been already obtained in (9.39), and rewritten in an alternative form in (9.40). Therefore, we have to complete the equilibrium equations by including the resultant **F** and the moment $\mathbf{M}_M$ (right-hand side terms). Of course, now we have to include the effects of the dampers, which are sensitive to the roll time rate $\dot{\phi}$.

We call c_ϕ the global damping coefficients with respect to roll, much like k_ϕ is the global stiffness with respect to roll. More precisely, as in (3.117), we have

$$k_\phi = k_{\phi_1} + k_{\phi_2} \quad \text{and} \quad c_\phi = c_{\phi_1} + c_{\phi_2} \tag{9.42}$$

Similarly, according to (10.55), we have the following global stiffness and global damping coefficient with respect to pitch

$$k_\theta = k_1 a_1^2 + k_2 a_2^2 \quad \text{and} \quad c_\theta = c_1 a_1^2 + c_2 a_2^2 \tag{9.43}$$

Therefore, the right-hand side terms to be inserted into the equilibrium equations (9.39) are as follows (cf. (3.94) and (3.95))

$$X = X_1 + X_2 - X_a$$

$$Y = Y_1 + Y_2$$

$$Z = Z_1 + Z_2 - (mg + Z_1^a + Z_2^a)$$

$$L_M = -k_\phi \phi - c_\phi \dot{\phi} + mg(h-d)\phi \qquad (9.44)$$

$$M_M = -k_\theta \theta - c_\theta \dot{\theta}$$

$$N_M = N = N_Y + N_X = (Y_1 a_1 - Y_2 a_2) + (\Delta X_1 t_1 + \Delta X_2 t_2)$$

9.7.2 Load Transfers

Having roll $\phi(t)$ and $\theta(t)$ as functions of time requires some other equations of the vehicle model developed in Chap. 3 to be updated. More precisely, we have to take dampers and inertia terms into account.

The lateral load transfers (3.144) now become

$$\Delta Z_1 t_1 = Y_1 q_1 + k_{\phi_1} \phi + c_{\phi_1} \dot{\phi}$$
$$\Delta Z_2 t_2 = Y_2 q_2 + k_{\phi_2} \phi + c_{\phi_2} \dot{\phi} \qquad (9.45)$$

which, if added, provide

$$\Delta Z_1 t_1 + \Delta Z_2 t_2 = (k_{\phi_1} + k_{\phi_2})\phi + (c_{\phi_1} + c_{\phi_2})\dot{\phi} + Y_1 q_1 + Y_2 q_2$$
$$= k_\phi \phi + c_\phi \dot{\phi} + Yd = k_\phi \phi + c_\phi \dot{\phi} + ma_y d \qquad (9.46)$$

since, as in (3.141), $Yd = Y_1 q_1 + Y_2 q_2$.

Combining (9.40), (9.44) and (9.46), we obtain

$$Y(h-d) + mg(h-d)\phi = k_\phi \phi + c_\phi \dot{\phi} + J_x \ddot{\phi} - J_{zx} \dot{r} \qquad (9.47)$$

which generalizes (3.157). With a little algebra, we can obtain also

$$\Delta Z_1 t_1 + \Delta Z_2 t_2 = ma_y h + mg(h-d)\phi - (J_x \ddot{\phi} - J_{zx} \dot{r}) \qquad (9.48)$$

which generalizes (3.106).

For the longitudinal load transfer ΔZ we can follow a similar line of reasoning, thus obtaining

$$\Delta Z = -\frac{Xh + J_y \ddot{\theta}}{l} = -\frac{ma_x h + J_y \ddot{\theta}}{l} \qquad (9.49)$$

which generalizes (3.104).

The main difference with respect to the model developed in Chap. 3, and summarized on Sect. 3.14, is that load transfers now depend explicitly on the angular accelerations of the vehicle body.

9.7.3 Constitutive (Tire) Equations

Taking explicitly into account the roll and pitch motions does not affect directly the tire equations. Therefore, the analysis developed in Chap. 3 applies entirely.

9.7.4 Congruence (Kinematic) Equations

The congruence equations listed in Sect. 3.14.7 can be employed even when the vehicle model has the roll and pitch degrees of freedom. Actually, according to Fig. 9.8, the lateral velocities of the front and rear axles should be, respectively

$$v_1 = v + ra_1 + (d - q_1)\dot{\phi} \quad \text{and} \quad v_2 = v - ra_2 + (d - q_2)\dot{\phi} \tag{9.50}$$

that is they include small contributions due to the different heights of the vehicle invariant point M and the two track invariant points M_1 and M_2. However, the additional terms are really very small, and hence can be neglected.

9.8 Steady-State and Transient Analysis

Obviously, including the roll and pitch motions into the vehicle model has very little, if any, influence on the vehicle steady-state behavior. We should not forget that the steady-state roll angle was part of the analyses carried out in Chaps. 3–8. On the other hand, the transient behavior, in particular when entering or exiting a curve, can be rather different.

9.9 Exercise

9.9.1 Roll Motion and Camber Variation

Camber variations $\Delta\gamma_i$ are strictly related to roll motion ϕ, and affect quite a bit the vehicle handling behavior. This topic was addressed in Sect. 3.10.3, where the first order relationships were provided.

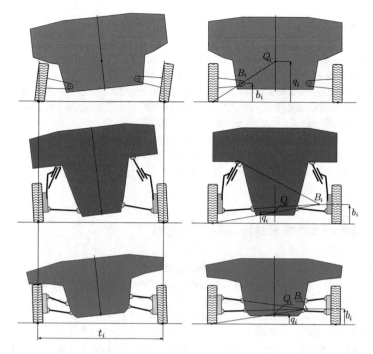

Fig. 9.10 Roll motion and camber variations (front view)

In particular, here we are interested in the equation

$$\Delta\gamma_i \simeq -\left(\frac{q_i - b_i}{b_i}\right)\phi \tag{9.51}$$

With the aid of a ruler and a protractor, check this equation for the three cases of Fig. 9.10.

Solution

In all three cases, the roll angle $\phi = 6$ deg. Top to bottom, we have $(q_i - b_i)/b_i$ equal to about -1.14, 0.6, and 0.36, respectively. Therefore, the corresponding camber angles should be -6.8 deg, 3.6 deg, and 2.1 deg. Indeed, direct measurements confirm these results.

9.10   Summary

The vehicle orientation has been defined by means of the yaw-pitch-roll elemental rotations. Then, to define the vehicle position, a careful analysis of what happens when the vehicle rolls has been performed. The key result is the definition of the Vehicle Invariant Point (VIP) as the best option for monitoring the vehicle position, and also for defining the lateral velocity and acceleration.

VIP allows for a simple and systematic analysis of the vehicle three-dimensional dynamics. Among other things, it has been shown that the well known roll-axis, as the axis about which the vehicle rolls, is nonsense.

9.11   List of Some Relevant Concepts

Section 9.2—finite rotations are not commutative;
Section 9.2—yaw, pitch, and roll are the three elemental rotations commonly and conveniently employed in vehicles;
Section 9.5.1—track invariant points belong to the vehicle body;
Section 9.5.2—vehicle invariant point (VIP) belongs to the vehicle body and it is the point best suited to represent the vehicle position, lateral velocity, and lateral acceleration;
Section 9.5.2—roll motion is better explained without any recourse to the roll axis;
Section 9.7.2—load transfers depend also on angular accelerations.

9.12   Key Symbols

a_1	distance of G from the front axle
a_2	distance of G from the rear axle
$\mathbf{a}_G$	acceleration of G
$\mathbf{a}_M$	acceleration of M
d	height of the no-roll axis below G
g	gravity acceleration
G	center of mass
h	height of G
L_M	first component of the torque with respect to M
$\mathbf{J}$	inertia tensor
J_x, J_y, J_z	moments of inertia
$\mathbf{K}_G^r$	angular momentum with respect to G
k_θ	total pitch stiffness
k_ϕ	total roll stiffness
k_{ϕ_i}	global roll stiffness of ith axle

l	wheelbase
m	mass
m_n	unsprung mass
m_s	sprung mass
M	vehicle invariant point (VIP)
M_i	track invariant point (TIP)
M_M	second component of the torque with respect to M
N	yaw moment
N_M	third component of the torque with respect to M
p	first component of $\boldsymbol{\Omega}$ in S_3
q	second component of $\boldsymbol{\Omega}$ in S_3
Q	no-roll center
q_1	height of the front no-roll center
Q_1	front no-roll center
q_2	height of the rear no-roll center
Q_2	rear no-roll center
r	third component of $\boldsymbol{\Omega}$ in S_3
$\mathbf{R}_i$	rotation matrix
S_i	Cartesian reference system
t_1	front track
t_2	rear track
u	longitudinal velocity
v	lateral velocity
$\mathbf{V}_G$	velocity of G
X	longitudinal force
Y	lateral force
Y_i	lateral force on the ith axle
Z	vertical force
Z_i	vertical load on ith axle
Z_i^0	static vertical load on ith axle
Z_i^a	aerodynamic vertical load on ith axle
ΔZ	longitudinal load transfer
ΔZ_i	lateral load transfer on ith axle
θ	pitch angle
ϕ	roll angle
ψ	yaw angle
$\boldsymbol{\Omega}$	angular velocity of the vehicle body

References

1. Bastow D, Howard G, Whitehead JP (2004) Car suspension and handling, 4th edn. SAE International, Warrendale
2. Blundell M, Harty D (2004) The multibody systems approach to vehicle dynamics. Butterworth-Heinemann
3. Dixon JC (1991) Tyres, suspension and handling. Cambridge University Press, Cambridge
4. Ellis JR (1994) Vehicle handling dynamics. Mechanical Engineering Publications, London
5. Font Mezquita J, Dols Ruiz JF (2006) La Dinámica del Automóvil. Editorial de la UPV, Valencia
6. Gillespie TD (1992) Fundamentals of vehicle dynamics. SAE International, Warrendale
7. Innocenti C (2007) Questioning the notions of roll center and roll axis for car suspensions. In: Deuxieme Congres International Conception et Modelisation des Systemes Mecaniques. Monastir
8. Meirovitch L (1970) Methods of analytical dynamics. McGraw-Hill, New York
9. Milliken WF, Milliken DL (1995) Race car vehicle dynamics. SAE International, Warrendale
10. Shim T, Velusamy PC (2011) Improvement of vehicle roll stability by varying suspension properties. Veh Syst Dyn 49(1–2):129–152
11. Wong JY (2001) Theory of ground vehicles. Wiley, New York

Chapter 10
Ride Comfort and Road Holding

Real roads are far from flat. Even freshly paved highways have small imperfections that interact with the vehicle dynamics by exciting vehicle vertical vibrations.

The capability to smooth down road imperfections affects both the *comfort* and the *road holding* of the vehicle. Improving comfort means, basically, limiting the vertical acceleration fluctuations of the vehicle body and hence of passengers. Improving road holding means, among other things, limiting the fluctuations of the vertical force that each tire exchanges with the road.[1] The main parameters that affect both comfort and road holding are the suspension *stiffness* and *damping*.

The study of the vibrational behavior of a vehicle going straight at constant speed on a *bumpy road* is called *ride* [1, 3, 5, 6, 9, 12]. More precisely, ride deals with frequencies in the range 0.25–25 Hz for road cars, a bit higher for race cars. Tires can, among other things, absorb small road irregularities at high frequency because of their vertical elasticity and low mass. However, for frequencies below 3 Hz the tires have little influence and can be considered as rigid. Therefore, the burden to absorb bigger bumps goes to the vehicle suspensions.

While when studying the handling of a vehicle we were also interested in the suspension geometry, we focus here on *springs* and *dampers*. We look for criteria for selecting the right stiffness and the right amount of damping for each suspension.

Actually, this is only half the truth. Real suspensions have nonlinear springs and nonlinear dampers, whose features cannot be reduced to a single number like in the linear case. However, suspensions with linear behavior are a good introduction to the study of ride and road holding.

The original version of this chapter was revised: Equations 10.13, 10.14, 10.15 and 10.18 has been updated. The correction to this chapter is available at https://doi.org/10.1007/978-3-031-06461-6_12

[1] Of course, we mean fluctuations due to road imperfections, not to load transfers.

M. Guiggiani, *The Science of Vehicle Dynamics*,
https://doi.org/10.1007/978-3-031-06461-6_10

439

Fig. 10.1   Schematics for spring, damper and inerter

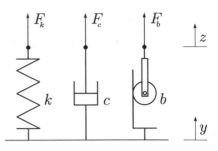

Although standard suspension systems are based on two components—springs and dampers (shock absorbers)—there is a third component that can turn out to be useful in some cases. It is the so-called *inerter*. The inerter is a device that provides a force proportional to the relative *acceleration* between its attachment points, much like a linear damper provides a force proportional to the relative velocity and a linear spring a force proportional to the relative displacement (Fig. 10.1)

$$
\begin{aligned}
F_k &= k(z - y) \\
F_c &= c(\dot{z} - \dot{y}) \\
F_b &= b(\ddot{z} - \ddot{y})
\end{aligned}
\tag{10.1}
$$

The inerter was missing indeed, till quite recently [11]. A typical inerter incorporates a flywheel which rotates in proportion to the relative displacement between its two ends. So far, it has been employed in some Formula cars. We will show how it can improve, in some cases, the car road holding.

10.1   Vehicle Models for Ride and Road Holding

We are mostly interested in the vehicle vertical motion. To keep our ride analysis quite simple, we assume that the vehicle goes *straight* and at *constant speed*. Therefore, there are no handling and/or performance implications here. The ride analysis comes into play because of the *uneven road*. Actually, we ask for a very peculiar road, albeit uneven. It must have exactly the same profile for both wheels of the same axle, thus not inducing roll motion at all. That means that we can rely on a two-dimensional model.

The vehicle models set up for handling and performance are not suitable for ride. We need to develop a tailored model like, e.g., the four-degree-of-freedom model shown in Fig. 10.2. In this model there are three rigid bodies:

- the sprung mass m_s (with moment of inertia J_y w.r.t. its center of gravity G_s), which has vertical motion z_s and pitch motion θ;
- the front unsprung mass m_{n_1}, which has only vertical (hop) motion y_1;

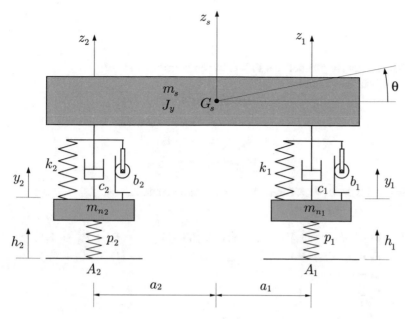

Fig. 10.2 Four-degree-of-freedom model to study ride and road holding

- the rear unsprung mass m_{n_2}, which has only vertical (hop) motion y_2.

Also shown in Fig. 10.2 are the two suspension springs, with stiffnesses k_1 and k_2, and two dampers, with damping coefficients c_1 and c_2, along with two springs p_1 and p_2 to model the tire vertical stiffnesses. Again, to keep the analysis simple, we assume that all these components have *linear* behavior. This is a very unrealistic hypothesis since real suspensions are designed to have hardening stiffness and are equipped with dampers with more resistance during the extension cycle than the compression cycle.

Inerters, with inertances b_1 and b_2, are also shown in Fig. 10.2. They have been used sparingly and only in some race cars. They are included for greater generality.

The vehicle model shown in Fig. 10.2 has four degrees of freedom. Points A_1 and A_2 are the centers of the front axle contact patches and of the rear axle contact patches, respectively. The two functions $h_1(t)$ and $h_2(t)$ are the road profiles as "felt" by the car, that is through the tires [4].

The sprung mass has two degrees of freedom z_s and θ. Alternatively, we could use, e.g., the vertical displacements z_1 and z_2. All displacements and rotations are absolute and taken from the static equilibrium position of the vehicle. We are investigating the oscillations with respect to the equilibrium position, that is the configuration the vehicle would have on a perfectly flat road.

The vehicle model shown in Fig. 10.2 is governed by three sets of equations, as usual:

1. congruence equations:

$$z_1 = z_s + a_1 \theta$$
$$z_2 = z_s - a_2 \theta \qquad (10.2)$$

that is a purely geometrical link between coordinates;
2. equilibrium equations:

$$m_s \ddot{z}_s = F_1 + F_2$$
$$J_y \ddot{\theta} = F_1 a_1 - F_2 a_2$$
$$m_{n_1} \ddot{y}_1 = N_1 - F_1$$
$$m_{n_2} \ddot{y}_2 = N_2 - F_2 \qquad (10.3)$$

that is a link between forces or couples and accelerations;
3. constitutive equations:

$$F_1 = -k_1(z_1 - y_1) - c_1(\dot{z}_1 - \dot{y}_1) - b_1(\ddot{z}_1 - \ddot{y}_1) = -(F_{k_1} + F_{c_1} + F_{b_1})$$
$$F_2 = -k_2(z_2 - y_2) - c_2(\dot{z}_2 - \dot{y}_2) - b_2(\ddot{z}_2 - \ddot{y}_2) = -(F_{k_2} + F_{c_2} + F_{b_2})$$
$$N_1 = -p_1(y_1 - h_1)$$
$$N_2 = -p_2(y_2 - h_2) \qquad (10.4)$$

which model springs, dampers and inerters.

By F_1 e F_2 we mean the vertical forces exchanged between the sprung mass and the two unsprung masses, respectively. By N_1 e N_2 we mean the forces exchanged by each axle with the road. All forces must be intended as perturbations with respect to the static equilibrium position. That is why the weight was not included in the equations.

Combining the above sets of equations, we end up with a system of four linear differential equations with constant coefficients. They are the governing equations of this vehicle model

$$\mathbf{M}\,\ddot{\mathbf{w}} + \mathbf{C}\,\dot{\mathbf{w}} + \mathbf{K}\,\mathbf{w} = \mathbf{h} \qquad (10.5)$$

where $\mathbf{w} = \mathbf{w}(t) = \big(z_s(t),\ \theta(t),\ y_1(t),\ y_2(t)\big)$ is the coordinate vector, and $\mathbf{h} = \mathbf{h}(t) = \big(0,\ 0,\ p_1 h_1(t),\ p_2 h_2(t)\big)$ is the road excitation. We also have the mass matrix $\mathbf{M}$

$$\mathbf{M} = \mathbf{M}_m + \mathbf{M}_b = \begin{bmatrix} m_s & 0 & 0 & 0 \\ 0 & J_y & 0 & 0 \\ 0 & 0 & m_{n_1} & 0 \\ 0 & 0 & 0 & m_{n_2} \end{bmatrix} + \begin{bmatrix} b_1 + b_2 & b_1 a_1 - b_2 a_2 & -b_1 & -b_2 \\ b_1 a_1 - b_2 a_2 & b_1 a_1^2 + b_2 a_2^2 & -b_1 a_1 & b_2 a_2 \\ -b_1 & -b_1 a_1 & b_1 & 0 \\ -b_2 & b_2 a_2 & 0 & b_2 \end{bmatrix} \qquad (10.6)$$

the damping matrix $\mathbf{C}$

$$\mathbf{C} = \begin{bmatrix} c_1 + c_2 & c_1 a_1 - c_2 a_2 & -c_1 & -c_2 \\ c_1 a_1 - c_2 a_2 & c_1 a_1^2 + c_2 a_2^2 & -c_1 a_1 & c_2 a_2 \\ -c_1 & -c_1 a_1 & c_1 & 0 \\ -c_2 & c_2 a_2 & 0 & c_2 \end{bmatrix} \qquad (10.7)$$

and the stiffness matrix $\mathbf{K}$

$$\mathbf{K} = \mathbf{K}_k + \mathbf{K}_p = \begin{bmatrix} k_1 + k_2 & k_1 a_1 - k_2 a_2 & -k_1 & -k_2 \\ k_1 a_1 - k_2 a_2 & k_1 a_1^2 + k_2 a_2^2 & -k_1 a_1 & k_2 a_2 \\ -k_1 & -k_1 a_1 & k_1 & 0 \\ -k_2 & k_2 a_2 & 0 & k_2 \end{bmatrix} + \begin{bmatrix} 0 & 0 & 0 & 0 \\ 0 & 0 & 0 & 0 \\ 0 & 0 & p_1 & 0 \\ 0 & 0 & 0 & p_2 \end{bmatrix} \tag{10.8}$$

A linear four-degree-of-freedom system is quite simple in principle, but also quite cumbersome to be dealt with analytically without the aid of a computer. Therefore, for educational purposes, it is useful to simplify this model further. The basic idea is to extract two models, both with two degrees of freedom. One model to study free vibrations and the other model to study forced vibrations. The two models are virtually obtained by cutting off the unnecessary parts (gray lines in Fig. 10.3) from the four-degree-of-freedom system.

The sprung mass m_s is always much higher than the total unsprung mass $m_n = m_{n_1} + m_{n_2}$. Typically we have $m_s \simeq 10 m_n$. Moreover, tire stiffness is, except in Formula cars, much higher than the suspension stiffness. Typically, $p_i = 6 - 12\, k_i$. Therefore, the tires have little influence on the free vibrations and can be considered as rigid, as done in Fig. 10.3 (top). In Formula cars we have $p_i = 1 - 2\, k_i$.

On the other hand, the road disturbances involve also high frequencies, and tire stiffness has to be taken into account. For studying forced vibrations, the vehicle is then split into two half-car models, as in Fig. 10.3 (bottom), where

$$m_{s_1} = m_s \frac{a_2}{l} \quad \text{and} \quad m_{s_2} = m_s \frac{a_1}{l} \tag{10.9}$$

Instead of the half-car model, it is customary to use the *quarter car model*, which is like the half-car model with all quantities divided by two.

Both models are rather crude approximations, but nevertheless they can provide very useful insights on how to choose the springs and dampers (and, just in case, the inerters as well).

10.2   Quarter (Full) Car Model

The quarter car model is shown in Fig. 10.4. For simplicity we dropped the subscript in all quantities. The model consists of a sprung mass m_s connected via the primary suspension to the unsprung mass m_n of the axle. The suspension is supposed to have linear behavior with stiffness k and damping coefficient c. An inerter, with inertance b, is also included. The tire vertical elasticity is represented again by a linear spring p. The tire damping is so small that can be neglected.

Quite contrary to common practice, here we prefer not to split the car into four corners, neither into two halves. Instead, we retain the sprung mass of the *whole vehicle* as m_s. Consistently, we have to include in k the total stiffness of the four

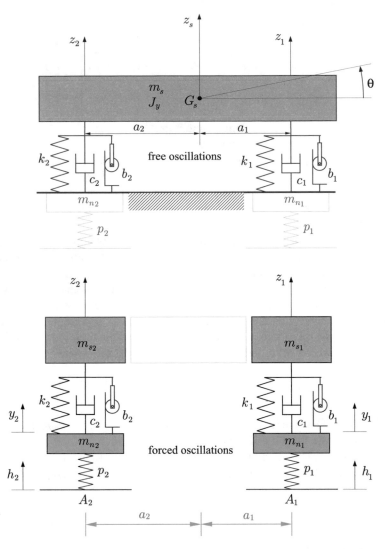

Fig. 10.3 "Extraction" of two-degree-of-freedom models to study free vibrations (top) and forced vibrations (bottom). Gray lines show the dropped parts

suspensions and so on. This approach has several advantages over the classical quarter car model:

- we deal with only one model instead of two (front and rear);
- we do not have to arbitrarily split the mass into front mass and rear mass;
- we are not tempted to use misleading concepts like the front/rear natural frequency.

Therefore, we are actually using a *full car model*. We still call it quarter car model just for the sake of uniformity with other books.

Fig. 10.4 Quarter car model
(better, full car model)

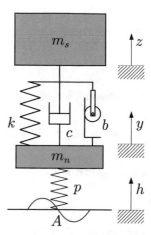

This quarter (full) car model is mainly used to study the vibrational behavior of
the vehicle when travelling on an uneven road. Therefore, the lowermost part of p
receives from the road a sinusoidal displacement $h(t) = H \cos \Omega t$. Someone may
object that real roads are not sinusoidal in shape. However, any road profile $g(x)$ of
length L can be expressed by its Fourier series [4]

$$g(x) = \sum_{n=0}^{\infty} \left[d_n \sin \left(\frac{2\pi n}{L} x \right) + e_n \cos \left(\frac{2\pi n}{L} x \right) \right], \qquad (10.10)$$

that is as an infinite sum of trigonometric functions. Fortunately, it is possible to take
only the first n terms without missing too much information. If the vehicle travels with
speed u, the Fourier term with spatial period L/n acts as a forcing displacement of
frequency $f_n = nu/L$. Therefore, the frequency of the excitation depends, obviously,
on the speed of the vehicle.

Because of the assumed *linearity* of the quarter car model, we can take advantage
of the *superposition principle*, and "feed" the system with one Fourier term at a time.
Should the system be nonlinear, this trick would be meaningless and we could no
longer apply a simple sinusoidal forcing function.

The quarter car model is a damped two-degree-of-freedom system. We employ as
coordinates the vertical displacement z of the sprung mass and the vertical displacement y (hop) of the unsprung mass. The road surface vertical displacement $h(t)$ can
be derived from the road surface profile and the car's speed. The equations of motion
of the quarter car model are readily obtained from Fig. 10.4 (recommended), or as a
special case of the equations given in Sect. 10.1

$$\begin{aligned}
m_s \ddot{z} &= -b(\ddot{z} - \ddot{y}) - c(\dot{z} - \dot{y}) - k(z - y) \\
m_n \ddot{y} &= -b(\ddot{y} - \ddot{z}) - c(\dot{y} - \dot{z}) - k(y - z) - p(y - h)
\end{aligned} \qquad (10.11)$$

where, as already stated, $h(t) = H \cos \Omega t$ is the excitation due to the road asperities. The same equations in matrix notation become

$$\mathbf{M}\ddot{\mathbf{w}} + \mathbf{C}\dot{\mathbf{w}} + \mathbf{K}\mathbf{w} = \mathbf{h} \tag{10.12}$$

with mass matrix $\mathbf{M}$

$$\mathbf{M} = \mathbf{M}_m + \mathbf{M}_b = \begin{bmatrix} m_s & 0 \\ 0 & m_n \end{bmatrix} + \begin{bmatrix} b & -b \\ -b & b \end{bmatrix} = \begin{bmatrix} m_s + b & -b \\ -b & m_n + b \end{bmatrix} \tag{10.13}$$

damping matrix $\mathbf{C}$

$$\mathbf{C} = \begin{bmatrix} c & -c \\ -c & c \end{bmatrix} \tag{10.14}$$

and stiffness matrix $\mathbf{K}$

$$\mathbf{K} = \mathbf{K}_k + \mathbf{K}_p = \begin{bmatrix} k & -k \\ -k & k \end{bmatrix} + \begin{bmatrix} 0 & 0 \\ 0 & p \end{bmatrix} = \begin{bmatrix} k & -k \\ -k & k+p \end{bmatrix} \tag{10.15}$$

We are mainly interested in the steady-state response, that is in the particular integral of the system of differential equations (10.11). In a case like this, it can be expressed as

$$\begin{aligned} z(t) &= Z \cos(\Omega t + \varphi) \\ y(t) &= Y \cos(\Omega t + \psi) \end{aligned} \tag{10.16}$$

that is in oscillations with the same angular frequency Ω of the excitation, but also with nonzero phases φ and ψ.

The mathematical analysis is much simpler if complex numbers are employed. The forcing function is therefore given as

$$h(t) = H(\cos \Omega t + i \sin \Omega t) = He^{i\Omega t} \tag{10.17}$$

with $H \in \mathbb{R}$. The steady-state solution is

$$\begin{aligned} z(t) &= Z\,[\cos(\Omega t + \varphi) + i \sin(\Omega t + \varphi)] = Ze^{i(\Omega t+\varphi)} = Ze^{i\varphi}e^{i\Omega t} = \mathsf{Z}e^{i\Omega t} \\ y(t) &= Y\,[\cos(\Omega t + \psi) + i \sin(\Omega t + \psi)] = Ye^{i(\Omega t+\psi)} = Ye^{i\psi}e^{i\Omega t} = \mathsf{Y}e^{i\Omega t} \end{aligned} \tag{10.18}$$

where $\mathsf{Z} = Ze^{i\varphi}$ and $\mathsf{Y} = Ye^{i\psi}$ are complex numbers with modulus Z and Y, and phases φ and ψ.

Inserting these expressions into (10.11) and dropping $e^{i\Omega t}$ provides the following algebraic system of equations in the complex unknowns Z and Y

$$\begin{cases} [(k - b\Omega^2) - m_s\Omega^2 + ic\Omega]Z - [(k - b\Omega^2) + ic\Omega]Y & = 0 \\ -[(k - b\Omega^2) + ic\Omega]Z + [p + (k - b\Omega^2) - m_n\Omega^2 + ic\Omega]Y & = pH \end{cases}$$
$$(10.19)$$

whose solution is

$$\frac{Z}{H} = \frac{p\,[(k - b\Omega^2) + ic\Omega]}{[(k - b\Omega^2) - m_s\Omega^2 + ic\Omega][p + (k - b\Omega^2) - m_n\Omega^2 + ic\Omega] - [(k - b\Omega^2) + ic\Omega]^2}$$

$$= p\,\frac{[(k - b\Omega^2) + ic\Omega]}{d(\Omega^2) + ic\Omega\,e(\Omega^2)} = \mathsf{G}_z(\Omega) \qquad (10.20)$$

and

$$\frac{Y}{H} = p\,\frac{[(k - b\Omega^2) - m_s\Omega^2 + ic\Omega]}{d(\Omega^2) + ic\Omega\,e(\Omega^2)} = \mathsf{G}_y(\Omega) \qquad (10.21)$$

where, for compactness,

$$d(\Omega^2) = m_s m_n \Omega^4 - \{[p + (k - b\Omega^2)]m_s + (k - b\Omega^2)\,m_n\}\Omega^2 + p\,(k - b\Omega^2)$$
$$e(\Omega^2) = p - (m_s + m_n)\Omega^2$$
$$(10.22)$$

The non-dimensional complex functions $\mathsf{G}_z(\Omega)$ and $\mathsf{G}_y(\Omega)$, given in (10.20) and (10.21), can be directly employed to obtain the steady-state solution

$$z(t) = H\,\mathsf{G}_z(\Omega)e^{i\Omega t}$$
$$y(t) = H\,\mathsf{G}_y(\Omega)e^{i\Omega t} \qquad (10.23)$$

From a practical point of view, we are mostly interested in the *amplitude* of these oscillations as functions of Ω

$$\frac{Z}{H} = \frac{|Z|}{H} = p\sqrt{\frac{(k - b\Omega^2)^2 + c^2\Omega^2}{d^2(\Omega^2) + c^2\Omega^2\,e^2(\Omega^2)}} = |\mathsf{G}_z(\Omega)| \qquad (10.24)$$

$$\frac{Y}{H} = \frac{|Y|}{H} = p\sqrt{\frac{[(k - b\Omega^2) - m_s\Omega^2]^2 + c^2\Omega^2}{d^2(\Omega^2) + c^2\Omega^2\,e^2(\Omega^2)}} = |\mathsf{G}_y(\Omega)| \qquad (10.25)$$

However, the phases can be obtained as well

$$\tan\varphi = \frac{\mathrm{Im}(Z)}{\mathrm{Re}(Z)} \qquad \tan\psi = \frac{\mathrm{Im}(Y)}{\mathrm{Re}(Y)} \qquad (10.26)$$

The amplitude of the vertical accelerations of the sprung and unsprung masses are given by $\Omega^2 Z$ and $\Omega^2 Y$, respectively.

Due to the oscillations, there are fluctuations in the vertical force exchanged by the tires with the road. More precisely, we have a sinusoidal force $Ne^{i\Omega t}$ superimposed on the constant force due to weight and, possibly, to aerodynamic downforces. From the quarter (full) car model of Fig 10.4 we get

$$\mathsf{N}e^{i\Omega t} = p(h - y) = p(H - \mathsf{Y})e^{i\Omega t} \tag{10.27}$$

From (10.21), we obtain the amplitude N as a function of the angular frequency Ω

$$
\begin{aligned}
\frac{N}{pH} = \frac{|\mathsf{N}|}{pH} &= \left| \frac{m_s m_n \Omega^4 - (m_s + m_n)\Omega^2[(k - b\Omega^2) + ic\Omega]}{d(\Omega^2) + ic\Omega\, e(\Omega^2)} \right| \\
&= \Omega^2 \sqrt{\frac{[m_s m_n \Omega^2 - (k - b\Omega^2)(m_s + m_n)]^2 + c^2 \Omega^2 (m_s + m_n)^2}{d^2(\Omega^2) + c^2 \Omega^2\, e^2(\Omega^2)}}
\end{aligned}
\tag{10.28}
$$

10.2.1   The Inerter as a Spring Softener

It is worth noting that all these expressions include the term $k - b\Omega^2$. This is the key to understand the *inerter* (also called J-Damper). It is pretty much like having a system whose suspension stiffness is sensitive to the frequency Ω of the excitation. At low frequencies $k - b\Omega^2 \simeq k$, but at high frequencies $k - b\Omega^2 \ll k$. The inertance b acts as a *spring softener*. This is a very interesting feature in Formula cars, with high aerodynamic loads, because we can use very stiff springs, thus limiting the spring deflection due to variable aerodynamic downforces, but at the same time the car will be able to absorb the high frequency road asperities, as if it were equipped with not-so-stiff springs. We will elaborate this idea quantitatively and in more detail in Sect. 10.3.3.

10.2.2   Quarter Car Natural Frequencies and Modes

A linear two-degree-of-freedom vibrating system, damped or not, has two natural modes, each one associated with its natural frequency.

To obtain these two modes, we consider the homogeneous counterpart of the system of differential equations (10.12)

$$\mathbf{M}\,\ddot{\mathbf{w}}_o + \mathbf{C}\,\dot{\mathbf{w}}_o + \mathbf{K}\,\mathbf{w}_o = \mathbf{0} \tag{10.29}$$

We seek a solution like

$$\mathbf{w}_o = \mathbf{x}e^{\mu t} \tag{10.30}$$

which, when inserted into (10.29), yields

$$e^{\mu t}(\mu^2 \mathbf{M} + \mu \mathbf{C} + \mathbf{K})\mathbf{x} = \mathbf{0} \tag{10.31}$$

The four values of μ that make (10.30) truly a solution are the roots of the characteristic equation

$$\det(\mu^2 \mathbf{M} + \mu \mathbf{C} + \mathbf{K}) = 0 \tag{10.32}$$

In an underdamped vibrating system, the four μ are complex numbers, complex conjugates in pairs

$$\mu_1 = -\zeta_1\omega_1 + i\omega_1\sqrt{1-\zeta_1^2} \qquad \mu_3 = \bar{\mu}_1 = -\zeta_1\omega_1 - i\omega_1\sqrt{1-\zeta_1^2}$$
$$\mu_2 = -\zeta_2\omega_2 + i\omega_2\sqrt{1-\zeta_2^2} \qquad \mu_4 = \bar{\mu}_2 = -\zeta_2\omega_2 - i\omega_2\sqrt{1-\zeta_2^2} \tag{10.33}$$

where $0 \le \zeta_i < 1$ are the damping ratios (or damping factors), and ω_i are close to the natural angular frequencies ω_{u_i} of the undamped system.[2] The two natural angular frequencies of the *damped* system (i.e., of the quarter car model) are

$$\omega_{d_i} = \omega_i\sqrt{1-\zeta_i^2} \tag{10.34}$$

Once the four μ_i have been obtained, we can go back to (10.31) and obtain the corresponding generalized eigenvectors $\mathbf{x}_l \in \mathbb{C}^2$, again complex conjugates in pairs. The general solution of (10.29) is given as linear combination of complex exponential functions

$$\mathbf{w}_o(t) = \gamma_1 \mathbf{x}_1 e^{(-\zeta_1\omega_1 + i\omega_{d_1})t} + \bar{\gamma}_1 \bar{\mathbf{x}}_1 e^{(-\zeta_1\omega_1 - i\omega_{d_1})t}$$
$$+ \gamma_2 \mathbf{x}_2 e^{(-\zeta_2\omega_2 + i\omega_{d_2})t} + \bar{\gamma}_1 \bar{\mathbf{x}}_2 e^{(-\zeta_2\omega_2 - i\omega_{d_2})t} \tag{10.35}$$

As an introduction to the general case, it is useful to study first two very special cases, that is $c = 0$ and $c = \infty$.

10.2.2.1 Undamped Quarter Car Model

Setting $c = 0$ makes the quarter car model a completely *undamped* two-degree-of-freedom vibrating system.

According to the expression of $d(\omega^2)$ in (10.22), the two natural angular frequencies ω_{u_1} and ω_{u_2} of the undamped system are the solutions of the algebraic equation

$$m_s m_n \omega^4 - \{[p + (k - b\omega^2)]m_s + (k - b\omega^2)\, m_n\}\omega^2 + pk = 0 \tag{10.36}$$

that is

---

[2] Let $\omega_{u_1} < \omega_{u_2}$, then $\omega_{u_1} < \omega_1$ and $\omega_2 < \omega_{u_2}$. Moreover, $\omega_{u_1}\omega_{u_2} = \omega_1\omega_2$.

$$\omega_{u_{1,2}}^2 = \frac{k(m_n + m_s) + (b + m_s)p}{2m_n m_s + 2b(m_n + m_s)}$$

$$\pm \frac{\sqrt{-4k[m_n m_s + b(m_n + m_s)]p + [k(m_n + m_s) + (b + m_s)p]^2}}{2m_n m_s + 2b(m_n + m_s)}$$

(10.37)

which, if there is no inerter b, simplifies into

$$\omega_{u_{1,2}}^2 = \frac{k(m_n + m_s) + m_s p \pm \sqrt{-4k(m_n m_s)p + [k(m_n + m_s) + m_s p]^2}}{2m_n m_s}$$

$$= \frac{1}{2}\left[\frac{p+k}{m_n} + \frac{k}{m_s} \pm \sqrt{\left(\frac{p+k}{m_n} - \frac{k}{m_s}\right)^2 + \frac{4k^2}{m_n m_s}}\right]$$

(10.38)

As already stated, in all road cars we have $m_s \gg m_n$ and $p \gg k$. Therefore, we can take the first-order Taylor expansion approximation of (10.38) for small values of m_n and k

$$\omega_{u_1}^2 \simeq \frac{kp}{(p+k)m_s} \quad \text{and} \quad \omega_{u_2}^2 \simeq \frac{p+k}{m_n}$$

(10.39)

In most cases, this very simple formulas provide very accurate estimates of the natural frequencies of the undamped quarter car model. For instance, with the data reported in the caption of Fig. 10.7, we get the following values using first the exact formula and then the approximate one

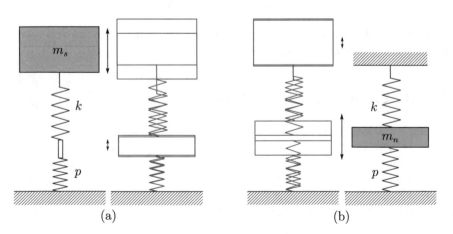

(a)   (b)

Fig. 10.5 One-degree-of-freedom systems for the *approximate* evaluation of the two natural frequencies of the *undamped* quarter car model (road cars only)

$$f_{u_1} = \frac{\omega_{u_1}}{2\pi} = 1.254 \text{ Hz} \simeq 1.255 \text{ Hz}$$

$$f_{u_2} = \frac{\omega_{u_2}}{2\pi} = 12.64 \text{ Hz} \simeq 12.63 \text{ Hz}$$

(10.40)

The results are almost identical. Typically, in road cars, $f_{u_2}/f_{u_1} \simeq 10$.

Of course there is a clear physical interpretation. The two approximate natural frequencies (10.39) would be the exact natural frequencies of the two one-degree-of-freedom systems shown in Fig. 10.5. Indeed, as also shown in Fig. 10.5, the two natural modes of the *undamped* quarter car model are very peculiar. For instance, again with the same data, the first mode, the one with $f_{u_1} = 1.2$ Hz, has $z(t) = 8.9y(t)$, whereas the second mode, with $f_{u_2} = 12.6$ Hz, has $z(t) = -y(t)/89$. That is, they look pretty much as if, in each mode, only one mass at the time were oscillating. Nonetheless, it is conceptually erroneous to think that sprung and unsprung mass have different natural frequencies [10, p. 104].

A Formula 1 car exhibits similar figures, although with some noteworthy differences. The undamped system has $f_{u_1} \simeq 5$ Hz with $z(t) = 2.5y(t)$, and $f_{u_2} \simeq 32$ Hz with $z(t) = -y(t)/25$.

It is very important to know that while the first natural mode is quite insensitive to damping, the second natural mode is very damping dependent. For instance, in a road car having what will be called the *optimal damping* c_{opt}, the first mode has $f_1 = 1.21$ Hz, which is very close to $f_{u_1} = 1.25$ Hz with no damping. Moreover, the amplitude of $z(t)$ is about 8.4 times the amplitude of $y(t)$, pretty much like in the undamped case. The second mode, on the other hand, has $f_{u_2} = 11.1$ Hz instead of $f_2 = 12.6$ Hz with no damping. But the most striking difference is that the amplitude of $y(t)$ is only about 12 times the amplitude of $z(t)$, instead of about 90 times, as it was with no damping. This is to say that we should not extrapolate results obtained with no damping to the real case, when there is a lot of damping because of the dampers.

10.2.2.2   Quarter Car Model with Stuck Damper

The other theoretical case is $c = \infty$, pretty much like having a stuck damper. The system behaves like an undamped one-degree-of-freedom system with one mass $m_s + m_n$ on top of a spring p (Fig. 10.6). There is only one natural frequency

$$\omega_c = \sqrt{\frac{p}{(m_s + m_n)}}$$

(10.41)

At first, it may appear a bit strange that $c = \infty$ leads to an undamped system. The effect of such an high value of c is to stick m_s and m_n together, thus leaving only the undamped oscillation with stiffness p.

Fig. 10.6 Quarter car model
with stuck damper, that is
with $c = \infty$

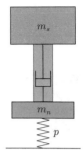

10.3   Damper Tuning

The quarter car model can now be used as a tool for the selection of the damping coefficient c of the damper. Of course, we have first to set up our goal. Typically, in road cars we are interested in minimizing the amplitude $\Omega^2 Z$ of the vertical acceleration $\ddot{z} = \Omega^2 Z e^{i\Omega t}$ of the sprung mass, thus optimizing the passenger comfort. On the other hand, in race cars we are more interested in minimizing the amplitude N of the oscillating part of the vertical force $N e^{i\Omega t}$, thus improving road holding.

10.3.1   Optimal Damper for Comfort

To select the right amount of damping to optimize passenger comfort, let us plot the normalized acceleration amplitude $\Omega^2 Z/H$ versus the angular frequency Ω of the road excitation. This is done in Fig. 10.7 for some values of c, including the two extreme cases $c = 0$ and $c = \infty$. The figure was obtained with $m_s = 1000$ kg, $m_n = 100$ kg, $k = 70$ kN/m and $p = 560$ kN/m, that is with $m_s = 10 m_n$ and $p = 8k$.

The plot for $c = 0$ and the plot for $c = \infty$ have four common points, marked by O, A, B and C in Fig. 10.7. Obviously, all other curves, for any value of $0 < c < \infty$, must pass through the same points.

The best curve, and hence the best value of the damping coefficient c, is perhaps the one with *horizontal tangent at point* A. It is a good compromise, as suggested in 1950 by Bourcier de Carbon [2]. As also shown in Fig. 10.7, lower or higher values of c would yield less uniform curves.

To obtain this optimal value c_{opt}, we have to impose that the derivative at A be zero

$$\left. \frac{\partial\big(\Omega^2 Z(c, \Omega)\big)}{\partial\Omega} \right|_{\Omega=\Omega_A} = 0 \qquad (10.42)$$

where $Z = Z(c, \Omega)$ is given in (10.24). The result is the sought *optimal damping coefficient* c_{opt}

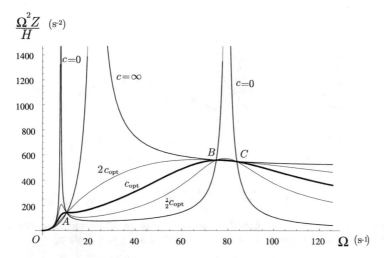

Fig. 10.7 Amplitude of the vertical acceleration of the sprung mass in a typical road car ($m_s = 1000$ kg, $m_n = 100$ kg, $k = 70$ kN/m and $p = 560$ kN/m)

$$c_{\text{opt}} = \sqrt{\frac{m_s k}{2}} \sqrt{\frac{p + 2k}{p}} \qquad (10.43)$$

where the second square root is quite close to one. With the data used to draw Fig. 10.7 we get $c_{\text{opt}} = 5916.08 \times 1.118 = 6614.38\,\text{N s/m}$. With this value of the damping coefficient, we have that the two natural modes of the quarter car model have, respectively, $\zeta_1 = 0.34$ and $\omega_{d_1} = \omega_1\sqrt{1 - \zeta_1^2} = 8.1\,\text{rad/s}$ for the first mode, and $\zeta_2 = 0.44$ and $\omega_{d_2} = \omega_2\sqrt{1 - \zeta_2^2} = 77.0\,\text{rad/s}$ for the second mode. We see that both modes are underdamped ($\zeta_i < 1$), but with a far from negligible amount of damping. A vehicle engineer should always bear in mind that the damping ratio ζ_1 of the first mode is usually something between 0.3 and 0.4 in road cars.

Another observation is in order here. Although the two values of ζ_i are quite similar, the time-rate decaying of the two modes, which depend on $\zeta_i \omega_i$, are drastically different because the two ω_i are quite far apart. For instance, in one second the amplitude of the first mode drops from 1 to $e^{-0.34 \times 8.61} = 0.05$, while that of the second mode drops to $e^{-0.44 \times 85.7} = 10^{-17}$. Quite a big difference.

It is worth noting that c_{opt} does not depend on the unsprung mass m_n. Therefore, it is not necessary to change the dampers when, for instance, mounting light alloy wheel rims. On the other hand, stiffer springs do require harder dampers.

Saying that m_n does not affect c_{opt} does not imply that the unsprung mass has no influence at all. The comfort performances for three different values of the ratio m_n/m_s are shown in Fig. 10.8. The lower the unsprung mass, the better, because the resulting curve is more uniform.

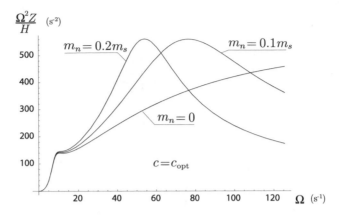

Fig. 10.8 Amplitude of the vertical acceleration of the sprung mass for three values of the unsprung mass (road car with $m_s = 1000$ kg, $c = c_{opt}$, $k = 70$ kN/m and $p = 560$ kN/m)

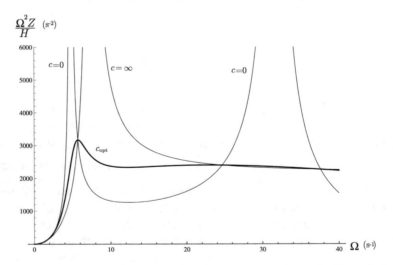

Fig. 10.9 Amplitude of the vertical acceleration of the sprung mass in a typical Formula 1 car

The formula for the optimal value of the damping coefficient here obtained perhaps works to get a close to optimal damping coefficient for a Formula 1 car as well. Figure 10.9 is the counterpart of Fig. 10.7. We see that the two Figures are quite different, but the c_{opt} curve is probably the best.

10.3.2 Optimal Damper for Road Holding

Needless to say that we need high vertical loads to have high friction forces. When the road is not flat, the vertical force fluctuations may impair road holding. Therefore, we are interested in how to determine the best damper tuning to counteract these force fluctuations as much as possible. The quarter car model can be usefully employed to this end. We have already obtained in (10.28) the expression of the amplitude of the sinusoidal component of the vertical load. Of course, it is superimposed on the vertical load due to weight, load transfers and, possibly, aerodynamic downforces.

The plot of the normalized amplitude $N/(pH)$ versus Ω is shown in Fig. 10.10 for several values of the damping coefficient c. As before, there are the curves for the extreme cases $c = 0$ and $c = \infty$. In this case there are only three fixed points O, $\hat{A}$ and $\hat{B}$. The curve corresponding to $c = \frac{1}{2}c_{\text{opt}}, c_{\text{opt}}, 2\,c_{\text{opt}}$ are also shown in Fig. 10.10. As before, we have assumed $m_s = 1000$ kg, $m_n = 100$ kg, $k = 70$ kN/m and $p = 560$ kN/m, that is $m_s = 10m_n$ and $p = 8k$.

The curve for $c = c_{\text{opt}}$ is not as good as it was with respect to comfort. For road holding optimization in road cars, it is better to use higher values of the damping coefficient c, that is $c > c_{\text{opt}}$.

Reducing the unsprung masses is very beneficial for road holding, as shown in Fig. 10.11. We see that the lower the unsprung mass, the lower the vertical force amplitude, and hence the better the road holding. Therefore, using light alloy wheels is certainly a way to improve road holding.

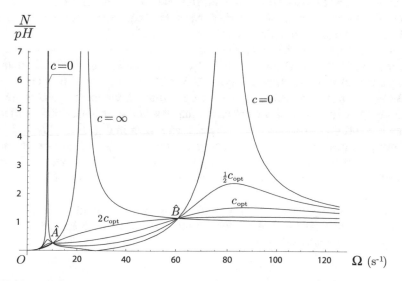

Fig. 10.10 Amplitude of the sinusoidal vertical load for a road car ($m_s = 1000$ kg, $m_n = 100$ kg, $k = 70$ kN/m and $p = 560$ kN/m)

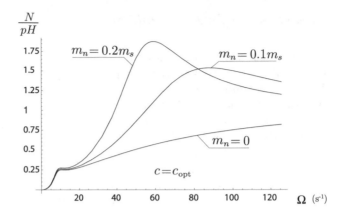

Fig. 10.11 Amplitude of the sinusoidal vertical load for a road car for three values of the unsprung mass and $c = c_{opt}$

10.3.3   The Inerter as a Tool for Road Holding Tuning

Formula cars, and Formula 1 cars in particular, have aerodynamic devices that provide fairly high downforces at high speed. These devices are most efficient if kept at constant distance from the road surface. To reduce the spring deflections under variable aerodynamic loads, very stiff springs have to be used. However, stiff springs are not very good to absorb road irregularities. Here is where the *inerter* comes into play. It works as a sort of spring softener at high frequencies, while being almost irrelevant with respect to static or slowly varying loads.

Let us have a look at the counterpart of Fig. 10.10 for, e.g., a Formula 1 car. The plot of $N/(pH)$ versus Ω for a Formula 1 car is shown in Fig. 10.12. Interestingly enough, the value of c_{opt} is optimal indeed. Any other value would be worse.

We are interested in increasing the spring stiffness k without impairing the suspension capability to filter down road irregularities. Unfortunately, simply stiffening the springs brings a worse plot of $N/(pH)$, as shown in Fig. 10.13 (dashed line). However, the inerter can help in balancing the stiffer spring, and, in fact we end up with a much better plot (thick solid line in Fig. 10.13). Typically, we can increase the stiffness by 10–20%, with an inertance of 25–100 kg per wheel in a Formula Indy car.

It is worth noting that in ordinary road cars the inerter would not be beneficial. This is due to the totally different values of mass, stiffnesses, etc. Indeed, Fig. 10.10 and Fig. 10.12 are very different.

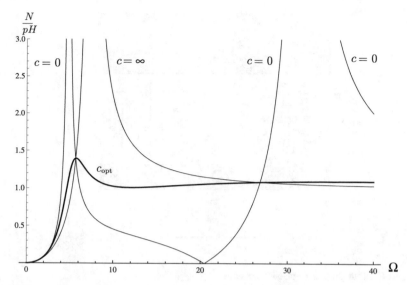

Fig. 10.12 Amplitude of the sinusoidal vertical load for a typical Formula 1 car

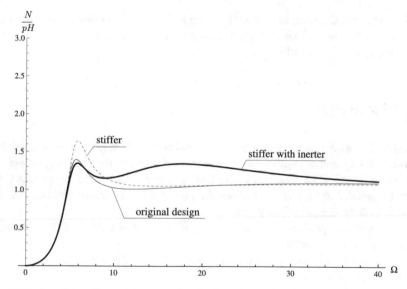

Fig. 10.13 Beneficial effect of the inerter in a Formula 1 car with stiffer springs

10.4 More General Suspension Layouts

More complex suspension layouts are possible. Some of them can be obtained by setting to zero some components in Fig. 10.14. Of course, never forget that dampers and inerters always need a spring in parallel to work properly.

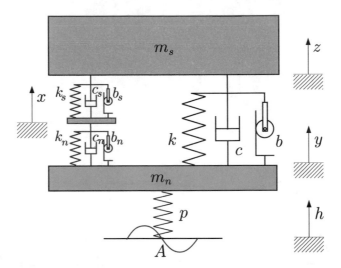

Fig. 10.14 Quite general suspension scheme

It is kind of interesting to note that setting $c_s = b_n = 0$ is not equivalent to $c_n = b_s = 0$. Finding the optimal configuration is not an easy task, but could in some cases turns out to be very rewarding.

10.5   Road Profiles

In probability theory, stationary ergodic process is a random process which exhibits both stationarity and ergodicity. In essence this implies that the random process will not change its statistical properties with time and that its statistical properties (such as the theoretical mean and variance of the process) can be deduced from a single, sufficiently long sample of the process.

Road elevation profiles are stationary ergodic processes. This allows for fairly simple statistical treatment.

The Fourier transform $F(\omega)$ is a very powerful tool to obtain the frequency feature of a given function $f(x)$

$$F(\omega) = \int_{-\infty}^{+\infty} f(x) e^{-i\omega x} dx \qquad (10.44)$$

The function $F(\omega) \in \mathbb{C}$ is precisely the frequency spectrum of $f(x)$.

We cannot apply directly the Fourier transform to a given road profile $g(x) \in \mathbb{R}$ because it does not tend to zero when $x \to \pm\infty$. However, we can introduce the spatial autocorrelation function $R_g(\tau)$ defined by

$$R_g(\tau) = \lim_{L \to \infty} \frac{1}{L} \int_{-L/2}^{+L/2} g(x)g(x+\tau)dx \tag{10.45}$$

where L is the length of the road with profile $g(x)$, and then compute its *power spectral density* (PSD) as its Fourier transform

$$S_g(s) = \int_{-\infty}^{+\infty} R(\tau)e^{-is\tau}d\tau \tag{10.46}$$

The power spectral density is measured in $m^2/$(cycles/m), if g is in meters and s is in cycles/m. Therefore, s is the spatial frequency.

If the vehicle travels at constant speed u, we can switch from the profile $g(x)$ to the time history $h(t)$ by means of the simple formula $h(t) = g(ut)$. The PSD $S_h(f)$, measured in m/Hz, of $h(t)$ can be obtained from $S_g(s)$ using

$$S_h(f) = \frac{S_g(f/u)}{u} \tag{10.47}$$

In general, if we know the PSD $S_h(f)$ of the excitation $h(t)$ and the frequency gain $G_z(\Omega)$ of the linear system at hand, we can easily obtain the PSD of the system response $z(t)$ as

$$S_z(f) = |G_z(2\pi f)|^2 S_h(f) \tag{10.48}$$

where, as well known, $\Omega = 2\pi f$.

For instance, the PSD $S_a(f)$ of the vertical acceleration $\ddot{z}$ of the sprung mass of the quarter car model is

$$S_a(f) = |(2\pi f)^2 G_z(2\pi f)|^2 S_h(f), \tag{10.49}$$

with $G_z(\Omega) = G_z(2\pi f)$ given in (10.20).

There is experimental evidence that the PSD of road profiles has a typical trend: the amplitude diminishes rapidly with the spatial frequency s. An often employed empirical formula for this behavior is

$$S_g(s) = Bs^{-k} \tag{10.50}$$

Unfortunately, there is not much agreement on the value of the exponent k. Typically it ranges between 2 and 4, including fractional values. The constant B characterizes the roughness of the road profile. The smoother the profile, the lower B. It is worth noting that the units to measure B are affected by the value of the exponent k.

According to (10.47), the counterpart of (10.50) in terms of time frequencies is

$$S_h(f) = Bu^{k-1}f^{-k} \tag{10.51}$$

which, obviously, shows that increasing the vehicle speed brings an increment in the PSD of the excitation.

10.6   Free Vibrations of Road Cars

The quarter car model looks at each axle as if it were alone. But it is not. Cars have two axles, and both take part in the vehicle body oscillations. Moreover, when we obtained the optimal value c_{opt} of the damping coefficient in (10.43) by means of the quarter car model, that was a function of the suspension stiffness k, beside the sprung mass m_s and the tire vertical stiffness p. But how was the stiffness k set? We do not have much freedom about m_s and p, and we may assume both of them as given for a certain kind of vehicle. But the stiffness k can be selected quite freely, for both front and rear axles.

Free oscillations are what happens right after the car has hit an isolated bump or hole. Since road cars usually do not employ the inerter, we use the even simpler two-degree-of-freedom model shown in Fig. 10.15, instead of the model of Fig. 10.3. As already discussed, we can safely consider the tires as rigid. The tires are indeed much stiffer than the springs, and at low frequencies (1–2 Hz) the unsprung masses oscillate very little. Moreover, the mode with higher natural frequency decays almost instantaneously, as already shown.

The analysis of the model of Fig. 10.15 will provide useful hints for the selection and tuning of the front and rear stiffnesses k_1 and k_2.

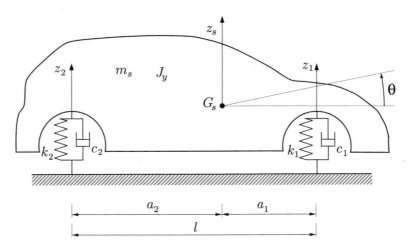

Fig. 10.15 Two-degree-of-freedom system for bounce and pitch analysis (rigid tires)

10.6.1 Governing Equations

To obtain all relevant equations for the two-degree-of-freedom vehicle model under investigation we follow the same path as in Sect. 10.1. We have (Fig. 10.15)

1. congruence equations:

$$z_1 = z_s + a_1 \theta$$
$$z_2 = z_s - a_2 \theta \tag{10.52}$$

that is a purely geometrical link between coordinates;
2. equilibrium equations:

$$m_s \ddot{z}_s = F_1 + F_2$$
$$J_y \ddot{\theta} = F_1 a_1 - F_2 a_2 \tag{10.53}$$

that is a link between forces or couples and accelerations; and
3. constitutive equations:

$$F_1 = -k_1 z_1 - c_1 \dot{z}_1$$
$$F_2 = -k_2 z_2 - c_2 \dot{z}_2 \tag{10.54}$$

When combined all together, they provide the governing equations

$$m_s \ddot{z}_s = -k_1(z_s + a_1\theta) - c_1(\dot{z}_s + a_1\dot{\theta}) - k_2(z_s - a_2\theta) - c_2(\dot{z}_s - a_2\dot{\theta})$$
$$J_y \ddot{\theta} = [-k_1(z_s + a_1\theta) - c_1(\dot{z}_s + a_1\dot{\theta})]a_1 - [-k_2(z_s - a_2\theta) - c_2(\dot{z}_s - a_2\dot{\theta})]a_2 \tag{10.55}$$

that can also be written in matrix notation as

$$\mathbf{M}\ddot{\mathbf{w}}_o + \mathbf{C}\dot{\mathbf{w}}_o + \mathbf{K}\mathbf{w}_o = \mathbf{0} \tag{10.56}$$

where $\mathbf{w}_o = (z_s, \theta)$. Formally, they look like (10.5), except for being homogeneous now. The 2×2 matrices are

$$\mathbf{M} = \begin{bmatrix} m_s & 0 \\ 0 & J_y \end{bmatrix} \tag{10.57}$$

$$\mathbf{C} = \begin{bmatrix} c_1 + c_2 & c_1 a_1 - c_2 a_2 \\ c_1 a_1 - c_2 a_2 & c_1 a_1^2 + c_2 a_2^2 \end{bmatrix} \tag{10.58}$$

and

$$\mathbf{K} = \begin{bmatrix} k_1 + k_2 & k_1 a_1 - k_2 a_2 \\ k_1 a_1 - k_2 a_2 & k_1 a_1^2 + k_2 a_2^2 \end{bmatrix} \tag{10.59}$$

As well known, the solutions of (10.56) are in the form $\mathbf{w}_o(t) = \mathbf{x}e^{\mu t}$, with μ and $\mathbf{x}$ such that

$$(\mu^2 \mathbf{M} + \mu \mathbf{C} + \mathbf{K})\mathbf{x} = \mathbf{0} \tag{10.60}$$

Quite surprisingly, it is common practice in the vehicle dynamics community to discard damping when studying free oscillations of a vehicle. Most books do that. But why?

Actually, vehicles have a lot of damping (in the quarter car model we obtained damping ratios ζ_i in the range 0.3–0.5). Perhaps they are the most damped system in mechanical engineering, and a good engineer cannot discard something which is not negligible at all. A rationale for neglecting damping should be provided, as a minimum. Unfortunately, in most cases there is just a sentence stating that damping will be neglected.

Free oscillations of undamped systems are much more predictable than those of a general damped system. Moreover, through modal analysis they can be treated as a collection of single-degree-of-freedom oscillators. But, we insist, vehicles are not undamped. They are very damped systems.

Fortunately, there is a way to have a damped system behave pretty much like an undamped system: it must have *proportional damping* (also called Rayleigh damping). Modes of proportionally damped systems preserve the simplicity of the real normal modes as in the undamped case, as we are going to discuss in a while.

10.6.2 Proportional Viscous Damping

The definition of proportional viscous damping is (e.g., [8, p. 522])

$$\mathbf{C} = \alpha\,\mathbf{M} + \beta\,\mathbf{K} \tag{10.61}$$

that is the damping matrix must be a linear combination of the mass and stiffness matrices, for suitable constants α and β.

Systems with proportional viscous damping have exactly the same mode shapes as the corresponding undamped systems. This is the key property.

The proof is quite simple. Inserting (10.61) into (10.56) and assuming, as usual, $\mathbf{w}_o(t) = \mathbf{x}e^{\mu t}$, we get

$$(\mu^2 + \mu\alpha)\mathbf{M}\,\mathbf{x} + (\mu\beta + 1)\mathbf{K}\,\mathbf{x} = \mathbf{0} \tag{10.62}$$

that is

$$\left(\frac{\mu^2 + \mu\alpha}{\mu\beta + 1}\right)\mathbf{M}\,\mathbf{x} = -\mathbf{K}\,\mathbf{x} \tag{10.63}$$

With respect to the general case (10.60), we have only two matrices instead of three. And it makes quite a big difference in the physical behavior of the vehicle, as will be shown hereafter.

Now, letting

$$\lambda = \frac{\mu^2 + \mu\alpha}{\mu\beta + 1} \quad \text{and} \quad \mathbf{A} = -\mathbf{M}^{-1}\mathbf{K} \tag{10.64}$$

we end up with *exactly the same eigenvalue problem as the undamped system*

$$\mathbf{A}\,\mathbf{x} = \lambda\,\mathbf{x} \tag{10.65}$$

which provides two real eigenvalues λ_1 and λ_2, and the corresponding *real* eigenvectors $\mathbf{x}_1$ and $\mathbf{x}_2$.

Solving the first equation in (10.64) with $\lambda = \lambda_1$, we obtain μ_1 and $\mu_3 = \bar{\mu}_1$. Similarly, solving with $\lambda = \lambda_2$ we obtain μ_2 and $\mu_4 = \bar{\mu}_2$. Therefore, we have apparently four μ_j and only two eigenvectors $\mathbf{x}_j$. The point is that the eigenvectors have real components, and hence coincide with their complex conjugates. Strictly speaking, we have two couples of identical eigenvectors.

The general solution, that is the free oscillations, for proportional damping (and hence also for no damping, which is just a special case of proportional damping) is[3]

$$\mathbf{w}_o(t) = \mathbf{x}_1\left(\gamma_1 e^{\mu_1 t} + \gamma_3 e^{\mu_3 t}\right) + \mathbf{x}_2\left(\gamma_2 e^{\mu_2 t} + \gamma_4 e^{\mu_4 t}\right) \tag{10.66}$$

Often, this equivalent expression, which only involves real quantities, is more convenient (cf. (7.238))

$$\mathbf{w}_o(t) = \chi_1\,\mathbf{x}_1 e^{-\zeta_1\omega_1 t}\sin(\omega_{d_1} t + \varphi_1) + \chi_2\,\mathbf{x}_2 e^{-\zeta_2\omega_2 t}\sin(\omega_{d_2} t + \varphi_2) \tag{10.67}$$

where

$$\mu_1 = -\zeta_1\omega_1 + i\,\omega_{d_1} \quad\text{and}\quad \mu_2 = -\zeta_2\omega_2 + i\,\omega_{d_2} \tag{10.68}$$

As usual in systems with proportional damping, ζ_j are the damping factors and ω_j are exactly the angular frequencies of the corresponding undamped system, while $\omega_{d_j} = \omega_j\sqrt{1 - \zeta_j^2}$ are the angular frequencies of the proportionally damped system.[4] The undamped system has $\lambda = \mu^2$, and hence

$$\omega_j = \sqrt{-\lambda_j} \quad\text{and}\quad \zeta_j = 0 \tag{10.69}$$

The four unknown constants depend on the four initial conditions.

The undamped and proportionally damped systems share almost everything, except the μ_j's. The really relevant aspect is that the eigenvectors $\mathbf{x}_j$ are exactly the same. This is the possible justification for "neglecting" the damping when studying the free oscillations of a vehicle. But the vehicle must be designed to have proportional damping, indeed. And a vehicle engineer should be well aware of this requirement.

---

[3] The quarter car model is a two-degree-of-freedom system whose damping is certainly not proportional. It is worth comparing (10.66) with the more general (10.35)

[4] The two natural frequencies of this model are not, of course, the two natural frequencies of the quarter car model. Another look at Fig. 10.3 should clarify the matter.

10.6.3   Vehicle with Proportional Viscous Damping

Looking at the three matrices (10.57), (10.58) and (10.59) for the case at hand, we see that the matrix $\mathbf{C}$ and the matrix $\mathbf{K}$ share the very same structure. Therefore, the only way to have proportional damping in a vehicle is to set $\alpha = 0$ and select springs and dampers such that

$$\beta = \frac{c_1}{k_1} = \frac{c_2}{k_2} \qquad (10.70)$$

thus having in (10.61) $\mathbf{C} = \beta\,\mathbf{K}$. This can be done fairly easily and cheaply.

From (7.220) we obtain

$$\lambda_{1,2} = -\frac{1}{2J_y m_s}\Bigg[J_y(k_1 + k_2) + m_s(k_1 a_1^2 + k_2 a_2^2) $$
$$\mp \sqrt{[J_y(k_1 + k_2) + m_s(k_1 a_1^2 + k_2 a_2^2)]^2 - 4 J_y m_s (a_1 + a_2)^2 k_1 k_2}\,\Bigg] $$
$$(10.71)$$

and the corresponding eigenvectors

$$\mathbf{x}_{1,2} = \Bigg(\frac{1}{2(k_1 a_1 - k_2 a_2)m_s}\Bigg[J_y(k_1 + k_2) - m_s(k_1 a_1^2 + k_2 a_2^2) $$
$$\mp \sqrt{[J_y(k_1 + k_2) + m_s(k_1 a_1^2 + k_2 a_2^2)]^2 - 4 J_y m_s (a_1 + a_2)^2 k_1 k_2}\,\Bigg], \ 1 \Bigg) $$
$$(10.72)$$

More compactly

$$\mathbf{x}_1 = (Z_{s_1}, 1) = \left(\frac{z_{s_1}(t)}{\theta_1(t)}, 1 \right) \qquad \text{and} \qquad \mathbf{x}_2 = (Z_{s_2}, 1) = \left(\frac{z_{s_2}(t)}{\theta_2(t)}, 1 \right) \qquad (10.73)$$

which means that the free oscillations are the linear combination of the two natural modes

$$z_s(t) = \chi_1 Z_{s_1} e^{-\zeta_1 \omega_1 t} \sin(\omega_{d_1} t + \varphi_1) + \chi_2 Z_{s_2} e^{-\zeta_2 \omega_2 t} \sin(\omega_{d_2} t + \varphi_2) = z_{s_1}(t) + z_{s_2}(t)$$
$$\theta(t) = \chi_1 e^{-\zeta_1 \omega_1 t} \sin(\omega_{d_1} t + \varphi_1) + \chi_2 e^{-\zeta_2 \omega_2 t} \sin(\omega_{d_2} t + \varphi_2) = \theta_1(t) + \theta_2(t)$$
$$(10.74)$$

The time histories for each mode are shown in Fig. 10.16. In each mode, the two coordinates move in a *synchronous way*. This is the key feature of systems with proportional damping.

Fig. 10.16 Time histories for bounce (top) and pitch (bottom) in case of proportional damping (synchronous motion)

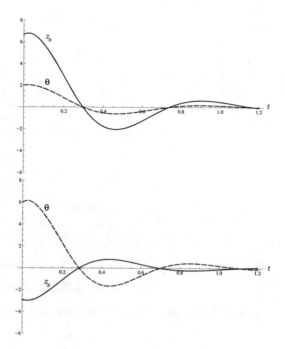

Each natural mode is an oscillation around a point P_i which has constantly zero vertical velocity. These points P_1 and P_2 are called *nodes*[5] and are defined as those points at which no vertical motion occurs when the system oscillates according to only one mode. Their position can be immediately obtained from (10.73). Each node P_j is at a horizontal distance d_j from G_s equal to Z_{s_j}, taken in the positive direction if Z_{s_j} is negative, and vice versa. In some sense, in a vehicle the eigenvectors can be visualized with a yardstick. This is not magic, it suffices to solve the equation

$$0 = \dot{z}_{s_j}(t) + d_j \dot{\theta}_j(t) \implies d_j = \frac{\dot{z}_{s_j}(t)}{\dot{\theta}_j(t)} = \frac{z_{s_j}(t)}{\theta_j(t)} = Z_{s_j} \qquad (10.75)$$

taking (10.73) into account.

The two natural modes and the corresponding nodes are shown in Fig. 10.17. Typically, the first mode, that is the one with lower natural frequency, has the node behind the rear axle. This mode is called *bounce*. The second mode has its node located ahead of G_s, near the front seat. This mode is called *pitch*.

Of course, it is not correct to speak of sprung natural frequencies at the front and at the rear of the car [10, p. 97].

We remark that *fixed nodes* are a prerogative of *proportional* damped systems. More general systems still have two natural modes, but in each mode the two coordinates $z_{s_j}(t)$ and $\theta_j(t)$ are no longer equal to zero simultaneously, i.e., the motion

---

[5] Other common names are motion centers or oscillation centers.

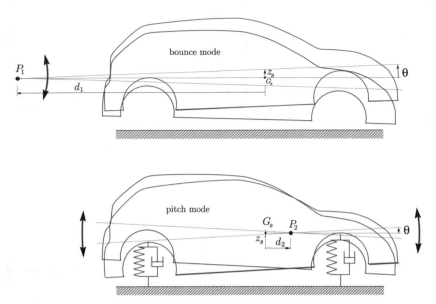

Fig. 10.17 Fixed nodes P_1 and P_2 of the two natural modes in case of proportional damping

is not synchronous. Therefore, their ratio $d_j(t)$ is a function of time and ranges from $-\infty$ to $+\infty$. At each time instant there is a different fixed point. We will discuss this topic further in Sect. 10.8.

10.6.4   Principal Coordinates

In a vehicle with proportional damping, the nodes P_1 and P_2 also mark where the *principal coordinates* z_b and z_p are, as shown in Fig. 10.18.

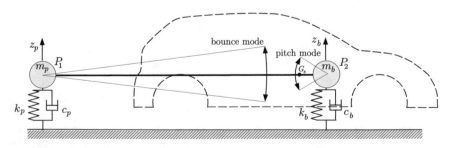

Fig. 10.18 Principal coordinates and equivalent system (proportional damping)

Let $\mathbf{S}$ be the matrix whose columns are the two eigenvectors (10.73), that is

$$\mathbf{S} = [\mathbf{x}_1 | \mathbf{x}_2]$$
(10.76)

The principal coordinates are equal to

$$\begin{bmatrix} z_b \\ z_p \end{bmatrix} = \mathbf{S}^{-1} \begin{bmatrix} z_s \\ \theta \end{bmatrix}$$
(10.77)

The key step is the diagonalization of the matrices. We have that

$$\begin{bmatrix} m_b & 0 \\ 0 & m_p \end{bmatrix} = \mathbf{S}^T \mathbf{M} \, \mathbf{S} \qquad \begin{bmatrix} c_b & 0 \\ 0 & c_p \end{bmatrix} = \mathbf{S}^T \mathbf{C} \, \mathbf{S} \qquad \begin{bmatrix} k_b & 0 \\ 0 & k_p \end{bmatrix} = \mathbf{S}^T \mathbf{K} \, \mathbf{S}$$
(10.78)

The system behaves precisely as if it were made up of two concentrated masses m_b and m_p, each one with its own spring k_b and k_p and damper c_b and c_p, respectively (Fig. 10.18). Obviously, we have that

$$2\zeta_1\omega_1 = \frac{c_b}{m_b} \qquad \omega_1^2 = \frac{k_b}{m_b}$$

$$2\zeta_2\omega_2 = \frac{c_p}{m_p} \qquad \omega_2^2 = \frac{k_p}{m_p}$$
(10.79)

10.6.5   Selection of Front and Rear Suspension Vertical Stiffnesses

In case of proportional damping, the shape of both modes (and hence the position of both nodes) depends on two nondimensional parameters: ρ and η.

The first parameter is the *dynamic index*

$$\rho = \frac{J_y}{m_s a_1 a_2}$$
(10.80)

It is a measure of how far the vehicle mass is distributed from its center of mass, with respect to the wheelbase. Of course, ρ depends on the whole vehicle architecture and it is very difficult to modify it. Usually, in ordinary road cars ρ ranges between 0.90 and 0.97 (Fig. 10.19). Cars with $\rho > 1$ must be like in Fig. 10.20, that is with the wheelbase much shorter than the whole vehicle length.

Another extremely important parameter is the ratio η

$$\eta = \frac{k_1 a_1}{k_2 a_2}$$
(10.81)

Fig. 10.19 A modern car with $0.9 < \rho < 1$

Fig. 10.20 An old car with $\rho > 1$

which characterizes how the axle stiffnesses relate to each other. Well tuned modern cars must have $\eta \simeq 0.95$.

Parameter η has a simple physical meaning. Just look at it as

$$\eta = \frac{a_1/k_2}{a_2/k_1} \tag{10.82}$$

It is the ratio between the static deflection at the rear $mga_1/(lk_2)$ and the static deflection at the front $mga_2/(lk_1)$.[6]

For a deeper comprehension of the possible effects of these two parameters ρ and η, we analyze the model of Fig. 10.15 in some special cases, before addressing how to tune the suspension stiffnesses in the general case.

For simplicity, we consider here the undamped system, whose governing equations are

$$m_s \ddot{z}_s + (k_1 + k_2)\, z_s + (k_1 a_1 - k_2 a_2)\, \theta = 0$$
$$J_y \ddot{\theta} + (k_1 a_1 - k_2 a_2)\, z_s + (k_1 a_1^2 + k_2 a_2^2)\, \theta = 0 \tag{10.83}$$

10.6.5.1 Case 1: $\eta = 1$

If the suspension stiffnesses are selected such that $\eta = 1$, that is

$$k_1 a_1 = k_2 a_2 \tag{10.84}$$

---

[6] Unfortunately, this physical interpretation often leads to the misconception that there are a front natural frequency and a rear natural frequency [6, p. 175].

the two equations in (10.83) become uncoupled. Both matrices are diagonal, which means that z_s and θ are the principal coordinates. The two undamped natural angular frequencies are

$$\omega_1 = \sqrt{\frac{k_1 + k_2}{m_s}}, \qquad \omega_2 = \sqrt{\frac{k_1 a_1^2 + k_2 a_2^2}{J_y}} \qquad (10.85)$$

Their ratio is equal, in this case, to the square root of the dynamic index

$$\left(\frac{\omega_1}{\omega_2}\right)^2 = \frac{J_y}{m_s a_1 a_2} = \rho \qquad (10.86)$$

The two eigenvalues are simply (cf. (10.72))

$$\mathbf{x}_1 = (1, 0) \quad \text{and} \quad \mathbf{x}_2 = (0, 1) \qquad (10.87)$$

Therefore, the bounce mode is a pure vertical translation and the pitch mode is a rotation around $G_s = P_2$.

10.6.5.2 Case 2: $\rho = 1$

Now, let us assume that a vehicle has $\rho = 1$, that is

$$J_y = m a_1 a_2 \qquad (10.88)$$

In this case the two principal coordinates are the vertical displacements z_1 and z_2 given in (10.2) and in Fig. 10.15, that is the displacements of the vehicle body at the two axles. After a little algebra, it is possible to rewrite the governing equations as

$$\begin{aligned} m_{s_1}\ddot{z}_1 + k_1 \, z_1 = 0, \\ m_{s_2}\ddot{z}_2 + k_2 \, z_2 = 0, \end{aligned} \qquad (10.89)$$

where

$$m_{s_1} = m_s \frac{a_2}{a_1 + a_2} \qquad m_{s_2} = m_s \frac{a_1}{a_1 + a_2} \qquad (10.90)$$

The undamped natural frequencies are

$$\omega_1 = \sqrt{\frac{k_1}{m_{s_1}}} \qquad \omega_2 = \sqrt{\frac{k_2}{m_{s_2}}} \qquad (10.91)$$

Their ratio is, in this case, equal to the square root of η

$$\left(\frac{\omega_1}{\omega_2}\right)^2 = \frac{k_1 \, a_1}{k_2 \, a_2} = \eta \qquad (10.92)$$

The two eigenvectors in the original coordinates z_s and θ are (cf. (10.72))

$$\mathbf{x}_1 = (a_2, 1) \quad \text{and} \quad \mathbf{x}_2 = (-a_1, 1) \qquad (10.93)$$

The nodes are precisely over the front axle and the rear axle, as expected. Otherwise, z_1 and z_2 would not be the principal coordinates.

10.6.5.3   Case 3: $\eta = 1$ and $\rho = 1$

But what happens if we set both η and ρ equal to one? From (10.86) and (10.92) we obtain that

$$\left(\frac{\omega_1}{\omega_2}\right)^2 = 1 \qquad (10.94)$$

that is the two undamped modes have exactly the same natural frequency.

The analysis of the shape of the two modes is more tricky. Apparently there is a paradox: the modes obtained before for $\eta = 1$ are not consistent with those obtained for $\rho = 1$, and vice versa. Which prevails? There is only one way out. Any point can be a node, that is, any vector $\mathbf{x}$ is an eigenvector. This happens because the matrix $\mathbf{A} = -\mathbf{M}^{-1}\mathbf{K}$ is like the identity matrix $\mathbf{I}$, times a suitable constant.

A vehicle designed to have $\eta = \rho = 1$ would have a very unpredictable behavior. As a matter of fact, a real vehicle could fulfill this condition only approximately. Therefore, the two nodes would be quite randomly located. Certainly, not a desirable behavior.

10.7   Tuning of the Suspension Stiffnesses

So far we have obtained the following results about the vehicle free oscillations:

1. tires can be considered as rigid;
2. damping should be proportional to the corresponding stiffness;
3. the two natural frequencies of the undamped system are very close to the natural frequencies of the proportionally damped system;
4. the shape of the modes of the undamped system are exactly equal to the shape of the modes of the proportionally damped system.

Now we can proceed to discuss how to choose k_1 and k_2. There are basically two requirements for road cars:

- both natural frequencies must fall in the range 1.0–1.5 Hz;
- the pitch mode should have its node located at about the front seat.

The first rule comes from the observation that oscillations at 1.0–1.5 Hz are quite comfortable for human beings. The second rule is an attempt to reduce the pitch motion of the driver. Pitch is typically more annoying than bounce.

As already stated, the value of ρ cannot be modified, unless the vehicle is completely redesigned. Modern road cars have (Fig. 10.19)

$$\rho \simeq 0.95 \tag{10.95}$$

To locate the pitch node on the front seat we can act on η, that is on the relative stiffnesses. Usually, a good value is

$$\eta \simeq 0.95 \tag{10.96}$$

With both η and ρ slightly lower than one, and with proportional damping, the car damped oscillations are like in Fig. 10.17, with the pitch node near the front seat and the bounce node quite far away behind the car. This is usually acknowledged as comfortable behavior.

For completeness, we provide the general formulas to compute the horizontal distances d_i from G_s of the nodes of bounce and pitch (Fig. 10.17), as functions of η and ρ

$$d_i = (a_1 + a_2)\frac{\rho - \eta + (\eta - 1)(\rho + 1)\chi \pm \sqrt{[(\eta - 1)(\rho - 1)\chi + \eta + \rho]^2 - 4\eta\rho}}{2(\eta - 1)} \tag{10.97}$$

where $\chi = a_2/(a_1 + a_2)$. Positive values of d_i means toward the rear axle, and vice-versa.

10.7.1 Optimality of Proportional Damping

Summing up, for a good suspension design we have found that we should fulfill these requirements

- $c_j \simeq c_{\text{opt}}$;
- $c_1/k_1 = c_2/k_2$ (proportional damping);
- $\eta \simeq 0.95$ (if also $\rho \simeq 0.95$).

But do they conflict with each other or not? Let us develop this point.

Optimal damping requires (cf. (10.43))

$$c_1 \simeq \sqrt{\frac{m_{s_1} k_1}{2}} \quad \text{and} \quad c_2 \simeq \sqrt{\frac{m_{s_2} k_2}{2}} \tag{10.98}$$

where $m_{s_1} = m_s a_2/l$ and $m_{s_2} = m_s a_1/l$. At the same time, proportional damping requires $c_1/k_1 = c_2/k_2 = \beta$, which combined with the former expression means

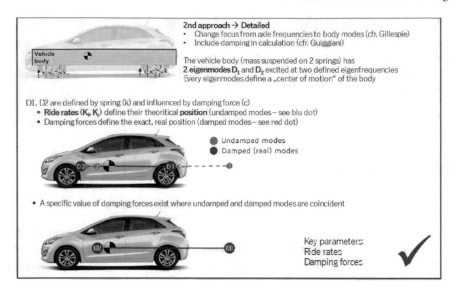

Fig. 10.21 Industrial application of proportional damping

$$\sqrt{\frac{m_{s_1}k_1}{2}}\frac{1}{k_1} \simeq \sqrt{\frac{m_{s_2}k_2}{2}}\frac{1}{k_2} \tag{10.99}$$

that is

$$\sqrt{\frac{m_s a_2}{k_1}} \simeq \sqrt{\frac{m_s a_1}{k_2}} \implies k_1 a_1 \simeq k_2 a_2 \implies \eta \simeq 1 \tag{10.100}$$

Therefore, we see that these three requirements do not conflict with each other.

We have insisted many times about having a vehicle with springs and dampers tuned to have proportional damping. The importance of this feature has been confirmed by Hyundai Motor Europe Technical Center GmbH in the communication entitled "A new R&H evaluation methodology applied during Hyundai i30 development", presented by Antonino Pizzuto at 2017 VI-grade Users Conference (Fig. 10.21).

As shown in Fig. 10.17, fixed nodes are a prerogative of proportionally damped systems. This is the outcome of having synchronous motion of both degrees of freedom in each natural mode, as shown in Fig. 10.16.

10.7.2   A Numerical Example

Crunching numbers helps a lot to grasp what we are really doing.

Let a vehicle have these features:

- sprung mass $m_s = 1000\,\text{kg}$ and moment of inertia $J_y = 1620\,\text{kgm}^2$;
- $a_1 = 1.2\,\text{m}$ and $a_2 = 1.5\,\text{m}$;
- axle vertical stiffnesses $k_1 = 31500\,\text{N/m}$ and $k_2 = 28000\,\text{N/m}$;
- proportional damping with $\beta = c_1/k_1 = c_2/k_2 = 0.0936\,\text{s}$.

We obtain immediately the dynamic index

$$\rho = \frac{J_y}{m_s a_1 a_2} = \frac{1620}{1800} = 0.9 \tag{10.101}$$

and the ratio

$$\eta = \frac{k_1 a_1}{k_2 a_2} = \frac{31.5 \times 1.2}{28.0 \times 1.5} = 0.9. \tag{10.102}$$

Both ρ and η are lower than one, although $k_1 > k_2$.

The matrix A is

$$\mathbf{A} = - \begin{bmatrix} 59.5 & -4.2 \\ -2.592 & 66.89 \end{bmatrix} \tag{10.103}$$

with eigenvalues

$$\lambda_1 = -58.24\,\text{s}^{-2} \qquad \lambda_2 = -68.15\,\text{s}^{-2} \tag{10.104}$$

and eigenvectors

$$\mathbf{x}_1 = (3.336, 1) \qquad \mathbf{x}_2 = (-0.486, 1) \tag{10.105}$$

The bounce mode has its node 3.336 m behind G_s, and hence $3.33 - 1.50 = 1.83$ m behind the rear axle (Fig. 10.17). The pitch mode has its node 0.486 m ahead of G_s.

Should the system be undamped, the natural frequencies would be

$$f_1 = \frac{\sqrt{-\lambda_1}}{2\pi} = 1.21\,\text{Hz} \qquad f_2 = \frac{\sqrt{-\lambda_2}}{2\pi} = 1.31\,\text{Hz} \tag{10.106}$$

These frequencies could be estimated by means of the simple formulas (10.85). The approximate values are $f_1 \simeq 1.23\,\text{Hz}$ and $f_2 \simeq 1.30\,\text{Hz}$, quite close to the exact ones although $\eta \neq 1$.

With proportional damping, we have to solve (10.64)

$$\mu^2 - \beta \lambda_i \mu - \lambda_i = 0 \tag{10.107}$$

with $\beta = c_1/k_1 = c_2/k_2 = 0.0936$ s, thus getting

$$\mu_{1,3} = -2.73458 \pm i\,7.12481\,\text{s}^{-1} \qquad \mu_{2,4} = -3.19975 \pm i\,7.60983\,\text{s}^{-1} \tag{10.108}$$

From the imaginary part we obtain the natural frequencies of the damped system

$$f_{s1} = \frac{\text{Im}(\mu_1)}{2\pi} = 1.13 \text{ Hz} \qquad f_{s2} = \frac{\text{Im}(\mu_3)}{2\pi} = 1.21 \text{ Hz} \qquad (10.109)$$

They are about 10% lower than those of the undamped system. Both fall within the acceptable range.

The bounce and pitch modes have $\zeta_1 = 0.36$ and $\zeta_2 = 0.39$, respectively. There is quite a lot of damping indeed.

If, just to see what happens, we set $J_y = 1980 \text{ kgm}^2$, thus having $\rho = 1.1$, we get that the bounce mode has $f_1 = 1.24$ Hz and its node located 2.93 m ahead of G_s, while the pitch mode has $f_2 = 1.16$ Hz and its node located at 0.67 m behind G_s. As expected, many things have been inverted, like the node positions and the frequency order.

10.8   Non-Proportional Damping

A vehicle with non-proportional damping has, in each natural mode, non-synchronous motion of the two degrees of freedom, as shown in Fig. 10.22, where the front damping coefficient has been reduced by 10%, while the rear damping coefficient has been increased by 10%. Also shown in Fig. 10.22 are the plots of $d_1(t)$ and $d_2(t)$, that is the time-varying positions of the nodes w.r.t. G_s

$$0 = \dot{z}_{s_j}(t) + d_j(t)\dot{\theta}_j(t) \implies d_j(t) = \frac{\dot{z}_{s_j}(t)}{\dot{\theta}_j(t)} \qquad (10.110)$$

These positions are functions of time and cycle from zero (when $\dot{z}_s = 0$) to $\pm\infty$ (when $\dot{\theta} = 0$). Therefore, the vehicle still has two modes, but their shapes are somehow mixed up. They are not so neatly different as they are with proportional damping. It is no longer possible to define the principal coordinates.

Actually, in some sense, both modes share some fundamental features. In both modes there are time instants in which $\dot{z}_s = 0$ with $\dot{\theta} \neq 0$, and hence the vehicle body is rotating around G_s, and other time instants in which $\dot{\theta} = 0$ with $\dot{z} \neq 0$, and hence the vehicle body is having a pure vertical translation.

Also observe that, differently from (10.75), the ratio in (10.110) do not extend to the ratio of coordinates.

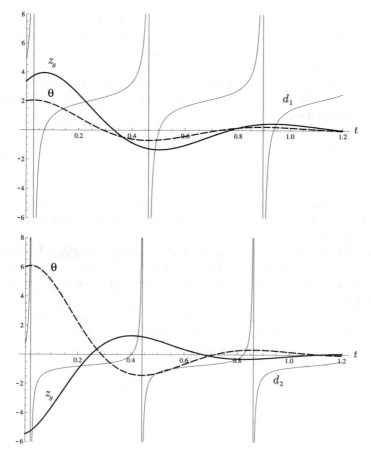

Fig. 10.22 Time histories for bounce (top) and pitch (bottom) in case of non-proportional damping (non-synchronous motion)

10.9   Interconnected Suspensions

So far we have employed the model of Fig. 10.15. Implicitly, we have considered it to be quite a general model for studying the ride of a two-axle vehicle. But it is not. Let us address the problem from a fresh point of view.

Still using z_s and θ as coordinates, a more general form of the equations of motion (10.83) for a linear two-degree-of-freedom undamped system are

$$
\begin{aligned}
m_s\,\ddot{z}_s + k_{zz}\,z_s + k_{z\theta}\,\theta &= 0 \\
J_y\,\ddot{\theta} + k_{\theta z}\,z_s + k_{\theta\theta}\,\theta &= 0
\end{aligned}
\tag{10.111}
$$

where $k_{z\theta} = k_{\theta z}$.

Each stiffness has a clear physical meaning. Let us impose a pure translation z_s to the system, that is with $\theta = 0$. The system reacts with a force $-k_{zz}\,z_s$ and a couple $-k_{\theta z}\,z_s$. Similarly, imposing a pure rotation around G_s, the system reacts with a force $-k_{z\theta}\,\theta$ and a couple $-k_{\theta\theta}\,\theta$.

In general, any 2×2 stiffness matrix is characterized by three coefficients. But in the system of Fig. 10.15 we have only two parameters, namely k_1 and k_2. Therefore the following equations

$$
\begin{aligned}
k_1 + k_2 &= k_{zz} \\
k_1 a_1 - k_2 a_2 &= k_{z\theta} \\
k_1 a_1^2 + k_2 a_2^2 &= k_{\theta\theta}
\end{aligned}
\tag{10.112}
$$

may not all be fulfilled. As anticipated, the scheme of Fig. 10.15 is not as general as it may seem at first. We need a suspension layout with three springs, although we still have only two axles.

Interconnected suspensions are the solution to this apparent paradox. A very basic scheme of interconnected suspensions is shown in Fig. 10.23. Its goal is to explain the concept, not to be a solution to be adopted in real cars (although, it was actually employed many years ago).

To understand how it works, first suppose the car bounces, as in Fig. 10.24. The springs contained in the floating device F get compressed, thus stiffening both axles. On the other hand, if the car pitches, as in Fig. 10.25, the floating device F just

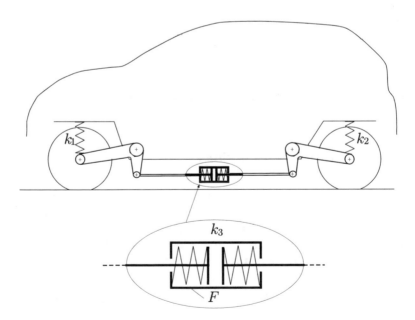

Fig. 10.23 Schematic for interconnected suspensions

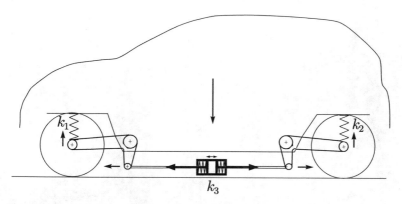

Fig. 10.24 Interconnected suspensions activated when bouncing

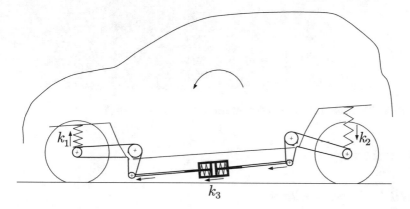

Fig. 10.25 Interconnected suspensions not activated when pitching

translates longitudinally, without affecting the suspension stiffnesses. This way we have introduced the third independent spring k_3 in our vehicle.

Obviously, hydraulic interconnections are much more effective, but the principle is the same. We have an additional parameter to tune the vehicle oscillatory behavior.

Although only a few cars have longitudinal interconnection, almost all cars are equipped with torsion (anti-roll) bars, and hence they have transversal interconnection. An example is shown in Fig. 10.26.

Using interconnected suspensions may lead to non-proportional damping, if proper counteractions are not taken, that is if the floating device F adds a stiffness k_3 without also adding a damping coefficient c_3.

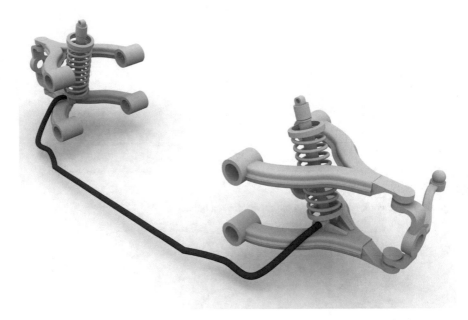

Fig. 10.26 Transversal interconnection by means of the anti-roll bar [7]

10.10   Exercises

10.10.1   Playing with η

By means of (10.97) *it is fairly easy to locate the nodes of bounce and pitch modes, in a vehicle with damping proportional to stiffness. Assuming $a_1 + a_2 = 2.6$ m and $\chi = a_2/(a_1 + a_2) = 0.5$, find d_1 and d_2 in the following cases:*

1. $\eta = 0.95$, $\rho = 0.95$;
2. $\eta = 0.99$, $\rho = 0.95$;
3. $\eta = 0.9$, $\rho = 0.95$;
4. $\eta = 0.6$, $\rho = 0.95$;
5. $\eta = 0.3$, $\rho = 0.95$.

Before jumping at the solution, try to figure out what the outcome can be.

Solution

To visualize and understand the results it is recommended to refer to Fig. 10.17

1. $d_1 = 3.06$ m, $d_2 = -0.52$ m;
2. $d_1 = 13.06$ m, $d_2 = -0.12$ m;
3. $d_1 = 2.03$ m, $d_2 = -0.79$ m;

4. $d_1 = 1.40\,\text{m}$, $d_2 = -1.14\,\text{m}$;
5. $d_1 = 1.33\,\text{m}$, $d_2 = -1.21\,\text{m}$.

10.10.2 Playing with ρ

By means of (10.97) *it is fairly easy to locate the nodes of bounce and pitch modes, in a vehicle with damping proportional to stiffness. Assuming $a_1 + a_2 = 2.6\,\text{m}$ and $\chi = a_2/(a_1 + a_2) = 0.5$, find d_1 and d_2 in the following cases:*

1. $\eta = 0.95$, $\rho = 0.95$;
2. $\eta = 0.95$, $\rho = 0.9$;
3. $\eta = 0.95$, $\rho = 0.8$;
4. $\eta = 0.95$, $\rho = 1$;
5. $\eta = 0.95$, $\rho = 1.1$;
6. $\eta = 1.1$, $\rho = 1.1$.

Before jumping at the solution, try to figure out what the outcome can be.

Solution

To visualize and understand the results it is recommended to refer to Fig. 10.17

1. $d_1 = 3.06\,\text{m}$, $d_2 = -0.52\,\text{m}$;
2. $d_1 = 5.35\,\text{m}$, $d_2 = -0.28\,\text{m}$;
3. $d_1 = 10.27\,\text{m}$, $d_2 = -0.13\,\text{m}$;
4. $d_1 = 1.3\,\text{m}$, $d_2 = -1.3\,\text{m}$;
5. $d_1 = 0.34\,\text{m}$, $d_2 = -5.41\,\text{m}$;
6. $d_1 = 3.29\,\text{m}$, $d_2 = -0.56\,\text{m}$.

Quite interesting the comparison between the first and the last cases.

10.11 Summary

In this chapter, the ride behavior of vehicles has been investigated. To keep the analysis very simple, two two-degree-of-freedom models have been formulated. The first, called quarter car model, has been used for determining the right amount of damping to have good comfort and/or road holding when the vehicle travels on a bumpy road (forced oscillations). In this framework, the inerter has been also introduced and discussed.

Free oscillations have been studied assuming the tires are perfectly rigid. The importance of proportional damping has been highlighted. This analysis has given indications on how to select spring stiffnesses.

Interconnected suspensions have been mentioned to show how to have a very general stiffness matrix.

10.12   List of Some Relevant Concepts

Section 10.1—the inerter is a device that provides a force proportional to the relative acceleration between its attachment points;

Section 10.2—the quarter car model is mainly used to study the vibrational behavior of a vehicle travelling on an uneven road;

Section 10.2.1—the inertance acts as a spring softener at high frequencies;

Section 10.3—the quarter car model is a tool for the selection of the damping coefficient of the dampers;

Section 10.6.2—systems with proportional viscous damping have exactly the same mode shapes as the corresponding undamped systems;

Section 10.6.3—only vehicles with proportional viscous damping have simple bounce and pitch modes.

10.13   Key Symbols

a_1	distance of G from the front axle
a_2	distance of G from the rear axle
b	inertance
c	damping coefficient
$\mathbf{C}$	damping matrix
c_i	damping coefficient
c_{opt}	optimal damping
f	frequency (Hz)
G	center of mass
h	vertical displacement
H	amplitude of the excitation
i	imaginary unit
l	wheelbase
J_y	moment of inertia
k	stiffness
$\mathbf{K}$	stiffness matrix
k_i	stiffness
$\mathbf{M}$	mass matrix
m_n	unsprung mass
m_s	sprung mass
N	amplitude of the vertical load on the tire
p	tire vertical stiffness

x	coordinate
$\mathbf{x}$	eigenvector
y	coordinate
Y	amplitude (real)
Y	amplitude (complex)
z	coordinate
z_b	principal coordinate (bounce)
z_p	principal coordinate (pitch)
z_s	vertical coordinate of G
Z	amplitude (real)
Z	amplitude (complex)
α	coefficient
β	coefficient
γ_i	coefficient
ζ_i	damping ratio
η	$k_1 a_1 / (k_2 a_2)$
θ	pitch rotation
λ	eigenvalue
μ	complex exponent
ρ	dynamic index
ω_i	natural angular frequency
ω_{d_i}	natural angular frequency of the damped system
Ω	angular frequency of the excitation

References

1. Bastow D, Howard G, Whitehead JP (2004) Car suspension and handling, 4th edn. SAE International, Warrendale
2. Bourcier de Carbon C (1950) Theorie mathématique et réalisation pratique de la suspension amortie des véhicules terrestres. In: 3rd Congres Technique de l'Automobile, Paris
3. Dixon JC (1991) Tyres suspension and handling. Cambridge University Press, Cambridge
4. Dixon JC (2009) Suspension geometry and computation. Wiley, Chichester
5. Font Mezquita J, Dols Ruiz JF (2006) La Dinámica del Automóvil. Editorial de la UPV, Valencia
6. Gillespie TD (1992) Fundamentals of vehicle dynamics. SAE International, Warrendale
7. Longhurst C (2013) https://www.carbibles.com/guide-to-car-suspension/
8. Palm WJ III (2007) Mechanical vibration. Wiley
9. Popp K, Schiehlen W (2010) Ground vehicle dynamics. Springer, Berlin
10. Seward D (2014) Race car design. Palgrave, London
11. Smith MC (2002) Synthesis of mechanical networks: the inerter. IEEE Trans Autom Control 47:1648–1662
12. Wong JY (2001) Theory of ground vehicles. Wiley, New York

Chapter 11
Tire Models

The global mechanical behavior of the wheel with tire has been addressed in Chap. 2. Basically, we have first found a way to describe the kinematics of a wheel with tire. This effort has led to the definition of the tire *slips*, as quantities that measure how far a tire is from pure rolling conditions. Then, the forces and couples that a tire receives from the road have been defined. The final step has been to investigate experimentally the link between these kinematic parameters and forces/couples.

In Chap. 2 no attempt was made to analyze what happens in the contact patch. That is, how the forces and couples are built by the elementary actions that arise at each point of the contact patch. This kind of analysis, however, is quite relevant for a real comprehension of the subtleties of vehicle setup.

In this chapter, what happens in the contact patch will be investigated by means of the so-called *brush model*. Great care will be devoted to clearly stating the assumptions on which this model is based. Moreover, the investigation will also cover the transient tire behavior. The final results are really interesting and enlightening.

11.1 Brush Model Definition

The *brush model* is perhaps the simplest *physical* tire model, yet it is quite significant and interesting. It is a tool to analyze qualitatively what goes on in the contact patch and to understand why the global mechanical behavior of a wheel with tire is, indeed, like in Figs. 2.20–2.25. Due to its simplicity,[1] the brush model is not always able to provide quantitative results. However, it is of great help in grasping some of the fundamental aspects of tire mechanics.

---

[1] Actually, the formulation presented here of the brush model is quite general, and hence it is a bit involved.

The Magic Formula, discussed in Sect. 2.11, provides curves that fit fairly well the experimental results, while the brush model attempts to describe the complex interaction between the tire and the road and how forces are generated. They are complementary approaches.

Basically, in the brush model, a belt equipped with infinitely many flexible bristles (the thread) is wrapped around a cylindrical rigid body (the rim), which moves on a flat surface (the roadway). In a well defined area (the contact patch), the tips of the bristles touch the ground, thus exchanging with the road normal pressures p and tangential stresses $\mathbf{t}$, provided the bristles also have a horizontal deflection $\mathbf{e}$. Each bristle is undistorted ($\mathbf{e} = \mathbf{0}$) when it enters the contact patch. A schematic of the brush model is shown in Fig. 11.1.

The brush model, as any mathematical model, relies on very many assumptions, more or less realistic. An attempt is made to clearly establish all of them, so that the impact of possible improvements can be better appreciated.

For generality, the model is first formulated for *transient conditions*. Steady-state behavior follows as a special case. The simplest case of translational slip only is discussed in Sect. 11.5.

Sound extensions of the brush model here presented have been recently published: [11–14].

11.1.1   Roadway and Rim

The *brush model*, like the tire, is something that connects the rim to the road. The roadway is assumed to be perfectly *flat*, like a geometric plane. The rim is modelled like a non-rolling cylindrical rigid body moving on the road, carrying on its outer surface a belt equipped with infinitely many flexible bristles (like a brush), which touch the road in the contact patch (Figs. 11.1 and 11.2). To simulate the rolling of the wheel, the belt slides on the rigid body with speed V_r (i.e., the rolling velocity defined in (2.50)).

11.1.2   Shape of the Contact Patch

As shown in Fig. 11.3, the contact patch $\mathcal{P}$ is assumed to be a convex, simply connected region. Therefore, it is quite different from a real contact patch, like the one in Fig. 2.4, which usually has lugs and voids.

It is useful to define a reference system $\hat{\mathsf{S}} = (\hat{x}, \hat{y}, \hat{z}; D)$ attached to the contact patch, with directions $(\mathbf{i}, \mathbf{j}, \mathbf{k})$ and origin at point D. Usually D is the center of the contact patch, as in Fig. 11.3. Directions $(\mathbf{i}, \mathbf{j}, \mathbf{k})$ resemble those of Fig. 2.6, in the sense that $\mathbf{k}$ is perpendicular to the road and $\mathbf{i}$ is the direction of the wheel pure rolling.

More precisely, the contact patch is defined as the region between the *leading edge* $\hat{x} = \hat{x}_0(\hat{y})$ and the *trailing edge* $\hat{x} = -\hat{x}_0(\hat{y})$, that is

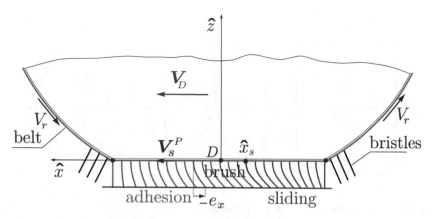

Fig. 11.1 Schematic of the brush model during braking (very important figure)

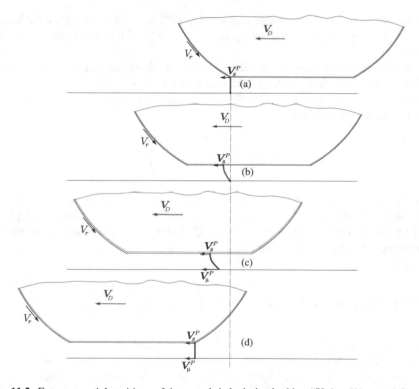

Fig. 11.2 Four sequential positions of the same bristle during braking ($|\mathbf{V}_D| > V_r$): **a** undeformed at the leading edge; **b** with growing deflection due to the tip stuck to the ground; **c** with lowering deflection with tip sliding on the ground; **d** undeformed at the trailing edge, with tip sliding on the ground

$$\mathcal{P} = \{(\hat{x}, \hat{y}) : \hat{x} \in [-\hat{x}_0(\hat{y}), \hat{x}_0(\hat{y})], \hat{y} \in [-b, b]\} \qquad (11.1)$$

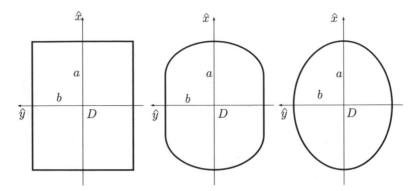

Fig. 11.3 Possible simple shapes of the contact patch

It is assumed for simplicity that the *shape* and *size* of the contact patch are not affected by the operating conditions, including the camber angle γ. Of course, this is not true in real tires.

For mathematical convenience, the contact patch is assumed here to be either a *rectangle*, centered at D, of length $2a$ and width $2b$ (Fig. 11.3, left), or an *ellipse*, again with axes of length $2a$ and $2b$ (Fig. 11.3, right). In the first case we have $\hat{x}_0 = a$, whereas in the second case

$$\hat{x}_0(\hat{y}) = \sqrt{a^2 \left(1 - \frac{\hat{y}^2}{b^2}\right)} \tag{11.2}$$

Typical values for a and b are in the range $0,04$–$0,08$ m. The rectangular shape is not a bad approximation of the contact patch of car tires (Fig. 2.4), while the elliptical one is better for motorcycle tires (Fig. 11.5). Occasionally, also a rounded rectangular contact patch is considered, as in Fig. 11.12.

11.1.3 Pressure Distribution and Vertical Load

Figures 11.4 and 11.5 show a typical pressure distribution as measured in a real motionless tire. The average ground pressure in the tire contact patch, considered as a single region, is not much higher than the tire inflation pressure. Of course there are high peaks near the tread edges.

A very simple pressure distribution $p(\hat{x}, \hat{y})$ on the contact patch $\mathcal{P}$, which roughly mimics the experimental results, may be *parabolic* along $\hat{x}$ and *constant* along $\hat{y}$

$$p = p(\hat{x}, \hat{y}) = p_0(\hat{y}) \frac{(\hat{x}_0(\hat{y}) - \hat{x})(\hat{x}_0(\hat{y}) + \hat{x})}{\hat{x}_0(\hat{y})^2} \tag{11.3}$$

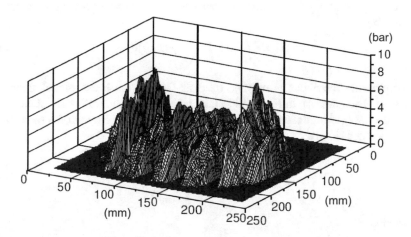

Fig. 11.4 Experimental results: pressure distribution for a motionless motorcycle tire [5]

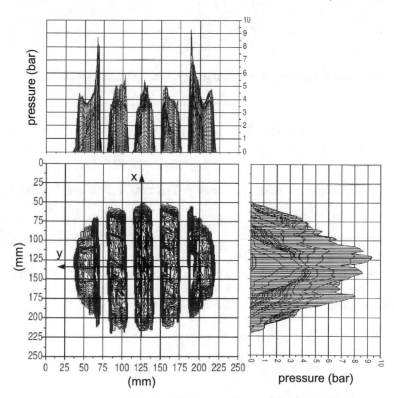

Fig. 11.5 Experimental results: contact patch and envelope of pressure distribution for a motionless motorcycle tire [5]

where $p_0(\hat{y}) = p(0, \hat{y})$ is the pressure peak value. The corresponding vertical load is given by

$$F_z = \int_{-b}^{b} d\hat{y} \int_{-\hat{x}_0(\hat{y})}^{\hat{x}_0(\hat{y})} p(\hat{x}, \hat{y}) d\hat{x} \tag{11.4}$$

Other pressure distributions may be used as well in the brush model, including nonsymmetric ones like in Fig. 2.43 to take into account the rolling resistance.

On a *rectangular* contact patch $\hat{x}_0(\hat{y}) = a$. Equation (11.3), with the same p_0 for all $\hat{y}$, becomes simply

$$p = p(\hat{x}, \hat{y}) = p_0 \left[1 - \left(\frac{\hat{x}}{a} \right)^2 \right] \tag{11.5}$$

and hence

$$F_z = \int_{-b}^{b} d\hat{y} \int_{-a}^{a} p(\hat{x}, \hat{y}) d\hat{x} = \frac{2}{3} p_0 2a 2b \tag{11.6}$$

which yields

$$p_0 = \frac{3}{2} \frac{F_z}{(2a)(2b)} \tag{11.7}$$

On an *elliptical* contact patch formulas (11.2) and (11.3) provide

$$p = p(\hat{x}, \hat{y}) = p_0 \left[1 - \frac{\hat{x}^2}{a^2 \left(1 - \frac{\hat{y}^2}{b^2} \right)} \right] \tag{11.8}$$

again with the same peak value p_0 for any $\hat{y}$. The total vertical load is given by

$$F_z = \int_{-b}^{b} d\hat{y} \int_{-\hat{x}_0(\hat{y})}^{\hat{x}_0(\hat{y})} p(\hat{x}, \hat{y}) d\hat{x} = \frac{1}{6} \pi p_0 2a 2b \tag{11.9}$$

The aspect ratio a/b of the contact patch is mainly determined by the shape of the tire. However, if F_z and p_0 are kept fixed, the product ab, and hence the total area of the contact patch, is not affected by the aspect ratio, as shown in Fig. 11.6.

11.1.4   Force–Couple Resultant

Exactly like in (2.14), the tangential stresses $\mathbf{t}(\hat{x}, \hat{y}, t)$ exerted by the road on the tire at each point of the contact patch yield a tangential force $\mathbf{F}_t$

Fig. 11.6 Footprints with different aspect ratio, but equal area

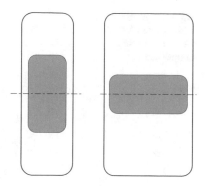

$$\mathbf{F}_t(t) = F_x\,\mathbf{i} + F_y\,\mathbf{j} = \int_{-b}^{b} d\hat{y} \int_{-\hat{x}_0(\hat{y})}^{\hat{x}_0(\hat{y})} \mathbf{t}(\hat{x}, \hat{y}, t)d\hat{x} \qquad (11.10)$$

and a vertical moment M_z^D with respect to point D

$$M_z^D(t)\,\mathbf{k} = \int_{-b}^{b} d\hat{y} \int_{-\hat{x}_0(\hat{y})}^{\hat{x}_0(\hat{y})} (\hat{x}\,\mathbf{i} + \hat{y}\,\mathbf{j}) \times \mathbf{t}(\hat{x}, \hat{y}, t)d\hat{x} \qquad (11.11)$$

All inertial effects, of any nature, are neglected.

11.1.5 Elastic Compliance of the Tire Carcass

Under pure rolling steady-state conditions, that is $F_x = F_y = M_z = 0$, let the position of D be the same of the point of virtual contact C, defined in (2.35) and in Fig. 2.14. We recall that, owing to the geometrical effect of camber γ, point C may not coincide with the origin O of the reference system defined in Sect. 2.4.1. However, in a car the camber angle is very small and hence $O \simeq C$.

Under general operating conditions, points D and C may have different positions on the road, mainly due to the *elastic compliance* of the carcass, as can be seen in Fig. 2.10. Therefore, as shown in Fig. 11.7

$$CD = \mathbf{q}(t) = q_x(t)\,\mathbf{i} + q_y(t)\,\mathbf{j} \qquad (11.12)$$

To approximately model the lateral and longitudinal compliance of the carcass, it has been assumed that the contact patch (with its reference system $\hat{\mathbf{S}}$) can have small rigid displacements q_x and q_y with respect to the rim, without changing its orientation. A linear relationship between $\mathbf{F}_t$ and $\mathbf{q}$ is the simplest option

$$\mathbf{F}_t = \mathbf{W}\mathbf{q} \qquad (11.13)$$

Fig. 11.7 Model of the
contact patch for taking into
account the carcass
compliance

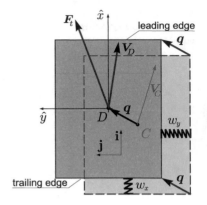

that is, using the vector components

$$F_x = w_x q_x(t) \quad \text{and} \quad F_y = w_y q_y(t) \tag{11.14}$$

if

$$\mathbf{W} = \begin{bmatrix} w_x & 0 \\ 0 & w_y \end{bmatrix} \tag{11.15}$$

with constant *carcass stiffnesses* w_x and w_y. Typically, $w_y \ll w_x$. Therefore, $\mathbf{F}_t$ and $\mathbf{q}$ are not parallel vectors.

The displacements q_x and q_y are usually quite small (i.e., $|q_x|, |q_y| \ll a$) and hence they can be neglected with respect to some phenomena, as will be discussed.

More advanced tire models may also include small rigid rotations of the contact patch [10], or employ the stretched string approach to model the carcass flexibility [1, 7, 9].

11.1.6 Friction

Let $V_\mu^P = |\mathbf{V}_\mu^P|$ be the magnitude of the *sliding velocity* $\mathbf{V}_\mu^P$, that is the velocity with respect to the road of a generic bristle tip with root at point $P = (\hat{x}, \hat{y})$,[2] and μ the *local friction coefficient*.[3]

Fairly general rules for *adhesion* and *sliding* between the bristle tip and the road are as follows

---

[2] For the very first time we look at the kinematics of points in the contact patch.
[3] Not to be confused with the global friction coefficients (2.87) and (2.89).

$$V_\mu^P = 0 \qquad\qquad \Longleftrightarrow \qquad |\mathbf{t}| < \mu\, p \qquad \text{(adhesion)} \qquad (11.16)$$

$$\mathbf{t} = -\mu\, p\, \frac{\mathbf{V}_\mu^P}{V_\mu^P} \qquad \Longleftrightarrow \qquad V_\mu^P \neq 0 \qquad \text{(sliding)} \qquad (11.17)$$

Equation (11.17) simply states that, at sliding, $\mathbf{t}$ and $\mathbf{V}_\mu^P$ have opposite direction and $|\mathbf{t}| = \mu p$. As a matter of fact, the ratio $\mathbf{V}_\mu^P / V_\mu^P$ is just a unit vector directed like the sliding velocity.

If thermal effects are neglected, μ may reasonably depend on the local value of the pressure p and of V_μ^P

$$\mu = \mu(p, V_\mu^P) \qquad (11.18)$$

It is common practice to call $\mu_0 = \mu(p, 0)$ the *coefficient of static friction* and $\mu_1 = \mu(p, V_\mu^P \neq 0)$ the *coefficient of kinetic friction*. In the present analysis, to keep it simple, we assume μ_0 and μ_1 to be *constant* all over the contact patch

$$\mu_0 = (1 + \chi)\mu_1, \quad \text{with } \chi > 0 \qquad (11.19)$$

thus discarding all dependencies on p and V_μ^P, except the switch from μ_0 to μ_1. Typically, $\mu_0 \approx 1, 2\mu_1$, that is $\chi \approx 0, 2$. More advanced friction models can be found, e.g., in [2, 3, 10].

11.1.7 Constitutive Relationship

The brush model owes its name to this section. It is indeed the constitutive relation that makes it possible to think of this model as having a moving belt equipped with infinitely many *independent* flexible bristles (Fig. 11.1).

Each massless bristle, while traveling in the contact patch, may have a horizontal deflection $\mathbf{e}(\hat{x}, \hat{y}, t) = e_x\,\mathbf{i} + e_y\,\mathbf{j}$ (Fig. 11.1). The key point is to assume that this deflection $\mathbf{e}(\hat{x}, \hat{y})$ does depend solely on the tangential stress $\mathbf{t}(\hat{x}, \hat{y}, t) = t_x\,\mathbf{i} + t_y\,\mathbf{j}$ at the very same point in the contact patch. In other words, each bristle behaves independently of the others: the constitutive relation is purely *local*. It is quite a strong assumption. Not very realistic, but terribly useful to get a simple model.

Actually, a truly simple model requires three further assumptions. In addition to being *local*, the constitutive relation need to be *linear*, *isotropic* and *homogeneous*, that is simply

$$\mathbf{t}(\hat{x}, \hat{y}, t) = k\,\mathbf{e}(\hat{x}, \hat{y}, t) \qquad (11.20)$$

where k is the *bristle stiffness*. In practical terms, it is the local thread stiffness. Usually, k ranges between 30 and 60 MN/m^3.

A linear, but anisotropic and non homogeneous constitutive relation would be like

Fig. 11.8 **a** isotropic
behavior, **b** anisotropic
behavior

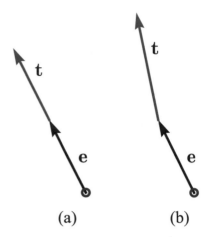

(a)   (b)

$$\begin{bmatrix} t_x \\ t_y \end{bmatrix} = \begin{bmatrix} k_{xx}(\hat{x}, \hat{y}) & k_{xy}(\hat{x}, \hat{y}) \\ k_{yx}(\hat{x}, \hat{y}) & k_{yy}(\hat{x}, \hat{y}) \end{bmatrix} \begin{bmatrix} e_x \\ e_y \end{bmatrix} \quad \text{that is} \quad \mathbf{t} = \boldsymbol{K}\mathbf{e} \qquad (11.21)$$

with $k_{xy} = k_{yx}$ and often equal to zero. It is anisotropic if $k_{xx} \neq k_{yy}$. It is non homo-geneous, if k's depend on their position $(\hat{x}, \hat{y})$ in the contact patch.

As shown in Fig. 11.8 and according to (11.20), isotropy implies that $\mathbf{t}$ and $\mathbf{e}$ always have the same direction.

It is worth noting that in (11.20) (and also in (11.21)) all quantities, including $\mathbf{t}$ and $\mathbf{e}$, are associated with the coordinates of the root, not of the tip of the bristle. Much like in the classical theory of linear elasticity, we are assuming that the problem can be safely formulated with reference to the undeformed state. This is reasonable provided the bristle deflections $\mathbf{e}$ are small, that is $|\mathbf{e}| \ll a$, which is usually the case.

11.1.8 Kinematics

We can define two fundamental *global* motions in the kinematics of a tire in contact with the ground (Fig. 11.9):

1. the continuous flow of undeformed rubber tread in the contact patch (due to the wheel rolling);
2. the motion of the contact patch with respect to the road.

The superposition of these two motions leads to what we call here the *skating velocity field* of the roots of the bristles.

For an in-depth discussion of some related topics, like the definitions of the *translational slip vector* $\boldsymbol{\sigma}$ and of the *spin slip* φ, we refer to Sect. 2.8.

Fig. 11.9 Kinematics of the
brush model (traction)

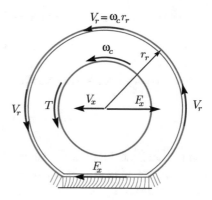

11.1.8.1 Belt Flow

As shown in Figs. 11.1 and 11.2, the first motion is modelled by assuming that the
belt (i.e., the root of each bristle) moves with respect to the rim with a velocity equal
to minus the *rolling velocity* $\mathbf{V}_r$

$$\mathbf{V}_r = V_r\,\mathbf{i} = \omega_c r_r\,\mathbf{i} \tag{11.22}$$

as defined in (2.50). This flow is always along parallel lines directed like $-\mathbf{i}$ in the
reference system $\hat{\mathsf{S}}$.

It is worth noting that in the brush model the rolling velocity may change in time
($V_r = V_r(t)$), but it must be the same at all points of the contact patch (it is a *global
parameter*).

This property makes it possible to define a sort of global *rolling distance* $s(t)$

$$s(t) = \int_0^t V_r(t)\mathrm{d}t \quad\text{that is}\quad \frac{\mathrm{d}s}{\mathrm{d}t} = V_r(t) \tag{11.23}$$

If $V_r > 0$, the function $s(t)$ is one-to-one. It will be shown that, in some cases, the
use of s as the independent variable is more convenient than the use of time t.

As already stated, the forefront border of the contact patch is called the leading
edge. It is very important to realize that it is through the leading edge that *undeformed*
rubber tread enters the contact patch (Fig. 11.1).

11.1.8.2 Motion of the Contact Patch

As shown in Fig. 11.7, the second fundamental motion is modelled by considering
the contact patch $\mathcal{P}$ as a rigid region that moves with respect to the road.

The velocity $\mathbf{V}_c = \mathbf{V}_C$, of the point of virtual contact C (see Sect. 2.7.2) is, by
definition, the *travel velocity* $\mathbf{V}_c$, defined in (2.55) for a real wheel.

From (2.57) and (2.66) it follows that

$$\mathbf{V}_s = \mathbf{V}_c - \mathbf{V}_r = V_r \boldsymbol{\sigma} \qquad (11.24)$$

where $\mathbf{V}_s = V_r \boldsymbol{\sigma}$ is the *slip velocity*, and $\boldsymbol{\sigma}$ is the *translational slip vector*

$$\boldsymbol{\sigma} = \frac{\mathbf{V}_s}{V_r} = \frac{\mathbf{V}_c - \mathbf{V}_r}{V_r} \qquad (11.25)$$

defined in Sect. 2.8 for real wheels.

Differentiating (11.12), we obtain the velocity $\mathbf{V}_D$ of the center D of the contact patch (Fig. 11.7)

$$\mathbf{V}_D = \mathbf{V}_C + \dot{\mathbf{q}} = \mathbf{V}_c + \dot{q}_x \, \mathbf{i} + \dot{q}_y \, \mathbf{j} \qquad (11.26)$$

Therefore, the generic point $\hat{P} = (\hat{x}, \hat{y})$ of the contact patch $\mathcal{P}$ (not of the tire) has a velocity $\mathbf{V}_{\hat{P}}$ equal to

$$\mathbf{V}_{\hat{P}} = \mathbf{V}_D + \omega_{s_z} \, \mathbf{k} \times D\hat{P} = \mathbf{V}_c + \dot{\mathbf{q}} + \omega_{s_z} \, \mathbf{k} \times (\hat{x} \, \mathbf{i} + \hat{y} \, \mathbf{j}) \qquad (11.27)$$

where ω_{s_z} is the slip yaw rate, as defined in (2.61). In the brush model there is slip yaw rate ω_{s_z} only within the contact patch, as if it were entirely due to the camber angle γ. Of course, if $\omega_{s_z} = 0$, we have $\mathbf{V}_{\hat{P}} = \mathbf{V}_D$.

11.1.8.3 Skating Velocity Field of the Bristle Roots

The combination of these two global motions yields the *local kinematics*, that is the motion of each *bristle root*.

The *root* of the bristle (momentarily) at point $\hat{P} = (\hat{x}, \hat{y}) = (\hat{x}_b(t), \hat{y})$ of the contact patch has a velocity $\mathbf{V}_s^P$ with respect to the ground given by the superimposition of the two global motions

$$
\begin{aligned}
\mathbf{V}_s^P(\hat{x}, \hat{y}, t) &= \mathbf{V}_{\hat{P}}(\hat{x}, \hat{y}, t) - \mathbf{V}_r(t) \\
&= \left(\mathbf{V}_c + \omega_{s_z} \, \mathbf{k} \times (\hat{x} \, \mathbf{i} + \hat{y} \, \mathbf{j}) + \dot{\mathbf{q}} \right) - \mathbf{V}_r \\
&= (\mathbf{V}_c - \mathbf{V}_r) + \omega_{s_z} \, \mathbf{k} \times (\hat{x} \, \mathbf{i} + \hat{y} \, \mathbf{j}) + \dot{\mathbf{q}} \\
&= \mathbf{V}_s + (\hat{x} \, \mathbf{j} - \hat{y} \, \mathbf{i})\omega_{s_z} + \dot{\mathbf{q}} \\
&= V_r[\boldsymbol{\sigma} - (\hat{x} \, \mathbf{j} - \hat{y} \, \mathbf{i})\varphi] + \dot{\mathbf{q}}
\end{aligned}
\qquad (11.28)
$$

where φ is the spin slip

$$\varphi = -\frac{\omega_{s_z}}{V_r} \qquad (11.29)$$

as in (2.65).

The velocity $\mathbf{V}_s^P$ of each bristle *root* is called here the *skating velocity*.[4] It is the velocity of the root of a bristle with respect to the road, not to be confused with the sliding velocity $\mathbf{V}_\mu^P$ of the bristle tip. Perhaps, a look at Fig. 11.18 can be useful to clarify the matter.

11.1.9 Brush Model Slips

The skating velocity field (11.28) of the brush model depends on the *translational slip* σ (11.25) and on the *spin slip* φ (11.29), exactly like in Sect. 2.8 for the rim of a real wheel with tire. This is no coincidence, as the kinematics of the brush model has been built around these slips.

However, the brush model behavior is better described if some other non-dimensional vectorial slips are defined.

11.1.9.1 Skating Slips

Equation (11.28) suggests to define the field of *skating slips* $\varepsilon(\hat{x}, \hat{y}, t)$

$$\varepsilon = \frac{\mathbf{V}_s^P}{V_r}$$
$$= \sigma - (\hat{x}\,\mathbf{j} - \hat{y}\,\mathbf{i})\varphi + \frac{\dot{\mathbf{q}}}{V_r} \tag{11.30}$$

Equation (11.28) can now be rewritten as

$$\mathbf{V}_s^P = V_r \varepsilon \tag{11.31}$$

Quite a compact formula.

A peculiar feature of the field of skating slips is that, whenever the spin slip $\varphi \neq 0$, they are *local*, in the sense that each point in the contact patch has its own $\varepsilon = \varepsilon(\hat{x}, \hat{y}, t)$. The bristle roots behave according to $\varepsilon(\hat{x}, \hat{y}, t)$.

11.1.9.2 Steady-State Skating Slips

Since it is very common to analyze the brush model assuming steady-state conditions ($\dot{\mathbf{q}} = \mathbf{0}$), it is convenient to define, at each point $(\hat{x}, \hat{y})$ in the contact patch, the field of *steady-state skating slips* $\boldsymbol{\lambda}(\hat{x}, \hat{y})$

$$\boldsymbol{\lambda}(\hat{x}, \hat{y}) = \sigma - (\hat{x}\,\mathbf{j} - \hat{y}\,\mathbf{i})\varphi \tag{11.32}$$

---

[4] The use of the practical slip κ would not have provided an equally neat formula.

11.1.9.3   Transient Translational Slip

By setting $\varphi = 0$ in (11.30), we can define the *transient translational slip* $\rho(t)$

$$\rho(t) = \sigma(t) + \frac{\dot{\mathbf{q}}(t)}{V_r(t)} \tag{11.33}$$

We see that ρ is a *global* quantity, like σ. They are the same at all points in the contact patch. However, unlike σ, it is not defined in a real wheel with tire, because it involves the velocity of the carcass deformation $\dot{\mathbf{q}}$.

Of course, both $\lambda(\hat{x}, \hat{y})$ and $\rho(t)$ are special cases of $\varepsilon(\hat{x}, \hat{y}, t)$

$$\varepsilon(\hat{x}, \hat{y}, t) = \lambda(\hat{x}, \hat{y}) + \frac{\dot{\mathbf{q}}(t)}{V_r} = \rho(t) - (\hat{x}\,\mathbf{j} - \hat{y}\,\mathbf{i})\varphi \tag{11.34}$$

We will see shortly that in the transient brush model the bristle roots behave according to ε, whereas the rim, by definition, behaves according to λ. This is the key to understand the physical meaning of ε.

11.1.10   Sliding Velocity of the Bristle Tips

To study the possible *sliding* of each bristle *tip* on the ground, let us consider the bristle root with coordinates $(\hat{x}, \hat{y}) = (\hat{x}_b(t), \hat{y})$.

According to (11.28), its root moves with respect to the road with a skating velocity $\mathbf{V}_s^P(\hat{x}, \hat{y}, t)$ (Figs. 11.1 and 11.2).

At the same time, the bristle has a deflection $\mathbf{e}(\hat{x}, \hat{y}, t)$, and hence, by definition, its tip has a velocity with respect to the root[5]

$$\dot{\mathbf{e}} = \frac{\mathrm{d}\mathbf{e}}{\mathrm{d}t} = \dot{\mathbf{e}}(\hat{x}, \hat{y}, t) \tag{11.35}$$

Therefore, the (possible) *sliding velocity* $\mathbf{V}_\mu^P$ of a bristle tip with respect to the road is given by the sum of these *two* vectorial contributions

$$\mathbf{V}_\mu^P(\hat{x}, \hat{y}, t) = \mathbf{V}_s^P + \dot{\mathbf{e}} \tag{11.36}$$

However, exactly like in fluid dynamics, it is more convenient to take a so-called Eulerian approach,[6] which provides

$$\dot{\mathbf{e}} = \frac{\mathrm{d}\mathbf{e}(\hat{x}_b(t), \hat{y}, t)}{\mathrm{d}t} = \frac{\partial \mathbf{e}}{\partial \hat{x}}\frac{\mathrm{d}\hat{x}_b}{\mathrm{d}t} + \frac{\partial \mathbf{e}}{\partial t} = -\mathbf{e}_{,\hat{x}}V_r + \mathbf{e}_{,t} \tag{11.37}$$

---

[5] The total time derivative is evaluated within $\hat{\mathsf{S}}$, that is as if $\mathbf{i}$ and $\mathbf{j}$ were fixed.

[6] As reported in [16, p. 4], this approach is actually due to d'Alembert.

since $d\hat{y}/dt = 0$ and where, for brevity, $\mathbf{e}_{,\hat{x}} = \partial\mathbf{e}/\partial\hat{x}$ and $\mathbf{e}_{,t} = \partial\mathbf{e}/\partial t$. The rationale of this last formula is that, again like in fluid dynamics, it is easier to keep fixed the observation point, rather than follow each bristle.

Combining (11.36) and (11.37), the *sliding velocity* $\mathbf{V}_\mu^P$ of a bristle tip with respect to the road is

$$\mathbf{V}_\mu^P(\hat{x}, \hat{y}, t) = \mathbf{V}_s^P + \dot{\mathbf{e}}$$
$$= V_r\varepsilon - V_r\mathbf{e}_{,\hat{x}} + \mathbf{e}_{,t} \qquad (11.38)$$
$$= V_r(\varepsilon - \mathbf{e}_{,\hat{x}}) + \mathbf{e}_{,t}$$

Of course, there is *adhesion* (i.e., no sliding) between the tip and the road if and only if $\mathbf{V}_\mu^P = \mathbf{0}$, like in (11.16) (see also Fig. 11.2a and b).

11.1.11 Summary of Relevant Velocities

A number of velocities, either global or local, have been defined or recalled in this section. It is perhaps useful to list all of them:

1. rolling velocity: $\mathbf{V}_r$, global;
2. sliding velocity: $\mathbf{V}_\mu^P$, local;
3. travel velocity: $\mathbf{V}_C = \mathbf{V}_c$, global;
4. carcass deformation velocity: $\dot{\mathbf{q}}$, global;
5. velocity of the center D: $\mathbf{V}_D$, global;
6. slip velocity: $\mathbf{V}_s$, global;
7. velocity of a generic point of the footprint: $\mathbf{V}_{\hat{p}}$, local;
8. skating velocity: $\mathbf{V}_s^P$, local;
9. bristle deflection velocity: $\dot{\mathbf{e}}$, local.

11.2 General Governing Equations of the Brush Model

The brush model has been completely defined in the previous section. A schematic was shown in Fig. 11.1. Its most distinguishing feature is that each bristle behaves independently of the others.

The fundamental governing equations for the *transient* behavior are to be obtained by combining all the relationships given in the brush model definition. Of course, the goal is

$$\text{rim kinematics} \qquad \Longleftrightarrow \qquad \text{force and moment} \qquad (11.39)$$

like in (2.85).

Therefore, we assume as given the following parameters:

- the shape of the contact patch (rectangular, elliptical, etc.);
- the size of the contact patch: a and b;
- the pressure distribution $p(\hat{x}, \hat{y})$;
- the grip coefficients $\mu_0 = (1 + \chi)\mu_1$;
- the bristle stiffness k;
- the carcass stiffnesses w_x and w_y;

and the following kinematic input functions:

- the rolling speed $V_r(t)$;
- the translational slip $\boldsymbol{\sigma}(t)$;
- the spin slip $\varphi(t)$.

We consider as unknown the functions $\mathbf{e}(\hat{x}, \hat{y}, t)$ and $\mathbf{q}(t)$, that is the field of bristle deflections and the longitudinal and lateral deflections of the carcass. Of course, the differential equations have to be supplied with suitable *initial conditions* on the whole contact patch and *boundary conditions* at the leading edge.

That said, let us dig into equations (relax, they look awful at first, but after a while their interplay will start to fascinate you, maybe…).

According to (11.16) and (11.38), and as exemplified in Fig. 11.2b, wherever there is *adhesion* between the tip and the road, the deflection $\mathbf{e}$ must change with the following time rate

$$\dot{\mathbf{e}} + \mathbf{V}_s^P = \mathbf{0} \qquad \Longleftrightarrow \qquad |k\mathbf{e}| < \mu_0\, p \qquad \text{(adhesion)} \qquad (11.40)$$

This is a complicated way to say simply that the tip does not move with respect to the road, while its root does.

The bristle tip starts *sliding* as soon as the friction limit is reached ($|\mathbf{t} = k\mathbf{e}| = \mu_0 p$). In some sense, adhesion has a higher priority than sliding.

Switching from adhesion to sliding means that the governing equation changes abruptly into (11.17), which, owing to (11.20) and (11.38), is equivalent to

$$k\mathbf{e} = -\mu_1 p\, \frac{\dot{\mathbf{e}} + \mathbf{V}_s^P}{|\dot{\mathbf{e}} + \mathbf{V}_s^P|} \qquad \Longleftrightarrow \qquad |\dot{\mathbf{e}} + \mathbf{V}_s^P| > 0 \qquad \text{(sliding)} \qquad (11.41)$$

This vectorial differential equation states that, whenever there is sliding, we have $k|\mathbf{e}| = \mu_1 p$, and the vectors $\mathbf{t} = k\mathbf{e}$ and $\mathbf{V}_\mu^P = \dot{\mathbf{e}} + \mathbf{V}_s^P$ have the same, unknown, direction.

Let us expand these observations. Sliding means that the deflection $\mathbf{e}$ is a vector whose intensity is equal to $\mu_1 p/k$, and is directed like the local sliding velocity $\mathbf{V}_\mu^P$. To fulfill simultaneously these two requirements there must be a nice interplay between $\mathbf{V}_s^P$ and $\dot{\mathbf{e}}$.

According to (11.37) and (11.38), Eqs. (11.40) and (11.41) can be recast as follows, where $\boldsymbol{\varepsilon} = \boldsymbol{\rho} - (\hat{x}\,\mathbf{j} - \hat{y}\,\mathbf{i})\varphi$

$$\mathbf{e}_{,\hat{x}} - \mathbf{e}_{,t}/V_r - \varepsilon = \mathbf{0} \qquad\qquad \Longleftrightarrow \qquad k|\mathbf{e}| < \mu_0\, p \qquad\qquad \text{(adhesion)}$$

$$(11.42)$$

$$k\mathbf{e} = \mu_1 p\, \frac{\mathbf{e}_{,\hat{x}} - \mathbf{e}_{,t}/V_r - \varepsilon}{|\mathbf{e}_{,\hat{x}} - \mathbf{e}_{,t}/V_r - \varepsilon|} \qquad \Longleftrightarrow \qquad |\mathbf{e}_{,\hat{x}} - \mathbf{e}_{,t}/V_r - \varepsilon| > 0 \quad \text{(sliding)}$$

$$(11.43)$$

with given boundary conditions at the leading edge $\hat{x} = \hat{x}_0(\hat{y})$

$$\mathbf{e}(\hat{x}_0(\hat{y}), \hat{y}, t) = \mathbf{0} \tag{11.44}$$

and initial conditions

$$\mathbf{e}(\hat{x}, \hat{y}, 0) = \mathbf{e}_0(\hat{x}, \hat{y}) \tag{11.45}$$

This is a two-state system, in the sense that only one partial differential equation applies at each point of the contact patch: we can either have adhesion or sliding, but not both (or none). By definition, adhesion means $|\mathbf{V}_\mu^P| = 0$ and the differential equation (11.43) of sliding is indeed meaningless.

A closer look shows that we have a different two-state system for any value of $\hat{y}$. Indeed, the spatial derivatives in (11.42) and (11.43) are only with respect to $\hat{x}$, that is in the direction $\mathbf{i}$ of the rolling velocity $V_r\, \mathbf{i}$. The rubber flows along parallel lines that do not interact to each other (in this model!).

However, the problem formulation needs an additional vectorial equation since $\dot{\mathbf{q}}$ is unknown, and so is $\rho(t) = \sigma + \dot{\mathbf{q}}/V_r$. Differentiating (11.13) with respect to time and taking (11.33) into account provides

$$\dot{\mathbf{F}}_t = \mathbf{W}\dot{\mathbf{q}} = \mathbf{W}\,(\rho - \sigma)V_r \tag{11.46}$$

Also useful is to insert the constitutive relationship (11.20) into (11.10) and then differentiate with respect to time

$$\dot{\mathbf{F}}_t = \frac{d}{dt}\left(k \int_{-b}^{b} d\hat{y} \int_{-\hat{x}_0(\hat{y})}^{\hat{x}_0(\hat{y})} \mathbf{e}\, d\hat{x}\right) = k \int_{-b}^{b} d\hat{y} \int_{-\hat{x}_0(\hat{y})}^{\hat{x}_0(\hat{y})} \mathbf{e}_{,t} d\hat{x} \tag{11.47}$$

Combining (11.46) and (11.47) yields the missing additional governing equation

$$k \int_{-b}^{b} d\hat{y} \int_{-\hat{x}_0(\hat{y})}^{\hat{x}_0(\hat{y})} \mathbf{e}_{,t} d\hat{x} = \mathbf{W}\dot{\mathbf{q}} = \mathbf{W}\,(\rho - \sigma)V_r \tag{11.48}$$

Summing up, the behavior of the *transient* brush model, that is the functions $\mathbf{e}(\hat{x}, \hat{y}, t)$ and $\rho(t)$, for given boundary conditions $\mathbf{e}(\hat{x}_0(\hat{y}), \hat{y}, t) = \mathbf{0}$ at the leading edge and initial conditions $\mathbf{e}(\hat{x}, \hat{y}, 0) = \mathbf{e}_0(\hat{x}, \hat{y})$ and $\rho(0) = \rho_0$, is completely defined by the *governing equations* (11.42) or (11.43), and (11.48).

Actually, a somehow more compact formulation of the very same problem can be obtained employing, instead of time t, the rolling distance s, defined in (11.23). Since there is a one-to-one correspondence between t and s, that is $t = t(s)$, and all time derivatives in the brush model are divided by $V_r(t) = \mathrm{d}s/\mathrm{d}t$, the general governing equations can be reformulated in terms of $\mathbf{e}(\hat{x}, \hat{y}, s)$ in the following way, with $\varepsilon = \rho - (\hat{x}\,\mathbf{j} - \hat{y}\,\mathbf{i})\varphi$

$$\mathbf{e},_{\hat{x}} - \mathbf{e},_s - \varepsilon = \mathbf{0} \qquad \Longleftrightarrow \qquad k|\mathbf{e}| < \mu_0\,p \qquad \text{(adhesion)} \quad (11.49)$$

$$k\mathbf{e} = \mu_1 p\,\frac{\mathbf{e},_{\hat{x}} - \mathbf{e},_s - \varepsilon}{|\mathbf{e},_{\hat{x}} - \mathbf{e},_s - \varepsilon|} \qquad \Longleftrightarrow \qquad |\mathbf{e},_{\hat{x}} - \mathbf{e},_s - \varepsilon| > 0 \quad \text{(sliding)} \quad (11.50)$$

along with

$$k \int_{-b}^{b} \mathrm{d}\hat{y} \int_{-\hat{x}_0(\hat{y})}^{\hat{x}_0(\hat{y})} \mathbf{e},_s\,\mathrm{d}\hat{x} = \mathbf{W}\frac{\mathrm{d}\mathbf{q}}{\mathrm{d}s} = \mathbf{W}(\rho - \sigma) \qquad (11.51)$$

where $\mathbf{e},_s = \partial\mathbf{e}/\partial s$. This formulation shows that the rolling velocity $V_r(t)$ does not have any influence on the behavior of the brush model with respect to the rolling distance s. The main reason is that all inertial effects have been neglected, as in (2.24).

Either in terms of t or s, this is quite a difficult mathematical problem if tackled in its full generality. Indeed, the *transient* behavior of a real wheel with tire (cf. (2.18)) is a rather difficult matter.

Fortunately, the brush model becomes much simpler under steady-state conditions, as discussed in Sect. 11.3. However, to deal with the simplest (and most classical) brush model you have to wait till Sect. 11.5, where the spin slip is set equal to zero and there is only translational slip. With a rectangular contact patch, as in Sect. 11.5.1, the whole model can be worked out analytically. Notwithstanding the very many simplifying assumptions, it is still an interesting and significant model.

11.2.1 Data for Numerical Examples

Almost all figures from here onwards in this chapter have been obtained with the following numerical values:

$$
\begin{array}{lll}
a = 7.5\,\mathrm{cm} & b = 5.6\,\mathrm{cm} & r_r = 25\,\mathrm{cm} \\[2pt]
\mu_0 = 1 & \chi = 0.2 & p_0 = 0.3\,\mathrm{MPa} \\[2pt]
k = 30\,\mathrm{MN/m^3} & w_x = 500\,\mathrm{KN/m} & w_y = 125\,\mathrm{KN/m}
\end{array}
\qquad (11.52)
$$

11.3 Brush Model Steady-State Behavior

The main, and most common, simplification is assuming the brush model to be in steady-state conditions (Fig. 11.1). Therefore, by definition, the field of bristle deflections $\mathbf{e}$ and the carcass deformation $\mathbf{q}$ are both time independent.

These conditions can be formulated as

- $\mathbf{e}_{,t} = \mathbf{0}$, and hence $\mathbf{e} = \mathbf{e}(\hat{x}(t), \hat{y})$, with no explicit time dependence;
- $\dot{\mathbf{q}} = \mathbf{0}$, which means that $\rho = \sigma$ is an input quantity for the tire model and it is constant in time.

The problem is substantially simpler, since the only unknown function is the field of bristle deflections $\mathbf{e}(\hat{x}, \hat{y})$, and both the adhesion and sliding zones are governed by first-order *ordinary* differential equations, with respect to the variable $\hat{x}$.

More in detail, the skating slip ε, defined in (11.30), becomes the steady-state skating slip λ, defined in (11.32)

$$\lambda(\hat{x}, \hat{y}) = \frac{\mathbf{V}_s^P(\hat{x}, \hat{y}, t)}{V_r(t)} = \sigma - (\hat{x}\mathbf{j} - \hat{y}\mathbf{i})\varphi \qquad (11.53)$$

with *constant* translational slip σ and *constant* spin slip φ. Therefore, the skating slip $\varepsilon = \lambda$ is a given, purely kinematic quantity, a known input to the model. It is worth noting that $\mathbf{V}_s^P$ and V_r may be still time dependent, but their ratio λ is not. According to (11.37), the total time derivative of the deflection of each bristle tip is given by

$$\frac{\dot{\mathbf{e}}}{V_r(t)} = -\frac{\partial \mathbf{e}}{\partial \hat{x}} = -\mathbf{e}'(\hat{x}, \hat{y}) \qquad (11.54)$$

where $\mathbf{e}'$ was introduced to stress that, in the brush model, $\hat{y}$ is more a parameter than a variable. Whenever $\mathbf{e}' \neq \mathbf{0}$, the bristle deflection changes as the bristle root changes its position with respect to the footprint (Fig. 11.2).

The sliding velocity (11.38) of each bristle tip becomes

$$\mathbf{V}_\mu^P(\hat{x}, \hat{y}, t) = V_r(\lambda - \mathbf{e}') \qquad (11.55)$$

Again, the ratio $\mathbf{V}_\mu^P(\hat{x}, \hat{y}, t)/V_r(t)$ is not time dependent.

11.3.1 Steady-State Governing Equations

According to (11.53) and (11.54), in the steady-state case the governing equations (11.42) and (11.43) of the brush model become (cf. [1, p. 761], [9, p. 83])

$$\mathbf{e}' - \boldsymbol{\lambda} = \mathbf{0} \qquad\qquad \Longleftrightarrow \qquad k|\mathbf{e}| < \mu_0\, p \qquad \text{(adhesion)} \qquad (11.56)$$

$$k\mathbf{e} = \mu_1 p\, \frac{\mathbf{e}' - \boldsymbol{\lambda}}{|\mathbf{e}' - \boldsymbol{\lambda}|} \qquad \Longleftrightarrow \qquad |\mathbf{e}' - \boldsymbol{\lambda}| > 0 \qquad \text{(sliding)} \qquad (11.57)$$

where $\mathbf{e}' = \mathbf{e}_{,\hat{x}}$.

These first-order differential equations in the unknown function $\mathbf{e}(\hat{x}, \hat{y})$, along with the boundary conditions at the leading edge, completely describe the behavior of the brush model.[7] Indeed, in this case the other Eq. (11.48) simply states $\rho = \sigma$.

As already remarked, this is a two-state system, since at each point there is, obviously, either adhesion or sliding. To distinguish between the solutions in the *adhesion* and in the *sliding* regions, we will use the symbols $\mathbf{e}_a$ and $\mathbf{e}_s$, respectively.

11.3.2   Adhesion and Sliding Zones

Each bristle, which behaves independently of the others, is *undeformed* when it enters the contact patch through the leading edge $\hat{x}_0(\hat{y})$. Its tip sticks to the ground (Fig. 11.2a) and, due to the skating velocity $\mathbf{V}_s^P$ between the bristle root and the road, a deflection $\mathbf{e}$ immediately starts to build up (Fig. 11.2b), along with a tangential stress $\mathbf{t} = k\mathbf{e}$. The physical interpretation of the adhesion equation $\mathbf{e}' = \boldsymbol{\lambda}$ is that the growth of the bristle deflection is completely and solely ruled by the wheel kinematics. It is not affected directly by the bristle stiffness k, neither by the pressure distribution.

On the other hand, the physical interpretation of the sliding equation is that the tangential stress $\mathbf{t}$ is always directed like the sliding velocity

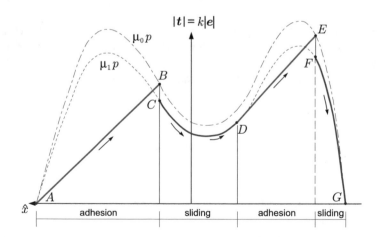

Fig. 11.10   Adhesion and sliding zones in the case $\lambda = \sigma = $ const. (very unusual pressure distribution!)

---

[7] More convenient governing equations for the sliding state are given in (11.62) and (11.63).

To better understand the roles played by adhesion and sliding, we refer to Fig. 11.10, where a fairly unusual pressure pattern has been depicted.

11.3.2.1 Adhesion

At first there is adhesion, and Eq. (11.56) applies with initial condition $\mathbf{e}_a = \mathbf{0}$ at $\hat{x}_0$ (point A in Fig. 11.10). A simple integration provides the behavior of the bristle deflection $\mathbf{e}_a$ in the adhesion zone

$$
\begin{aligned}
\mathbf{e}_a(\hat{x}, \hat{y}) &= \int_{\hat{x}_0}^{\hat{x}} \mathbf{e}' d\hat{x} = \int_{\hat{x}_0}^{\hat{x}} \boldsymbol{\lambda} d\hat{x} = \int_{\hat{x}_0}^{\hat{x}} [\boldsymbol{\sigma} - \varphi(\hat{x}\,\mathbf{j} - \hat{y}\,\mathbf{i})] d\hat{x} \\
&= -\boldsymbol{\sigma}\,(\hat{x}_0 - \hat{x}) + \varphi \left[\frac{(\hat{x}_0 - \hat{x})(\hat{x}_0 + \hat{x})}{2}\,\mathbf{j} - \hat{y}(\hat{x}_0 - \hat{x})\,\mathbf{i} \right]
\end{aligned}
\tag{11.58}
$$

It is worth noting that this expression is linear with respect to $\boldsymbol{\sigma}$ and φ. Moreover, it is not affected directly by the pressure distribution.

The magnitude of $\mathbf{e}_a$ is given by

$$
|\mathbf{e}_a| = \sqrt{\mathbf{e}_a \cdot \mathbf{e}_a} = (\hat{x}_0 - \hat{x})\sqrt{(\sigma_x + \varphi\hat{y})^2 + \left(\sigma_y - \varphi\frac{\hat{x}_0 + \hat{x}}{2}\right)^2}
\tag{11.59}
$$

Expressions (11.58) and (11.59) simplify considerably if $\varphi = 0$, that is $\boldsymbol{\lambda} = \boldsymbol{\sigma} =$ const.

Line A–B in Fig. 11.10 shows an example of linear growth ($\boldsymbol{\lambda} = \boldsymbol{\sigma}$). According to (11.56), the adhesion state is maintained as far as $k|\mathbf{e}_a| < \mu_0 p$, that is up to $\hat{x}_s = \hat{x}_s(\boldsymbol{\sigma}, \varphi, \hat{y})$ (point B in Fig. 11.10) where

$$
|\mathbf{t}| = k|\mathbf{e}_a(\hat{x}_s, \hat{y})| = \mu_0 p(\hat{x}_s, \hat{y})
\tag{11.60}
$$

In the proposed model, as soon as the static friction limit is reached at point $\hat{x} = \hat{x}_s$, the following sudden change in the deflection (massless bristle) occurs

$$
\mathbf{e}_s(\hat{x}_s, \hat{y}) = \frac{\mu_1}{\mu_0} \mathbf{e}_a(\hat{x}_s, \hat{y})
\tag{11.61}
$$

Therefore, at the transition from adhesion to sliding the deflection preserves its direction, but with a sudden reduction in magnitude (line B–C in Fig. 11.10).

11.3.2.2 Sliding

The sliding state starts with $\mathbf{e}_s(\hat{x}_s, \hat{y})$ as initial condition and evolves according to (11.57), that is to a system of two nonlinear first-order ordinary differential equations. However, (11.57) can be recast in a simpler, more convenient form

$$\mathbf{e}_s \cdot \mathbf{e}_s = \left(\frac{\mu_1 p}{k}\right)^2$$

$$(\mathbf{e}_s \times (\boldsymbol{\lambda} - \mathbf{e}'_s)) \cdot \mathbf{k} = 0 \tag{11.62}$$

that is, using components

$$e_x^2 + e_y^2 = \left(\frac{\mu_1 p}{k}\right)^2$$

$$e_x(\lambda_y - e'_y) = e_y(\lambda_x - e'_x) \tag{11.63}$$

which is a *differential–algebraic* system. Indeed, the sliding state requires:

- the magnitude of the tangential stress **t** to be equal to the kinetic coefficient of friction times the pressure (curved line C–D in Fig. 11.10)
- the direction of **t** (and hence of $\mathbf{e}_s$) to be the same as that of the sliding velocity $\mathbf{V}_\mu^P = V_r(\boldsymbol{\lambda} - \mathbf{e}'_s)$.

These are precisely the two conditions stated by (11.62) or (11.63).

Although, in general, the exact solution cannot be obtained by analytical methods, some features of the solution are readily available.

Let **s** be a unit vector directed like the sliding velocity $\mathbf{V}_\mu^P$, that is such that

$$\mathbf{V}_\mu^P = |\mathbf{V}_\mu^P|\mathbf{s} \tag{11.64}$$

or, equivalently, $\mathbf{t} = -|\mathbf{t}|\mathbf{s}$ and $\mathbf{e} = -|\mathbf{e}|\mathbf{s}$.

As well known, for any unit vector we have $\mathbf{s} \cdot \mathbf{s}' = 0$, where $\mathbf{s}' = \partial \mathbf{s}/\partial \hat{x}$. Therefore, $\mathbf{m} = \mathbf{s}'/|\mathbf{s}'|$ is a unit vector orthogonal to **s** (and hence to $\mathbf{V}_\mu^P$), and the skating slip $\boldsymbol{\lambda}$ can be expressed as

$$\boldsymbol{\lambda} = (\boldsymbol{\lambda} \cdot \mathbf{s})\mathbf{s} + (\boldsymbol{\lambda} \cdot \mathbf{m})\mathbf{m} \tag{11.65}$$

Moreover, according to (11.57)

$$\mathbf{e}_s = -\frac{\mu_1 p}{k}\mathbf{s} \quad \Longrightarrow \quad \mathbf{e}'_s = -\frac{\mu_1 p'}{k}\mathbf{s} - \frac{\mu_1 p}{k}\mathbf{s}' \tag{11.66}$$

Combining (11.64)–(11.66) we get

$$\frac{\mathbf{V}_\mu^P}{V_r} = \frac{|\mathbf{V}_\mu^P|}{V_r}\mathbf{s} = \boldsymbol{\lambda} - \mathbf{e}'_s$$

$$= (\boldsymbol{\lambda} \cdot \mathbf{s})\mathbf{s} + (\boldsymbol{\lambda} \cdot \mathbf{m})\mathbf{m} + \frac{\mu_1 p'}{k}\mathbf{s} + \frac{\mu_1 p}{k}|\mathbf{s}'|\mathbf{m} \tag{11.67}$$

$$= \left(\boldsymbol{\lambda} \cdot \mathbf{s} + \frac{\mu_1 p'}{k}\right)\mathbf{s}$$

which shows which terms actually contribute to the sliding velocity $\mathbf{V}_\mu^P$.

In most cases, the sliding regime is preserved up to the trailing edge, that is till the end of the contact patch. However, it is interesting to find the conditions that can lead the bristle to switch back to adhesion (like point D in Fig. 11.10). From (11.67) it immediately arises that

$$|\mathbf{V}_\mu^P| = 0 \qquad \Longleftrightarrow \qquad k\boldsymbol{\lambda} \cdot \mathbf{s} + \mu_1 p' = 0 \tag{11.68}$$

Since $\mathbf{s}$ depends on the solution $\mathbf{e}_s$ of the algebraic-differential system of Eq. (11.63), this condition has to be checked at each numerical integration step.

The governing equation (11.57) of the sliding state deserves some further discussion. The "annoying" term $(\boldsymbol{\lambda} - \mathbf{e}_s')/|\boldsymbol{\lambda} - \mathbf{e}_s'|$ is simply equal to $\boldsymbol{\lambda}/|\boldsymbol{\lambda}|$ if $\mathbf{e}_s$ and $\boldsymbol{\lambda}$ are parallel vectors. This observation may suggest the following approximate approach to (11.57)

$$k\mathbf{e}_f = -\mu_1 p \frac{\boldsymbol{\lambda} - \mathbf{e}_f'}{|\boldsymbol{\lambda}|}$$
$$k\tilde{\mathbf{e}}_s = -\mu_1 p \frac{\mathbf{e}_f}{|\mathbf{e}_f|} \tag{11.69}$$

First we solve two *separate linear* differential equations (not a system) for the two components of the "fictitious" deflection $\mathbf{e}_f$. Then, we obtain the *approximate* deflection $\tilde{\mathbf{e}}_s$ in the sliding region as a vector with magnitude $\mu_1 p/k$ and directed like $\mathbf{e}_f$. We remind that linear first-order differential equations can always be solved by integration (see, e.g., [17, p. 410]).[8] In many cases $\tilde{\mathbf{e}}_s$ is a very good approximation of $\mathbf{e}_s$.

An even simpler, less accurate, but often employed idea is to assume that the governing equation in the sliding state is just an algebraic equation

$$k\hat{\mathbf{e}}_s = -\mu_1 p \frac{\boldsymbol{\lambda}}{|\boldsymbol{\lambda}|} \tag{11.70}$$

Therefore, we allow a sudden discontinuity of the direction of the deflection at the transition from adhesion to sliding.[9] This is not correct, but very appealing because of its simplicity. Of course, as already mentioned, (11.70) is exact if $\mathbf{e}_s$ and $\boldsymbol{\lambda}$ happen to be parallel throughout the whole sliding region, that is if $\varphi = 0$ and hence $\boldsymbol{\lambda} = \boldsymbol{\sigma}$.

---

[8] The solution of $y' + f(x)y = g(x)$ is

$$y(x) = \exp\left(-\int^x f(t)dt\right)\left[\int^x \exp\left(\int^z f(t)dt\right)g(z)dz + C\right]$$

[9] This approach can be found in [4].

11.3.3 Force–Couple Resultant

The solution of the steady-state brush model shows whether there is adhesion or sliding at each point of the contact patch $\mathcal{P}$ and provides the corresponding bristle deflection $\mathbf{e}_a(\hat{x}, \hat{y})$ or $\mathbf{e}_s(\hat{x}, \hat{y})$. Therefore, the tangential stress $\mathbf{t}$ at each point of $\mathcal{P}$ is

$$\mathbf{t}(\hat{x}, \hat{y}) = \begin{cases} \mathbf{t}_a = k\mathbf{e}_a(\hat{x}, \hat{y}) & \text{(adhesion)} \\ \mathbf{t}_s = k\mathbf{e}_s(\hat{x}, \hat{y}) & \text{(sliding)} \end{cases} \tag{11.71}$$

Like in (2.14) and (11.10), the tangential force $\mathbf{F}_t = F_x\,\mathbf{i} + F_y\,\mathbf{j}$ that the road applies on the tire model is given by the integral of $\mathbf{t}$ over the contact patch

$$\mathbf{F}_t(\boldsymbol{\sigma}, \varphi) = \int_{-b}^{b} d\hat{y} \int_{-\hat{x}_0(\hat{y})}^{\hat{x}_0(\hat{y})} \mathbf{t}(\hat{x}, \hat{y}) d\hat{x} \tag{11.72}$$

which is a function, among other things, of the global slips $\boldsymbol{\sigma}$ and φ.[10]

It may be convenient to use the nondimensional or *normalized* tangential force $\mathbf{F}_t^n$ and its components [8]

$$\mathbf{F}_t^n = F_x^n\,\mathbf{i} + F_y^n\,\mathbf{j} = \frac{\mathbf{F}_t}{F_z} = \frac{F_x\,\mathbf{i} + F_y\,\mathbf{j}}{F_z} \tag{11.73}$$

Of course, under whichever operating condition of the brush model, we always have $|\mathbf{F}_t^n| < \mu_0$. It is quite interesting to find the combination of σ_x, σ_y and φ which provides the highest possible value. Equations (2.87) and (2.89) address a similar issue in an experimental context.

The overall moment of the tangential stresses with respect to point D is given by

$$M_z^D(\boldsymbol{\sigma}, \varphi)\,\mathbf{k} = \int_{-b}^{b} d\hat{y} \int_{-\hat{x}_0(\hat{y})}^{\hat{x}_0(\hat{y})} (\hat{x}\,\mathbf{i} + \hat{y}\,\mathbf{j}) \times \mathbf{t}(\hat{x}, \hat{y}) d\hat{x} \tag{11.74}$$

However, in general, we are more interested in the vertical moment (usually called self-aligning torque) M_z, that is the moment with respect to the origin O of S. According to (11.12) and (11.14), we have to take into account the effects of the carcass compliance and of camber (Fig. 2.14) to locate D with respect to O

$$\begin{aligned} M_z(\gamma, \boldsymbol{\sigma}, \varphi) &= M_z^D - F_x(c_r(\gamma) + q_y) + F_y q_x \\ &= M_z^D - F_x\left(c_r(\gamma) + \frac{F_y}{w_y}\right) + F_y\frac{F_x}{w_x} \\ &= M_z^D - F_x c_r(\gamma) + F_x F_y \frac{w_y - w_x}{w_x w_y} \end{aligned} \tag{11.75}$$

---

[10] Since the tangential force is constant in time, it is possible to exploit its dependence on the given slips.

11.3.4 *Examples of Tangential Stress Distributions*

To gain insights into the steady-state brush model behavior, we will address some particular cases. Some of them can be solved analytically, while others require a numerical approach.

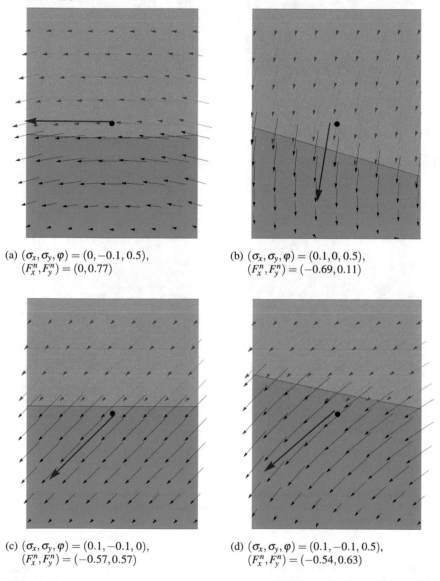

(a) $(\sigma_x, \sigma_y, \varphi) = (0, -0.1, 0.5),$
    $(F_x^n, F_y^n) = (0, 0.77)$

(b) $(\sigma_x, \sigma_y, \varphi) = (0.1, 0, 0.5),$
    $(F_x^n, F_y^n) = (-0.69, 0.11)$

(c) $(\sigma_x, \sigma_y, \varphi) = (0.1, -0.1, 0),$
    $(F_x^n, F_y^n) = (-0.57, 0.57)$

(d) $(\sigma_x, \sigma_y, \varphi) = (0.1, -0.1, 0.5),$
    $(F_x^n, F_y^n) = (-0.54, 0.63)$

Fig. 11.11 Examples of tangential stress distributions in rectangular contact patches. Also shown the line separating the adhesion region (top) and the sliding region (bottom), and the components of the normalized tangential force. Values of φ are in m^{-1}

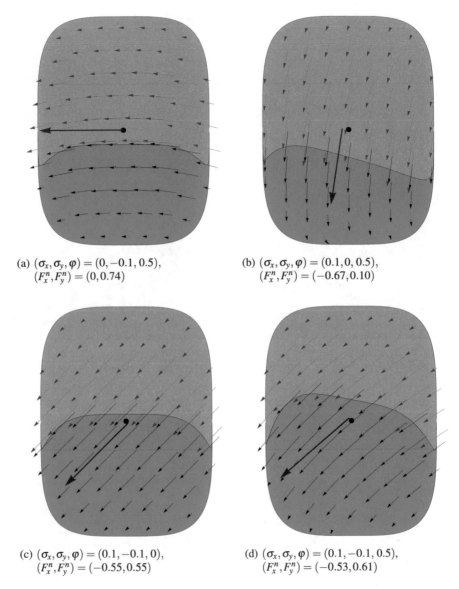

(a) $(\sigma_x, \sigma_y, \varphi) = (0, -0.1, 0.5)$,
$(F_x^n, F_y^n) = (0, 0.74)$

(b) $(\sigma_x, \sigma_y, \varphi) = (0.1, 0, 0.5)$,
$(F_x^n, F_y^n) = (-0.67, 0.10)$

(c) $(\sigma_x, \sigma_y, \varphi) = (0.1, -0.1, 0)$,
$(F_x^n, F_y^n) = (-0.55, 0.55)$

(d) $(\sigma_x, \sigma_y, \varphi) = (0.1, -0.1, 0.5)$,
$(F_x^n, F_y^n) = (-0.53, 0.61)$

Fig. 11.12 Examples of tangential stress distributions in rounded rectangular contact patches. Also shown the line separating the adhesion region (top) and the sliding region (bottom), and the components of the normalized tangential force. Values of φ are in m^{-1}

The shape of the contact patch is taken to be rectangular or elliptical, although it would not be much more difficult to deal with more realistic shapes, like the one in the center of Fig. 11.3.

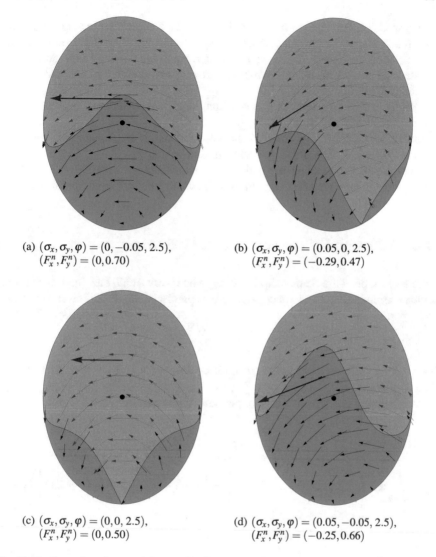

(a) $(\sigma_x, \sigma_y, \varphi) = (0, -0.05, 2.5)$,
$(F_x^n, F_y^n) = (0, 0.70)$

(b) $(\sigma_x, \sigma_y, \varphi) = (0.05, 0, 2.5)$,
$(F_x^n, F_y^n) = (-0.29, 0.47)$

(c) $(\sigma_x, \sigma_y, \varphi) = (0, 0, 2.5)$,
$(F_x^n, F_y^n) = (0, 0.50)$

(d) $(\sigma_x, \sigma_y, \varphi) = (0.05, -0.05, 2.5)$,
$(F_x^n, F_y^n) = (-0.25, 0.66)$

Fig. 11.13 Examples of tangential stress distributions in elliptical contact patches. Also shown the line separating the adhesion region (top) and the sliding region(s) (bottom), and the components of the normalized tangential force. Values of φ are in m^{-1} and correspond to a camber angle of about $38°$

Figure 11.11, obtained with the data listed in (11.52), shows the tangential stress pattern in *rectangular* contact patches, along with the adhesion and sliding regions, for four combinations of $(\sigma_x, \sigma_y, \varphi)$. The corresponding values of the normalized longitudinal and lateral forces are also reported. As typical in car tires, the value of φ is small.

Exactly the same combinations of slips, but for *rounded rectangular* contact patches, are shown in Fig. 11.12.

In Fig. 11.13, four cases for *elliptical* contact patches are shown. In these cases, the spin slip φ is quite high, as typical in motorcycle tires.

As expected, large values of φ make the stress distributions strongly non-parallel, thus reducing the value of the maximum achievable resultant tangential force.

11.4 Adhesion Everywhere (Linear Behavior)

If the magnitude of the skating slip λ is everywhere very small, then there is adhesion almost everywhere in the contact patch. More precisely, small skating slips means

$$|\lambda| \ll \frac{\mu_0 p_0}{2ak} \tag{11.76}$$

that is $|\lambda| < 0,03$ on a dry paved road. Of course, we are still dealing with steady-state conditions.

According to (11.58) and (11.72), the tangential force is

$$
\begin{aligned}
\mathbf{F}_t(\sigma, \varphi) = F_x \mathbf{i} + F_y \mathbf{j} &= \int_{-b}^{b} d\hat{y} \int_{-\hat{x}_0(\hat{y})}^{\hat{x}_0(\hat{y})} k \mathbf{e}_a(\hat{x}, \hat{y}) d\hat{x} \\
&= \int_{-b}^{b} d\hat{y} \int_{-\hat{x}_0}^{\hat{x}_0} k \left(-\sigma(\hat{x}_0 - \hat{x}) + \varphi \left[\frac{(\hat{x}_0 - \hat{x})(\hat{x}_0 + \hat{x})}{2} \mathbf{j} - \hat{y}(\hat{x}_0 - \hat{x}) \mathbf{i} \right] \right) d\hat{x} \\
&= -C_\sigma \sigma + C_\varphi \varphi \mathbf{j} \\
&= -C_\sigma \sigma_x \mathbf{i} - (C_\sigma \sigma_y - C_\varphi \varphi) \mathbf{j}
\end{aligned}
\tag{11.77}
$$

which, as expected, is *linear* in both σ and φ. This is a crude approximation of the real tire behavior, unless all the force components are very small.

It is worth noting that the longitudinal force F_x is a function of σ_x only, whereas the lateral force F_y depends on both σ_y and φ.

The coefficient C_σ may be called *slip stiffness*. In the isotropic brush model, C_σ is the same for any direction of the tangential force, that is for any combination of σ_x and σ_y. Moreover, in the brush model

$$C_\sigma = C_\alpha = C_{\kappa_x} \tag{11.78}$$

where C_α and C_{κ_x} were defined in (2.88) and (2.86).

The coefficient C_φ is the *spin stiffness* for the lateral force. Owing to the symmetric shape of the contact patch, the spin slip does not contribute to the longitudinal force.

It is possible to insert (2.75) and (2.76), that is the practical slip components, into (11.77), but the resulting function is no longer linear

$$\mathbf{F}_t(\kappa, \varphi) = F_x \mathbf{i} + F_y \mathbf{j} = -C_\sigma \frac{\kappa_x \mathbf{i} + \kappa_y \mathbf{j}}{1 - \kappa_x} + C_\varphi \varphi \mathbf{j} \tag{11.79}$$

Once again, the practical slip does not do a good job.

As shown in (2.68),

$$\varphi = -\frac{\sin \gamma (1 - \varepsilon_r)}{r_r} \tag{2.68'}$$

if the yaw rate ω_z is zero or at least negligible (as discussed at Sect. 2.10), the spin slip φ becomes a function of the camber angle γ only (besides F_z). In this case, we can define the *camber stiffness* C_γ

$$C_\gamma = -\frac{C_\varphi}{r_r}(1 - \varepsilon_r) < 0 \tag{11.80}$$

and obtain $(\sin \gamma \approx \gamma)$

$$\mathbf{F}_t(\sigma, \gamma) = F_x \mathbf{i} + F_y \mathbf{j} = -C_\sigma(\sigma_x \mathbf{i} + \sigma_y \mathbf{j}) + C_\gamma \gamma \mathbf{j} \tag{11.81}$$

Typically, $F_z/C_\gamma \approx 1$ for a motorcycle tire. Quite often, $-C_\sigma \sigma_y \mathbf{j}$ is called *cornering force* and $C_\gamma \gamma \mathbf{j}$ is called *camber force* (or camber thrust). Obviously, only under the very strong assumption of adhesion all over the contact patch, that is for very small values of the skating slip λ, we have two separate and independent contributions to the lateral force.

Under the same conditions and according to (11.75) we can compute the vertical moment with respect to the center D of the contact patch

$$M_z^D(\sigma_y, \varphi)\mathbf{k} = \int_{-b}^{b} d\hat{y} \int_{-\hat{x}_0(\hat{y})}^{\hat{x}_0(\hat{y})} (\hat{x}\,\mathbf{i} + \hat{y}\,\mathbf{j}) \times k\mathbf{e}_a(\hat{x}, \hat{y})d\hat{x}$$
$$= \left(C_{M_\sigma}\sigma_y + C_{M_\varphi}\varphi\right)\mathbf{k} = -F_y t_c\,\mathbf{k} \tag{11.82}$$

where t_c is the *pneumatic trail* with respect to the contact center D. The last expression states quite a remarkable fact: that $F_y = 0$ means $M_z^D = 0$ as well. The minus sign makes $t_c > 0$ under standard operating conditions.

Combining (11.75), (11.77) and (11.82) we obtain the vertical moment with respect to point O

$$M_z(\gamma, \sigma, \varphi) = C_{M_\sigma}\sigma_y + C_{M_\varphi}\varphi + C_\sigma \sigma_x \left[c_r(\gamma) + \frac{w_x - w_y}{w_x w_y}(-C_\sigma \sigma_y + C_\varphi \varphi) \right] \tag{11.83}$$

For a *rectangular* contact patch (i.e., $x_0(\hat{y}) = a$) we have

$$C_\sigma = 4ka^2b \tag{11.84}$$

and

$$C_\varphi = C_{M_\sigma} = \frac{a}{3}C_\sigma \qquad C_{M_\varphi} = \frac{b^2}{3}C_\sigma \qquad C_\gamma = -\frac{a(1 - \varepsilon_r)}{3r_r}C_\sigma \tag{11.85}$$

Typically, $C_\gamma \ll |C_\sigma|$. From (11.77), (11.82) and (11.85) we can obtain the pneumatic trail t_c for a rectangular contact patch

$$t_c = \frac{\sigma_y a + \varphi b^2}{3\sigma_y - \varphi a} \tag{11.86}$$

Special, but quite important cases are:
$\varphi = 0$, which yields

$$t_c = \frac{a}{3} \tag{11.87}$$

and $\sigma_y = 0$, that yields

$$t_c = -\frac{b^2}{a} \tag{11.88}$$

For an *elliptical* contact patch the algebra is a bit more involved. The final expression of the tangential force $\mathbf{F}_t$ is exactly like in (11.77), but with the following stiffnesses

$$C_\sigma = \frac{8}{3}ka^2b \quad \text{and} \quad C_\varphi = C_{M_\sigma} = \frac{3\beta a}{32}C_\sigma \tag{11.89}$$

We recall that the contact patch has length $2a$ and width $2b$. The product ab, and hence also the area of the contact patch, are determined by the vertical load F_z and the tire inflation pressure p_0, as obtained in (11.6) and (11.9). However, in the expressions (11.84) and (11.89) of the slip stiffness C_σ there appear the term $a^2b = a(ab)$. That means that the aspect ratio a/b of the footprint does affect C_σ. The reason for this dependence is promptly explained with the aid of Fig. 11.14.

If we compare Fig. 11.14a and b, we see that, for given slip angle α, the longer the footprint, the higher the final deflection of the bristles. This phenomenon is partly compensated by the fact that bristle deflections act on a wider strip in case (b). As predicted by (11.84), the net result is that tire (a) has a slip stiffness C_σ twice as much as tire (b). In other words, to obtain the same lateral force F_y from the wider tire we need to double the slip angle, as shown in Fig. 11.14c.

All tires in Fig. 11.14 share the same F_z, p_0 and k.

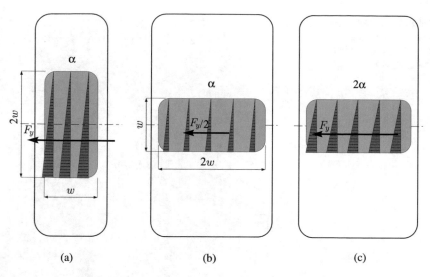

Fig. 11.14 Comparison of tires with the same F_z, p_0 and k, but different width

11.5 Translational Slip Only ($\sigma \neq \mathbf{0}, \varphi = 0$)

The investigation of the steady-state behavior of the brush model is much simpler if there is no spin slip φ. Indeed, the first two figures in this chapter referred to the case of pure braking. It is not a bad idea to go back and have another look.

According to (11.28), if $\varphi = 0$ and $\dot{\mathbf{q}} = \mathbf{0}$ all points in the contact patch $\mathcal{P}$ have the same skating velocity $\mathbf{V}_s^P = \mathbf{V}_s$. Therefore, from (11.53) we obtain that the skating slip λ is equal to the translational slip σ

$$\lambda = \sigma \tag{11.90}$$

and the governing equation (11.56) in the *adhesion* region becomes (Fig. 11.15)

$$\mathbf{e}_a' = \sigma = \text{ const.} \tag{11.91}$$

whose solution, which is a *linear* function of $\hat{x}$, is readily obtained as a special case of (11.58)

$$\mathbf{e}_a(\hat{x}, \hat{y}) = -(\hat{x}_0(\hat{y}) - \hat{x})\sigma = -(\hat{x}_0(\hat{y}) - \hat{x})\sigma \mathbf{s} \tag{11.92}$$

As shown in Fig. 11.15, all vectors $\mathbf{e}_a$ have the *same direction* $\mathbf{s} = \sigma/\sigma$, with $\sigma = |\sigma|$. Moreover, they grow linearly in the adhesion region. A look at Fig. 11.15 should clarify the matter.

The physical interpretation of these equations is simply that in the adhesion region everything is ruled by the kinematics of the wheel.

Like in (11.60), the adhesion state is maintained up to $\hat{x}_s = \hat{x}_s(\sigma, \hat{y})$, which marks the point where the friction limit is reached (Fig. 11.15)

Fig. 11.15 Linear pattern in the adhesion region and parabolic pattern in the sliding region

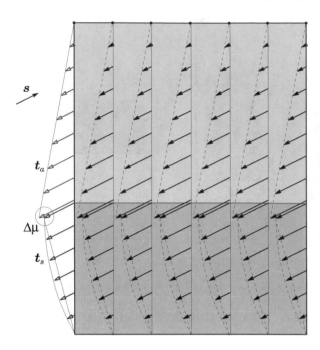

$$k|\mathbf{e}_a(\hat{x}_s, \hat{y})| = k\sigma\,(\hat{x}_0(\hat{y}) - \hat{x}_s) = \mu_0 p(\hat{x}_s, \hat{y}) \tag{11.93}$$

For the parabolic pressure distribution (11.3) we obtain

$$\hat{x}_s(\sigma, \hat{y}) = \hat{x}_0(\hat{y}) \left[\frac{k\hat{x}_0(\hat{y})}{\mu_0 p_0(\hat{y})}\,\sigma - 1 \right] \tag{11.94}$$

It is worth noting that, if $\varphi = 0$, the line separating the adhesion and the sliding regions depends solely on the magnitude σ of the slip. It is not affected by the direction $\mathbf{s}$ of $\boldsymbol{\sigma}$. However, this separating line depends on the shape of the contact patch. It is a straight line for a rectangular footprint, as in Fig. 11.15. For an elliptical contact patch, Fig. 11.16 shows the lines between adhesion and sliding for a sequence of growing values of σ.

At $\hat{x}_s$ the friction coefficient switches from μ_0 to its kinetic value μ_1, and the sliding state starts according to (11.61), that is with a *parabolic* pattern (Fig. 11.15)

$$k\mathbf{e}_s(\hat{x}_s, \hat{y}) = -\mu_1 p(\hat{x}_s, \hat{y})\mathbf{s} \tag{11.95}$$

The really important aspect is that sliding begins with the bristle deflection $\mathbf{e}_a$ that has already the *same direction* $\mathbf{s}$ as $\boldsymbol{\lambda} = \boldsymbol{\sigma}$. Therefore, also $\mathbf{e}'_s$ is directed like $\mathbf{s}$, and the governing equation (11.57) (or (11.62)) for the *sliding* region becomes simply

$$k\mathbf{e}_s(\hat{x}, \hat{y}) = -\mu_1 p(\hat{x}, \hat{y})\mathbf{s} \tag{11.96}$$

Fig. 11.16 Lines separating
the adhesion region (top) and
the sliding region (bottom)
for $\sigma =$
$(0.01, 0.05, 0.10, 0.15, 0.20, 0.266)$
and $\varphi = 0$. Pressure
distribution as in (11.8)

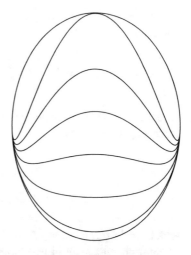

which is no longer a differential equation. Actually this is already the definition of
$\mathbf{e}_s$ in the sliding region.

Equations (11.92) and (11.96) provide the complete solution for this case. There-
fore, the tangential stress $\mathbf{t}$ at each point of the contact patch $\mathcal{P}$ is given by (Fig. 11.15)

$$
\mathbf{t}(\hat{x}, \hat{y}) = \begin{cases} \mathbf{t}_a = -t_a \mathbf{s} = -(\hat{x}_0(\hat{y}) - \hat{x})\sigma k \, \mathbf{s}, & \text{(adhesion)} \\ \mathbf{t}_s = -t_s \mathbf{s} = -\mu_1 p(\hat{x}, \hat{y}) \, \mathbf{s}, & \text{(sliding)} \end{cases} \tag{11.97}
$$

where $\mathbf{s} = \sigma/\sigma$, $t_a = |\mathbf{t}_a|$ and $t_s = |\mathbf{t}_s|$. Actually, as in Fig. 11.17, we have assumed
that, for any $\hat{y}$, a single adhesion region $(\hat{x}_s(\sigma, \hat{y}) \leq \hat{x} \leq \hat{x}_0(\hat{y}))$ is followed by a
single sliding region $(-\hat{x}_0(\hat{y}) \leq \hat{x} < \hat{x}_s(\sigma, \hat{y}))$, as it is normally the case. However,
as shown in Fig. 11.10 for a fairly unrealistic pressure distribution, it is possible, at
least in principle, to have multiple regions.

Summing up, we have the following features (Figs. 11.15, 11.17 and 11.18):

- the tangential stress $\mathbf{t}$ is directed like σ, with opposite sign;
- t_a grows linearly in the adhesion region;
- t_s follows the $\mu_1 p$ parabolic pattern in the sliding region;
- both t_a and t_s are not affected by the direction of σ;
- the higher σ, the steeper the growth of t_a and hence the closer the transition point
  $\hat{x}_s$ to the leading edge $\hat{x}_0$.

All these features can be appreciated in Figs. 11.19 and 11.20, which show the
tangential stress pattern, as predicted by the brush model, in rectangular and elliptical
contact patches under pure *translational* slip σ. It is worth remarking that in each
contact patch all arrows are parallel to each other.

The global tangential force $\mathbf{F}_t = F_x \mathbf{i} + F_y \mathbf{j}$ that the road applies to the tire model
is given by the integral of $\mathbf{t}$ on the contact patch, like in (11.72). Of course, here the
analysis will provide $\mathbf{F}_t(\sigma, 0)$. Since all tangential stresses $\mathbf{t}$ have the same direction

−s, the computation simply amounts to integrating |t| (shaded area in Figs. 11.17 and 11.18)

$$\mathbf{F}_t = -\mathbf{s}\, F_t(\sigma) = -\mathbf{s}\left[\int_{-b}^{b} d\hat{y} \int_{\hat{x}_s(\sigma,\hat{y})}^{\hat{x}_0(\hat{y})} t_a(\sigma, \hat{x}, \hat{y}) d\hat{x} + \int_{-b}^{b} d\hat{y} \int_{-\hat{x}_0(\hat{y})}^{\hat{x}_s(\sigma,\hat{y})} t_s(\hat{x}, \hat{y}) d\hat{x} \right]$$
(11.98)

where $F_t = |\mathbf{F}_t|$. The two components, that is the longitudinal force F_x and the lateral force F_y, are given by

$$F_x = F_x(\sigma_x, \sigma_y) = -\frac{\sigma_x}{\sigma} F_t(\sigma),$$
$$F_y = F_y(\sigma_x, \sigma_y) = -\frac{\sigma_y}{\sigma} F_t(\sigma)$$
(11.99)

which imply $\sigma_x/F_x = \sigma_y/F_y$.

Summing up, in the brush model with $\varphi = 0$, the magnitude $F_t(\sigma)$ of the tangential force $\mathbf{F}_t$ depends on the magnitude $\sigma = \sqrt{\sigma_x^2 + \sigma_y^2}$ of the translational slip, and the vectors $\mathbf{F}_t$ and σ have the same direction, but opposite signs

$$\mathbf{F}_t = -\frac{\sigma}{\sigma} F_t(\sigma)$$
(11.100)

In practical applications, it is a good idea to employ the Magic Formula for $F_t(\sigma)$ in (11.99), since it follows better the real tire behavior.

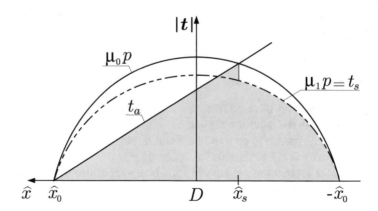

Fig. 11.17 Typical pattern of the tangential stress in the adhesion region (left) and in the sliding region (right)

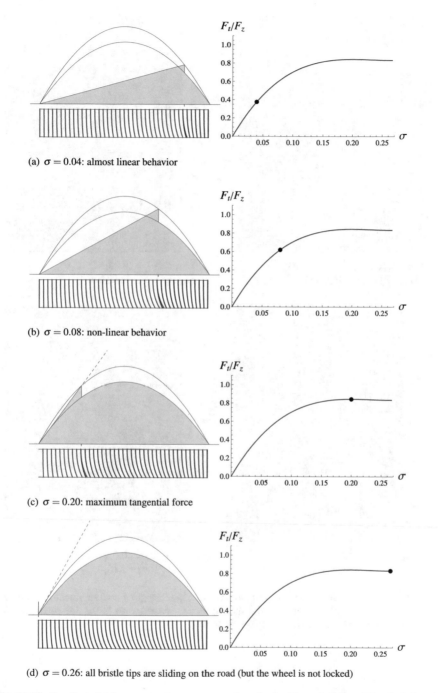

(a) $\sigma = 0.04$: almost linear behavior

(b) $\sigma = 0.08$: non-linear behavior

(c) $\sigma = 0.20$: maximum tangential force

(d) $\sigma = 0.26$: all bristle tips are sliding on the road (but the wheel is not locked)

Fig. 11.18 Brush model for rectangular contact patch under braking conditions. The shaded area (left) is proportional to the global tangential force F_t, marked by a point on the plot (right). Green bristles have the tip stuck to the ground, red bristles have the tip sliding on the ground

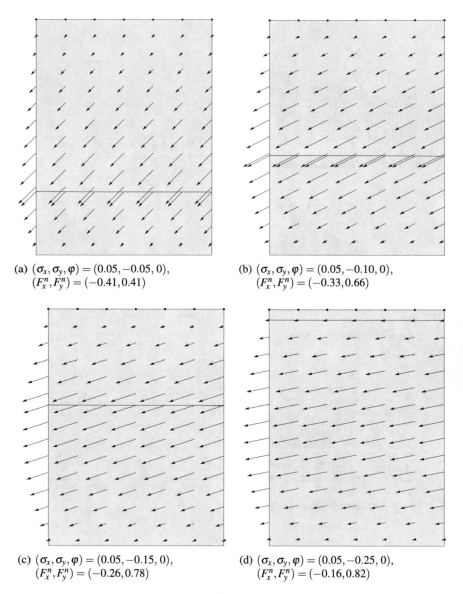

(a) $(\sigma_x, \sigma_y, \varphi) = (0.05, -0.05, 0)$,
$(F_x^n, F_y^n) = (-0.41, 0.41)$

(b) $(\sigma_x, \sigma_y, \varphi) = (0.05, -0.10, 0)$,
$(F_x^n, F_y^n) = (-0.33, 0.66)$

(c) $(\sigma_x, \sigma_y, \varphi) = (0.05, -0.15, 0)$,
$(F_x^n, F_y^n) = (-0.26, 0.78)$

(d) $(\sigma_x, \sigma_y, \varphi) = (0.05, -0.25, 0)$,
$(F_x^n, F_y^n) = (-0.16, 0.82)$

Fig. 11.19 Examples of tangential stress distributions in rectangular contact patches under *pure translational slip* σ. All arrows have the same direction. Also shown is the line separating the adhesion region (top) and the sliding region (bottom)

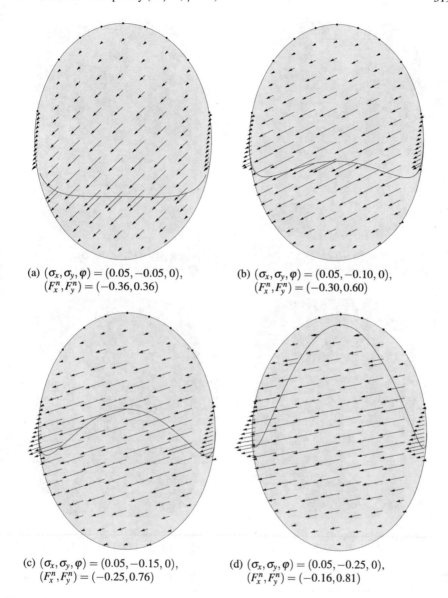

(a) $(\sigma_x, \sigma_y, \varphi) = (0.05, -0.05, 0)$,
$(F_x^n, F_y^n) = (-0.36, 0.36)$

(b) $(\sigma_x, \sigma_y, \varphi) = (0.05, -0.10, 0)$,
$(F_x^n, F_y^n) = (-0.30, 0.60)$

(c) $(\sigma_x, \sigma_y, \varphi) = (0.05, -0.15, 0)$,
$(F_x^n, F_y^n) = (-0.25, 0.76)$

(d) $(\sigma_x, \sigma_y, \varphi) = (0.05, -0.25, 0)$,
$(F_x^n, F_y^n) = (-0.16, 0.81)$

Fig. 11.20 Examples of tangential stress distributions in elliptical contact patches under *pure translational slip* σ. All arrows have the same direction. Also shown is the line separating the adhesion region (top) and the sliding region (bottom)

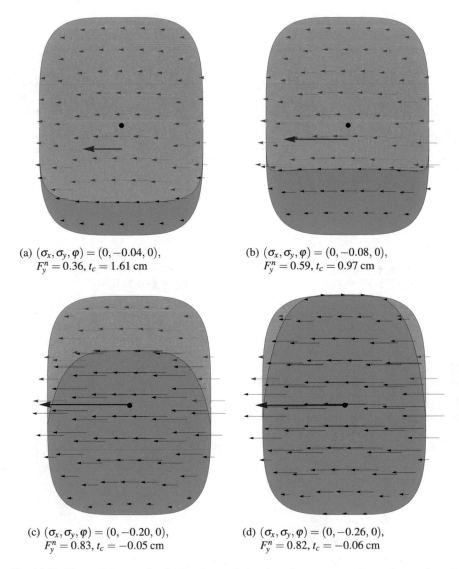

(a) $(\sigma_x, \sigma_y, \varphi) = (0, -0.04, 0)$,
   $F_y^n = 0.36$, $t_c = 1.61$ cm

(b) $(\sigma_x, \sigma_y, \varphi) = (0, -0.08, 0)$,
   $F_y^n = 0.59$, $t_c = 0.97$ cm

(c) $(\sigma_x, \sigma_y, \varphi) = (0, -0.20, 0)$,
   $F_y^n = 0.83$, $t_c = -0.05$ cm

(d) $(\sigma_x, \sigma_y, \varphi) = (0, -0.26, 0)$,
   $F_y^n = 0.82$, $t_c = -0.06$ cm

Fig. 11.21 Tangential stress distributions in rounded rectangular contact patches under pure lateral slip σ_y. Also shown the global tangential force. Values of σ as in Fig. 11.18

Partial derivatives can be readily obtained from (11.99)

$$
\begin{aligned}
-\frac{\partial F_x}{\partial \sigma_x} &= \frac{\partial}{\partial \sigma_x}\left(\frac{\sigma_x}{\sigma} F_t(\sigma)\right) = \left(\frac{\sigma_x}{\sigma}\right)^2 \left(F_t' - \frac{F_t}{\sigma}\right) + \frac{F_t}{\sigma} \\
-\frac{\partial F_x}{\partial \sigma_y} &= \frac{\partial}{\partial \sigma_y}\left(\frac{\sigma_x}{\sigma} F_t(\sigma)\right) = \left(\frac{\sigma_x \sigma_y}{\sigma^2}\right)\left(F_t' - \frac{F_t}{\sigma}\right)
\end{aligned}
\tag{11.101}
$$

Those of F_y simply need interchanging x and y.

Equation (11.74) provides the vertical moment M_z^D with respect to point D. However, it can be considerably simplified in the case of $\varphi = 0$. As a matter of fact, we see from (11.97) that $\mathbf{t}(\hat{x}, \hat{y}) = \mathbf{t}(\hat{x}, -\hat{y})$[11] and hence

$$
M_z^D(\sigma_x, \sigma_y) = -\frac{\sigma_y}{\sigma}\left[\int_{-b}^{b} d\hat{y} \int_{\hat{x}_s(\sigma,\hat{y})}^{\hat{x}_0(\hat{y})} \hat{x}\, t_a(\sigma, \hat{x}, \hat{y}) d\hat{x} + \int_{-b}^{b} d\hat{y} \int_{-\hat{x}_0(\hat{y})}^{\hat{x}_s(\sigma,\hat{y})} \hat{x}\, t_s(\hat{x}, \hat{y}) d\hat{x}\right]
\tag{11.102}
$$

It may be convenient to recast this equation in the following form

$$
M_z^D(\sigma_x, \sigma_y) = \frac{\sigma_y}{\sigma} F_t(\sigma)\, t_c(\sigma) = -F_y(\sigma_x, \sigma_y)\, t_c(\sigma)
\tag{11.103}
$$

which is, indeed, the definition of the *pneumatic trail* t_c, that is the (signed) distance from the contact center D of the line of action of the lateral force $F_y\, \mathbf{j}$. As shown in Fig. 11.21, a positive t_c stands for a lateral force behind D, which is the standard case.

11.5.1   Rectangular Contact Patch

Assuming a rectangular shape (Fig. 11.3) essentially means setting $\hat{x}_0(\hat{y}) = a$ as the equation of the leading edge. Therefore, any dependence on $\hat{y}$ disappears and the problem becomes one-dimensional, that is $\mathbf{e}_a = \mathbf{e}_a(\hat{x})$ and $\mathbf{e}_s = \mathbf{e}_s(\hat{x})$.

As shown in Fig. 11.19, in this case the line between the adhesion and the sliding regions is simply a *straight line* directed like $\mathbf{j}$

$$
\hat{x}_s(\sigma) = a\left(\frac{ka}{\mu_0 p_0}\sigma - 1\right) = a\left(2\frac{\sigma}{\sigma_s} - 1\right)
\tag{11.104}
$$

where

$$
\sigma_s = \frac{2\mu_0 p_0}{ka} = \frac{3\mu_0 F_z}{C_\sigma} = \frac{\mu_0}{k}|p'(a)|
\tag{11.105}
$$

with the slip stiffness C_σ defined in (11.78).

---

[11] If, as usual, also $\hat{x}_0(\hat{x}, \hat{y}) = \hat{x}_0(\hat{x}, -\hat{y})$ and $p(\hat{x}, \hat{y}) = p(\hat{x}, -\hat{y})$.

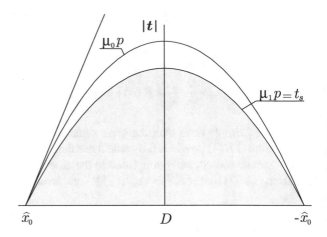

Fig. 11.22 Tangential stress if $\sigma = \sigma_s$ (total sliding)

The physical interpretation of σ_s is promptly obtained. If $\sigma \geq \sigma_s$, regardless of the direction of $\boldsymbol{\sigma}$, there is sliding on the whole rectangular contact patch, that is $\hat{x}_s = a$. For instance, with the numerical values of (11.52) on Sect. 11.3, we have $\sigma_s = 0.27$, that is a fairly low value.

At first, it may be surprising to have full sliding without wheel locking (i.e., $\sigma = \infty$). The phenomenon is explained in Fig. 11.22 (and also in Fig. 11.18d): to have total sliding it suffices that the straight line to be tangent to the upper parabola at the leading edge. The value (11.105) of σ_s predicted by the brush model is therefore quite "weak", in the sense that it is very much affected by the assumed pressure distribution. However, the existence of full sliding without (necessarily) wheel locking is an important result.

Also interesting is to observe that

$$\sigma_s C_\sigma = 3\mu_0 F_z \tag{11.106}$$

A simple formula that shows the strong relationship between σ_s and C_σ: the stiffer the tire, the smaller σ_s. The quantity μ_1 plays no role.

Application of (11.98) with $\hat{x}_0 = a$ and $\hat{x}_s(\sigma)$ as in (11.104) (and hence $0 \leq \sigma \leq \sigma_s$), provides the expression of the magnitude F_t of the tangential force

$$F_t = F_t(\sigma) = C_\sigma \sigma \left[1 - \frac{\sigma}{\sigma_s} \left(\frac{1 + 2\chi}{1 + \chi} \right) + \left(\frac{\sigma}{\sigma_s} \right)^2 \left(\frac{1 + 3\chi}{3(1 + \chi)} \right) \right] \tag{11.107}$$

where $\mu_0 = (1 + \chi)\mu_1$ as in (11.19). In this model and under these specific operating conditions, $F_t(\sigma)$ is a polynomial function of σ, whose typical behavior is shown in Fig. 11.23, along with its linear approximation ("good" only up to $\sigma \approx 0.03$). From Fig. 11.24 we can also appreciate how the adhesion and sliding regions contribute separately to build up the total tangential force.

Fig. 11.23 Magnitude F_t of the tangential force as a function of σ, and corresponding linear approximation

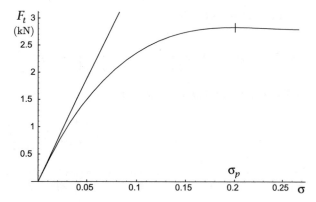

Fig. 11.24 Contributions to F_t (*solid line*) of the adhesion region (*long-dashed line*) and of the sliding region (*short-dashed line*)

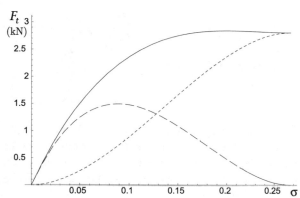

The derivative of $F_t(\sigma)$ is

$$F_t'(\sigma) = \frac{dF_t}{d\sigma} = C_\sigma \left[1 - 2\frac{\sigma}{\sigma_s}\left(\frac{1+2\chi}{1+\chi}\right) + \left(\frac{\sigma}{\sigma_s}\right)^2 \left(\frac{1+3\chi}{1+\chi}\right) \right] \quad (11.108)$$

which, among other things, clearly provides the important result

$$\left.\frac{dF_t}{d\sigma}\right|_{\sigma=0} = C_\sigma \quad (11.109)$$

that clarifies why C_σ is called slip stiffness.

As expected, the force with total sliding is

$$F_t(\sigma_s) = \mu_1 F_z \quad (11.110)$$

since all tangential stresses **t** have the same direction.

The peak value of F_t is

$$F_t^{\max} = F_t(\sigma_p) = \mu_0 \left[\frac{4 - 3(\mu_1/\mu_0)}{[3 - 2(\mu_1/\mu_0)]^2} \right] F_z = \mu_1 \left[1 + \frac{4\chi^3}{(3\chi + 1)^2} \right] F_z = \mu_p F_z$$

$$(11.111)$$

and it is achieved at $\sigma = \sigma_p$ (Fig. 11.23)

$$\sigma_p = \frac{1 + \chi}{1 + 3\chi} \sigma_s \qquad (11.112)$$

Typically, as in Fig. 11.23, good tires have low values of σ_p. In this model, the global friction coefficient μ_p is given by (cf. (2.87) and (2.89))

$$\mu_p = \frac{F_t^{\max}}{F_z} = \mu_1 \left[1 + \frac{4\chi^3}{(3\chi + 1)^2} \right] \qquad (11.113)$$

which means that, as expected

$$\mu_1 < \mu_p \ll \mu_0 \qquad (11.114)$$

For instance, if $\mu_0 = 1.2\mu_1$, we have $F_t^{\max} = 0.84\mu_0 F_z = 1.013\mu_1 F_z$, that is a value only marginally higher than $F_t(\sigma_s)$. Indeed, as shown in Fig. 11.27 (and also in Fig. 11.18c), the mechanics of the tire makes it very difficult to have tangential stresses close to $\mu_0 p$. In practical terms, attempts at increasing μ_1 are more worthwhile than those at increasing μ_0.

It may be interesting to fit the curve of $F_t(\sigma)$ shown in Fig. 11.23 by means of the Magic Formula $y(x)$ given in (2.90). According to Sect. 2.11, the four unknown coefficients can be obtained by matching the peak value $y_m = F_t(\sigma_p) = 2.84 \, \text{kN}$, the asymptotic value $y_a = F_t(\sigma_s) = 2.80 \, \text{kN}$, the slope at the origin $y'(0) = C_\sigma = 37.8 \, \text{kN} = \text{rad}$ and the abscissa of the peak value $x_m = \sigma_p = 0.2$. The resulting coefficients are $B = 12.1$, $C = 1.10$, $D = 2.835 \, \text{kN}$ and f $E = -3.63$. The comparison is shown in Fig. 11.25. The agreement between the two curves is quite poor. Particularly unacceptable is the initial increase of the slope, which is never found in experimental curves (cf. Figs. 2.20 and 2.22). Indeed, $E < -(1 + C^2/2)$ and hence $y'''(0) > 0$.

A better agreement is shown in Fig. 11.26, where the asymptotic value was arbitrarily lowered to $y_a = 0.7F_t(\sigma_s)$, thus obtaining $B = 8.81, C = 1.51, D = 2.84 \, \text{kN}$ and $E = 0.1$. The lesson to be learnt is, perhaps, that the Magic Formula may occasionally provide unexpected results and, therefore, should be used with care (Fig. 11.27).

Going back to the brush model, the explicit expressions of $F_x(\sigma_x, \sigma_y)$ and $F_y(\sigma_x, \sigma_y)$, that is of the longitudinal and lateral components, can be obtained by inserting (11.107) into (11.99). Figure 11.28 illustrates the combined effect of σ_x and σ_y. Quite remarkable is the effect on the slope at the origin, that is on the generalized slip stiffness $\widetilde{C}_\sigma$. From (11.101) and (11.107) it follows that

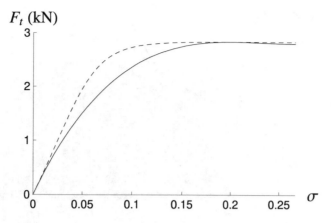

Fig. 11.25 Brush model curve (*solid line*) and the corresponding classical fitting by the Magic Formula (*dashed line*)

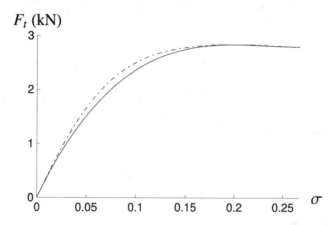

Fig. 11.26 Brush model curve (*solid line*) and another possible fitting by the Magic Formula (*dot-dashed line*)

$$\widetilde{C}_\sigma(\sigma_y) = -\left.\frac{\partial F_x}{\partial \sigma_x}\right|_{\sigma_x=0} = C_\sigma\left[1 - \frac{|\sigma_y|}{\sigma_s}\frac{1+2\chi}{1+\chi} + \left(\frac{\sigma_y}{\sigma_s}\right)^2\frac{1+3\chi}{3(1+\chi)}\right] \quad (11.115)$$

and, interchanging x and y

$$\widetilde{C}_\sigma(\sigma_x) = -\left.\frac{\partial F_y}{\partial \sigma_y}\right|_{\sigma_y=0} = C_\sigma\left[1 - \frac{|\sigma_x|}{\sigma_s}\frac{1+2\chi}{1+\chi} + \left(\frac{\sigma_x}{\sigma_s}\right)^2\frac{1+3\chi}{3(1+\chi)}\right] \quad (11.116)$$

Of course $\widetilde{C}_\sigma(0) = C_\sigma$. This stiffness reduction has strong practical implications on the handling behavior of vehicles. σ_s was defined in (11.105).

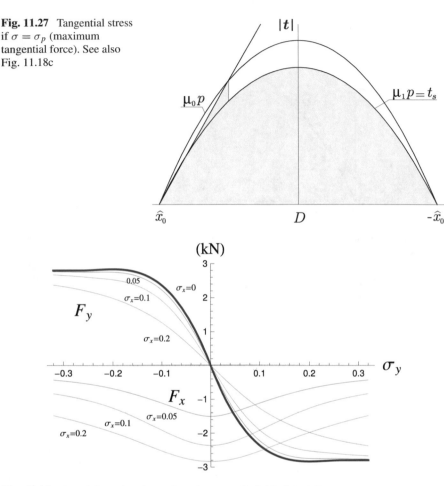

Fig. 11.27 Tangential stress if $\sigma = \sigma_p$ (maximum tangential force). See also Fig. 11.18c

Fig. 11.28 F_y and F_x as functions of σ_y, for $\sigma_x = (0,\ 0.05,\ 0.1,\ 0.2)$

It should be observed that the generalized cornering stiffness $\widetilde{C}_\alpha(\sigma_x)$ is no longer equal to $\widetilde{C}_\sigma(\sigma_x)$ (cf. (11.78))

$$\widetilde{C}_\alpha(\sigma_x) = (1 + \sigma_x)\widetilde{C}_\sigma(\sigma_x) \tag{11.117}$$

whereas $\widetilde{C}_{\kappa_x}(\sigma_y) = \widetilde{C}_\sigma(\sigma_y)$.

Another useful plot is the one shown in Fig. 11.29. For any combination of (σ_x, σ_y), a point in the plane (F_x, F_y) is obtained such that $\sigma_x/\sigma_y = F_x/F_y$. All these points fall within a circle of radius $F_t^{\max}$, usually called the *friction circle*. Lines with constant σ_y are also drawn in Fig. 11.29. Because of the symmetry of this tire model, lines with constant σ_x are identical, but rotated of 90° around the origin, as shown in Fig. 11.38b.

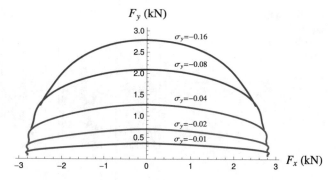

Fig. 11.29 Friction circle with lines at constant σ_y

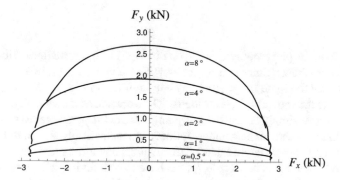

Fig. 11.30 Friction circle of Fig. 11.29, but with lines at constant α

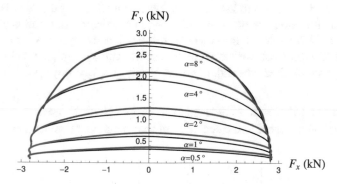

Fig. 11.31 Comparison

Fig. 11.32 Same slip angle α, but different σ_y if: **a** $\sigma_x = 0$, **b** $\sigma_x < 0$ (driving), **c** $\sigma_x > 0$ (braking)

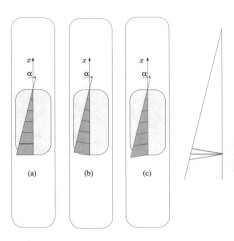

More often, the plot employed is the one in Fig. 11.30, where lines with constant slip angle α are drawn. Since α is a function of σ_x *and* σ_y (Eq. (2.78)), the two plots contain exactly the same information. While the lines in Fig. 11.29 are symmetric with respect to the vertical axis, lines in Fig. 11.30 are not, as shown in Fig. 11.31. The asymmetry arises simply because the slip angle is not the parameter to be used for a neat description of the tire mechanics. Indeed, as schematically shown in Fig. 11.32, the bristles may have different lateral deformations under the same slip angle.

As already mentioned on Sect. 2.9, tires have to be built in such a way to provide the maximum tangential force $\mathbf{F}_t$ *in any direction* with small slip angles α, as shown in Fig. 11.33. This is a fundamental requirement for a wheel with tire to behave almost like a wheel, that is to have a *directional capability*. In other words, while $\mathbf{F}_t$ can have any direction, the travel velocity $\mathbf{V}_c$ must undergo just small deviations α. According to (2.78), this condition will be fulfilled if and only if the tire exhibits the peak value of $\mathbf{F}_t$ for small values of the theoretical slip σ_p, typically below 0.2. On the contrary, in a locked wheel the two vectors $\mathbf{F}_t$ and $\mathbf{V}_c$ always point in opposite directions.

Equation (11.102), with $\hat{x}_0 = a$ and $\hat{x}_s$ as in (11.104), provides the vertical moment M_z^D with respect to the center D of the rectangular contact patch

$$M_z^D(\sigma_x, \sigma_y) = \sigma_y C_\sigma \frac{a}{3} \left[1 - 3\frac{\sigma}{\sigma_s}\frac{1+2\chi}{1+\chi} + 3\left(\frac{\sigma}{\sigma_s}\right)^2 \frac{1+3\chi}{1+\chi} - \left(\frac{\sigma}{\sigma_s}\right)^3 \frac{1+4\chi}{1+\chi} \right]$$

$$= \frac{\sigma_y}{\sigma} F_t(\sigma) t_c(\sigma) = -F_y(\sigma_x, \sigma_y) t_c(\sigma)$$

$$(11.118)$$

where t_c is the pneumatic trail. The typical behavior of M_z^D is shown in Fig. 11.34.

However, under combined slip conditions, to obtain M_z with respect to point O it is necessary to take into account the *carcass compliance*, according to (11.75). The

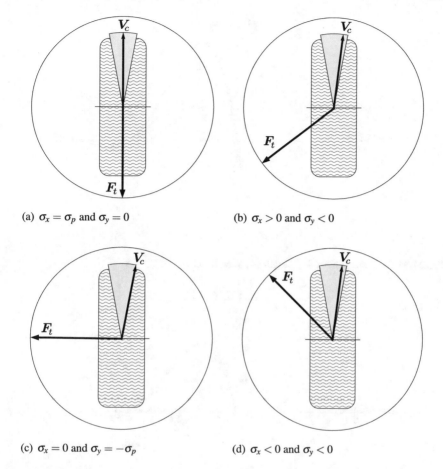

(a) $\sigma_x = \sigma_p$ and $\sigma_y = 0$ (b) $\sigma_x > 0$ and $\sigma_y < 0$

(c) $\sigma_x = 0$ and $\sigma_y = -\sigma_p$ (d) $\sigma_x < 0$ and $\sigma_y < 0$

Fig. 11.33 Typical relationships between the tangential force $\mathbf{F}_t$ and the travel velocity $\mathbf{V}_c$ for a tire with the same theoretical slip $\sigma = \sqrt{\sigma_x^2 + \sigma_y^2} = \sigma_p$, but different σ_x/σ_y

typical behavior of $M_z(\sigma_x, \sigma_y)$ is shown in Fig. 11.35. The difference with Fig. 11.34 is quite relevant.

Also of practical interest may be the plots of M_z versus F_x (Fig. 11.36) and of F_y versus $-M_z$ (Fig. 11.37), this one being often called *Gough plot* if $\sigma_x = 0$.

The three functions $F_x(\sigma_x, \sigma_y)$, $F_y(\sigma_x, \sigma_y)$ and $M_z(\sigma_x, \sigma_y)$ can be seen as the parametric equations of a three-dimensional surface that fully describes, at constant vertical load F_z, the tire mechanical behavior. Such surface is shown in Fig. 11.38a, along with its three projections, which are precisely like Figs. 11.28, 11.36 and 11.37, respectively. The surface in Fig. 11.38a is called here the *tire action surface*.

As already mentioned, a wheel with tire can be called a wheel because:

1. the tire action surface is regular, in the sense that it does not fold onto itself, for a limited set of values (σ_x, σ_y). It has therefore a limited contour and, hence,

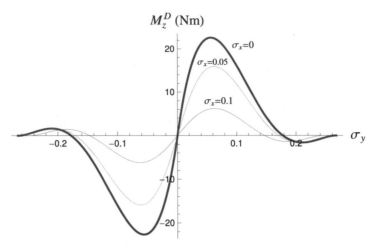

Fig. 11.34 Vertical moment M_z^D versus σ_y, at constant σ_x

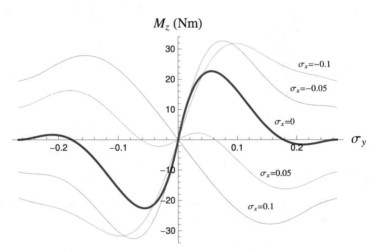

Fig. 11.35 Vertical moment M_z versus σ_y, at constant σ_x and $\gamma = 0$

the slip angle α is always quite low, according to (2.78). The goal of ABS [15] is to avoid wheel locking and also to keep $|\alpha|$ very low, thus maintaining the *directional capability* of the wheels;

2. the vertical moment M_z is always moderate. A wheel must provide forces applied not far from the center of the contact patch.

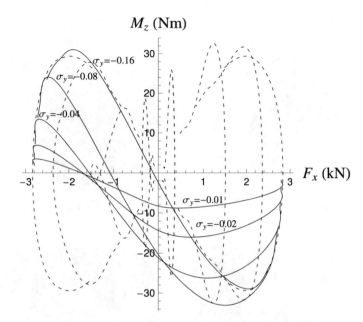

Fig. 11.36 Vertical moment M_z versus longitudinal force F_x, with lines at constant σ_y (*solid*) and constant σ_x (*dashed*: ± 0.01, ± 0.05, ± 0.1, ± 0.2)

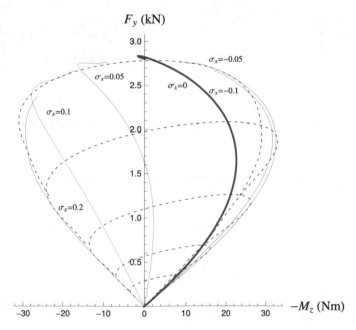

Fig. 11.37 Lateral force F_y versus vertical moment M_z, with lines at constant σ_x (*solid*) and constant σ_y (*dashed*: -0.01, -0.02, -0.04, -0.08, -0.16)

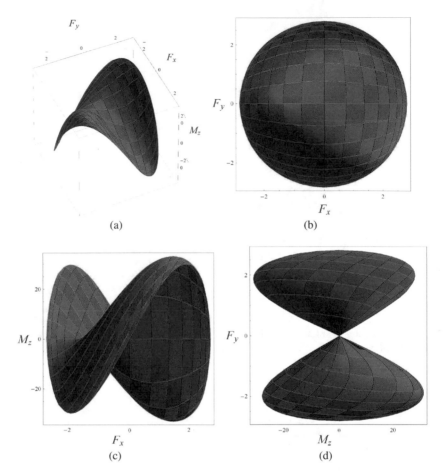

Fig. 11.38 *Tire action surface*, and its three projections (forces in kN and moments in Nm). Also shown lines at constant σ_x (blue) and constant σ_y (black)

11.5.2 Elliptical Contact Patch

Assuming an elliptical shape (Fig. 11.3) essentially means setting $\hat{x}_0(\hat{y})$ according to (11.2). As shown in Figs. 11.16 and 11.20, in this case the line between the adhesion and the sliding regions is *curved*. Its explicit equation is obtained inserting (11.2) into (11.94). To have sliding on the whole elliptical contact patch, a very high value of σ is necessary (Fig. 11.16).

Application of (11.98) with suitable $\hat{x}_0(\hat{y})$ and $\hat{x}_s(\sigma, \hat{y})$ provides the expression of the magnitude F_t of the tangential force

$$F_t = F_t(\sigma) = C_\sigma \sigma \left[1 - \frac{18\text{\ss}}{64} \frac{\sigma}{\sigma_s} \left(\frac{1+2\chi}{1+\chi} \right) + \frac{12}{45} \left(\frac{\sigma}{\sigma_s} \right)^2 \left(\frac{1+3\chi}{1+\chi} \right) \right]$$

$$(11.119)$$

where C_σ was obtained in (11.89) and σ_s is as in (11.105), although it has no special meaning in this case. Again, $F_t(\sigma)$ is a polynomial function of σ, whose typical behavior is much like in Fig. 11.23, but with a less evident peak.

11.6   Wheel with Pure Spin Slip ($\sigma = \mathbf{0}$, $\varphi \neq 0$)

The investigation of the behavior of the brush model becomes much more involved if there is *spin slip* φ. Even if $\sigma = \mathbf{0}$, the problem in the sliding region has to be solved in full generality according to the governing equations (11.63). Therefore, numerical solutions have to be sought.

The definition of φ was given in (2.65) and is repeated here

$$\varphi = -\frac{\omega_z + \omega_c \sin \gamma \, (1 - \varepsilon_r)}{\omega_c \, r_r} \qquad (2.65')$$

It involves ω_z, $\sin \gamma$, ε_r, ω_c and r_r. However, in most applications spin slip means camber angle γ, since $\omega_z / \omega_c \approx 0$. Figure 11.39 reports an example of the relationship between γ and φ, if $\varepsilon_r = 0$ (motorcycle tire), $r_r = 0.25$ m and $\omega_z = 0$.

Large values of φ are attained only in motorcycles.[12] Therefore, in this section the analysis is restricted to *elliptical* contact patches. Figure 11.40 shows the almost linear growth of the (normalized) lateral force $F_y^n(0, \varphi) = F_y^n(\varphi) = F_y / F_z$, even for very large values of the spin slip. A similar pattern can be observed in Fig. 11.41 for the vertical moment $M_z^D = M_z$. In both cases, the main contribution comes from the adhesion regions.

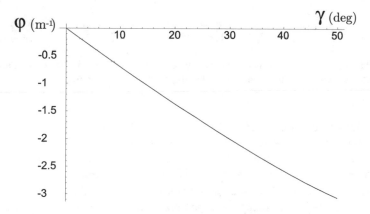

Fig. 11.39 Relationship between the camber angle γ and the spin slip φ, if $\omega_z = 0$, $\varepsilon_r = 0$, and $r_r = 0.25$ m

---

[12] More generally, in tilting vehicles, which may have three wheels, like *MP3* by *Piaggio*, or even four.

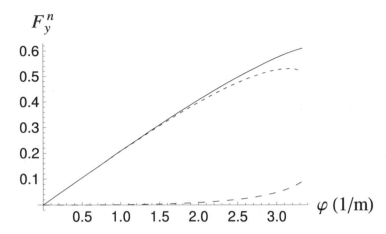

Fig. 11.40 Normalized lateral force versus spin slip (*solid line*). Also shown is the contribution of the adhesion zone (*short-dashed line*) and of the sliding zone (*long-dashed line*)

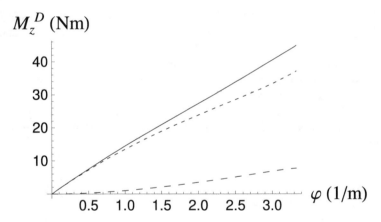

Fig. 11.41 Vertical moment versus spin slip (*solid line*). Also shown is the contribution of the adhesion zone (*short-dashed line*) and of the sliding zone (*long-dashed line*)

The lateral force plotted in Fig. 11.40 is precisely what is usually called the *camber force*, that is the force exerted by the road on a tire under pure spin slip.

Some examples of tangential stress distributions are shown in Fig. 11.42. They are quite informative. There is adhesion along the entire central line, and the stress has a parabolic pattern. The value of φ does not affect the direction of the arrows in the adhesion region, but only their magnitude. Even at $\varphi = 3.33 \, \mathrm{m}^{-1}$, i.e. a very high value, the two symmetric sliding regions have spread only on less than half the contact patch.

Another important observation is that there are longitudinal components of the tangential stress, although the longitudinal force $F_x = 0$. In some sense, these components are wasted, and keeping them as low as possible is a goal in the design of real tires.

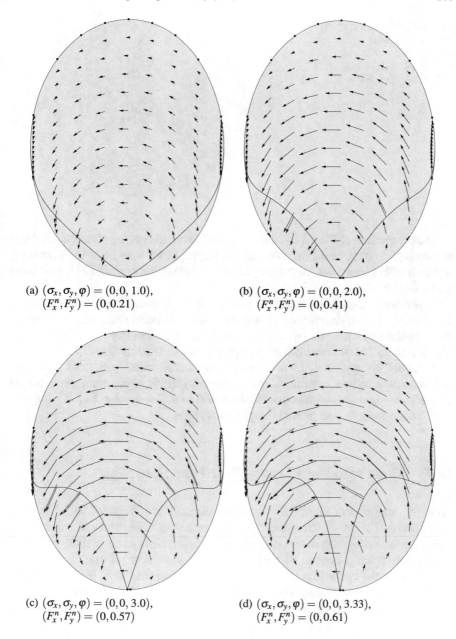

(a) $(\sigma_x, \sigma_y, \varphi) = (0, 0, 1.0)$,
$(F_x^n, F_y^n) = (0, 0.21)$

(b) $(\sigma_x, \sigma_y, \varphi) = (0, 0, 2.0)$,
$(F_x^n, F_y^n) = (0, 0.41)$

(c) $(\sigma_x, \sigma_y, \varphi) = (0, 0, 3.0)$,
$(F_x^n, F_y^n) = (0, 0.57)$

(d) $(\sigma_x, \sigma_y, \varphi) = (0, 0, 3.33)$,
$(F_x^n, F_y^n) = (0, 0.61)$

Fig. 11.42 Examples of tangential stress distributions in elliptical contact patches under *pure spin slip* φ. Also shown is the line separating the adhesion region (top) and the two sliding regions (bottom). Values of φ are in m^{-1}

Fig. 11.43 Elliptical contact
patch with inverted
proportions.
$(\sigma_x, \sigma_y, \varphi) = (0, 0, \ 3.33)$,
$(F_x^n, F_y^n) = (0, 0.36)$

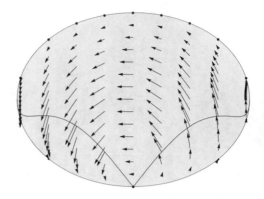

The comparison of Figs. 11.42d and 11.43 gives an idea of the effect of the shape
of the contact patch. In the second case the lengths of the axes have been inverted,
while all other parameters are unchanged. Nevertheless, the normalized lateral force
is much lower (0.36 vs 0.61).

In the brush model developed here, the lateral force and the vertical moment
depend on φ, but not directly on γ. Therefore, there is no distinction between operating conditions with the same spin slip φ, but different camber angle γ as in Fig. 2.19.
This is a limitation of the model with respect to what stated on Sect. 2.10.

It should be appreciated that a cambered wheel under pure spin slip cannot be in free rolling conditions. According to (2.109), there must be a torque
$\mathbf{T} = M_z \sin \gamma \, \mathbf{j}_c = T \, \mathbf{j}_c$ with respect to the wheel axis. Conversely, $T = 0$ requires a
longitudinal force F_x and hence a longitudinal slip σ_x.

11.7 Wheel with Both Translational and Spin Slips

From the tire point of view, there are fundamentally two kinds of vehicles: cars,
trucks and the like, whose tires may operate at relatively large values of translational
slip and small values of spin slip, and motorcycles, bicycles and other tilting vehicles,
whose tires typically operate with high camber angles and small translational slips.
In both cases, the interaction between σ and φ in the mechanics of force generation
is of great practical relevance. The tuning of a vehicle often relies on the right balance
between these kinematical quantities.

11.7.1 Rectangular Contact Patch

Rectangular contact patches mimic those of car tires. Therefore, we will address the
effect of just a bit of spin slip on the lateral force of a wheel mainly subjected to lateral
slips. The goal is to achieve the highest possible value of F_y^n. Unfortunately, it is not
possible to obtain analytical results and a numerical approach has to be pursued.

Fig. 11.44 Rectangular
contact patch under pure spin
slip (arrows magnified by a
factor 5 with respect to the
other figures).
$(\sigma_x, \sigma_y, \varphi) = (0, 0, 0.21)$,
$(F_x^n, F_y^n) = (0, 0.06)$

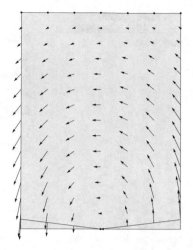

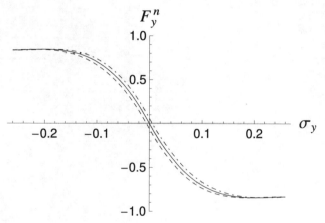

Fig. 11.45 Normalized lateral force F_y^n versus σ_y, for $\varphi = 0$ (*solid line*), $\varphi = -0.21\,\mathrm{m}^{-1}$ (*dashed line*), $\varphi = 0.21\,\mathrm{m}^{-1}$ (*dot-dashed line*). Rectangular contact patch and $\sigma_x = 0$ in all cases

A rectangular contact patch under pure spin slip (arrows magnified by a factor 5) is shown in Fig. 11.44. The global effect is a small lateral force, usually called camber force.

Indeed, as shown in Fig. 11.45, the effect of a small amount of spin slip φ is, basically, to translate *horizontally* the curve of the lateral force versus σ_y.[13] However, the peak value is also affected, as more clearly shown in Fig. 11.46. By means of a trial-and-error procedure it has been found, in the case at hand, that $\varphi = 0.21\,\mathrm{m}^{-1}$ does indeed provide the highest positive value of F_y^n. In general, car tires need just a

---

[13] Of course, the effect cannot be to "add" the camber force, that is to translate the curve vertically.

Fig. 11.46 Detail of
Fig. 11.45 showing different
peak values

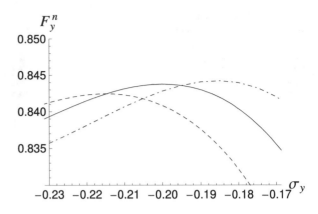

Fig. 11.47 Rectangular
contact patch under lateral
and spin slips. $(\sigma_x, \sigma_y, \varphi) =$
$(0, -0.185, 0.21)$,
$(F_x^n, F_y^n) = (0, 0.84)$

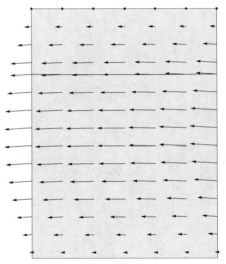

few degrees of camber to provide the highest lateral force as a function of the lateral
slip σ_y (Fig. 11.47).

Such small values of spin slip have very little influence on the longitudinal force
generation.

11.7.2 Elliptical Contact Patch

Elliptical contact patches mimic those of motorcycle tires. Therefore, in this case we
will study the effect of just a bit of lateral slip σ_y on the lateral force of a cambered
wheel. Again, the goal is to achieve the highest possible value of F_y^n.

The large effect of even a small amount of σ_y on the normalized lateral force F_y^n
as a function of φ is shown in Fig. 11.48. However, this is quite an expected result

after (11.89). Consistently, also the vertical moment M_z^D changes a lot under the influence of small variations of σ_y (Fig. 11.49).

Figures 11.50 and 11.51 provide a pictorial representation of the tangential stress in two relevant cases, that is those that yield the highest lateral force. Quite remarkably, a 10% higher value of F_y^n is achieved in case (b) with respect to case (a). In general, a little σ_y has a great influence on the stress distribution in the contact patch. Conversely, the same lateral force can be obtained by infinitely many combinations (σ_y, φ). This is something most riders know intuitively. Obviously, $F_x = 0$ in all cases of Figs. 11.50 and 11.51.

Under these operating conditions, according to (2.78), the slip angle α never exceeds two degrees. Therefore, the wheel has excellent directional capability.

It should be observed that the larger value of F_y^n of case (b) in Fig. 11.50 is associated with a smaller value of M_z^D. Basically, it means that the tangential stress distribution in the contact patch is *better organized* to yield the lateral force, without wasting much in the vertical moment (mainly due to useless longitudinal stress components). The comparison shown in Fig. 11.50c confirms this conclusion.

A lateral slip in the "wrong" direction, like in Fig. 11.50d, yields a reduction of the lateral force and an increase of the vertical moment.

As reported in Figs. 11.48 and 11.49, there are particular combinations of (σ_y, φ) which provide either $F_y^n = 0$ or $M_z^D = 0$. The stress distributions in such two cases are shown in Fig. 11.52.

Fig. 11.48 Elliptical contact patch: normalized lateral force versus spin slip, at different values of lateral slip

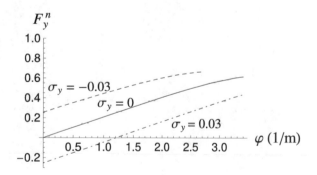

Fig. 11.49 Elliptical contact patch: vertical moment versus spin slip, at different values of lateral slip

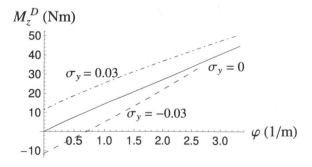

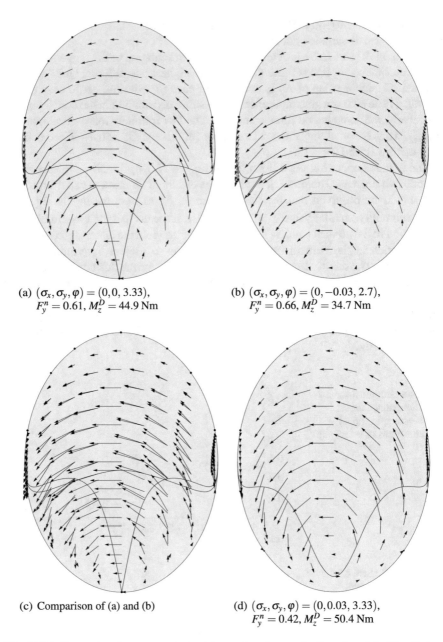

(a) $(\sigma_x, \sigma_y, \varphi) = (0, 0, 3.33)$,
    $F_y^n = 0.61$, $M_z^D = 44.9$ Nm

(b) $(\sigma_x, \sigma_y, \varphi) = (0, -0.03, 2.7)$,
    $F_y^n = 0.66$, $M_z^D = 34.7$ Nm

(c) Comparison of (a) and (b)

(d) $(\sigma_x, \sigma_y, \varphi) = (0, 0.03, 3.33)$,
    $F_y^n = 0.42$, $M_z^D = 50.4$ Nm

Fig. 11.50 Comparison between contact patches under **a** large spin slip only and **b** still quite large spin slip with the addition of a little of lateral slip. Case **d** shown for completeness. Values of φ are in m^{-1}

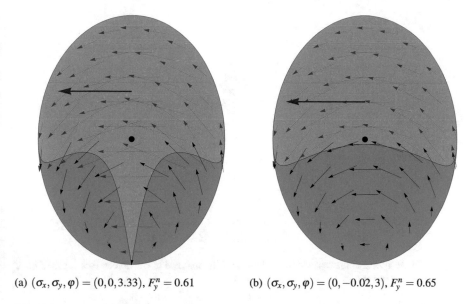

(a) $(\sigma_x, \sigma_y, \varphi) = (0,0,3.33),\, F_y^n = 0.61$

(b) $(\sigma_x, \sigma_y, \varphi) = (0,-0.02,3),\, F_y^n = 0.65$

Fig. 11.51 Normalized lateral force in elliptical contact patches under **a** large spin slip only and **b** still quite large spin slip with the addition of a little of lateral slip

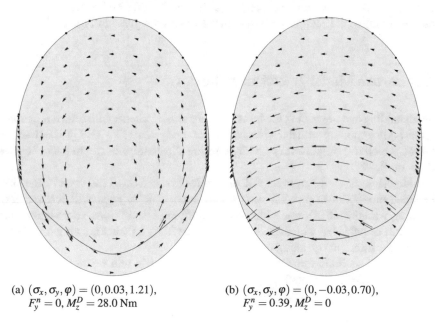

(a) $(\sigma_x, \sigma_y, \varphi) = (0,0.03,1.21),$
$F_y^n = 0,\, M_z^D = 28.0\,\mathrm{Nm}$

(b) $(\sigma_x, \sigma_y, \varphi) = (0,-0.03,0.70),$
$F_y^n = 0.39,\, M_z^D = 0$

Fig. 11.52 Special cases: **a** zero lateral force and **b** zero vertical moment

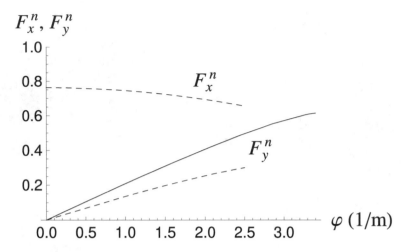

Fig. 11.53 Elliptical contact patch: normalized longitudinal and lateral forces versus spin slip, at $\sigma_x = 0$ (*solid line*) and $\sigma_x = -0.15$ (*dashed lines*)

The interaction of longitudinal slip σ_x and spin slip φ yields the effects reported in Fig. 11.53 on the longitudinal and lateral forces. A fairly high value $\sigma_x = -0.15$ has been employed. Examples of stress distributions are given in Fig. 11.54.

11.8   Brush Model Transient Behavior

Understanding and describing the transient behavior of wheels with tires has become increasingly important with the advent of electronic systems like ABS [15] or traction control, which may impose very rapidly varying slip conditions (up to tens of cycles per second).

Addressing the problem in its full generality like in Sect. 11.2, even in the simple brush model, looks prohibitive (but not impossible to good will researchers). However, with the aid of some additional simplifying assumptions, some interesting results can be achieved which, at least, give some hints on what is going on when a tire is under transient operating conditions.

In the next sections some simplified transient models will be developed. In all cases, inertia effects are totally neglected.

11.8.1   Transient Models with Carcass Compliance Only

A possible way to partly generalize the steady-state brush model discussed in Sect. 11.3 is to relax only the second condition of Sect. 11.3, while still retaining the first one, that is:

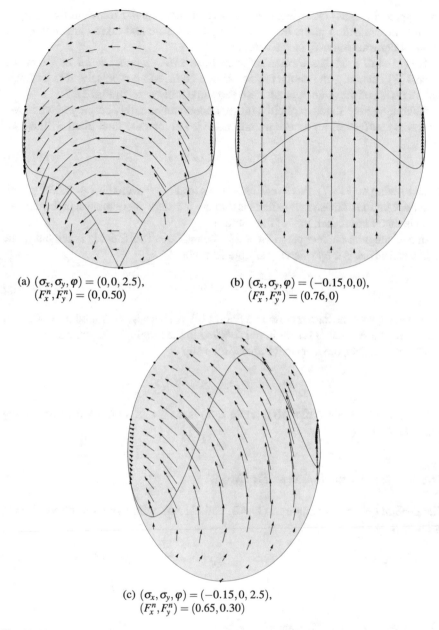

(a) $(\sigma_x, \sigma_y, \varphi) = (0, 0, 2.5)$,
    $(F_x^n, F_y^n) = (0, 0.50)$

(b) $(\sigma_x, \sigma_y, \varphi) = (-0.15, 0, 0)$,
    $(F_x^n, F_y^n) = (0.76, 0)$

(c) $(\sigma_x, \sigma_y, \varphi) = (-0.15, 0, 2.5)$,
    $(F_x^n, F_y^n) = (0.65, 0.30)$

Fig. 11.54 Examples of tangential stress distributions: **a** pure spin slip φ, **b** pure longitudinal slip σ_x and **c** both φ and σ_x. Values of φ are in m^{-1}

- $\mathbf{e}_{,t} = \mathbf{0}$, which means $\mathbf{e} = \mathbf{e}(\hat{x}, \hat{y})$, with no time dependence;
- $\dot{\mathbf{q}} \neq \mathbf{0}$, which means that $\rho(t) \neq \sigma(t)$.

This approach, which leads to some simple and very popular transient tire models, discards the transition in the bristle deflection pattern $\mathbf{e}$ and takes care only of the transient deformation $\mathbf{q}(t)$ of the carcass.

This kind of models are often referred to as *single contact point* transient tire models [9]. Actually, the contact is not at one point. More precisely, it is assumed that all points of the contact patch have the same motion, as in Fig. 11.7.

Although rarely stated explicitly, these models can be safely employed whenever the carcass stiffnesses w_x and/or w_y are much lower than the total tread stiffness k_t

$$w_i \ll k_t, \qquad i = x, y \tag{11.120}$$

Indeed, owing to (11.47), this condition allows for $\dot{\mathbf{F}}_t \neq \mathbf{0}$ even if $\mathbf{e}_{,t} \approx \mathbf{0}$. The physical interpretation of these inequalities is that the transient phenomenon in the contact patch is much faster than that of the carcass.

In a rectangular contact patch $2a \times 2b$, the total tread stiffness k_t is related to the local tread stiffness k by this very simple formula

$$k_t = 4abk \tag{11.121}$$

For instance, with the data reported on Sect. 11.3, we have $w_x = k_t$ and $w_y = 0.25k_t$. Therefore, we see that (11.120) in *not fulfilled* in the longitudinal direction!

In these models, the transient translational slip

$$\boldsymbol{\rho}(t) = \boldsymbol{\sigma}(t) + \dot{\mathbf{q}}(t)/V_r(t) \tag{11.122}$$

is an unknown function, like $\mathbf{q}(t)$, while $\boldsymbol{\sigma}(t)$ is, as usual, an input function, along with $\varphi(t)$ and $V_r(t)$.

11.8.1.1   Transient Nonlinear Tire Model

The general governing equations (11.42) and (11.43), with the assumption $\mathbf{e}_{,t} = \mathbf{0}$, become

$$\mathbf{e}' - \boldsymbol{\varepsilon} = \mathbf{0} \qquad \Longleftrightarrow \qquad k|\mathbf{e}| < \mu_0\, p \qquad \text{(adhesion)} \tag{11.123}$$

$$k\mathbf{e} = -\mu_1 p\, \frac{\mathbf{e}' - \boldsymbol{\varepsilon}}{|\mathbf{e}' - \boldsymbol{\varepsilon}|} \qquad \Longleftrightarrow \qquad |\mathbf{e}' - \boldsymbol{\varepsilon}| > 0 \qquad \text{(sliding)} \tag{11.124}$$

where $\boldsymbol{\varepsilon} = \boldsymbol{\rho} - (\hat{x}\,\mathbf{j} - \hat{y}\,\mathbf{i})\varphi$ and $\mathbf{e}' = \mathbf{e}_{,\hat{x}}$.

These equations are *formally identical* to the governing equations (11.56) and (11.57) of the steady-state case. Both cases share the assumption $\mathbf{e}_{,t} = \mathbf{0}$. Therefore, the whole analysis developed in Sect. 11.3 holds true in this case as well, with the important difference that $\boldsymbol{\rho} = \boldsymbol{\sigma} + \dot{\mathbf{q}}/V_r(t)$ has to replace any occurrence of $\boldsymbol{\sigma}$, since now $\dot{\mathbf{q}} \neq \mathbf{0}$.

Of particular importance is to understand that the global tangential force $\mathbf{F}_t = \mathbf{F}_t(\rho, \varphi)$ is exactly the *same function* of (11.72). For instance, in a *rectangular* contact patch with $\varphi = 0$ the magnitude of $\mathbf{F}_t$ is given by a formula identical to (11.107), that is

$$F_t = F_t(\rho(t)) = C_\sigma \rho \left[1 - \frac{\rho}{\sigma_s} \left(\frac{1 + 2\chi}{1 + \chi} \right) + \left(\frac{\rho}{\sigma_s} \right)^2 \left(\frac{1 + 3\chi}{3(1 + \chi)} \right) \right] \quad (11.125)$$

with $\rho = |\rho|$.

Consequently, the components $F_x(\rho_x, \rho_y)$ and $F_y(\rho_x, \rho_y)$ of F_t are

$$F_x = -\frac{\rho_x}{\rho} F_t(\rho), \qquad F_y = -\frac{\rho_y}{\rho} F_t(\rho) \quad (11.126)$$

Of course, $\rho = \rho_x \mathbf{i} + \rho_y \mathbf{j}$. The partial derivatives are given by (11.101), again with ρ replacing σ.

Since $\rho(t) = \sigma(t) + \dot{\mathbf{q}}(t)/V_r(t)$, the transient slip $\rho(t)$ is an unknown function and an additional vectorial equation is necessary (it was not so in the steady-state case, which had $\dot{\mathbf{q}} = \mathbf{0}$). The key step to obtain the missing equation is getting $\dot{\mathbf{F}}_t$ and inserting it into (11.46), as already done in Sect. 11.2 for the general case.

The simplification with respect to the transient general case, as already stated, is that here $\mathbf{F}_t(\rho, \varphi)$ is a *known* function and hence

$$\begin{cases} \dot{F}_x = \dfrac{\partial F_x}{\partial \rho_x}\dot{\rho}_x + \dfrac{\partial F_x}{\partial \rho_y}\dot{\rho}_y + \dfrac{\partial F_x}{\partial \varphi}\dot{\varphi} = w_x V_r(\rho_x - \sigma_x) \\[2mm] \dot{F}_y = \dfrac{\partial F_y}{\partial \rho_x}\dot{\rho}_x + \dfrac{\partial F_y}{\partial \rho_y}\dot{\rho}_y + \dfrac{\partial F_y}{\partial \varphi}\dot{\varphi} = w_y V_r(\rho_y - \sigma_y) \end{cases} \quad (11.127)$$

is a system of linear differential equations with nonconstant coefficients in the unknown functions $\rho_x(t)$ and $\rho_y(t)$. In general, it requires a numerical solution. The influence of the spin slip rate $\dot{\varphi}$ is negligible and will be discarded from here onwards.

Generalized relaxation lengths can be defined in (11.127)

$$\begin{aligned} s_{xx}(\rho_x, \rho_y) &= -\frac{\partial F_x}{\partial \rho_x}\frac{1}{w_x}, & s_{xy}(\rho_x, \rho_y) &= -\frac{\partial F_x}{\partial \rho_y}\frac{1}{w_x} \\[2mm] s_{yx}(\rho_x, \rho_y) &= -\frac{\partial F_y}{\partial \rho_x}\frac{1}{w_y}, & s_{yy}(\rho_x, \rho_y) &= -\frac{\partial F_y}{\partial \rho_y}\frac{1}{w_y} \end{aligned} \quad (11.128)$$

where the minus sign is there to have positive lengths. System (11.127) can be rewritten as

$$\begin{cases} -s_{xx}\dot{\rho}_x - s_{xy}\dot{\rho}_y = V_r(\rho_x - \sigma_x) \\ -s_{yx}\dot{\rho}_x - s_{yy}\dot{\rho}_y = V_r(\rho_y - \sigma_y) \end{cases} \quad (11.129)$$

In [9, p. 346] this kind of model is called *nonlinear single point*.

In classical handling analysis, only lateral slips are supposed to be significant. The model becomes simply

$$- s_{yy} \dot{\rho}_y = V_r(\rho_y - \sigma_y) \tag{11.130}$$

11.8.1.2 Transient Linear Tire Model

The simplest version of (11.127) assumes a linear function $\mathbf{F}_t(\rho) = -C_\sigma(\rho_x \mathbf{i} + \rho_y \mathbf{j})$, like in (11.77). Accordingly, Eq. (11.127) become

$$\begin{aligned} -C_\sigma \dot{\rho}_x &= w_x V_r(\rho_x - \sigma_x) \\ -C_\sigma \dot{\rho}_y &= w_y V_r(\rho_y - \sigma_y) \end{aligned} \tag{11.131}$$

often conveniently rewritten as

$$\begin{aligned} s_x \dot{\rho}_x + V_r \rho_x &= V_r \sigma_x \\ s_y \dot{\rho}_y + V_r \rho_y &= V_r \sigma_y \end{aligned} \tag{11.132}$$

where the positive constants

$$s_x = \frac{C_\sigma}{w_x} \quad \text{and} \quad s_y = \frac{C_\sigma}{w_y} \tag{11.133}$$

are called, respectively, longitudinal and lateral *relaxation lengths*. With the data listed in (11.52), we have $s_x = a$ and $s_y = 4a$: as expected the lateral relaxation length is much higher than the longitudinal one. The two equations in (11.132) are now uncoupled, which simplifies further this model, called *linear single point*.

If we compare the linear and the non linear single point models, we see that $s_x \geq s_{xx}$ and $s_y \geq s_{yy}$.

Consistently with the assumption of linear tire behavior, inserting $\mathbf{F}_t = -C_\sigma \rho$ into (11.132) leads to the most classical *transient linear tire model* ($i = x, y$)

$$s_i \dot{F}_i + V_r(t) F_i = -V_r(t) C_\sigma \sigma_i(t) \tag{11.134}$$

that is, to nonhomogeneous linear first-order differential equations [6]. It is worth noting that (11.132) and (11.134) are perfectly equivalent.

The simplest, canonical, case is with *constant* V_r, which makes the equations with *constant* coefficients. The homogeneous counterpart of (11.134) has solution

$$F_i^O(t) = Ae^{-\frac{V_r}{s_i} t} \tag{11.135}$$

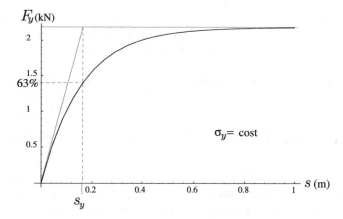

Fig. 11.55 Lateral force asymptotic response to a step change of σ_y and measurement of the relaxation length s_y

If also σ_i is *constant*, a particular solution F_i^p is simply

$$F_i^p = -C_\sigma \sigma_i \tag{11.136}$$

Therefore, in this case the general solution of (11.134), with initial condition $F_i(0) = 0$, is

$$F_i(t) = F_i^O(t) + F_i^p = -C_\sigma \sigma_i \left(1 - e^{-\frac{V_r}{s_i}t}\right) \tag{11.137}$$

In Fig. 11.55 this solution is plotted using the travel distance $s = V_r t$ instead of time. Also shown is how to experimentally measure the relaxation length. Just take the value of s that makes 63% of the asymptotic value of the force. This is much more reliable than trying to use the tangent in the origin of a noisy experimental signal.

Also interesting is the particular solution if $\sigma_i(t) = \sigma_0 \sin(\omega t)$ (the homogeneous solution decays very rapidly)

$$F_i^p(t) = -\frac{C_\sigma \sigma_0}{\sqrt{1 + (\omega s_i / V_r)^2}} \sin\left(\omega t - \arctan(\omega s_i / V_r)\right) \tag{11.138}$$

It is worth noting how the term $\omega s_i / V_r$ affects both the amplitude (reducing it) and the phase shift. The tire force is delayed with respect to the sinusoidal input.

However, this completely linear model provides acceptable results only if the tire slips are very small. An alternative, a little better approach is presented hereafter.

11.8.1.3 Transient Semi-nonlinear Tire Model

It is very common in vehicle dynamics to combine the linear equations (11.132) for $\rho(t)$ with a nonlinear function for the tangential force, like, e.g., $\mathbf{F}_t = -(\rho/\rho)F_t(\rho)$, as in (11.125). Things are a bit mixed up, but the allure of simplicity is quite powerful.

Indeed, the differential equations in (11.127) are much more involved than those in (11.132), while combining (11.127) with a nonlinear function for $\mathbf{F}_t$ is fairly straightforward. In [9, p. 345] this kind of model is called *semi-nonlinear single point*.

Results are much more realistic than those provided by the linear model.

11.8.2   Transient Model with Carcass and Tread Compliance

If the carcass and tread stiffnesses are comparable, that is if (11.120) does not hold, the effects of $\mathbf{e}_{,t}$ should also be taken into account, particularly under severe transient conditions. Therefore, both conditions listed at Sect. 11.3 are relaxed, that is:

- $\mathbf{e}_{,t} \neq \mathbf{0}$;
- $\dot{\mathbf{q}} \neq \mathbf{0}$.

To keep the formulation rather simple, while still grasping the main phenomena, it is useful to work under the following simplifying assumptions:

1. rectangular shape of the contact patch, which means $x_0(\hat{y}) = a$;
2. no spin slip φ;
3. either pure longitudinal slip σ_x or pure lateral slip σ_y, but not both;
4. $\mu_0 = \mu_1$, that is both equal to μ.

It is worth noting that complete adhesion in the contact patch is *not* assumed (cf. [9, p. 220]). Like in Sect. 11.2, boundary conditions at the leading edge and initial conditions on the whole contact patch need to be supplied, that is

$$\mathbf{e}(a, \hat{y}, t) = 0, \quad \text{and} \quad \mathbf{e}(\hat{x}, \hat{y}, 0) = \mathbf{0} \quad (11.139)$$

Nonzero initial conditions are possible, but may lead to more involved formulations.

Like in Sect. 11.5.1, the first two simplifying assumptions (rectangular shape and no φ), along with zero initial conditions, make $\mathbf{e}$, and actually the whole formulation, *not dependent* on $\hat{y}$. That means that we have to deal with ordinary differential equations, instead of partial differential equations. The additional effect of the third assumption ($\sigma_x \sigma_y = 0$) is to have $\rho(t)$, $\mathbf{q}$, and $\mathbf{e}(\hat{x}, t)$ with only one nonzero component (i.e., directed like either $\mathbf{i}$ or $\mathbf{j}$). That means that we have to deal with scalar functions, not vectorial functions. The fourth assumption ($\Delta \mu = 0$) makes all function continuous.

With $\varphi = 0$, the first general governing equation (11.42) (adhesion region) becomes

$$V_r \mathbf{e}_{,\hat{x}} - \mathbf{e}_{,t} = V_r \rho \quad (11.140)$$

which is a nonhomogeneous transport equation in the unknown function $\mathbf{e}(\hat{x}, t) = \mathbf{e}_a(\hat{x}, t)$. The tangential stress in the adhesion region is given by $\mathbf{t}_a(\hat{x}, t) = k\mathbf{e}_a$.

The adhesion state starts at the leading edge $\hat{x} = a$ and is maintained up to $\hat{x} = \hat{x}_s(t)$, which marks, at time t, the moving point where the friction limit is reached

$$k|\mathbf{e}_a(\hat{x}_s(t), t)| = \mu p(\hat{x}_s(t)) \qquad (11.141)$$

and hence where the sliding region begins.

Exactly like in (11.96), the onset of sliding is with the bristle deflection that has the *same direction* as $\mathbf{e}_a(\hat{x}_s(t), t)$. Therefore, the governing equation (11.43) for the *sliding* region becomes simply

$$\mathbf{t}_s(\hat{x}) = k\mathbf{e}_s(\hat{x}) = \mu p(\hat{x}) \frac{\mathbf{e}_a(\hat{x}_s(t), t)}{|\mathbf{e}_a(\hat{x}_s(t), t)|}, \qquad \text{with} \ -a \leq \hat{x} < \hat{x}_s(t) \qquad (11.142)$$

which is already the definition of $\mathbf{e}_s$ and hence of $\mathbf{t}_s$. It is important to note that in the sliding region the bristle deflections $\mathbf{e}_s$ *do not depend on time* and, therefore, are *known*. It is the *moving transition point* $\hat{x}_s(t)$ that has to be found as a function of time.

The global tangential force $\mathbf{F}_t(t) = F_x \mathbf{i} + F_y \mathbf{j}$ that the road applies to the tire model is given by the integral of $\mathbf{t} = k\mathbf{e}$ on the contact patch, like in (11.72), with all tangential stresses $\mathbf{t}$ having the *same direction*

$$\mathbf{F}_t(t) = -\mathbf{s} \, F_t(t) = k \left[2b \int_{\hat{x}_s(t)}^{a} \mathbf{e}_a(\hat{x}, t) d\hat{x} + 2b \int_{-a}^{\hat{x}_s(t)} \mathbf{e}_s(\hat{x}) d\hat{x} \right] \qquad (11.143)$$

Since also $\rho(t) = \sigma(t) + \dot{\mathbf{q}}(t)/V_r(t)$ is unknown, an additional equation is necessary. Exactly like in (11.47), it is obtained by differentiating $\mathbf{F}_t(t)$. But here, owing to the simplifying assumptions, some further steps can be carried out,[14] thus getting

$$\dot{\mathbf{F}}_t = 2bk \int_{\hat{x}_s(t)}^{a} \mathbf{e}_{,t} d\hat{x} = 2bkV_r \int_{\hat{x}_s(t)}^{a} (\mathbf{e}_{,\hat{x}} - \rho) d\hat{x}$$
$$= 2bkV_r \left[-\mathbf{e}(\hat{x}_s(t), t) - (a - \hat{x}_s(t)) \rho(t) \right] \qquad (11.144)$$

since $\mathbf{e}(a, t) = 0$. This result can be inserted into (11.48) to get the sought for equation

$$-2bk \left[\mathbf{e}(\hat{x}_s(t), t) + (a - \hat{x}_s(t)) \rho(t) \right] = \mathbf{W} \left[\rho(t) - \sigma(t) \right] \qquad (11.145)$$

where $\mathbf{W}$ is a diagonal matrix, as in (11.15).

Summing up, the problem is therefore governed by either of the two following (formally identical) systems of differential–algebraic equations, with suitable boundary and initial conditions

---

[14] The crucial aspects are: $\mathbf{e}_s$ not depending on time, $\mathbf{e}_a(\hat{x}_s, t) = \mathbf{e}_s(\hat{x}_s)$.

$$\begin{cases} V_r\, e_{x,\hat{x}} - e_{x,t} = V_r \rho_x, \quad \hat{x}_s(t) < \hat{x} < a \\ k|e_x(\hat{x}_s(t),t)| = \mu p(\hat{x}_s(t)) \\ \rho_x(t) = \dfrac{w_x \sigma_x(t) - 2bk\, e_x(\hat{x}_s(t),t)}{w_x + 2bk(a - \hat{x}_s(t))} \\ e_x(a,t) = 0 \\ e_x(\hat{x},0) = 0 \end{cases} \qquad \begin{cases} V_r\, e_{y,\hat{x}} - e_{y,t} = V_r \rho_y \\ k|e_y(\hat{x}_s(t),t)| = \mu p(\hat{x}_s(t)) \\ \rho_y(t) = \dfrac{w_y \sigma_y(t) - 2bk\, e_y(\hat{x}_s(t),t)}{w_y + 2bk(a - \hat{x}_s(t))} \\ e_y(a,t) = 0 \\ e_y(\hat{x},0) = 0 \end{cases}$$

$$\tag{11.146}$$

where, possibly, $V_r = V_r(t)$. Zero initial conditions imply that

$$\rho_i(0) = \frac{w_i \sigma_i(0)}{w_i + 4abk} \tag{11.147}$$

It is quite counterintuitive that if we apply a step function to $\sigma(t)$, we obtain $\rho_i(0) \neq 0$.

This model can be called *nonlinear full contact patch*.

It should be remarked that, unlike the commonly used approaches described in the previous section, the proposed model accounts not only for the transient deformation of the carcass (i.e., $\dot{\mathbf{q}} \neq \mathbf{0}$), but also for the transient behavior of the bristle deflection pattern (i.e., $\mathbf{e}_{,t} \neq \mathbf{0}$). It will be shown that this last effect may be far from negligible in some important cases, particularly in braking/driving wheels. More precisely, the larger any of the ratios

$$\theta_x = \frac{w_x}{k_t} \qquad \theta_y = \frac{w_y}{k_t} \tag{11.148}$$

where $k_t = 4abk$ is the tread stiffness, the more relevant the effect of the bristle deflection in that direction. Since $w_x \gg w_y$, the transient behavior in the bristle deflection pattern has more influence when the wheel is subject to time-varying longitudinal slip. For instance, with the data reported on Sect. 11.3, we have $\theta_x = 1$ and $\theta_y = 0.25$. In practical terms, bristle transient pattern has some relevance in ABS systems and also in launch control systems.

11.8.3   Model Comparison

The proposed models for the transient behavior of tires are compared on a few numerical tests. The goal is to show the range of applicability and to warn about employing a model without really understanding its capabilities.

In particular, three models of increasing complexity are compared:

- semi-nonlinear single point, (11.132) with (11.125);
- nonlinear single point, (11.127);
- nonlinear full contact patch, (11.146).

The linear single point model is not considered because of its limitations.

Of course, all the simplifying assumptions listed at the beginning of Sect. 11.8.2 have to be fulfilled. Therefore, tests are performed with the data listed in (11.52), except for $\chi = 0$, and under either pure longitudinal slip or pure lateral slip. Moreover, a *rectangular* contact patch and *parabolic* pressure distribution is assumed.

All models are tested applying step functions to either σ_x or σ_y, the step values being -0.21 and -0.07. In all cases, the index i means either x (longitudinal) or y (lateral) direction.

The *first model* (semi-nonlinear single contact point, Sect. 11.8.1.2) takes into account only the carcass compliance and employs a *constant relaxation length* s_i, with $i = x, y$. This model is by far the most popular model for the transient behavior of tires, if limited to pure lateral conditions. According to (11.132), the model is defined by

$$\begin{cases} s_i \dot{\rho}_i + V_r \rho_i = V_r \sigma_i \\ \rho_i(0) = 0 \end{cases} \tag{11.149}$$

where $s_i = C_\sigma/w_i$, with $C_\sigma = 4ka^2b$ as in (11.84). Once the function $\rho_i(t)$ has been obtained, the global tangential force is given by the nonlinear function

$$F_i(\rho_i) = -C_\sigma \rho_i \left[1 - \frac{|\rho_i|}{\sigma_s} + \frac{1}{3} \left(\frac{\rho_i}{\sigma_s} \right)^2 \right] \tag{11.150}$$

much like in (11.99) with (11.107).

The *second model* (nonlinear single contact point, Sect. 11.8.1) is similar, but employs a *nonconstant relaxation length*, as in (11.127)

$$\begin{cases} -\dfrac{F_i'(\rho_i)}{w_i} \dot{\rho}_i + V_r \rho_i = V_r \sigma_i \\ \rho_i(0) = 0 \end{cases} \tag{11.151}$$

where (cf. (11.108) with $\chi = 0$)

$$F_i'(\rho_i) = -C_\sigma \left[1 - 2\frac{|\rho_i|}{\sigma_s} + \left(\frac{\rho_i}{\sigma_s} \right)^2 \right] \tag{11.152}$$

is the derivative of (11.150). A numerical solution is usually required. As in the first model, the function $\rho_i(t)$ is then inserted into (11.150) to obtain the longitudinal/lateral force.

The *third model* (nonlinear full contact patch, Sect. 11.8.2) takes into account both the carcass and tread compliances, as in (11.146)

$$\begin{cases} V_r\, e_{i,\hat{x}} - e_{i,t} = V_r \rho_i \\ k|e_i(\hat{x}_s(t), t)| = \mu p(\hat{x}_s(t)) \\ \rho_i(t) = \dfrac{w_i \sigma_i(t) - 2bk\, e_i(\hat{x}_s(t), t)}{w_i + 2bk(a - \hat{x}_s(t))} \\ e_i(a, t) = 0 \\ e_i(\hat{x}, 0) = 0 \end{cases} \tag{11.153}$$

To obtain a numerical solution, an iterative method can be employed. First make an initial guess for $\rho_i^{(0)}(t)$ (for instance $\rho_i^{(0)}(t) = (\sigma_i(t) + \rho_i^s(t))/2$, where $\rho_i^s(t)$ is the solution of (11.149)). By means of the first equation, numerically obtain $e_x^{(0)}(\hat{x}, t)$, and then, using the second equation, evaluate the function $\hat{x}_s^{(0)}$. At this stage, the first iteration can be completed by computing $\rho_i^{(1)}(t)$ by means of the third equation. The whole procedure has to be repeated (usually 5 to 15 times) until convergence is attained.

Once a good approximation of $e_i(\hat{x}, t)$ and $\hat{x}_s(t)$ (and also of $\rho_i(t)$) has been computed, the tangential force can be obtained from the following integral over the contact patch

$$F_i(t) = 2bk \left[\int_{\hat{x}_s(t)}^{a} e_i(\hat{x}, t)\mathrm{d}\hat{x} + \mu \operatorname{sign}(e_i(\hat{x}_s(t), t)) \int_{-a}^{\hat{x}_s(t)} p(\hat{x})\mathrm{d}\hat{x} \right] \qquad (11.154)$$

11.8.4 Selection of Tests

A *step change* in the input (forcing) function $\sigma_i(t)$ works well to highlight the differences between the three models. With the data of (11.52), except $\chi = 0$, the static tangential force (11.150) has maximum magnitude for $\sigma = 0.266$. To test the models in both the (almost) linear and nonlinear ranges, a small ($\sigma_i = -0.07$) and a large ($\sigma_i = -0.21$) step have been selected. Since $w_x = 4w_y$, both longitudinal and lateral numerical tests are performed.

In all cases, results are plotted versus the rolling distance s, instead of time, thus making $V_r(t)$ irrelevant.

11.8.5 Longitudinal Step Input

The longitudinal force $F_x(s)$, as obtained from the three tire models with step inputs $\sigma_x = -0.07$ and $\sigma_x = -0.21$, is shown in Fig. 11.56. Because of the high value of the longitudinal carcass stiffness w_x (equal to the tread stiffness k_t), the transient phenomenon is quite fast. Indeed, in the first model (*dashed line*) the relaxation length $s_x = 7.5 \, \mathrm{cm}$.

Quite remarkably, the three models provide very different results for $s < 0.25 \, \mathrm{cm}$, thus showing that the selection of the transient tire model may be a crucial aspect in vehicle dynamics, particularly when considering vehicles equipped with ABS.

Obviously, all models converge to the same asymptotic (i.e, steady-state) value of F_x.

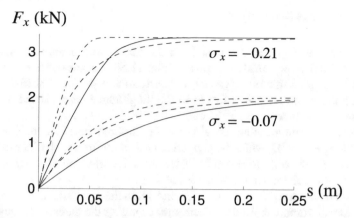

Fig. 11.56 Longitudinal force asymptotic response to a small and to a large step change of σ_x. Comparison of three tire models: semi-nonlinear single contact point (*dashed line*), nonlinear single contact point (*dot-dashed line*), nonlinear full contact patch (*solid line*)

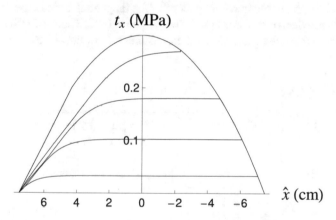

Fig. 11.57 Transient patterns of the tangential stress t_x in the contact patch (third model)

The behavior of the first model (*dashed lines*) is the same in both cases, except for a vertical scaling. This is not the case for the second model (*dot-dashed lines*) because of the nonconstant generalized relaxation length. The more detailed third model (*solid lines*) behaves in quite a peculiar way, thus confirming that the contribution of the transient tread deflection is far from negligible.

Figure 11.57 shows the transient pattern of the tangential longitudinal stress t_x in the contact patch as provided by the third model with $\sigma_x = -0.21$. It is worth noting how greatly, in the adhesion region, the pattern departs from the linear behavior of the static case (Fig. 11.17).

11.8.6 Lateral Step Input

The lateral force $F_y(s)$, as obtained from the three tire models with step inputs $\sigma_y = -0.07$ and $\sigma_y = -0.21$, is shown in Fig. 11.58. Because of the low value of the lateral carcass stiffness w_y (equal to one fourth of the tread stiffness k_t), the transient phenomenon is not as fast as in the longitudinal case. Indeed, in the first model the relaxation length $s_y = 30$ cm.

In this case, the three models provide not very different results in the linear range, that is with $\sigma_y = -0.07$, while they depart significantly from each other in the non-linear range, that is with $\sigma_y = -0.21$. Therefore, the selection of the transient tire model may be crucial in lateral dynamics as well.

It should be observed from Figs. 11.56 and 11.58 that the first and second models have the same "formal" behavior. Therefore, changing the carcass stiffness results only in a horizontal scaling. This is not true for the third model.

Obviously, all models converge to the same asymptotic (i.e, steady-state) value of F_y.

Figure 11.59 shows the transient pattern of the tangential lateral stress in the contact patch as provided by the third model with $\sigma_y = -0.21$. There are still differences with respect to the static case, although not as much as in Fig. 11.57.

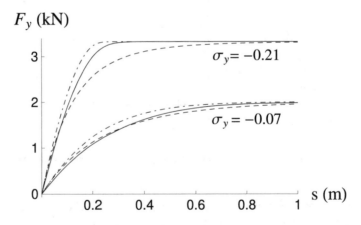

Fig. 11.58 Lateral force asymptotic response to a small and to a large step changes of σ_y. Comparison of three tire models: semi-nonlinear single contact point (*dashed line*), nonlinear single contact point (*dot-dashed line*), nonlinear full contact patch (*solid line*)

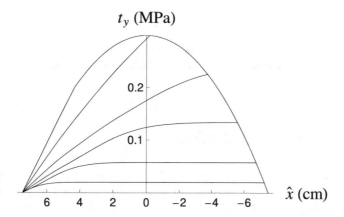

Fig. 11.59 Transient patterns of the tangential stress t_y in the contact patch (third model)

11.9 Exercises

11.9.1 Braking or Driving?

Figure 11.1 *shows a schematic of the brush model. Is it braking or driving? Explain why.*

Solution

It is obviously braking because all bristles are deflected backward. It is braking pretty much like in Fig. 11.18b: the sliding part is about 1/3 of the footprint.

11.9.2 Carcass Compliance

Do you expect the tire carcass to be more compliant in the longitudinal or in the lateral direction?

Solution

It is quite intuitive that it is more compliant in the lateral direction, at least in road tires (Fig. 11.7).

11.9.3 Brush Model: Local, Linear, Isotropic, Homogeneous

To which of these four properties does the brush model owe its name?

Solution

The answer is "local". Indeed, this is the main simplification. Not very realistic, but dramatically important to keep the model formulation amenable to an almost analytic treatment.

11.9.4 Anisotropic Brush Model

Assuming, as in (11.20), *the same tread stiffness k in both longitudinal and lateral directions may be not always correct. Try to figure out how to generalize the brush model to have different stiffnesses.*

Solution

The solution is already in (11.21). Just replace any occurrence of $k\mathbf{e}$ with $\mathbf{Ke}$, maybe with $\mathbf{K}$ being diagonal. Particularly important is to upgrade the comments around (11.96). Even under pure translational slip the transition from adhesion to sliding involves a change in the direction of the bristle deflections, as governed by (11.57).

11.9.5 Carcass Compliance 2

Does the carcass compliance affect the mechanical behavior of tire under steady-state conditions?

Solution

Well, yes and no. In this model the tangential force is not affected by the carcass compliance. On the other hand, the moments with respect to the origin O of the reference system S_w (Fig. 2.6) does indeed depend also on the carcass compliance, as shown in Fig. 11.35.

11.9.6 Skating Versus Sliding

What is the difference between skating velocity and sliding velocity?

Solution

Let us consider a bristle in the tire brush model. The *skating* velocity is the velocity of its *root* with respect to the ground. The *sliding* velocity is the velocity of its *tip* with respect to the ground.

11.9.7 Skating Slip

Is the skating slip ε local or global?

Solution

The skating slip is, in general, a local quantity. It becomes global if there is no spin slip φ. Moreover, we can observe that ε is a transient slip, since it takes into account also $\dot{\mathbf{q}}$.

11.9.8 Simplest Brush Model

Select the options to have the simplest brush model.

Solution

A fairly simple brush model is obtained with the following options (Fig. 11.15):

footprint shape: rectangular;
slips: translational;
working condition: steady-state.

11.9.9 Velocity Relationships

Find out under which operating conditions the following equations hold true:

1. $\mathbf{V}_D = \mathbf{V}_C$;
2. $\mathbf{V}_s^P = \mathbf{V}_s$;
3. $\mathbf{V}_s^P = -\dot{\mathbf{e}}$
4. $\rho = \sigma + \frac{\dot{\mathbf{q}}}{V_r}$;
5. $\varepsilon = \rho$;
6. $\mathbf{V}_s^P = \varepsilon V_r$;
7. $\mathbf{V}_s^P = \sigma V_r$;
8. $\mathbf{V}_s = \sigma V_r$.

Solution

1. $\dot{\mathbf{q}} = \mathbf{0}$;
2. $\dot{\mathbf{q}} = \mathbf{0}$ and $\omega_{s_z} = 0$;
3. adhesion region;
4. always;

5. $\varphi = 0$;
6. always;
7. $\dot{\mathbf{q}} = \mathbf{0}$ and $\varphi = 0$;
8. always.

11.9.10 Slip Stiffness Reduction

Figure 11.28 *shows the effects of the combined action of* σ_x *and* σ_y *on* F_x *and* F_y. *According to* (11.116) *and with the data in* (11.52), *evaluate the reduction of the slip stiffness* C_σ *(i.e., the slope in the origin) under the combined working conditions of Fig.* 11.28.

Solution

We know from Sect. 11.5.1 that in this case $\sigma_s = 0.27$. Moreover, we have $\chi = 0.2$. It is a simple calculation to find that $\widetilde{C}_\sigma(0.05) = 0.80\, C_\sigma$, $\widetilde{C}_\sigma(0.1) = 0.63\, C_\sigma$, $\widetilde{C}_\sigma(0.2) = 0.38\, C_\sigma$. These results show how strong is the interaction between σ_x and σ_y.

Let us do the same calculation, but with $\chi = 0.1$. First we observe that σ_s does not change. The results are as follows: $\widetilde{C}_\sigma(0.05) = 0.81\, C_\sigma$, $\widetilde{C}_\sigma(0.1) = 0.65\, C_\sigma$, $\widetilde{C}_\sigma(0.2) = 0.41\, C_\sigma$. Not a big difference.

11.9.11 Total Sliding

Can a non-locked wheel have all bristle tips sliding on the road surface?

Solution

Yes. Just have a look at Fig. 11.18d.

11.9.12 Spin Slip and Camber Angle

According to Fig. 11.39, *obtain the camber angle* γ *corresponding to spin slip* φ *equal to* −1, −2 *and* −3 m^{-1}.

Solution

The camber angles are 14.5, 30.0 and 48.6°, respectively. We see it is slightly non linear.

11.9.13   The Right Amount of Camber

In which figure it was shown that just a bit of camber can improve the maximum lateral force?

Solution

This topic was addressed in Fig. 11.46. See also Fig. 11.47. In motorcycle tire it is often the other way around: lot of camber and a little of lateral slip to adjust the lateral force to the required value.

11.9.14   Slip Stiffness

Let us consider tires all with the same F_z, p_0 and k. According to (11.84), the brush model predicts a lower slip stiffness for wider tires. Elaborate mathematically this concept with reference to Fig. 11.14.

Solution

Tire (a) has a footprint with length $2w$ and width w. Tires (b) and (c) have a footprint with length w and width $2w$. The area of the footprint is the same in all cases. From (11.84) we obtain that the narrower tire has C_σ twice as much as that of the wider tires.

11.10   Summary

In this chapter a relatively simple, yet significant, tire model has been developed. It is basically a brush model, but with some noteworthy additions with respect to more common formulations. For instance, the model takes care of the transient phenomena that occur in the contact patch. A number of figures show the pattern of the local actions within the contact patch (rectangular and elliptical).

11.11   List of Some Relevant Concepts

Section 11.1.9—the skating slip takes into account both transient translational slip and spin slip;
Section 11.3.2—each bristle is undeformed when it enters the contact patch;
Section 11.5—the analysis of the steady-state behavior of the brush model is quite simple if there is no spin slip;

Section 11.5.1—full sliding does not imply wheel locking;

Section 11.5.1—the slip angle α is not a good parameter for a neat description of tire mechanics;

Section 11.5.1—the tire action surface summarizes the tire characteristics under a constant vertical load;

Section 11.5.1—tires have to be built in such a way to provide the maximum tangential force in any direction with small slip angles. This is a fundamental requirement for a wheel with tire to have directional capability;

Section 11.5.2—the tire action surface summarizes the steady-state behavior of a tire;

Section 11.7.2—good directional capability of a wheel means small slip angles.

11.12   Key Symbols

a	longitudinal semiaxis of the contact patch
b	lateral semiaxis of the contact patch
C	point of virtual contact
C_γ	camber stiffness
C_σ	slip stiffness
C_φ	spin stiffness
D	center of the contact patch
$\mathbf{e}$	bristle deflection
$\mathbf{e}_a$	bristle deflection in the adhesion zone
$\mathbf{e}_s$	bristle deflection in the sliding zone
$\mathbf{F}_t$	tangential force
F_x	longitudinal component of $\mathbf{F}_t$
F_y	lateral component of $\mathbf{F}_t$
F_z	vertical load
k	bristle stiffness
M_z^D	vertical moment with respect to D
O	origin of the reference system
p	pressure
p_0	pressure peak value
$\mathbf{q}$	horizontal deformation of the carcass
q_x, q_y	components of $\mathbf{q}$
r_r	rolling radius
s_x, s_y	components of the relaxation length
$\mathbf{t}$	tangential stress
$\mathbf{V}_c$	travel velocity
$\mathbf{V}_r$	rolling velocity
$\mathbf{V}_s$	slip velocity
$\mathbf{V}_s^P$	skating velocity
$\mathbf{V}_\mu^P$	sliding velocity
w_x, w_y	carcass stiffnesses

α	slip angle
γ	camber angle
ε	skating slip
ε_r	camber reduction factor
λ	steady-state skating slip
μ	local friction coefficient
μ_0	coefficient of static friction
μ_1	coefficient of kinetic friction
ρ	transient translational slip
σ	theoretical slip vector
φ	spin slip
ω_{s_z}	slip yaw rate

References

1. Clark SK (ed) (1971) Mechanics of pneumatic tires. National Bureau of Standards, Washington
2. Deur J, Asgari J, Hrovat D (2004) A 3D brush-type dynamic tire friction model. Veh Syst Dyn 42:133–173
3. Deur J, Ivanovic V, Troulis M, Miano C, Hrovat D, Asgari J (2005) Extension of the LuGre tyre friction model related to variable slip speed along the contact patch length. Veh Syst Dyn 43(Supplement):508–524
4. Guiggiani M (2007) Dinamica del Veicolo. CittaStudiEdizioni, Novara
5. Hüsemann T (2007) Tire technology: simulation and testing. Technical report, Institut für Kraftfahrwesen, Aachen. http://www.ika.rwth-aachen.de/lehre/kfz-labor/2_tires_en.pdf
6. Kreyszig E (1999) Advanced engineering mathematics, 8th edn. Wiley, New York
7. Lugner P, Pacejka H, Plöchl M (2005) Recent advances in tyre models and testing procedures. Veh Syst Dyn 43:413–436
8. Milliken WF, Milliken DL (1995) Race car vehicle dynamics. SAE International, Warrendale
9. Pacejka HB (2002) Tyre and vehicle dynamics. Butterworth-Heinemann, Oxford
10. Pacejka HB, Sharp RS (1991) Shear force development by pneumatic tyres in steady state conditions: a review of modelling aspects. Veh Syst Dyn 20:121–176
11. Romano L, Sakhnevych A, Strano S, Timpone F (2019) A novel brush-model with flexible carcass for transient interactions. Meccanica 54(10):1663–1679. https://doi.org/10.1007/s11012-019-01040-0
12. Romano L, Bruzelius F, Jacobson B (2020) Brush tyre models for large camber angles and steering speeds. Veh Syst Dyn 1–52. https://doi.org/10.1080/00423114.2020.1854320
13. Romano L, Bruzelius F, Jacobson B (2021) Unsteady-state brush theory. Veh Syst Dyn 59(11):1643–1671. https://doi.org/10.1080/00423114.2020.1774625
14. Romano L, Timpone F, Bruzelius F, Jacobson B (2021) Analytical results in transient brush tyre models: theory for large camber angles and classic solutions with limited friction. Meccanica. https://doi.org/10.1080/00423114.2020.1774625
15. Savaresi SM, Tanelli M (2010) Active braking control systems design for vehicles. Springer, London
16. Truesdell C, Rajagopal KR (2000) An introduction to the mechanics of fluids. Birkhäuser, Boston
17. Zwillinger D (ed) (1996) CRC standard mathematical tables and formulae, 30th edn. CRC Press, Boca Raton

Correction to: The Science of Vehicle Dynamics

Correction to:
M. Guiggiani, *The Science of Vehicle Dynamics*,
https://doi.org/10.1007/978-3-031-06461-6

The book was inadvertently published with some low resolution figures and wrong equations. The corresponding chapters have been corrected as follows:

In Chapter 3, Figures 3.70 to 3.75 have been replaced with higher resolution figures.

In Chapter 7, Figures 7.12 to 7.19 have been replaced with higher resolution figures.

In Chapter 10, Equations 10.13, 10.14, 10.15 and 10.18 have been corrected.

The correction chapters and the book have been updated with the changes.

The updated version of these chapters can be found at
https://doi.org/10.1007/978-3-031-06461-6_3,
https://doi.org/10.1007/978-3-031-06461-6_7,
https://doi.org/10.1007/978-3-031-06461-6_10

Correction to: The Science of Vehicle Dynamics

Correction to:
Chapters 5 and 8 in: M. Guiggiani, *The Science of Vehicle Dynamics,* **https://doi.org/10.1007/978-3-031-06461-6**

In the original version of the book, the legends of figure 5.21 and 5.22 were swapped by mistake. Moreover, equation 8.64 was mistaken. The figure legends were corrected in chapter 5 and the equation was corrected in chapter 8. The chapters and the book were updated with this change.

The updated version of these chapters can be found at
https://doi.org/10.1007/978-3-031-06461-6_5
https://doi.org/10.1007/978-3-031-06461-6_8

Index